Green Energy and Technology

More information about this series at http://www.springer.com/series/8059

Zhongchao Tan

Air Pollution and Greenhouse Gases

From Basic Concepts to Engineering Applications for Air Emission Control

Zhongchao Tan
Department of Mechanical and
Mechatronics Engineering
University of Waterloo
Waterloo, ON
Canada

ISSN 1865-3529 ISSN 1865-3537 (electronic)
ISBN 978-981-287-211-1 ISBN 978-981-287-212-8 (eBook)
DOI 10.1007/978-981-287-212-8

Library of Congress Control Number: 2014950678

Springer Singapore Heidelberg New York Dordrecht London

Printed on acid-free paper

Springer is part of Springer Science+Business Media (www.springer.com)

Preface

Air emissions include air pollution emissions and greenhouse gas emissions. Effective air emission control requires multidisciplinary expertise in engineering, education, physics, chemistry, mathematics, medical science, psychology, agriculture, architecture, business management, economics, and politics. It is a difficult task for the author(s) of any single book to address all aspects of air emissions. The focus of this book is on engineering science and technology, upon which effective air emission control program must be built. It does not prescribe social, economic, and political factors that lie outside the scope of this book.

This book aims at senior undergraduate and graduate students with educational backgrounds in mechanical, chemical, and/or environmental engineering. It can also be used by professionals with similar training background. It focuses on the basic concepts and engineering applications of technologies for the control of air emissions resulted from fossil fuel combustion.

This book is divided into three parts. The general basic concepts introduced in Part I are necessary to the understanding of air emission engineering topics in Parts II and III. Part II presents the engineering applications of the principles introduced in Part I. Part III covers some emerging topics related to air emission engineering and they include carbon capture and storage, nanoaerosol, indoor air quality.

Following a brief introduction to air emission in Chap. 1, Chaps. 2–4 present the general basic properties of gases and aerosol particles. They are necessary to understand the formation and behavior of air emissions. Chapters 5 and 6 present basic principles for the separation of unwanted gases and particulates from the contaminated air. These are the principles for the related engineering applications in Parts II and III such as syngas cleaning, carbon capture, and flue gas cleaning.

Part II of the book introduces the strategies for precombustion (Chaps. 7 and 8), in-combustion (Chap. 9) and postcombustion (Chap. 10) air emission control, step by step, from a process point of view. While air dispersion model (Chap. 11) is a powerful tool for air quality assessment and impact prediction, air dispersion itself is also a measure for air emission control by dilution.

Part III includes special topics related to the scope of this book, but they do not fit into the process introduced above. Chapter 12 is devoted to carbon sequestration and storage, which are of increasing interest to the society. Although debates are still ongoing, it is time to summarize the techniques that have been developed for CO_2 capture and storage. There may be some overlapping between this chapter and the other parts of this book. Chapter 13 presents an emerging topic of air pollution, nanosized air pollution. Nanomaterials are now widely used in many industries, for example, improved combustion efficiency, environmental protection, health, and solar panel fabrication. The unique properties of nanoaerosol and its implications on monitoring and filtration technologies are covered. Indoor air quality is introduced in Chap. 14. Indoor air quality is related extensively to air pollution. The sources of indoor air pollutants are different from their outdoor counterparts, as are their control techniques. The last chapter is about air quality and air emission monitoring techniques. They are commonly needed in industrial practices, government standard enforcement, and research and development in a laboratory setting.

The seed from which this book has grown was the engineering lecture notes that I have developed over the last 10 years. More teaching materials are available at this link: http://tan.uwaterloo.ca/book.html. They include PowerPoint presentations, extra assignment problems, and the solutions to the practice problems. They will be updated without notice.

Many people have helped me in writing this book, and my sincere appreciation goes to Dr. Dongqing Li (University of Waterloo), Dr. Mark Rood (University of Illinois at Urbana-Champaign), Raheleh Givehchi, Jing Min, Ishpinder Kailey, and all the undergraduate and graduate students who have commented on the manuscripts.

It has been a pleasure working with the team at Springer in publishing this book. My thanks are due to Anand Jayaprakash, Ramesh Premnath, Udhaya Kumar, and those working behind the scenes.

Contents

Acronyms

AC	Activated Carbon
ACGIH	American Conference of Governmental and Industrial Hygienists
AFBC	Atmospheric Fluidized Bed Combustor
AFT	Adiabatic Flame Temperature
AQI	Air Quality Index
ASHRAE	American Society of Heating, Ventilating, and Air-conditioning Engineering
ASTM	American Society for Testing and Materials
B20	A mixture of 20 % of biodiesel with 80 % of petroleum diesel
BDC	Bottom Dead Center
BRI	Building Related Illness
CB_1	Coal Bug 1
CCS	Carbon Capture and Storage
CEM	Continuous Emission Monitoring
CFBC	Circulating Fluidized Bed Combustor
CLC	Chemical Looping Combustion
CLOU	Chemical Looping Combustion with Oxygen Uncoupling
CNT	Carbon Nanotube
CPC	Condensation Particle Counter
CRC	Chemical Rubber Company
CRT	Continuously Regenerating Trap
DI	Direct Injection
DMA	Differential Mobility Analyzer
DMT	Derjaguin, Muller, and Toporov
EDX	Energy-Dispersive X-ray Spectroscopy
EGR	Exhaust Gas Recirculation
EIA	Energy Information Administration
EOR	Enhanced Oil Recovery
EPA	Environmental Protection Agency
ESP	Electrostatic Precipitator/Precipitation

EU	European Union
FBC	Fluidized Bed Combustion
FGR	Flue Gas Recirculation
FGD	Flue Gas Desulfurization
FSI	Furnace Sorbent Injection
GHG	Greenhouse Gas
HAP	Hazardous Air Pollutant
HC	Hydrocarbon
HFCs	Hydro Fluorocarbons
HHV	Higher Heating Value
HTC	Hydrothermal Conversion
HVAC	Heating, Ventilating, and Air Conditioning
IAQ	Indoor Air Quality
IDI	Indirect Injection Engine
IEA	International Energy Agency
IGCC	Integrated Gasification Combined Cycle
IL	Ionic Liquid
IPCC	Intergovernmental Panel on Climate Change
ISO	International Organization for Standardization
IUPAC	International Union of Pure and Applied Chemistry
JKR	Johnson, Kendall, and Roberts
LCA	Life Cycle Analysis
LFL	Lower Flammable Limit
LHS	Left-Hand Side
LHV	Lower Heating Value
LIFAC	Limestone Injection into the Furnace and Activation of Calcium Oxide
LLB	Lurgi Lentjes Bischoff
MOF	Metal–Organic Framework
NAAQS	National Ambient Air Quality Standard
NGCC	Natural Gas Fired Combined Cycles
NIOSH	National Institute for Occupational Safety and Health
NREL	National Renewable Energy Laboratory
OSHA	Occupational Safety and Health Agency
PAH	Polycyclic Aromatic Hydrocarbon
PBL	Planetary Boundary Layer
PFBC	Pressurized Fluidized Bed Combustor
PFC	Perfluorinated Compound
PM	Particulate Matter
$PM_{2.5}$	Mass of particulate matter with particle diameters smaller than 2.5 μm
PM_{10}	Mass of particulate matter with particle diameters smaller than 10 μm
PM_x	Mass of particulate matter with particle diameters smaller than x μm
PSA	Pressure Swing Adsorption
PZ	Piperazine

R&D	Research and Development
RHS	Right-Hand Side
SBS	Sick Building Syndrome
SCR	Selective Catalytic Reduction
SEM	Scanning Electron Microscopy
SMPS	Scanning Mobility Particle Sizer
SNCR	Selective Non-Catalytic Reduction
Syngas	Synthesis Gas
TCC	Thermochemical Conversion
TDC	Top Dead Center
TEM	Transmission Electron Microscope
TGA	Thermo Gravimetric Analysis
THP	Total Heat Production
TLV	Threshold Limit Value
TLV-C	Threshold Limit Value-Ceiling
TLV-STEL	Short-Term Exposure Limit Threshold Limit Value
TLV-TWA	Time-Weighted Average Threshold Limit Value
TRS	Total Reduced Sulfur
UFL	Upper Flammable Limit
UN	The United Nations
US DOE	The United States Department of Energy
US EPA	The United States Environmental Protection Agency
USA	The United States of America
UV	Ultraviolet
VOC	Volatile Organic Compound
WGS	Water–Gas Shift
WHO	The World Health Organization
WWII	The Second World War
XRD	X-Ray Diffraction

Nomenclature

a	Packing material property (m^2/m^3)
a	Contact radius between particle and another surface (m)
A	Area (m^2)
A	Rate coefficient for reaction rate constant (Depends)
A_c	Cross-sectional area (m^2)
A_s	Filter porosity parameter (–)
A/F	Air fuel ratio in combustion (–)
b	Depth of the ESP plates (m)
B	Rate coefficients for reaction rate constant (Depends)
c_x	Molecular speed along x-direction (m/s)
c_y	Molecular speed along y-direction (m/s)
c_z	Molecular speed along z-direction (m/s)
c_{rms}	Root mean square speed (m/s)
$c_{p,a}$	Air heating capacity (J K/kg)
$\vec{c}$	A gas molecule velocity vector (m/s)
$\bar{c}$	Mean molecular speed (m/s)
$\vec{c}_{A/B}$	Average relative velocity (m/s)
$\bar{c}_i$	Mean thermal speed of the ions (m/s)
c_p	Specific heat of the fuel (kJ/pg)
C	Concentration of gas or air pollutant (kg/m^3)
C_e	Air pollutant concentration in air exiting the room (kg/m^3)
C_c	Cunningham correction factor (–)
C_D	Drag coefficient (–)
C_i	Concentration of the pollutant in entering the box in a box model (kg/m^3)
$C_{i\infty}$	Steady state air pollutant concentration when $t \rightarrow \infty$ (kg/m^3)
C_N	Normalized air pollutant concentration (kg/m^3)
C_N	Number of molecules per unit volume of the gas ($\#/m^3$)
C_o	Real concentration in the air stream (kg/m^3)
C_s	Air pollutant concentration in supply air (kg/m^3)
C_s	Mass of adsorbate per mass of adsorbent (kg/kg)

C_s Sampled concentration (kg/m^3)
C_{ss} The steady state concentration (kg/m^3)
C_{std} Concentration corrected to 12 % CO_2 in the stack gas (kg/m^3)
C_x Exit concentration in adsorption (kg/m^3)
C_u A coefficient (–)
$C_{\Delta P}$ Pressure drop coefficient (–)
d Diameter (m)
d_a Aerodynamic particle diameter (m)
d_{bed} Body diameter of granular filter bed (m)
d_c Collision diameter (m)
d_c Characteristic dimension (m)
d_e Equivalent volume diameter (m)
d_f Aerodynamic focusing orifice diameter (m)
d_f Fiber diameter (m)
d_p^m Maximum diameter of the focused particles (m)
d_p Particle diameter (m)
d_G Diameter of the granules (m)
d_{50} Cut size, particle size for which the separation efficiency is 50 % (m)
D Cyclone body diameter (m)
D Diameter of the tube (m)
D Characteristic length in Re (m)
D The diffusivity or diffusion coefficient of gas in air (m^2/s)
D_e Diameter of gas exit of a cyclone (m)
D_p Diffusivity of the particles in the gas (m^2/s)
e Coefficient of restitution (–)
e Enhanced absorption factor (–)
e_k Kinetic energy for one mole of an ideal gas (–)
E Electric field intensity (V/m)
E_{ad} Adhesion energy (J)
E_A Activation energy (J/mol)
f An empirical fitting factor (–)
f Correction factor to the plume rise due to stack downwash (–)
f_D Dimensionless Darcy friction factor (–)
f_v Vapor mass fraction at the droplet surface (–)
F Force (N)
F_B Buoyancy flux (m^4/s^3)
F_C Centrifugal force (N)
F_D Drag force (N)
F_M Momentum flux (m^4/s^2)
Fr Froude number (–)
g Gravitational acceleration, 9.81 (m/s^2)
G Mole flow rate of the gas mixture (mol/s)
$\bar{G}$ Mole flow rate of the solute-free carrier gas (mol/s)
Gr Gravity number (–)

h	Height or length of a packed bed filled with granules (m)
h	Sensible heat of air (kJ/kg of dry air)
h_e	Total enthalpy of the air exiting the room (J)
h_{fg}	Latent heat of vaporization of water (J/mol)
h_{fg}	Latent heat of vaporization of the liquid fuel (J/mol)
$h_{f,i}^{o}$	Heat of formation of substance i (J/mol)
h_s	Total enthalpy of the supply air (J)
$h(T)$	Molar enthalpy of a component at temperature T (J/mol)
H	Cyclone height of inlet (m)
H	Distance between two plates of ESP (m)
H	Hamaker constant (J)
H	Henry's law constant (Depends)
H	Length/height of the column (m)
H	Plume center height (m)
H	Total enthalpy (J)
H_c	Critical height of a settling chamber (m)
H_{ij}	Hamaker constant (J)
H_B	Height of the building (m)
HHV	Higher Heating Value (J/mol)
HTU	Height of transfer unit (–)
I	Electric current (A)
j_x	Number of molecule colliding with the wall per unit area per unit time (1/s m^2)
J	Flux of mass flow per unit area (kg/s m^2)
J_x	Collision per unit time per unit area (1/m^2)
k	Boltzmann constant, 1.3807×10^{-23} (J/K)
k	Chemical reaction coefficient corresponding to y_{ss} (mol/s)
k	Karman constant = 0.4 (–)
k	Permeability of the porous medium (m^2)
k	Permeability of a filter dust cake (m^2)
k	Rate constant of the reaction (Depends)
k	Thermal conductivity of the liquid droplet at the surface (W/(m K))
k_{AmH}	Reaction rate constant of CO_2 and the amine (m^3/mol s)
k_b	Rate constant for the backward/reverse reaction (Depends)
k_f	Rate constant for the forward reaction (Depends)
k_g'	Amine liquid film mass transfer coefficient (m/s)
k_P	Mechanical constant of the particle (–)
k_y	Gas phase mass transfer coefficient (mol/m^2 s)
k_x	Liquid phase mass transfer coefficient (mol/m^2 s)
k_1	Mass transfer coefficient (depends)
k_2	Mass transfer coefficient (depends)
K_f	Mechanical constant of the particle filter materials (–)
Kn	Knudsen number (–)
K_C	Concentration based chemical equilibrium constant (Depends)

K_E	Force constant of proportionality, 9×10^9 (Nm^2/C^2)
K_P	Partial pressure based equilibrium constant (Depends)
K_{P0}	Partial pressure based equilibrium constant under a reference condition (Depends)
K	Aerosol coagulation coefficient (–)
K_x	Mass transfer rate coefficient in adsorption (1/s)
K_x	Overall mass transfer coefficient for the liquid phase (mol/m^2 s)
K_y	Overall mass transfer coefficient for the gas phase (mol/m^2 s)
L	Length of the ESP plates (m)
L	Mole flow rate of the liquid (mol/s)
L	Obukhov Length (m)
L_B	Cyclone body length (m)
L_C	Cyclone cone length (m)
$\bar{L}$	Mole flow rate of the solute fee liquid (mol/s)
LHV	Lower Heating Value (J/mol)
m	Mass of a molecule (kg)
$\dot{m}$	Air emission rate by mass (kg/s)
$\dot{m}$	Mass flow rate of fluid (kg/s)
m_i^0	Molality (mol/kg)
m_l	Mass of the fuel droplet (kg)
m_w	Mass of water vapor in given air volume (kg)
$m_{w,s}$	Mass of water vapor required to saturate (kg)
$\dot{m}_v$	Liquid fuel droplet vaporization rate (kg/s)
M	Molar weight (kg/kmol)
M_{eq}	Equilibrium concentration in adsorption (kg/kg)
M_{max}	Maximum loading potential of adsorbate that can be loaded to per mass of adsorbent (kg/kg)
M_x	Flux of momentum along x-direction (1/s m^2)
n	Mole amount of gases (mol, kmol)
n	Number of ions (–)
n	Number of molecules (–)
n	Particle number concentration ($1/m^3$)
n'	Mole transfer of pollutant (mol/m^2 s)
n_c	Number of molecule colliding with the wall (–)
N	Number of aerosol particles (–)
N	Cumulative number of particles deposited per unit area of surface ($1/m^2$)
N_e	Number of turns the gas flow makes before turning upward in a cyclone
N_{i0}	Ion concentration in the charging zone ($1/m^3$)
N_{vdw}	Van der Waals number (J m^3/mol^2)
N_A	Avogadro constant, 6.022×0^{23} (1/mol)
NTU	Number of transfer unit (–)
p_{sat}	Saturation vapor partial pressure at the actual dry bulb temperature (Pa)
p_w	Vapor partial pressure (Pa)

$p(x)$ Probability function (–)
P Gas pressure (Pa)
P Potential energy (J)
P_i Partial pressure of gas i in the gas phase above the liquid (Pa)
P_p Penetration efficiency of particles (–)
$P^n_{o_2}$ Partial pressure of the oxidizer (Pa)
P^0_v The original vapor pressure (Pa)
Pe Peclet number (–)
Pe_m Modified Peclet number (–)
q Charge carried by particle (C)
$\dot{q}$ Total sensible heat transfer rate (J)
Q Heat added to the system (J)
Q Volumetric gas flow rate (m^3/s)
Q_e Volumetric flow rate of exhaust air (m^3/s)
Q_e Air exits in the room at a volumetric flow rate (m^3/s)
Q_i Volumetric flow rate of internal air cleaning (m^3/s)
Q_o Volumetric flow rate of fresh air (m^3/s)
Q_r Volumetric flow rate of recirculating air (m^3/s)
Q_s Intake (supply) fresh air volumetric flow rate (m^3/s)
Q_{sh} Sheath air flow rate (m^3/s)
r Radial position (m)
r_A Reaction rate (mol/m^3 s)
r_s Radius of a droplet (m)
r_{s0} Initial droplet diameter (m)
r_{EGR} EGR percentage (–)
r^* Dimensionless distance from the center of the fiber (–)
r_1 Inner radius of annular space (m)
r_2 Outer radius of annular space (m)
R Interception parameter (m)
R Universal ideal gas constant, 8.314 (J/mol K)
R Overall resistance to mass transfer (m/s)
R_x Liquid phase resistance to mass transfer (m/s)
R_y Gas phase resistance to mass transfer (m/s)
R^* Characteristic radius of two bodies in contact (m)
Re Reynolds number (–)
Re_p Particle Reynolds number (–)
Re_x Boundary layer Reynolds number at x (–)
RH Relative humidity (–)
S Internal source rate with respect to the air pollutant (Depends)
S Length of vortex finder of a cyclone (m)
$\dot{s}$ Pollutant production rate in the room (kg/s)
s_f Length of the uniform single fiber (m)
S_f Particle shape factor (–)
s_z Atmosphere stability indicator (–)

Stk	Stokes number (–)
Stk_m	Modified Stokes number (–)
Stk^*	Critical Stokes number (–)
t_e	The life time of fuel droplet (s)
t_x	Breakthrough time of an adsorption column (s)
T	Temperature (K)
T_a	Adiabatic flame temperature (K)
T_0	Initial temperature (K)
T_0	Reference temperature (K)
T_s	Sutherland's constant (K)
T_s	Surface temperature of the fuel droplet (K)
T_v	Vaporization temperature for the liquid fuel (K)
THP	Total heat production of occupants (J/s)
TLV	Threshold limit value of the air pollutant (Depends)
u	Velocity (m/s)
u_g	Magnitude of the mean fluid velocity (m/s)
u_s	Vapor speed leaving the surface of the fuel droplet (m/s)
u_∞	Air speed of the free steam (m/s)
u_*	Friction speed (m/s)
u_{10}	Wind speed at 10 m above the ground (m/s)
$\bar{u}_g$	Mean gas speed (m/s)
u_θ	Transverse/tangential speed (m/s)
U	Gas velocity (m/s)
U_0	Air speed approaching the filter fiber (m/s)
U_0	Free air stream velocity (m/s)
U_s	Sampling velocity (m/s)
U_∞	Air velocity approaching the filter (not the fiber) (m/s)
v	Speed of the air parcel (m/s)
v	Gas specific volume (m^3/kg)
v	Velocity on the streamline (m/s)
v_f	Average air velocity in the focusing orifice exit plane (m/s)
v_r	Radial speed of the particle (m/s)
v_s	Stack emission speed (m/s)
v_{TS}	Terminal settling velocity (m/s)
v_0	Initial velocity of particle in the still air (m/s)
v_{cr}	Particle critical velocity (m/s)
$\bar{v}_{im}$	Mean impact velocity in thermal rebound (m/s)
v_{im}	Impact velocity in thermal rebound (m/s)
v_θ	Transverse/tangential speed (m/s)
V	Total reactor volume (m^3)
V	Volume of the indoor pace (m^3)
V	Volume of gas (m^3)
V	Volume of the air parcel (m^3)
V	Voltage of ESP (V)

V_{az} Adsorption wave propagation speed (m/s)
V_E Terminal precipitating velocity of ESP (m/s)
$\bar{V}$ Average voltage on the inner collector rod (V)
w Water vapor mass fraction in the air (kg vapor/kg air)
w Humidity ratio of air (–)
w Specific humidity of air (–)
w_e Moisture content of the air exiting the room (–)
w_s Moisture content of the supply air (–)
$\dot{w}$ Moisture production rate in the room (kg/s)
W Weight of the adsorbent packed (kg)
W Work done by the system on the surroundings (J)
x Mole ratio in the bulk liquid phase (–)
x_c Critical distance for maximum plume rise (m)
x_i The equilibrium concentration of gas i in the liquid phase (mol/m^3)
x_F Remaining mass fraction of the solid fuel (–)
x^* Hypothetical mole fraction corresponding to x in the bulk fluid (–)
X Mole ratio of pollutant to the corresponding liquid (–)
X^* Hypothetical or equilibrium mole ratio in gas (–)
y Mole fraction of the target gas at gas–liquid interface (–)
y_i Mole fraction at the interface in gas phase (–)
y_{ss} Mole fraction of the target gas in the bulk liquid at steady state (–)
y^* Hypothetical mole fraction corresponding to y in the bulk fluid (–)
Y Mole ratio of pollutant to the corresponding gas (–)
Y Hydrodynamic factor (–)
Y^* Composite Young's modulus (N/m^2)
Y^* Hypothetical or equilibrium mole ratio in gas (–)
z Elevation (m)
z_0 Surface roughness height (m)
Z_e Mobility of the ions (m^2/V s)
z_{mix} Mixing height in atmosphere (m)
Z_e Equilibrium distance between the particle and another surface (0.4 nm) (m)
Z_p Electrical mobility of a particle (m^2/V s)
Z_p^* Optimum electrical mobility of a particle (m^2/V s)
α Solidification of a filter (–)
δ Boundary layer thickness (m)
δ Width of the adsorption wave (m)
δ_{AB} Jump distance in analyzing the aerosol coagulation (m)
Δh Plume rise (m)
Δh_m Maximum plume rise (m)
ΔH_R Enthalpy of reaction (J/mol)
Δt Time difference (s)
ΔP Pressure drop across the filter (a positive number) (Pa)
ΔU Internal energy in the system (J)
Δx Thickness of the porous media (m)

$\Delta\gamma$	Specific adhesion energy (J/m^2)
ε_r	Relative permittivity of the fiber (–)
ε_r	Relative permittivity or dielectric constant of the particle with respect to vacuum (–)
ε_0	Permittivity of a vacuum, 8.854 × 10^{-12} (C/V m)
η_{ad}	Particle adhesion efficiency (–)
η_D	Single fiber efficiency by diffusion (–)
η_{ip}	Single fiber efficiency by impaction (–)
η_{it}	Single fiber filtration by interception (–)
η_s	Sampling efficiency (–)
η_{sG}	Single granule efficiency (–)
η_{sf}	Total efficiency of a single fiber (–)
η_{ts}	Particle transport efficiency (–)
η_p	Separation efficiency for particles (–)
η_x	Breakthrough efficiency in adsorption (–)
θ	Angle (rad)
λ	Mean free path of gas (m)
μ	Kinetic (Dynamic) viscosity (Pa s, N s/m^2)
ρ	Density (kg/m^3)
ρ_a	Density of the surrounding air (kg/m^3)
ρ_b	Bulk density of the adsorbent (kg/m^3)
ρ_0	Ground level air density (kg/m^3)
ρ_p	Density of the air parcel (kg/m^3)
ρ_p	Particle density (kg/m^3)
ρ_r	Reaction rate ratio (–)
σ_g	Geometric standard deviation (–)
σ_y	Dispersion coefficient in the transverse (y) direction (m)
σ_z	Dispersion coefficient in the vertical (z) direction (m)
τ	Characteristic time for aerodynamic particle settling (s)
τ	Half-value time in aerosol coagulation (s)
τ	Shear stress (N/m^2)
τ_0	The ground level (surface) shear stress (m)
Φ	Equivalence ratio in combustion (–)
Φ	Latitude where the air is of concern (–)
$\emptyset(u)$	Cumulative probability function (–)
Ω	Angular speed of the earth, 7.25 × 10^{-5} rad/s
$[i]$	Mole concentration (mol/m^3)

Chapter 1
Air Emissions

1.1 Air

Air surrounds planet Earth and is held in place by gravity. Typical standard air is defined as dry air at sea level, where the temperature and pressure are 15 °C and 101,325 Pa, respectively. The composition of standard air is given in Table 1.1 [4]. By mole or volume, standard air contains 78.08 % nitrogen, 20.95 % oxygen, 0.93 % argon, 0.0314 % carbon dioxide, and trace amounts of other gases. These ratios change very little with location or time. Most of the time, however, atmospheric air is not dry, and there is a moisture content of about 1 %. The moisture content of the air presents in form of water vapor or liquid droplets or ice crystals.

1.2 Air Pollution and Greenhouse Gases

Air emissions include air pollution emissions and greenhouse gas (GHG) emissions. Air pollution emissions, which are commonly referred to as air pollutants, are species in the air that cause acute or chronic human health problems and negative environmental impacts. GHGs are not toxic unless at extremely high concentrations, but their higher-than-normal concentrations in the atmosphere may have resulted in global warming and climate change with possible catastrophic consequences on planet Earth.

Air pollution occurs when unwanted materials are added to the air, especially in abundance, resulting in effects on the environment and health. In principle, these unwanted materials may be anything in any phase, such as carbon dioxide (CO_2), liquid droplets, and solid dust particles. A material is labeled as an air pollutant only when it is against the interest of human beings. For example, sulfur dioxide (SO_2) is considered as an air pollutant due to its negative impact on the environment and human health, which will be elaborated on shortly.

Z. Tan, *Air Pollution and Greenhouse Gases*, Green Energy and Technology,
DOI 10.1007/978-981-287-212-8_1

Table 1.1 Pure dry air at sea level

Gas	Symbol	Percent by volume (%)
Nitrogen	N_2	78.084
Oxygen	O_2	20.9476
Argon	Ar	0.934
Carbon dioxide	CO_2	0.0314
Neon	Ne	0.001818
Methane	CH_4	0.0002
Helium	He	0.000524
Krypton	Kr	0.000114
Hydrogen	H_2	0.00005
Xenon	Xe	0.0000087

In another example, if there is over 78 % of nitrogen in the air, it is not considered as an air pollutant, because it does not impose any noticeable negative effect on human beings or the environment. On the other hand, although CO_2 has been a component of clean dry air (Table 1.1) before human history, it was not labeled as an GHG emission until recently, when scientists observed a strong correlation between the increased level of atmospheric CO_2 and global warming (or climate change), which could eventually eliminate the existence of human beings on this planet.

1.2.1 Air Pollution

Air emissions have natural or anthropogenic origins; however, this book focuses primarily on the latter, which include industrial activities and the burning of fossil fuels. These constituents of pollution have the potential to affect the majority of people in a region.

Air pollutants comprise primary and secondary air pollutants. Primary air pollutants are emitted directly from sources. They include, but are not limited to, particulate matter (PM), sulfur dioxide (SO_2), nitric oxides (NO_x), hydrocarbon (HC), volatile organic compounds (VOCs), carbon monoxide (CO), and ammonia (NH_3). Secondary air pollutants are produced by the chemical reactions of two or more primary pollutants or by reactions with normal atmospheric constituents. Examples of secondary air pollutants are ground level ozone, formaldehyde, smog, and acid mist.

Particulate matter is a mixture of solid particles and liquid droplets suspended in the air. In this book, particulate matter is interchangeable with aerosol, which is a suspension of solid or liquid particles in a gas. It is a two-phase system consisting of particles and the gas in which they are suspended. Particulate matter can be both primary pollutants and secondary pollutants that are sent directly into the atmosphere in the form of windblown dust and soil, sea salt spray, pollen, and spores. Other examples of PM are smoke, fumes, and haze.

For particulate matter where particle diameters are smaller than x micrometers, it is defined as PMx. Commonly used terms are PM_{10} and $PM_{2.5}$. Sometimes, particulate matter and aerosol is exchangeable. Monodisperse aerosols, in which all particles have the same size, can be produced in laboratory for use as test aerosols. In practice, engineers deal with polydisperse aerosols (i.e. suspended particles are in a wide range of sizes), and statistical measures should be used to characterize particle sizes. Aerosol Technology by William Hinds (2006) is one of the reference books for this subject.

Air pollutants other than PM present primarily as gases. Volatile organic compounds (VOCs) are chemicals that contain carbon and/or hydrogen and evaporate easily. VOCs are the main air emissions from the oil and gas industry, as well as indoor consumer products and construction materials, such as new fabrics, wood, and paints. VOCs have been found to be a major contributing factor to ground-level ozone, a common air pollutant, and a proven public health hazard. Sulfur dioxide (SO_2) and nitric oxides (NO_x) are two major gaseous air pollutants generated through combustion processes. Carbon monoxide (CO) and hydrocarbon (HC) are generated from incomplete combustion and are converted into CO_2 through a complete combustion process.

Secondary air pollutants are those formed through complex physical and/or chemical reactions, e.g., coagulation and condensation or photochemical reactions, respectively, e.g.:

$$SO_2 + NH_3 + VOCs \rightarrow \text{Aerosol} \tag{1.1}$$

Atmospheric physics and chemistry are so complicated that they are still not well understood. Most gaseous air emissions such as NO_x, SO_2, HC, VOCs, and ammonia (NH_3) are converted into PM. Readers are recommended to read *Atmospheric Chemistry and Physics* by Seinfeld and Pandis [20] for in-depth knowledge on this subject.

Air pollution is an evolving subject and inevitable, as the demand for energy increases. Air pollution really flourished with the Industrial Revolution and continues to grow with the human appetite for comfort and speed.

At first, the study of air pollution focused on recurring episodes of high levels of air pollution in areas surrounding industrial facilities, such as coal burning power plants and chemical refineries. These pollution episodes were accompanied by acute human sickness and the exacerbation of chronic illness. After the mid-twentieth century, when industrialized nations' economies recovered rapidly from World War II, many urban regions without heavy industrial facilities also began to experience high levels of photochemical smog and nitrogen oxides.

The twentieth century marked the beginning of the understanding that human activity was having deleterious effects upon the natural world, including human health and welfare. These effects included increasing pollution of air, water, and land by the byproducts of industrial activity, and the permanent loss of natural species of plants and animals through changes in laboratory settings, water usage, and human predations.

The topic of indoor air quality (IAQ) has become popular, due to the awareness of asthma and allergies triggered by indoor air pollutants such as mold. IAQ awareness also increased with the involvement of the United States Environmental Protection Agency. The energy crisis in the 1970s resulted in tighter building envelope, sealing, and insufficient ventilation. Most existing heating, ventilation, and air conditioning (HVAC) systems were designed for temperature control without consideration of air pollutant accumulations. As a result, IAQ degraded, and problems arose. Recent findings have demonstrated that indoor air is often more polluted than outdoor air in many developed countries (except for regions like Beijing, China, where outdoor air is extremely polluted), thereby causing a greater health concern as current lifestyles demand more time indoors.

The later twentieth and early twenty-first centuries saw a boom in nanotechnology. Nanotechnology has been tested for air quality remediation in such areas as noncatalytic combustion and photocatalytic oxidation of volatile organic compounds (VOCs). On the other hand, the environmental effects of nanotechnology are not well understood; and, concerns have recently begun to increase. The world is not ready for nanotechnology because "the future is coming sooner than it is expected" [15]. The effect of nanotechnology to air quality is still waiting for systematic studies to confirm its environmental effects. Scientific evidence is needed before definitive conclusions can be made.

1.2.2 Greenhouse Gases

Greenhouse gases (GHGs) include CO_2, methane (CH_4), hydrofluorocarbons (HFCs), perfluorinated compounds (PFCs), and others. Most of them have been produced by natural sources and present in the atmosphere for a long, long time. It is only recently that extra anthropogenic GHGs may have contributed to global climate change and extreme weather.

CO_2 emissions are sometime referred to carbon emissions and are largely due to the combustion of fuels in electric power generation, engines, building heating, and industrial plants. In addition to oil and gas industry, CH_4 emissions can also be produced by the biological degradation of biomass from agricultural activities and landfills. Industrial processes produce HFCs and PFCs. Chapter 13 discusses the capture and storage of CO_2.

1.3 Effects of Air Pollution and GHGs

Clean air is essential to humans and other living beings on Earth. One can survive without water for 2 days or food for 2 weeks, but very few can survive a few minutes without air. On a daily base, an average adult consumes 2 kg of water [25], 1 kg of food and 20 kg of air. However, polluted air impacts public health when the pollutant concentrations are high enough. Polluted air also affects the environment by lowering its visibility and damaging the other species and materials on the planet. Excessive GHGs in the atmosphere affect Earth's energy balance and, consequently, the climate and weather. The effects of air pollution and GHGs are introduced in the following subsections.

1.3.1 Health Effects of Air Pollution

Poor air quality is responsible for one out of eight global deaths [31]. Air pollution has become the single largest environment and health challenge to society. Air pollution has a direct negative effect on human health. Acute air pollution episodes can result in cardiac and respiratory diseases, bronchitis, pneumonia, and sometimes death. Historically, these conditions occurred in many places, including Muese River Valley, Belgium (1930), Donora, Pennsylvania, USA (1948), London, England (1952, 1962), New York City (1953) and Bhopal, Indian (1984). It is still happening: for example, the most recent outbreak of smog in many Chinese cities, which were responsible for the deaths of 1.2 million/year in China.

Air pollutants enter the human body (or that of any creature) mainly through the respiratory system, although they can also be ingested or absorbed through the pores of skin. Part of the inhaled air pollutants can be exhaled, but most reach the lungs, with some penetrating through the lungs and entering the circulation system. These contaminants can be then transported all over the body, with some chemical reactions forming new chemicals.

Toxic chemicals can interfere with normal body functions, resulting in many health problems. Some of these effects are acute, including eye irritation, headaches, and nausea, while others may be chronic and irreversible, such as organ damage, birth defects, heart disease, cancer, and even death.

Aerosol particles are deposited into the respiratory system at different locations according to particle sizes. Particles larger than 15 μm (in diameter, which is this book's default unless stated otherwise) are captured by small nasal hairs or mucus membranes. Particles smaller than 10 μm are trapped in the cilia, which are the body's last line of defense in the bronchioles before reaching the alveoli. The lower respiratory system is composed of the bronchi, bronchioles, and alveoli, where most pollutant exchange occurs. Alveoli contain millions of tiny air sacs. Most particles between 1 μm and 100 nm (=0.1 μm) settle in the alveoli. Alveolar tissue holds the particles in place for weeks to years. The toxins and chemicals adsorbed by the particulates are then dissolved and transported to the circulation system of the body.

As such, the potential chronic health effects of particulate matter are lung cancer, pulmonary emphysema, bronchitis, asthma, and other respiratory infections [10]. In addition, inhaled gaseous air pollutants may slow the action of the tiny cilia and result in the deep embedment of more particles in lung tissue.

Although human beings have mechanisms to remove and clear particles from the respiratory system, air pollutants can quickly overcome these natural defense mechanisms. Once the chemicals carried by the air pollutants are absorbed into the bloodstream, they can be transported to a distant organ or tissue to create a variety of adverse effects. The respiratory effects result in skin and eye irritation, inflammation, allergic reaction, cough, chest pain, bronchitis, pulmonary emphysema, lung cancer, decreased respiratory efficiency, diminished pulmonary circulation, and enlargement and weakening of the heart and blood vessels.

In the United States of America (USA), where air is relative clean on average, lung dysfunction, which is a result of exposure to air pollutants, ranks among top 10 most important occupational diseases and injuries. Consequently, these effects can result in the increase in medical costs and premature death. Individual reactions to air pollutants depend on the type of pollutant a person is exposed to, the degree of exposure, the individual's health status and genetics. People staying in offices are exposed to indoor air pollutants rather than those who work outdoors. Elevated exposure to air pollutants correlates to increases in emergency room visits, hospital admissions, and premature deaths.

Children and elders are more vulnerable to air pollution. Children are very sensitive to the effects of air pollution, as they breathe more rapidly and inhale more pollutants per body weight than adults do. Therefore, their lungs have a greater chance of exposure to harmful air pollutants. Fine particles can harm lung development and cause early childhood asthma. Polluted air may contribute to permanent lung damage during the development periods of children's lungs. Recent studies also indicate that air pollutants have a larger impact on people who are already ill than unhealthy ones.

The short-term health effects of air pollutants are quantified using an air quality index (AQI) [14]. Computation of the AQI requires multiple air pollutant concentrations, often in terms of the presence of five common pollutants—sulfur dioxide, nitrogen dioxide, carbon monoxide, suspended particulates, and ground-level ozone [26]. The calculated AQI values are divided into ranges, and each range is assigned a descriptor and a color code. A lower AQI value indicates better air quality.

In the USA AQI, for example, there are six levels of AQI, Good (0–50), Moderate (51–100), Unhealthy for Sensitive Groups (101–150), Unhealthy (151–200), Very unhealthy (201–300), and Hazardous (>300).

- **Good** indicates that the air quality is considered satisfactory and poses little or no risk.
- **Moderate** air quality is acceptable except for a very small number of individuals, such as those who are sensitive to ozone. These people may experience respiratory symptoms.

- **Unhealthy for sensitive groups** indicates that the air quality is not acceptable for those who are particularly sensitive to certain air pollutants. It is likely to affect at levels lower than that of the general public: for example, children and adults who are active outdoors and people with respiratory disease are at greater risk from exposure to ozone. However, the general public is fine with the air.
- **Unhealthy** AQI values correspond to the air that the majority of the public may begin to experience an adverse effect on health.
- **Very unhealthy** air triggers a health alert, meaning everyone may experience serious health effects.
- **Hazardous** AQI values trigger health warnings of emergency conditions.

1.3.2 Environmental Impact

Air pollution affects not only air quality, but also indirectly the quality of groundwater, land, vegetation, forest, and climate. Air pollutants contaminate the clouds first and then return to Earth together with precipitation. These precipitations carry air pollutants back to Earth and to any subjects with which they come in contact, such as land, water, and vegetation. As introduced shortly, acid rain is an excellent example of this case. Air pollutants, such as SO_2 and NO_x, cause direct damage to leaves of plants and trees when they enter leaves' stomata. Chronic exposure of leaves and needles to air pollutants can also break down the waxy coating that helps prevent excessive water loss and damage from diseases, pests, drought, and frost. "In the Midwestern United States crop losses of wheat, corn, soybeans, and peanuts from damage by ozone and acid deposition amount to about $5 billion a year" [17].

1.3.2.1 Low Visibility

Low visibility is a direct and practical perception of poor air quality. Low visibility occurs as a result of the scattering and absorption of light by air pollutants. Both primary and secondary particulate air pollutants contribute to visibility impairment. Humidity can worsen the low visibility by changing the fate and sizes of the particulate matter: for example, sulfates accumulate water and grow in size, becoming more effective in lowering the visibility.

Visibility is also reduced by ground-level ozone, which is a secondary air pollutant formed during the reaction between NO_x and VOCs. Sulfate and nitrate fine particles can suspend in air for a long period of time; and, more importantly, they give smog its yellowish-brown color and reduce the visibility. Particles falling on public buildings and vegetation also affect their appearance.

1.3.2.2 Acid Rain

There are wet forms of acidic pollutants found in rain, snow, fog, and cloud vapor. Most acid rain forms when NO_x and SO_2 are converted through oxidation and dissolution to nitric acid (HNO_3) and sulfuric acid (H_2SO_4), respectively, and also when ammonia gas (NH_3) is converted into ammonium (NH_4^+). Emissions of sulfur and nitrogen oxides to the atmosphere have increased since the Industrial Revolution and have recently become more prevalent in China, Eastern Europe, and Russia, where sulfur-containing coal is the major energy source.

Acid rain was first discovered in 1852 and has been extensively studied since the late 1960s [2, 19]. The most important gas that leads to acidification is sulfur dioxide. Emissions of nitrogen oxides are also of increasing importance, due to their contribution to ozone and nitric acid.

Acid rain has an adverse impact on ecology. It falls on forests, soils, and bodies of freshwater, killing off insect and aquatic life forms. Acid rain can seriously damage soil biology. While some microbes can quickly consume acids, others cannot tolerate much of it. The enzymes of the latter are changed by the acid and lose their capabilities. In addition, acid rain also removes nutrients and minerals, which are necessary to maintain a flourishing ecosystem. The depletion of minerals in the soil can slow the growth of food crops, plants, and forests. When the pH values are extremely low, the acid rain can cause a part of or an entire forest to die, mostly because the trees are weakened and become susceptible to harsh environments.

Acid rain can also corrode building materials and historical monuments. Sculptures commonly lose their sheen, because acid rain chemically reacts with the calcium compounds in the stones.

1.3.3 Greenhouse Gas Effects

Under normal conditions, solar radiation reaches the surfaces of the Earth after penetrating through its atmosphere. Some of the solar energy (heat) is absorbed, reflected or reradiated within the Earth's atmosphere. With the increases of greenhouse gases (GHGs) in the atmosphere, long-wave infrared (IR) heat radiation is reradiated and trapped within the Earth's atmosphere, resulting in the warming of the Earth's surface. Readers are referred to heat transfer textbooks for in-depth knowledge of solar radiation. Related courses are also offered to undergraduate students in most engineering schools.

The stratosphere of the Earth contains a layer of ozone gas (O_3) that normally absorbs most of the harmful, shortwave ultraviolet radiation (UVB) from the sun, protecting Earth's living organisms. However, this protective shield has become thinner and thinner. Recent data shows a trend of 3.4 % decrease per decade in the average total ozone over the northern hemisphere's mid-latitudes since 1979. As the ozone layer thins, more UVB radiation reaches the surface of the Earth, resulting in global warming and climate change. Ozone depletion is also expected to lead to increased skin cancer rates and suppression of the immune system. It also impacts the ecosystem by slowing down the growth of certain food plants.

Scientists had linked several substances associated with human activities to ozone depletion, including the use of chlorofluorocarbons (CFCs), halons, carbon tetrachloride, and methyl chloroform. These chemicals are emitted from such industrial processes as commercial air conditioners, refrigerators, and insulating foam.

Climate change due to GHG emissions has become a prominent issue on the global stage. Although there is a disagreement of the causes, it is doubtless that the global climate is changing. The global temperature has a very close correlation with the atmospheric CO_2 concentration. Over the past century, the atmospheric CO_2 concentration increased from $\sim$280 to 370 ppmv, with a rapid increase in last few decades. The global temperature has risen by 0.5 °C during this time period [11]. This change rate is unmatched in the past 1,000 years, and 11 of the 12 years from 1995 to 2006 were the warmest years in the instrumental record of global temperature since 1850. The temperature increase is widespread over the globe and is greater in cold and remote areas. Climate change is believed to be caused by human activity, primarily the burning of fossil fuels, resulting in a buildup of GHGs [13].

1.4 Roots of Air Pollution and GHGs

Air emissions originate from both natural and anthropogenic sources. Nature has been producing many materials that are now being considered as air emissions. For example, carbon monoxide (CO), sulfur dioxide (SO_2), and ash particles are abundant in volcanic emissions. Methane is naturally produced by the digestion of food by animals and the degradation of biomass. Radon gas from radioactive decay within the Earth's crust, and wildfires generate smoke and CO. However, these sources are not the major concern to society.

1.4.1 Anthropogenic Air Emissions

Anthropogenic sources are related to the production and combustion of different fuels. Human activities are responsible for the increases in ground-level O_3 and GHGs in recent years. About 95 % of NO_x (nitric oxides) from human activity come from the burning of fossil fuels in power plants, engines, homes, and industries.

Air emissions are the byproducts of energy production and consumption; for example, the combustion of methane in pure oxygen can be written in a simplified form as

$$CH_4 + O_2 \rightarrow CO_2 + H_2O + \text{Heat} \tag{1.2}$$

The main objective of fuel combustion is heat, which is used for other engineering applications. Unfortunately, CO_2 is released as an air emission from the combustion process, which is complicated and generates many pollutants. In reality, air instead of pure oxygen (as shown in Eq. (1.2)) is used for fuel combustion. Moreover, fuels are not purely carbon and hydrogen, and their compositions have to be analyzed before the potential pollutants can be predicted. A power plant can be used as a more realistic example. When burning coal to generate thermal energy to heat the water for steam, primary air emissions including CO_2, SO_2, NO_x, heavy metals, and particulates are produced and emitted from the stack into the atmosphere.

The emissions of air pollutants and GHGs from the stack depend on many factors, including but are not limited to the properties of the fuel, the combustion process, and downstream flue gas treatment devices. There is much more SO_2 in flue gas when high sulfur coal is used as the fuel instead of gasoline. Nonetheless, commercial natural gas is much cleaner than other fossil fuels because of it is free of ash and sulfur. Fossil energy consumption is the main source of air emissions. In the United States, more than 80 % of the CO_2 emissions are related to direct energy consumption [26, 27].

1.4.2 Growing Population and Energy Consumption

From a global point of view, the growing population and the desire of human beings in pursuit of high-speed transportation and comfortable lives are at the roots of the air emissions. There has been a worldwide growth in population in both developed and developing countries. However, the world population growth is mainly attributed to that in developing countries. The twentieth century saw the biggest increase in human population in history, due to medical advances and massive increases in agricultural productivity. The United Nations [24] estimated that the world's population was growing at the rate of 1.14 % or about 75 million people per year in 2012. By 2050, the world population is projected to be in the range of 7.4–10.6 billion.

Global emissions of CO_2, for example, also increased with the population [1], rising from 21.2 to 26.9 billion metric tons. There were nearly no increases in North American emissions, and a significant decrease occurred in Europe, whereas CO_2 emissions from Asia have doubled.

The growing global population demands more energy consumption to maintain the same level of lifestyle; and, for developing countries, the increasing population has a more significant effect as the life quality keeps improving. There is a minimum amount of energy needed to sustain human life. Fuel energy is needed for cooking in houses and restaurants, as well as in heating and/or cooling for houses and buildings. Much more energy than the minimum is consumed in developed countries, providing food, clothing, shelter, transportation, communications, lighting, materials, and numerous services for the entire population.

The end uses of energy consumption vary from nation to nation, and from sector to sector. In developed countries, such as the US and Canada, more energy is used for luxury. In 1995, the worldwide consumption rate was about 2 kW per capita, but the US consumed about 13 kW per capita, with the global rate. If energy usage is divided into four sectors—industrial, transportation, commercial services, and residential—these categories were responsible for the consumption of 36, 37, 16, and 21 %, respectively, of the total energy used (1996 data). Most recent data can be found in the annual International Energy Outlook in the following section.

1.4.3 International Energy Outlook

Global energy demand has been growing despite the increase in oil and gas prices. Worldwide energy consumption is projected to increase by 57 % from 2004 to 2030 [7]. Total energy demand in developing countries will rise by 95 %, compared to an increase of 24 % in developed countries. Fossil fuels will remain the major energy sources worldwide, with liquids remaining dominant due to their irreplaceable importance in the transportation and industrial sectors. However, their share of the world energy market in 2007s outlook is lessened in the projection, as other fuels replace liquids where possible outside those sectors. Nuclear power and renewable energy sources are expected to expand with the increases in fossil fuel prices (Fig. 1.1). Figure 1.1 is produced using the original data of website by searching "Key World Energy Statistics 2010" published by the International Energy Agency [12].

Natural gas consumption will rise on average by 1.9 % per year from 2004 to 2030. The price of gas is expected to rise continuously as demand increases, consequently making coal more cost-competitive, especially in electric power generation. As a result, coal, which is a major combustion polluter, will be the fastest growing energy source worldwide, producing even more air pollutants and greenhouse gases.

Nuclear power generation in developing countries is projected to increase by 4 % per year from 2004 to 2030, but will remain a small portion compared to other energy sources. Unconventional resources, such as biofuels and liquefied coal or gas, are expected to become increasingly competitive. Renewable energy sources are expected to continue to increase annually and become more competitive economically with higher fossil fuel prices and support from government incentives.

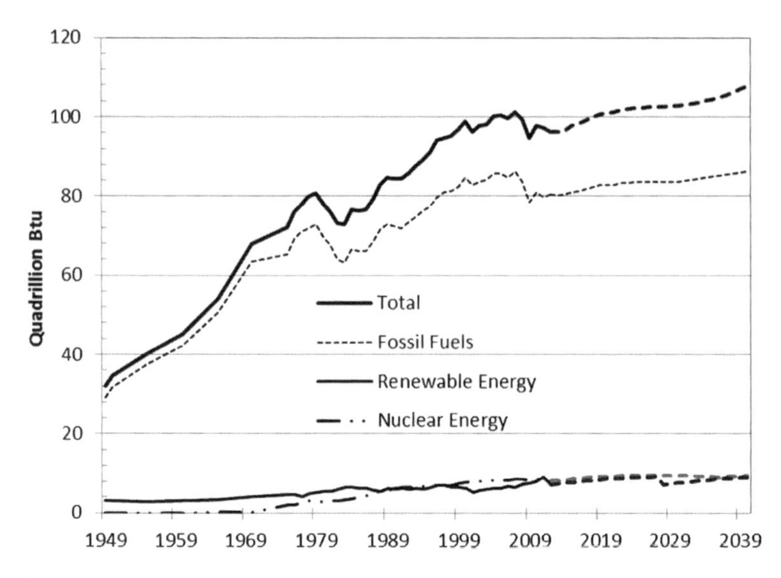

Fig. 1.1 US energy historical use and projection by fuel type (1949–2040)

1.4.4 Global Air Emissions

Before the 1980s, air pollution was mainly a local concern; and, strict national industrial controls in developed countries led to a significant reduction in air emissions and improved the air quality in urban areas. For example, the US Clean Air Act of 1970 was passed to address the increasing concerns of air pollution. Similar changes took place throughout the industrialized world at that time, and the ambient air quality has improved locally in developed countries. In a modern world, air emissions should be considered from a global perspective. While some people enjoy the comfort and speed of modern conveniences and luxuries provided by energy consumption, others are suffering from the resulting air emissions, with or without intention. There is no doubt that fossil fuel is and will remain the dominant source of global energy in the coming decades. The demand of fossil fuels is still growing, as is the output of air emissions.

Global air pollution continues to worsen due to the population growth and scant environmental restrictions in many nations. The situation is worse in the developing countries, especially for the emerging economies. Figure 1.2 shows the map of global air pollution using $PM_{2.5}$ as an indicator [28]. As seen in Fig. 1.2, recent rapid unsustainable economic development has also resulted in considerable air pollution in China. Many Chinese cities often seem wrapped in toxic gray shrouds, and visibility is extremely low most of the time. Ninety-nine percent of the city dwellers in China breathe air that is considered to be polluted by the European Union [16].

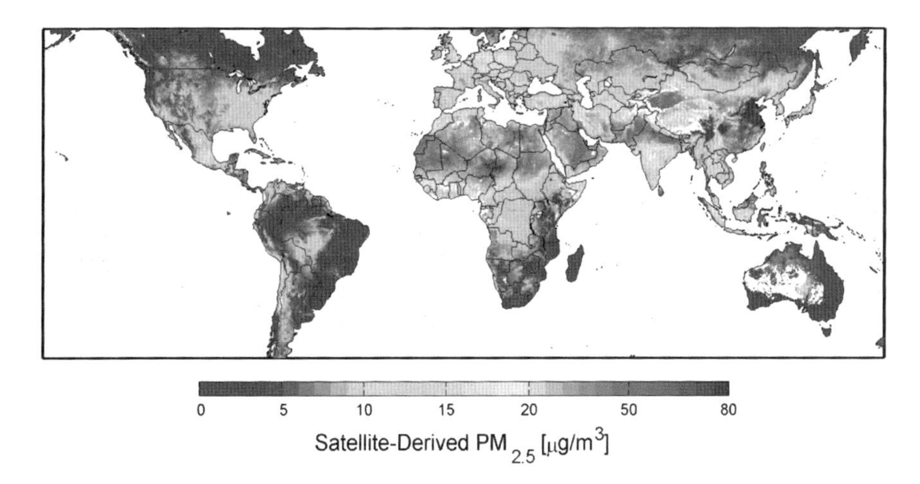

Fig. 1.2 Global satellite-derived $PM_{2.5}$ [28], reproduced with permission from environmental health perspectives)

Air pollution can be persistent, and many air pollutants travel with atmospheric turbulence and can thus affect areas far-remote from the source of the pollution. For example, the air pollution problems in China and India have affected not only their nearby neighbors, but also locations on the other side of the globe. Emissions of SO_2 and NO_x from China's coal-fired power plants fall as acid rain onto the territory of neighboring countries. Dust storms originating in China can reach the atmosphere of the United States. Some of the particulate pollution over Los Angeles, USA is generated in Asian countries.

Atmospheric circulation describes the global air circulation, which can take years to complete. Most of the motions of the atmosphere are actually horizontal as a result of uneven heating of the Earth's surface (most at the equator and least at the poles), the Earth's rotation (Coriolis force), and the influence of the ground and the sea [20]. While it is not within the scope of this book, global air circulation is the driving force for air pollutants emitted from a Chinese power plant to travel to the opposite side of the globe.

As a result of globalization, developing countries become polluted before and during their economy growth. Although the reasons behind it are subject to a debate, the fact is that newly developed countries, such as China and India, are much polluted as they grow. Differences among countries' environmental standards and the costs for remediation have caused the relocation of pollution-intensive industries from strictly controlled countries to those with few or no standards by creating "pollution havens" for developed countries. Air pollution in the United States has decreased steadily since the 1970s, but the opposite is the case in developing countries [6, 8].

Highly polluting manufacturing factories, such as chemicals, electronics, and automobiles, have shifted from developed countries to developing countries, partially due to less stringent emission control policies and weak implementation, in addition to the low labor cost. Countries that have a particularly high demand for environmental quality currently specialize in products that can be produced in an environmentally friendly manner. However, these countries overlook the protocol of other countries that manufacture products that produce pollution.

1.5 General Approaches to Air Emission Control

Currently, it is seemingly impossible to have a pollution-free environment without a cost. It is only reasonable to control the air quality to a certain level at an appropriate cost. To achieve this goal, all governments, industries, and the public have to work together and share the cost, more or less. The government's role is the making and enforcing of laws and regulations based on local affordability and implementation ability. Regulations change over time and at different locations, but they should be cost-effective and enforceable. It is meaningless to set a standard so high that cannot be met.

1.5.1 Air Emission and Air Quality Standards

Government agencies approach air pollution control by developing and enforcing air emission standards and air quality standards. Air emission taxes and cost–benefit standards are relatively newer approaches that are not yet well implemented and, as such, are excluded from this book.

In most regions, standards are implemented by using a combination of both air emission and air quality standards. In practice, air quality standards set up by governments define "clean air." Each standard is set by government agencies, and implementation protocols are also provided. Air emission and air quality standards are widely accepted and used in many countries and are elaborated in the following subsections. These standards are first introduced here. Chap. 15 covers the technical knowledge for the implementation of these standards, which require air quality monitoring and air emission monitoring.

1.5.1.1 Air Emission Standards

An air emission standard sets some maximum levels of air emissions from a facility. If emissions from all facilities meet the standards, it enables the lowest level of air emissions into the atmosphere and the best air quality possible. Air emission standards developed by government agencies (such as the US Environmental Protection Agency and Environment Canada) set the upper limits of air emissions from a source and also include the methodologies of meeting these standards.

One approach to the lowest emission level is the employment of the best technologies currently available to keep the equipment in a good operating condition. The corresponding regulations (or standards) have to specify the "best technology" equipment. Since the best technology currently available allows for the lowest emissions currently possible, the corresponding standard does not have to specify the emission rate from a facility or equipment installed with such devices. This "best technology" approach is a good application where accurate emission monitoring is challenging. For example, US state regulations for large underground storage tanks (UST) require the operators to install the best floating roofs available and the equipment has to be well maintained. Another example is the prohibition of manure spread in many local municipalities to minimize the odor dispersion to local communities. It is sometimes economically challenging for industry to keep up with advances in technology. Replacement of the entire air pollution control system of a power plant is not as easy as changing the diesel particulate filter under a truck.

Another approach to air emission standards does not require an operator to install the best technology; it is up to the operator to decide. The regulations set the maximum air emission rates for different air pollutants.

It is simple, but expensive, to implement air emission standards. In these standards, permitted emission rates and the standard test procedures are clearly described. Any trained professional can follow the standard procedure to determine whether the standards are met at a facility. As the best technology approach is generally expensive, it may not be always necessary to implement this standard for facilities in remote areas with low populations.

The flexibility of air emission standards is poor. It is impractical to shut down a utility plant that supplies electricity or heat to a large community, even if it cannot meet the local air emission standard. In this case, it is obvious that air emission standards may fail to control air pollution. Compared to air emission standards, air quality standards are much more flexible in addressing this political challenge.

1.5.1.2 Air Quality Standards

Air quality standards are developed based on extensive scientific study on and validation of the threshold values of the air pollutants. Due to the nature of scientific research, statistical errors are allowed. One well-known example of air quality standards is the US National Ambient Air Quality Standards (NAAQS), which have established limitations on air pollutants considered harmful to public health and the environment (with details available at http://www.epa.gov/air/criteria.html). These standards vary with location, time, and sector and they are subject to regular reviews and revisions, as needed.

Not all air pollutants are regulated by air quality standards. The US Clean Air Act requires the Environmental Protection Agency (EPA) to set National Ambient Air Quality Standards for six criteria air pollutants:

- ozone (O_3)
- particulate matter (PM)
- carbon monoxide (CO)
- sulfur dioxide (SO_2)
- nitric oxides (NO_x)
- lead (Pb)

These air pollutants are found all over the world and may come from primary and/or secondary sources. Of the six pollutants, particulate pollution and ground-level ozone are the most widespread health threats.

The US EPA calls these pollutants criteria air pollutants, because it regulates them by developing human health and/or environmentally based criteria for setting permissible levels. The set of limits based on human health is called primary standards. Secondary standards are another set of limits intended to prevent environmental and property damage. The primary standards are aimed at the protection of the public's health, and the secondary standards at welfare protection. The US EPA has set threshold values for six principal pollutants as listed in Table 1.2.

To implement air quality standards, regulators also provide instructions to quantify the air quality. If the measured pollutant concentrations are less than the threshold values specified in the standards, the air quality is considered acceptable in the near future and no action is needed. On the other hand, if the air quality does not meet the standard, air emissions from some sources have to be reduced in order to comply with the standards of concern. For example, during the 2008 Summer Olympics in Beijing, China, air quality in Beijing improved much more than expected, according to the government official media. Before the Games, it implemented drastic measures to reduce air pollution by allowing only one-third of Beijing's 3.3 million cars on the road on alternate days under an even-odd license plate system.

Air quality standards are flexible in that they offer governments different approaches to improving the air quality in a region. They can, however, be complex and sometimes difficult to enforce, unless there is only one single emission source in a region. Air quality is usually not a problem in this type of regions, due to the effective dispersion of the air. When the air quality standard is not met in populated regions where there are different mobile and stationary sources, it can be difficult to assign the responsibility, although sometimes it is obvious that automobiles and large plants are at the root of the problem. To make it more challenging, the atmosphere itself is responsible for the majority of secondary air pollutants, such as ground-level ozone and smog, which are the products of many complicated chemical reactions in the air. Lawyers are likely to make it even more complicated if there is a court action.

Table 1.2 National Ambient Air Quality Standards (NAAQS) for six principal pollutants

Pollutant		Primary/secondary	Averaging time	Level	Form
Carbon monoxide		Primary	8 h	9 ppm	Not to be exceeded more than once per year
			1 h	35 ppm	
Lead		Primary and secondary	Rolling 3 month average	0.15 μg/m^3	Not to be exceeded
Nitrogen dioxide		Primary	1 h	100 ppb	98th percentile, averaged over 3 years
		Primary and secondary	Annual	53 ppb	Annual mean
Ozone		Primary and secondary	8 h	0.075 ppm	Annual fourth-highest 8-hr daily maximum concentration, averaged over 3 years
Particulate matter	$PM_{2.5}$	Primary	Annual	12 μg/m^3	Annual mean, averaged over 3 years
		Secondary	Annual	15 μg/m^3	Annual mean, averaged over 3 years
		Primary and secondary	24 h	35 μg/m^3	98th percentile, averaged over 3 years
	PM_{10}	Primary and secondary	24 h	150 μg/m^3	Not to be exceeded more than once per year on average over 3 years
Sulfur dioxide		Primary	1 h	75 ppb	99 % percentile of 1-hr daily maximum concentrations, averaged over 3 years
		Secondary	3 h	0.5 ppm	Not to be exceeded more than once per year

Source http://www.epa.gov/air/criteria.html (accessed June 2014)

1.5.2 General Engineering Approaches to Air Emission Control

Before we elaborate on the technical approaches to air emission control, let us first take a look at the fate of air emissions from man-made combustion sources. As shown in Fig. 1.3, a fuel enters a combustion device (furnace or engine) and is oxidized and converted into different gases and particulates flue gas traveling through the duct and being discharged through the stack or tail pipe into the atmosphere. Chemical reactions continue, although at a much slower pace than that in the combustion device. These primary air pollutants are either partially or completely, depending on time, converted into secondary air pollutants. Both primary

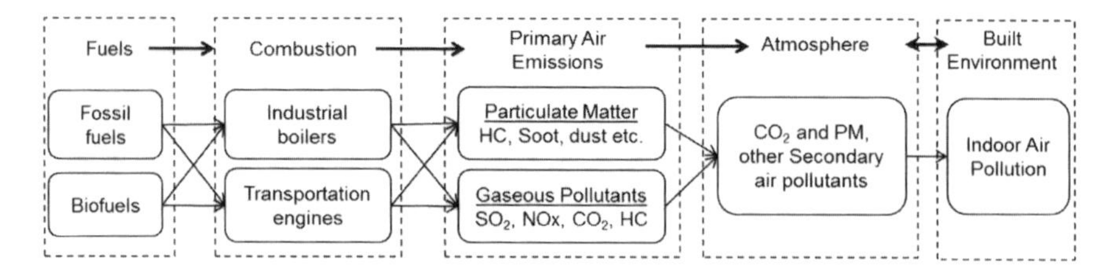

Fig. 1.3 Fate of combustion-related air emission control

and secondary air pollutants may enter built environments through building envelopes and/or ventilation systems, and there is also an exchange between indoor and outdoor environments.

According to the fate of the air emissions, technical approaches to effective air emission control can be classified by contaminant species, if a specific compound is tracked from its birth in combustion to air dispersion in the atmosphere. Alternatively, they can also be classified by the point of control, as in

- Pre-combustion approaches,
- In-combustion approaches, and
- Post-combustion approaches.

In this book, we follow the latter classification (pre-combustion, in-combustion, and post-combustion approaches), which is based on the location of air emission control approaches with each compound at the second level. Pre-combustion approaches are the most cost-effective, because elements that may be converted into air pollutants in a combustion process are taken out of the fuels. Examples include coal washing, crude oil refinery, and natural gas sweetening (more in-depth discussion later). Despite fuel cleaning prior to combustion, air emission forming elements are still in the fuel.

In-combustion approaches include combustion process modification. For example, lowering the combustion temperature can reduce the formation of nitric oxides (NO_x), while combustion at high temperature promotes complete combustion and consequently reduce the formation of hydrogen carbon (HC) and carbon monoxide. Injection of a calcium-based sorbent into the furnace can reduce the downstream concentration of sulfur dioxide (SO_2).

As a final attempt to capture air emissions, air pollutants and GHGs are separated from the post-combustion gases, namely flue gas or engine exhaust, with air cleaning devices. These air cleaning devices are usually designed for specific air emission species. Air pollutants penetrating through these devices are discharged into the atmosphere through the stacks of stationary facilities or the exhaust pipes of engines.

The last step is called air dispersion. Effective air dispersion helps reduce the immediate negative impact on local air quality; however, it does not reduce the total amount of air pollutants or GHGs entering the atmosphere.

1.6 Scope and Structure of This Book

Air emissions are a subject that requires multidisciplinary backgrounds in engineering, education, physics, chemistry, mathematics, medical science, psychology, agriculture, architecture, business management, economics, politics, and so on. It is difficult for any single book to address all aspects of air emissions.

This book is aimed at an advanced education for professionals, senior undergraduate, and graduate students with educational backgrounds in mechanical, chemical, and/or environmental engineering. It focuses on the scientific principles and applications of technologies to control air emissions resulting from fossil fuel combustion in stationary sources and internal combustion engines. The focus is on science and technology, upon which many effective air emission control program must be built. It does not purposely prescribe social, economic, and political factors that lie outside the scope of this book.

Adequate descriptions of fundamental principles and their engineering applications in air emission control form the main body of this book. The structure of this book generally tracks air emissions from their generation sources to their destination in the atmosphere and built environments. An air pollutant is generated from its source of combustion, either stationary or mobile, and ends up in the atmosphere. In between, there are multiple technical approaches to capture this air pollutant. These technical approaches are designed based on the fundamental principles related to air pollution control. Once air pollutants are emitted, they go through secondary reactions in the atmosphere, which are beyond the scope of this book.

Following a brief introduction to air emission in this chapter, Chaps. 2 and 4 cover the basic properties of gaseous and particulate pollutants, respectively. The combustion of fossil fuels produces pollutants that include oxides of carbon, sulfur and nitrogen, particulate matter, toxic metals, vapors, and ash. Most air emission control principles are applicable to a family of air pollutants rather than a single type. For example, absorption can be used for not only clean fuel production by cleaning the syngas, but also post-emission air cleaning. Sorption can be used for the separation of many gaseous pollutants from the air instead of only SO_x or VOCs. A good understanding of their properties is helpful in the design and evaluation of effective air emission control devices.

Chapter 3 gives a brief review of combustion chemistry and the formation of air pollutants and GHGs during combustion. The author found that many graduate students were not familiar with this subject, which is critical to a good understanding of air emissions. Effective in-combustion air emission control technologies to be covered in Part 2 of this book are developed from the principles introduced in this chapter. For example, NO_x formation is very temperature sensitive and low NO_x burners are developed based on this principle.

The last step of air pollution control is separation of air pollutants from air. The separations of gases and particulates follow different principles, which are introduced in Chaps. 5 and 6, respectively. These are the principles for the corresponding engineering applications for pre-combustion fuel cleaning and post-combustion air cleaning.

The general basic principles in Part 1 are necessary for the understanding of air emission engineering topics in Parts 2 and 3. Part 2 of the book introduces the strategies for pre-combustion (Chaps. 7 and 8), in-combustion (Chap. 9) and post-combustion (Chap. 10) air pollution control, step by step, from a process point of view. While air dispersion model is a powerful tool for air quality assessment and impact prediction, air dispersion itself is a measure for air emission control by dilution. It is elaborated in Chap. 11.

Part 3 includes special topics related to the scope of this book, but they do not closely fit into the process introduced above. Chapter 12 is devoted to carbon sequestration and storage, which are of increasing interest to society. Although debates are still ongoing, it is time to summarize the techniques that have been developed for GHG control, especially for CO_2 capture and storage. There may be some overlapping between this chapter and the other parts of this book.

Chapter 13 discusses an emerging topic of air pollution, nano air pollution. Nanomaterials are now widely used in many industries, for example, improved combustion efficiency, environmental protection, health, solar panel fabrication. Once these nanoparticles enter the air, they may have to be separated for the protection of the environment and health. Engineering approaches to nano air pollution control is the core emphasis of this chapter. Specifically, the properties of nano air pollution and its implications on monitoring and filtration technologies are covered.

Indoor air quality is introduced in Chap. 14 as it is also an important emerging topic. Indoor air quality is related extensively to air pollution. The direct sources are different from their outdoor counterparts, as are their control techniques.

The last chapter is about air quality and air emission monitoring techniques. They are commonly needed in industrial practices, government standard enforcement and research and development in a laboratory setting.

1.7 Units and Dimensions

The International System of Units (SI) is used in this book unless stated otherwise. Typical conversion factors are listed in Table A.3 in the Appendices. For example, the dimension of length (e.g., particle diameter) and mass may take different units and the conversions are as follows:

- 1 mm = 1000 micrometer (μm)
- 1 μm = 1000 (nm)
- 1 gram (g) = 1000 milligram (mg) = 1,000,000 microgram (μg)

Concentrations of air pollutants and GHGs can be presented in several dimensions, including mass per volume, volume ratio and mass ratio. Mass per volume is usually presented with units of kg/m^3, mg/m^3 or $\mu g/m^3$; and, volume and mass ratios are measured in units of parts per million (ppm) and percentages. At a low concentration, ppm is used more often than percentage to quantify the concentration of an air pollutant or GHG in the air. One frequently encountered unit conversion is between volume (or mass) percentage and ppmv (or ppmm):

$$x\% = 0.01 \times x \times 10^6 \text{ ppm} \ = 10000x \text{ (ppm)} \tag{1.3}$$

$$y \text{ ppm} = \frac{y}{10^6} = \frac{y}{10^4}\% \tag{1.4}$$

Sometimes, it is necessary to specify the concentration by volume or mass. Accordingly, *ppmv* and *ppmm* are used to express parts per million by *volume* and *mass*, respectively. The density of a species is needed in order to convert ppmv to ppmm:

$$x\,(\text{ppmm}) = y\,(\text{ppmv})\frac{\rho_{spieces}}{\rho_{mix}} \tag{1.5}$$

The conversion between the volume concentration and the mass concentration as determined by the density of the subject of concern can be explained using the following example (Example 1.1).

Example 1.1: Unit conversion The National Ambient Air Quality Standard for carbon monoxide (CO) is 35 ppmv measured over a one-hour averaging time. What is the equivalent concentration in (a) percentage, (b) mg/m^3 under standard condition when its density is 1.145 kg/m^3?

Solution 35 ppmv means that there is 35 m^3 of CO in one million m^3 of air. Therefore,

(a) by volume percentage, 35 ppmv is $\frac{35}{10000}\ \% = 0.0035\ \%$
(b) by mass concentration, 35 ppmv of CO in air becomes 1.145 $kg/m^3 \times 35/10^6 = 40.075 \times 10^{-6}\ kg/m^3 = 40.075\ mg/m^3$

Unless otherwise stated, the standard ambient condition is at a temperature of 25 °C and standard atmospheric pressure (1 atm or 1.013×10^5 Pa).

1.8 Practice Problems

1.8.1 Multiple Choice Problems

1. Which one of the following gases is NOT considered as a greenhouse gas

   a. Carbon dioxide
   b. Methane
   c. Water vapor
   d. Ammonia

2. Which one of the following is NOT one of the criteria air pollutants set by US National Ambient Air Quality Standards

   a. Ozone
   b. Carbon dioxide
   c. Sulfur dioxide
   d. Lead

3. Which air pollution control approach has the greatest simplicity with the least flexibility?

   a. Emission standards
   b. Air quality standards
   c. Emission taxes
   d. Both b and c above

1.8.2 Calculations

1. The national ambient air quality standard for ozone is 0.08 ppmv measured over an eight hour averaging time. What is the equivalent concentration in μg/m^3 at 25 °C?
2. The primary air quality standard for sulphur dioxide (SO_2) measured over 24 h averaging time is 0.14 ppmv. What is the equivalent concentration in μg/m^3 at 1 atm and 25 °C?
3. The primary air quality standard for nitrogen dioxide (NO_2) as an annual average is 100 μg/m^3. What is the equivalent concentration in ppmv at 1 atm and 25 °C?
4. The exhaust gas from automobile contains 8,000 ppmv of carbon monoxide (CO).

   a. What is the equivalent concentration in g/m^3 at 1 atm and 25 °C?
   b. What is the concentration in exhaust pipe in g/m^3 if it is at 220 °C and 1.2 atm?

5. The exhaust gas from automobile contains 20 μg/m^3 of lead.

   a. What is the equivalent concentration in ppmm at 1 atm and 25 °C?
   b. What is the concentration in exhaust pipe in μg/m^3 if it is at 200 °C and 1.1 atm?

6. At 25 °C and 1 atm pressure the ozone concentration at a monitoring site is 180 μg/m^3.

   a. What is the equivalent concentration in ppmm at 1 atm and 25 °C?
   b. Does this concentration exceeds the national ambient air quality standard of 0.08 ppmv (based on an 8 h average value)?

7. The dust concentration in an urban area is increased to a level of 100 μg/m^3 by wind storm. What is the equivalent concentration in grains per cubic foot at 1 atm and 25 °C?
8. Dry air at 25 °C has 78 % N_2, 21 % O_2, and 1 % Rn by volume. What is the concentration of each component in ppmv and μg/m^3?
9. The inhalable particle (PM_{10}) concentration in an industrial area is measured to be 5×10^{-5} grains/ft^3. What is the equivalent concentration in micrograms per cubic meter at 1 atm and 25 °C?
10. What is the resulting concentration of carbon monoxide in ppmv when 100 mL of carbon monoxide is mixed with 1,000,000 mL of air? Does this concentration exceeds the national ambient air quality standard of 9 ppmv (based on an 8 h average value)?
11. The national ambient air quality standard for nitrogen dioxide (NO_2) as an annual average is 100 μg/m^3. Assume that a person take in about 1 litre of air with every breath. How many grams of nitrogen dioxide does a person take in with every breath, if air contains 80 μg/m^3 of nitrogen dioxide?
12. The national ambient air quality standard for sulphur dioxide (SO_2) as an annual average is 80 μg/m^3. Assume that a person take in about 1 litre of air with every breath. How many molecules of sulphur dioxide does a person take in with every breath, if air contains 80 μg/m^3 of sulphur dioxide?
13. Calculate the weight of a dust particle having 1.5 μm diameter in air at 25 °C and 1 atm. Assume that density of particle is 1,000 kg/m^3 density.

References and Further Readings

1. Boden TA, Marland G, Andres RJ (2013) Global, regional, and national fossil-fuel CO_2 emissions. Carbon Dioxide Information Analysis Center, Oak Ridge National Laboratory, U. S. Department of Energy, Oak Ridge. doi:10.3334/CDIAC/00001_V2013, http://cdiac.ornl.gov/trends/emis/glo_2010.html
2. Bresser AHM, Salomons W (eds) (1990) Acidic precipitation, vols 1–5. Springer, New York
3. Cooper CD, Alley FC (2002) Air pollution control: a design approach, 3rd edn. Waveland Press, Inc., Long Grove
4. CRC (2013) US standard atmosphere. In: Lide DR (ed) Handbook of chemistry and physics, 94th edn. CRC Press, Boca Raton, pp 14–19. http://www.hbcpnetbase.com/
5. De Nevers N (2000) Air pollution control engineering, 2nd edn. McGraw–Hill Companies, New York
6. EIA (2004) International Energy Annual 2004 (May–July 2006). www.eia.doe.gov/iea
7. EIA (Energy Information Administration) (2007a) International Energy Outlook (IEO): May 2007
8. EIA (2007b) 2010–2030: EIA, system for the analysis of global energy markets
9. Ferris BJJ (1978) Health effects of exposure to low levels of regulated air pollutants: a critical review. J Air Pollut Control Assoc 28:482
10. Godish T (1997) Air quality. Lewis Publishers, USA
11. Houghton JT et al (eds) (1996) Climate change 1995: the science of climate change. IPCC, Cambridge University Press, Cambridge
12. IEA (2010) Key world energy statistics. International Energy Agency, Paris

13. IPCC (2013) Climate change 2013. In: Solomon S, Qin D, Manning M, Marquis M., Averyt K, Tignor MMB, Miller HL, Chen Z Jr (eds) The physical science basis. Cambridge University Press, 32 Ave. of the Americas, New York
14. Johnson DL, Ambrose SH, Bassett TJ, Bowen ML, Crummey DE, Isaacson JS, Johnson DN, Lamb P, Saul M, Winter-Nelson AE (1997) Meanings of environmental terms. J Environ Qual 26:581–589
15. Kenney J (2008) Nanotechnology: the future is coming sooner than you think. In: Fisher E, Selin S, Wetmore JM (eds) The yearbook of nanotechnology in society, volume I: presenting futures. Springer, Netherlands, pp 1–21
16. Khan J, Yardley J (2007) As China roars, pollution reaches deadly extremes. The New York Times, New York
17. Miller GT (1992) Living in the environment: an introduction to environmental science (Contemporary issues in crime and justice series), 7th edn. Wadsworth Publishing Company, Belmont
18. NRC (1981) Committee on indoor air pollutants: indoor pollutants, national research council. National Academy Press, Washington
19. Regens JL, Rycroft RW (1988) The acid rain controversy. University of Pittsburgh Press, Pittsburgh
20. Seinfeld JH, Pandis SN (2006) Atmospheric chemistry and physics: from air pollution to climate change, 2nd edn. John Wiley and Sons, Inc., New York
21. Shy CM, Goldsmith JR, Hackney JD, Lebowitz MD, Menzel DB (1978) Health effects of air pollution. American Lung Association, New York
22. Simmons WS, Rinne SP, Tesche NS, Weir BR (1985) Toxicology of fossil fuel combustion products, vols 1 and 2. EPRI EA—3920, EPRI, Palo Alto
23. Smith JB, Tirpak DA (1990) The Potential effects of global climate change on the United States. Hemisphere, New York
24. UN (2012) World Population Division Website. http://www.un.org/en/development/desa/population/publications/pdf/trends/WPP2012_Wallchart.pdf. Accessed Jan 2014
25. US EPA (2004) Estimated Per Capita Water Ingestion and Body Weight in the United States–An Update, (EPA-822-R-00-001). http://water.epa.gov/action/advisories/drinking/upload/2005_05_06_criteria_drinking_percapita_2004.pdf
26. US EPA (2008) Greenhouse gas inventory reports: inventory of U.S. Greenhouse Gas Emissions and Sinks: 1990-2006. USEPA #430-R-08-005. Available via http://www.epa.gov/climatechange/emissions/downloads/08_CR.pdf. Accessed July 2008
27. US EPA (2014) Inventory of U.S. Greenhouse gas emissions and sinks, EPA 430-R-14-003 website http://www.epa.gov/. 1200 Pennsylvania Avenue, N.W., Washington
28. van Donkelaar A, Martin RV, Brauer M, Kahn R, Levy R, Verduzco C, Villeneuve PJ (2010) Global Estimates of ambient fine particulate matter concentrations from satellite-based aerosol optical depth: development and application. Environ Health Perspect 118(6):847–855
29. Watson AY, Bates RR, Kennedy D (1988) Air pollution, the automobile and public health. National Academy Press, Washington
30. WHO (2005) WHO air quality guidelines for particulate matter, ozone, nitrogen, dioxide and sulfur dioxide; Global update 2005. (http://whqlibdoc.who.int/hq/2006/WHO_SDE_PHE_OEH_06.02_eng.pdf)
31. WHO (2013) International Agency for Research on Cancer (IARC) report on outdoor air pollution a leading environmental cause of cancer death. http://www.iarc.fr/en/media-centre/pr/2013/pdfs/pr223_E.pdf

Part I
Basic Concepts

Chapter 2
Basic Properties of Gases

The purpose of this chapter is to describe some basic properties of gases, which are applicable to not only air but also gaseous air emissions.

The terms gas and vapor are both used to describe the gaseous state of a substance. However, gas is primarily for a pure substance or mixture that exists in gaseous state under normal conditions. Vapor is used to describe a substance that is in gaseous state, which exists in liquid or solid state under normal conditions. A gas can be compressed above the atmospheric pressure and even nontoxic gases can be lethal when their concentrations are high enough to displace too much oxygen in the air. While too much vapor will result in the phase change from gas to liquid by condensation.

Gaseous air emissions can be divided into organic and inorganic types. Organic compound includes most chemicals based on a structure of carbon atoms. Organic air emissions include such gases as methane, but the majorities are vapors under normal conditions. Inorganic gaseous air emissions are primarily gases, except for mercury.

Despite the differences in gas and vapor, they also share some common properties. Therefore, within this text, gas is used to include vapor unless otherwise stated.

2.1 Gas Kinetics

Kinetic theory is also known as kinetic molecular theory or collision theory. Kinetic theory of gases attempts to explain macroscopic properties of gases, such as pressure, temperature, or volume, by considering their microscopic compositions and motion. In the kinetic theory of gases, the following assumptions are made

- Gas molecules are considered uniform spherical particles, each of which has a mass but negligible volume compared to the gas container.
- The number of molecules is large and thus their behaviors can be analyzed statistically.
- Gas molecules move rapidly, constantly, and randomly. The collision between the molecules and the wall is considered perfectly elastic and instantaneous.
- The average distance between the gas molecules is large compared to their size.

Z. Tan, *Air Pollution and Greenhouse Gases*, Green Energy and Technology,
DOI 10.1007/978-981-287-212-8_2

2.1.1 Speeds of Gas Molecules

An understanding of the gas properties requires a good understanding of the molecular velocities. In engineering dynamics analysis, we describe a particle velocity with its magnitude and its direction. Similarly, a gas molecule velocity vector $\vec{c}$ is described using its three directional components in a rectangular x-y-z coordinate system as

$$\vec{c} = c_x\hat{i} + c_y\hat{j} + c_z\hat{k} \tag{2.1}$$

Maxwell-Boltzmann distribution is the most commonly used for molecular speed distribution. The distribution of one-dimensional velocity $-\infty < c_i < \infty$ is

$$f(c_i) = \left(\frac{m}{2\pi kT}\right)^{3/2} \exp\left(-\frac{mc_i^2}{2kT}\right) \quad i = x, y, z \tag{2.2}$$

This is a normal distribution with a mean of 0 and a variance of kT/m. It also applies to the other two velocity components.

In engineering applications, total speeds of molecules are of more interest than their components. The Maxwell-Boltzmann distribution describes the probability of molecular speed [16].

$$f(c) = 4\pi c^2 \left(\frac{m}{2\pi kT}\right)^{3/2} \exp\left(-\frac{mc^2}{2kT}\right) \tag{2.3}$$

where the Boltzmann constant $k = 1.3807 \times 10^{-23}$(J/K), c is the molecular speed of a molecule, m is the mass of the molecule, and T is the temperature of the gas.

In air pollution, we are interested in the distributions of the molecular mean speed, root-mean-square speed and mean relative speed. As to be seen shortly, they are useful parameters in molecular kinetics that lead us to microscopic properties like pressure, viscosity, diffusivity, and so on. These speeds can be computed from the Maxwell–Boltzmann distribution of molecular speed described in Eq. (2.3).

The *mean molecular speed* ($\bar{c}$) is the mathematical average of the speed distribution and it can be calculated by integration

$$\bar{c} = \int_0^\infty cf(c)\mathrm{d}c = 4\pi\left(\frac{m}{2\pi kT}\right)^{3/2} \int_0^\infty \left[c^3 \exp\left(-\frac{mc^2}{2kT}\right)\right] \mathrm{d}c \tag{2.4}$$

In order to complete the integration, we need to know that

$$\int_0^\infty [x^3 \exp(-ax^2)]\mathrm{d}x = \frac{1}{2a^2} \quad (2.5)$$

For this specific problem, $a = -m/2kT$, and the integration term can be determined as

$$\int_0^\infty \left[c^3 \exp\left(-\frac{mc^2}{2kT}\right)\right] dc = \frac{1}{2(-m/2kT)^2} = 2\left(\frac{kT}{m}\right)^2 \quad (2.6)$$

Substituting Eq. (2.6) into Eq. (2.4) leads to

$$\bar{c} = \left(\frac{8kT}{\pi m}\right)^{1/2} \quad (2.7)$$

By similar approaches, we can get the root-mean-square speed (v_{rms}), which is the square root of the average squared speed:

$$c_{\mathrm{rms}} = \sqrt{\int_0^\infty c^2 f(c)\mathrm{d}c} = \sqrt{\frac{3kT}{m}} \quad (2.8)$$

Comparison between Eqs. (2.7) and (2.8) shows that $c_{\mathrm{rms}} > \bar{c}$ because c_{rms} contains a factor of 3 and $\bar{c}$ contains a factor of $8/\pi \approx 2.55$. This is resulted from the fact that greater speeds are weighted more heavily in the integration based on c^2.

Average relative velocity is another molecular speed needed in our analysis that follows. From engineering dynamics, we have learned that the relative velocity of any two molecules A and B which behave like particles is

$$\vec{c}_{A/B} = \vec{c}_A - \vec{c}_B \quad (2.9)$$

where $\vec{c}_{A/B}$ is the velocity of molecule A relative to molecule B (m/s), and the magnitude of the relative velocity is the square root of the scale product of itself:

$$\vec{c}_{A/B}^{\,2} = \vec{c}_{A/B} \times \vec{c}_{A/B} = (\vec{c}_A - \vec{c}_B) \times (\vec{c}_A - \vec{c}_B) = \vec{c}_A \times \vec{c}_A - 2\vec{c}_A \times \vec{c}_B + \vec{c}_B \times \vec{c}_B \quad (2.10)$$

Replacing the speeds in the above equation with the average speeds gives

$$\bar{c}_{A/B}^{\,2} = (\vec{c}_A \times \vec{c}_A)_{\mathrm{ave}} - 2(\vec{c}_A \times \vec{c}_B)_{\mathrm{ave}} + (\vec{c}_B \times \vec{c}_B)_{\mathrm{ave}} \quad (2.11)$$

Since A and B are randomly selected and they are independent on each other, the term $(\vec{c}_A \times \vec{c}_B)_{\text{ave}}$ results in a quantity of zero. Meanwhile, since A and B are randomly selected from the same population, where the molecules behave statistically the same, we also have $\bar{c}_A = \bar{c}_B = \bar{c}$. Then Eq. (2.11) becomes

$$\bar{c}^2_{A/B} = \bar{c}^2_A + \bar{c}^2_B = 2\bar{c}^2 \tag{2.12}$$

The relation between the magnitudes of the average relative speed and mean speed is

$$\bar{c}_{A/B} = \sqrt{2}\bar{c} = \left(\frac{16kT}{\pi m}\right)^{1/2}. \tag{2.13}$$

2.1.2 Avogadro Constant and Molar Weight

A gas volume contains a large number of molecules, which are treated as particles, in rapid motions. Mole amount is used to quantify the amount of molecules. In 1 mol of gas there are 6.022×10^{23} molecules. This is described using the Avogadro number or Avogadro constant

$$N_A = 6.022 \times 10^{23} \quad (1/\text{mol}) \tag{2.14}$$

Any gas can be characterized with its molar weight, which is the mass of 1 mol of the gas

$$M = N_A m \tag{2.15}$$

where m is the mass of a single molecule, and M is the molar weight of a gas with a unit of g/mol or kg/kmol. Molar weights of typical gases with known molecular formula can be determined by the corresponding number of atoms. For example, the molar weight of O_2 is 32 because there are two oxygen atoms in one oxygen molecule and each of the atom weight is 16 g/mol.

2.1.3 Gas Pressure

The pressure of a gas is resulted from the force exerted by gas molecules on the walls of the container due to the collision between the wall and molecules. Consider a cubic container having N gas molecules and the length of the container is l.

The linear momentum before and after the impact is $m\vec{c}_1$ and $m\vec{c}_2$, respectively, when a gas molecule collides with the wall of the container that is normal to the x coordinate axis and bounces back in the opposite direction. From the principle of impulse and linear momentum one has,

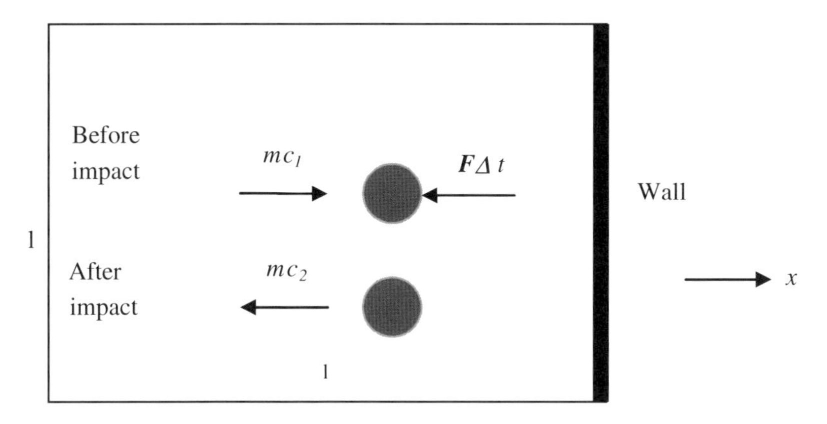

Fig. 2.1 Principle of impulse and linear momentum analysis on a single molecule

$$m\vec{c_1} + \vec{F}\Delta t = m\vec{c_2} \tag{2.16}$$

where the term $\vec{F}\Delta t$ is the impulse during the collision [9]. Both force and velocity are vector quantities defined by their magnitudes and directions. Along one of the rectangular coordinates, say x direction as depicted in Fig. 2.1, the above equation can be written in terms of magnitudes,

$$mc_{1x} - F_x\Delta t = -mc_{2x} \tag{2.17}$$

Reorganize it and one gets,

$$F_x\Delta t = m(c_{1x} + c_{2x}) \tag{2.18}$$

Since the impact between the surface and the gas molecules is elastic, we have $c_{1x} = c_{2x} = c_x$ and the above equation becomes

$$F_x\Delta t = 2mc_x \tag{2.19}$$

With a constant speed of c_x, the molecule will impact on the same wall once every $2l/c_x$ time units (for a round trip), where l is the distance between two opposite walls of the container. Then, the force on the wall produced by the same molecule along x direction is:

$$F_x = \frac{2mc_x}{2l/c_x} = \frac{mc_x^2}{l} \tag{2.20}$$

Now, consider all the N molecules in the container. The total force acting on the wall due to N molecules with the same mass of m is

$$F = \frac{m}{l}\left(\sum_{i=1}^{N} c_{ix}^2\right) \tag{2.21}$$

Recall the assumptions of kinetic theory, there are a large number of molecules moving randomly and constantly, therefore, Eq. (2.21) is applicable to any directions in the container. And the force on each wall can be considered same in magnitude. Consider the force acting only on one wall and the magnitude of the velocity can be calculated using

$$c_i^2 = c_{ix}^2 + c_{iy}^2 + c_{iz}^2 \tag{2.22}$$

Since x, y, and z are randomly chosen and the molecular motion is random and uniform along any direction, we get

$$c_{ix}^2 = c_{iy}^2 = c_{iz}^2 = \frac{1}{3}c_i^2 \tag{2.23}$$

Now, the total force exerted by the molecules on one wall described in Eq. (2.21) can be expressed in terms of the total speed instead of a single component of the velocity,

$$F = \frac{m}{3l}\left(\sum_{i=1}^{N} c_i^2\right) \tag{2.24}$$

According to the definition of root-mean-square speed described in Eq. (2.8),

$$Nc_{\text{rms}}^2 = \sum_{i=1}^{N} c_i^2 \tag{2.25}$$

Then the total force on the wall of the container can be written as:

$$F = \frac{Nmc_{\text{rms}}^2}{3l} \tag{2.26}$$

The resultant pressure, which is force per unit area of the wall, of the gas can then be written as

$$P = \frac{F}{A} = \frac{Nmc_{\text{rms}}^2}{3lA} = \frac{Nm}{3V}c_{\text{rms}}^2 \tag{2.27}$$

where A = the area of the wall on which force is exerted, and V = the volume of the container. Since Nm stands for the total mass of the gas, then the density of the gas is

$$\rho = Nm/V \tag{2.28}$$

Equation (2.27) can be further rewritten as

$$P = \frac{1}{3}\rho c_{\mathrm{rms}}^2 \tag{2.29}$$

This formula demonstrates the relationship between a macroscopic property (pressure) and a microscopic property (the root-mean-square speed). For example, the pressure of a gas is due to collisions between molecules and the wall.

2.1.4 Density and Specific Volume of a Gas

In an engineering practice, we hardly use or even care about the exact number of molecules in a volume. Rather we use the mole amount of gases, which is denoted as n and

$$n = N/N_A = Nm/M \tag{2.30}$$

Then, the gas density is described as

$$\rho = nN_A m/V = nM/V \tag{2.31}$$

The equation allows us to estimate the density of a gas with known molar weight by comparing with another gas with known density.

A term related to density is specific volume, which is the inverse of density

$$v = 1/\rho \tag{2.32}$$

It has a unit of volume per mass, for example, m^3/kg.

2.1.5 Ideal Gas Law and Dalton's Law

Air and typical gases of interest in air emissions are often considered as ideal gases. The ideal gas law governs the relationship between the pressure P, the volume V, and the temperature T of an ideal gas. It can be derived by continuing with the gas molecular kinetics.

Substituting Eq. (2.8) into Eq. (2.27) we can get, with Eq. (2.25 below),

$$PV = nRT \tag{2.33}$$

With $\rho = Nm/V$, Eq. (2.33) becomes

$$P = \frac{\rho RT}{M} \tag{2.34}$$

This is the so called ideal gas law, where n is the mole amount of the gas, and R = the universal gas constant and $R = 8314 \frac{\text{J}}{\text{kmol K}}$ or $8.314 \frac{\text{J}}{\text{mol}}$ K.

Example 2.1: Gas density calculation
Estimate dry air density at 0 °C and 1 atm using Eq. (2.34)

Solution
From Eq. (2.34) we have

$$\rho = \frac{PM}{RT} = \frac{101,325(\text{Pa}) \times 28.84(\text{kg/kmol})}{8,314\text{J}/(\text{kmol K}) \times 273\text{K}} = 1.29\,\text{kg/m}^3$$

The universal ideal gas constant is related to the Boltzmann constant k as,

$$R = kN_A \tag{2.35}$$

As such, the ideal gas law can be rewritten in terms of the Boltzmann constants

$$PV = nN_AkT = NkT \tag{2.36}$$

where $(N = nN_A)$ is the total number of molecules in the subject gas.

Dalton's law is an empirical law that was observed by John Dalton in 1801 and it is related to the ideal gas law. It is important to air emission studies in that gases in air emission engineering are often mixtures of multiple compounds.

Consider a mixture of gases, the mole number n of a gas mixture equals to the sum of the mole numbers of all its components.

$$n = \sum_{i=1}^{N} n_i \tag{2.37}$$

and the mole fraction, denoted as y_i, of any given species is

$$y_i = \frac{n_i}{n} \quad \text{and} \quad \sum_{i=1}^{N} y_i = 1 \tag{2.38}$$

For ideal gases under some conditions, mole fraction of any species is equal to its volume fraction. The molar weight of a mixture of ideal gases can be determined from the mole fraction of each compound and the corresponding molar weight using Eq. (2.39):

$$M = \sum_{i=1}^{N} y_i M_i \tag{2.39}$$

where M_i is the molar weight of each substance in the gas mixture.

Example 2.2: Molar weight of simplified air
Dry air can be approximated as a mixture of nitrogen and oxygen molecules where oxygen takes 21 % by volume. The approximate molar weights of nitrogen and oxygen molecules are 28 and 32 g/mol, respectively. Estimate the molar weight of standard dry air.

Solution
Using Eq. (2.39) we can get

$$\begin{aligned} M_{\text{air}} &= y_{N_2} M_{N_2} + y_{O_2} M_{O_2} \\ &= 0.79 \times 28 + 0.21 \times 32 = 28.84(\text{kg/kmol}) \end{aligned}$$

Assuming the gases are nonreactive, each individual gas in the mixture is also governed by the ideal gas law:

$$P_i V = n_i RT \tag{2.40}$$

where P_i is partial pressure of the gas compound i. Combining Eqs. (2.33) and (2.40) gives,

$$P_i = \frac{n_i}{n} P = y_i P \tag{2.41}$$

This relationship is also referred to as *Dalton's law*. It states that the total pressure of a mixture of nonreactive gases is equal to the sum of the partial pressures of all individual gases.

The partial pressure of a gas in a mixture is an important property that affects many engineering practices, for example, the solubility of a gas in liquid depends on the partial pressure of the gas (see Sect. 2.3).

Example 2.3: Partial pressure of an ideal gas
Table 2.1 shows the compositions of pure dry air at sea level. Using the volume percentage in this table, determine the partial pressures of nitrogen, oxygen, methane, and carbon dioxide in Pascal at sea level where the atmospheric pressure is 101.325 kPa.

Table 2.1 Mean free paths for ideal gases

Pressure (atm)	Temperature (K)	Mean free path (m)
1	300	7×10^{-8}
10^{-4}	300	7×10^{-4}
10^{-8}	300	7

Solution

Gas	Percent by volume (%)	Mole ratio	Partial pressure (Pa)
Nitrogen	78.084	0.78084	79,118.61
Oxygen	20.9476	0.209476	21,225.16
Carbon dioxide	0.0314	0.00314	318.16
Methane	0.0002	0.000002	0.203

Sometimes, we need to determine the amount of moisture or water vapor in the air. It is useful in air emission monitoring, characterization of the air cleaning efficiency, or simply quantification of indoor air quality. It can be expressed by either specific humidity or relative humidity of the air. Specific humidity quantifies the mass ratio of water molecules to dry air. The specific humidity, w, of air can be calculated using

$$w = \frac{m_w}{m_{\text{air}}} = \frac{n_w M_w}{n_{\text{air}} M_{\text{air}}} \tag{2.42}$$

Relative humidity of air is expressed as the ratio of the vapor partial pressure of the air to the saturation vapor partial pressure of the air at the actual dry bulb temperature.

$$RH = \frac{P_w}{P_{\text{sat}}} \tag{2.43}$$

where P_w = vapor partial pressure and P_{sat} = saturation vapor partial pressure at the actual dry bulb temperature. They shall carry the same unit to make RH dimensionless. Since vapor partial pressure is less than the saturation vapor partial pressure, we have $0 < RH < 1$. More practically, relative humidity can also be expressed as the ratio of actual mass of water vapor in a given air volume to the mass of water vapor required to saturate at this air.

$$RH = \frac{m_w}{m_{w,s}} \tag{2.44}$$

where m_w = mass of water vapor in the given air volume and $m_{w,s}$ = mass of water vapor required to saturate at this volume

Example 2.4: Vapor pressure
Relative humidity of air in a typical cool summer day in Canada is about 40 % and it is known that the saturation pressure at 21°C is 25 mbar. What is the corresponding vapor pressure in the air.

Solution
From Eq (2.43), we have

$$P_w = P_{\text{sat}} \times RH = 25.0\,\text{mbar} \times \ 40\ \% = 10\,\text{mbar}$$

2.1.6 Kinetic Energy of Gas Molecules

Combining Eqs. (2.35) and (2.8) leads to the root-mean-square speed of an ideal gas in terms of microscopic variables, molecular weight M and temperature T.

$$c_{\text{rms}} = \left(\frac{3RT}{N_a m}\right)^{1/2} = \left(\frac{3RT}{M}\right)^{1/2} \tag{2.45}$$

This equation can be used to determine c_{rms} of a known ideal gas at a certain temperature. This formula shows that root-mean-square speed of a gas is proportional to the square root of the temperature, so it increases with the increase in gas temperature.

The molecular kinetic energy can be determined from root-mean-square speed. Usually, it is quantified on a per mole base. The kinetic energy for 1 mol of ideal gas can be calculated as,

$$e_k = \frac{1}{2} N_a m c_{\text{rms}}^2 = \frac{3}{2} RT \tag{2.46}$$

This equation shows that the kinetic energy of an ideal gas depends only on its temperature. This implies that the molar kinetic energy of different gases is the same at the same temperature.

Example 2.5: Gas kinetic energy
Compute root-mean-square speeds and the kinetic energy of 1 mol of the following gases H_2, H_2O vapor, air, and CO_2 at standard temperature of 293 K.

Solution

Step 1. Determine the molar weight of the gases

Gas or Vapor	Molar weight (g/mol)
H_2	2
H_2O	18
Air	29
CO_2	44

Step 2. Compute the root-mean-square speed of the gas molecules using Eq. (2.45) and the kinetic energy of 1 mol of the gas using Eq. (2.46)

Gas or Vapor	Molecular weight (g/mol)	Equation (2.45) c_{rms} (m/s)	Equation (2.46) (J/mol)
H_2	2	1911.54	3654.0
H_2O	18	637.18	3654.0
Air	29	501.99	3654.0
CO_2	44	407.54	3654.0

It is apparent that the lighter molecules have the greater root-mean-square speeds to maintain the equal kinetic energy at the same temperature.

2.1.7 Gas Mean Free Path

The mean free path of a gas is the average distance traveled by gas molecule between the collisions. Gas mean free path affects the aerosol dynamics and consequently air sampling and cleaning technologies. Gas mean free path may be estimated from kinetic theory too.

Consider a gas with molecules with a uniform diameter of d. As shown in Fig. 2.2, the effective cross section for collision is $d_c = 2d$, and the cross section area can be calculated as $A = 4\pi d^2$.

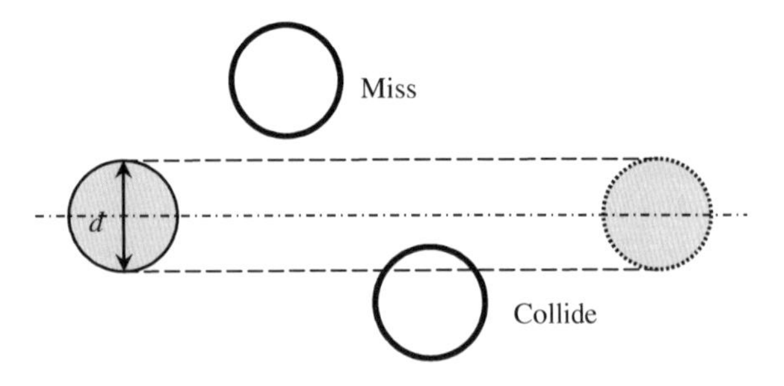

Fig. 2.2 Collision diameter of gas molecules

The number of collisions this molecule experiences during the time interval of Δt is determined by the number molecules colliding with the molecule in the swept volume, ΔV, which is

$$\Delta V C_N = \pi d^2 \bar{c}_{A/B} \Delta t C_N \tag{2.47}$$

where C_N is the number of molecules per unit volume of the gas ($\#/m^3$). The relative velocity is used because other molecules are also traveling within the space; it is described in Eq. (2.13):

$$\bar{c}_{A/B} = \sqrt{2}\bar{c}$$

The distance it travels in Δt is $\bar{c}\Delta t$, then the mean free path, which is the distance traveled divided by the number of collisions, can then be calculated as

$$\lambda = \frac{\bar{c}\Delta t}{\pi d^2 \bar{c}_{A/B} \Delta t C_N} = \frac{1}{\sqrt{2}\pi d^2 C_N} \tag{2.48}$$

There is no need to correct the average speed in the numerator for the calculation of distance traveled, which is supposed to be calculated using the average speed of the molecules.

Assuming ideal gas, the number concentration of the gas molecules can be calculated as

$$C_N = \frac{N}{V} = \frac{nN_a}{V} = \frac{P}{RT} N_a \tag{2.49}$$

where N is the total number of molecules in the container with a volume V. There is a simple relationship between the gas density and molecule number concentration as

$$\rho = C_N m \tag{2.50}$$

With Eq. (2.49) and Eq. (2.50), Eq. (2.48) becomes

$$\lambda = \frac{RT}{\sqrt{2}\pi N_a d^2 P} = \frac{M}{\sqrt{2}\pi N_a d^2 \rho} \tag{2.51}$$

Molar weight (M) and gas molecule diameter (d) are fixed for a gas at stable state, therefore, the mean free path depends only on the density of the gas (ρ), which depends on the pressure and temperature of the gas. For ambient conditions, the mean free path increases with increasing temperature or decreasing pressure.

Example 2.6: Air mean free path
For air, the molecule diameter is approximately 3.7×10^{-10} m. Determine the mean free path of air at sea level at 293 K. What is the value for air mean free path at an elevation of 2,000 m by assuming same temperature?

Solution
Air density at the sea level is $\rho_0 = 1.21$ kg/m^3. At the elevation of 2,000 m, air density becomes $\rho_{2000} = 0.95$ kg/m^3. The corresponding mean free paths are calculated using Eq. (2.51).

At sea level,

$$\lambda_0 = \frac{0.02884}{\sqrt{2}\pi \times 6.0221 \times 10^{23} \times (3.7 \times 10^{-10})^2} \times \frac{1}{1.21}$$
$$= 0.066 \times 10^{-6}\,\text{m} = 0.066\,\mu\text{m}$$

At 2,000 m above the sea level,

$$\lambda_{2000} = \frac{0.02884}{\sqrt{2}\pi \times 6.0221 \times 10^{23} \times (3.7 \times 10^{-10})^2} \times \frac{1}{0.95}$$
$$= 0.083 \times 10^{-6}\,\text{m} = 0.083\,\mu\text{m}$$

Conversion between the mean free paths under different conditions was given by Allen and Raabe [1], Cited by [14]

$$\frac{\lambda}{\lambda_0} = \frac{p_0}{p}\left(\frac{T}{T_0}\right)^2\left(\frac{T_0 + T_s}{T + T_s}\right) \tag{2.52}$$

where the parameters with subscript of 0 is for the standard condition and those without subscripts are for any other conditions. Examples of values of mean free path for air are given in Table 2.1.

2.1.8 Number of Collisions with Wall/Surface

By applying the kinetic theory, one can also calculate the number of collisions on the walls of a container. The quantitative analysis of the collision on the wall of the container is important for the study of the kinetic molecular theory of transport properties such as diffusion and viscosity.

Again, consider the scenario in Fig. 2.1, a cubic container with a wall area, A, and assume elastic impact between the molecules and the wall. During a small time interval of Δt, the distance that the molecule with the moving speed of c_x travel is $\Delta x = c_x \Delta t$ in the +x-direction if they do not collide with the wall. In another word, the molecules will collide with the wall if they are within a distance $\Delta x = c_x \Delta t$ from the wall.

Assume one molecule collides with the wall only once during Δt, then the number of collisions is the same as the number of molecules within the volume formed by Δx and A. Then, the number of collisions is expressed as the corresponding number of molecules that collide with the wall during Δt

$$n_c = C_N c_x A \Delta t \tag{2.53}$$

where C_N is the number of molecules per unit volume of gas, which is defined in Eq. (2.49). From this equation, we can get the number of molecules colliding with the wall per unit area per unit time ($1/\mathrm{s\ m^2}$)

$$j_x = \frac{n_c}{A\Delta t} = C_N c_x \tag{2.54}$$

Then, the total number of collisions considering the molecule speed distribution is determined by integration, assuming constant molecule number concentration C_N at steady state,

$$J_x = \int_0^\infty j_x f(c_x) \mathrm{d}c_x = C_N \int_0^\infty c_x f(c_x) \mathrm{d}c_x \tag{2.55}$$

Using the Maxwell-Boltzmann distribution described in Eq. (2.2), the integration part in Eq. (2.55) can be manipulated following

$$\int_0^\infty c_x f(c_x) \mathrm{d}c_x = \frac{1}{2} \int_{-\infty}^\infty |c_x| f(c_x) \mathrm{d}c_x = \frac{1}{2} \bar{c}_x \tag{2.56}$$

where $|\bar{c}_x|$ is the average of the absolute value of c_x. Then Eq. (2.55) becomes

$$J_x = \frac{1}{2} C_N \bar{c}_x \tag{2.57}$$

A step further from the Boltzmann distribution, we can relate $|\bar{c}_x|$ with average molecular speed that can be calculated from Eq. (2.2)

$$\bar{c}_x = \int_0^\infty c_x f(c_x) \mathrm{d}c_x = \int_0^\infty \left(\frac{m}{2\pi kT}\right)^{3/2} \exp\left(-\frac{mc_x^2}{2kT}\right) c_x \mathrm{d}c_x = \frac{1}{2} \bar{c} \tag{2.58}$$

Thus, the collision per unit time per unit area along x direction in terms of average molecular speed is

$$J_x = \frac{1}{4} C_N \bar{c} \tag{2.59}$$

Substitute Eq. (2.7) into this equation and we can get

$$J_x = C_N \sqrt{\frac{kT}{2\pi m}} \tag{2.60}$$

Considering the relationship between gas density and the molecule number concentration described in Eq. (2.50), the total number of collisions per unit time per unit area can also be expressed in terms of gas density as

$$J_x = \rho \sqrt{\frac{kT}{2\pi m^2}}. \tag{2.61}$$

2.1.9 Diffusivity of Gases

We can derive the diffusivity of a single gas by applying the preceding analysis of molecule collision on a surface to an imaginary cubic container that is formed by a distance of mean free path, 2λ. We can apply finite element analysis from $(x-\lambda)$ to $(x+\lambda)$. The concentration at x is C_N. Assuming a constant gas concentration gradient of $\frac{\mathrm{d}C_N}{\mathrm{d}x}$ from $(x-\lambda)$ to$(x+\lambda)$; the concentrations at $(x+\lambda)$ and $(x-\lambda)$ are $\left(C_N + \frac{\mathrm{d}C_N}{\mathrm{d}x}\lambda\right)$ and $\left(C_N - \frac{\mathrm{d}C_N}{\mathrm{d}x}\lambda\right)$, respectively.

Considerations of symmetry lead us to assert that the average number of particles traveling in a given direction $(\pm x, \pm y \text{ or } \pm z)$ will be one-sixth of the total, and thus the mean rate at which molecules crosses a plane is $\mathrm{N}\bar{c}/6$ per unit area in unit time. N is the total number of molecules in the container. This differs slightly from the exact result, although it is suitable for some simplified argument.

Statistically, 1/6 of the molecules at $(x+\lambda)$ will move along $-x$ direction. According to the definition of mean free path, these molecules leaving the plane $(x+\lambda)$ along $-x$ direction will reach plane x. Therefore, the number of molecules leaving plane $(x+\lambda)$ per second per unit area is

$$J_{\leftarrow x} = \frac{1}{6}\left(C_N + \frac{\mathrm{d}C_N}{\mathrm{d}x}\lambda\right)\bar{c}$$

Similarly, the number of molecules per second per unit area leaving plane $(x-\lambda)$ to plane x is

$$J_{\rightarrow x} = \frac{1}{6}\left(C_N - \frac{\mathrm{d}C_N}{\mathrm{d}x}\lambda\right)\bar{c}$$

Then, the net flux of molecules in the positive x direction (#/s m^2) is

$$J_x = J_{\to x} - J_{\leftarrow x} = -\frac{1}{3}\frac{\mathrm{d}C_N}{\mathrm{d}x}\lambda\bar{c} \tag{2.62}$$

Comparing this equation with the Fick's law of diffusion

$$J_x = -D\frac{\mathrm{d}C_N}{\mathrm{d}x} \tag{2.63}$$

gives the diffusivity or diffusion coefficient of gas in m^2/s:

$$D = \frac{1}{3}\lambda\bar{c} \tag{2.64}$$

Combining with Eqs. (2.7) and (2.51), we can get leads to

$$D = \frac{1}{3}\frac{RT}{\sqrt{2}\pi N_a d^2 P}\sqrt{\frac{8kT}{\pi m}} = \frac{2}{3\pi^{1.5}d^2}\frac{RT}{PN_a}\sqrt{\frac{RT}{M}}. \tag{2.65}$$

2.1.10 *Viscosity of a Gas*

Viscosity is a measure of the resistance of a fluid being deformed by either shear or extensional stress. Viscosity in gases arises principally from the molecular diffusion that transports momentum between layers of flow. The kinetic theory of gases allows accurate prediction of the viscosity of a gas.

Similar to the analysis for diffusivity, consider a laminar flow of gas above a horizontal plate. We can apply finite element analysis from $(x - \lambda)$ to $(x + \lambda)$. The concentration at x is C_N. Assuming a constant gas velocity gradient of $\frac{\mathrm{d}u}{\mathrm{d}x}$ from $(x - \lambda)$ to $(x + \lambda)$; the velocities at $(x + \lambda)$ and $(x - \lambda)$ are $\left(u + \frac{\mathrm{d}u}{\mathrm{d}x}\lambda\right)$ and $\left(u - \frac{\mathrm{d}u}{\mathrm{d}x}\lambda\right)$, respectively. The gas may be treated as layers perpendicular to the moving direction (say x axis). Also assume a constant gas molecule concentration C_N. In each layer, the steady flow gas velocity is also constant. Then, the rate of transport of momentum per unit area in the $+x$ and $-x$ directions, respectively, are

$$M_{\to x} = \frac{1}{6}C_N m\bar{c}\left(u - \frac{\partial u}{\partial x}\lambda\right)$$

$$M_{\leftarrow x} = \frac{1}{6}C_N m\bar{c}\left(u + \frac{\partial u}{\partial x}\lambda\right)$$

Table 2.2 Sutherland's constants and reference temperatures

Gas	Formula	T_s (K)	T_0 (K)	μ_0 (10^{-6} Pa s)
Hydrogen	H_2	72	293.85	8.76
Nitrogen	N_2	111	300.55	17.81
Oxygen	O_2	127	292.25	20.18
Air	–	120	291.15	18.27
Carbon dioxide	CO_2	240	293.15	14.8
Carbon monoxide	CO	118	288.15	17.2
Ammonia	NH_3	370	293.15	9.82
Sulfur dioxide	SO_2	416	293.65	12.54

where m is the mass of a single molecule. C_N is the molecule number concentration. The term $\left(\frac{1}{6}C_N m\bar{c}\right)$ stands for the average mass flow rate per unit area through plane x. This leads to a net flux of momentum in the $+x$ direction through the plane x as

$$M_x - -\frac{1}{3}C_N m\bar{c}\frac{\partial u}{\partial x}\lambda \tag{2.66}$$

Comparing with the definition of shear stress

$$\tau = -\mu\frac{\partial u}{\partial x} \tag{2.67}$$

we can get the kinetic viscosity of the gas as

$$\mu = \frac{1}{3}C_N m\bar{c}\lambda = \frac{1}{3}\rho\bar{c}\lambda \tag{2.68}$$

Combining this equation with Eqs. (2.7) and (2.51), we can get

$$\mu = \frac{1}{3}\rho\sqrt{\frac{8kT}{\pi m}}\frac{RT}{\sqrt{2}\pi N_a d^2 P} = \frac{2}{3}\frac{\sqrt{mkT}}{\pi^{1.5}d^2} \tag{2.69}$$

where $\mu =$ kinetic viscosity in Pa s or N s/m^2.

The effect of temperature on the dynamic viscosity of an ideal gas can also be calculated using the Sutherland's equation (Licht and Stechert 1944 cited by [12]).

$$\frac{\mu}{\mu_0} = \frac{T_0 + T_s}{T + T_s}\left(\frac{T}{T_0}\right)^{3/2} \tag{2.70}$$

where $\mu =$ viscosity in Pa.s at input temperature T, $\mu_0 =$ reference viscosity at reference temperature T_0, $T =$ input temperature, $T_0 =$ reference temperature, $T_s =$ Sutherland's constant. Values of Sutherland's constant T_s are taken from Crane [4].

The viscosities at different reference temperatures can be found in the handbook of CRC [5]; and some examples are listed Table 2.2. For temperatures between $0 < T < 555$ K, the Sutherland's constants and reference temperatures for some gases are listed. The maximum error is 10 %.

2.2 Gas Fluid Dynamics

2.2.1 Reynolds Number

Reynolds number of a fluid quantifies the relative importance of inertial forces (ρu) and viscous forces (μ/L) for a flow. Mathematically, it is described by

$$\mathrm{Re} = \frac{\rho u D}{\mu} = \frac{4\dot{m}}{\mu \pi D} \tag{2.71}$$

where u = magnitude of the mean fluid velocity in m/s, D = characteristic length in m, μ = dynamic viscosity of the fluid in $(N.s/m^2)$ or Pa.s, ρ = density of the fluid in kg/m^3, and $\dot{m}$ is the mass flow rate of the fluid.

The characteristic length depends on the flow condition, internal or external, and the cross section of the pipe for internal flow. For flow in a pipe, it is the hydraulic diameter of the pipe, and for flow over a body, the characteristic length is usually the length of the body. The flow is likely laminar if Re < 2,000 and turbulent for Re > 4,000 for either internal or external flows. In a boundary layer analysis (Sect. 2.2.3), the characteristic length is the distance measured from the leading edge where the boundary layer starts to develop.

2.2.2 Bernoulli's Equation

Bernoulli's equation is important to air emission analysis too. It is derived from the basic concept of conservation of mass and conservation of energy. Very briefly, consider a streamline of a moving fluid without heat transfer; Bernoulli's equation describes the relationship between the static pressure of the fluid, fluid velocity, and the elevation for a steady flow,

$$\frac{v^2}{2} + \int \frac{\Delta P}{\rho} + gz = \text{constant} \tag{2.72}$$

where v is the local velocity on the streamline (m/s), P is the absolute static pressure (Pa, N/m^2), ρ is the density of the fluid (kg/m^3), g is gravitational acceleration (9.81 m/s^2) and z is the elevation (m).

Assuming constant fluid density for the incompressible fluid, Eq. (2.72) becomes

$$P + \frac{1}{2}\rho v^2 + \rho g z = \text{constant} \tag{2.73}$$

where P represents the static pressure and $\rho v^2/2$ the velocity pressure. The consequent total pressure is $P + \frac{1}{2}\rho v^2$. This equation is applicable to most air emission related to analysis.

When $z_1 = z_2$, the pressure difference between two points along a streamline is related to the speeds at these points

$$P_1 - P_2 = \frac{1}{2}\rho\left(v_2^2 - v_1^2\right) \tag{2.74}$$

This simple equation finds important applications in air emission analysis such as the estimation of the resistance to airflow (pressure drop) in an air cleaning device. It is also the principle behind the design of a Pitot tube, which will be introduced later in Chap. 15.

2.2.3 Boundary Layer and Drag

When a gas flows around outside of a body, it produces a force on the body that tends to drag the body in the direction of the gas flow. There are two mechanisms behind this drag effect, one is the skin friction drag and another is form drag. To illustrate the skin drag, let us consider a flat surface with a sharp leading edge attacked by a uniform fluid flow (Fig. 2.3).

Denote the uniform free stream speed as u_∞ and set the coordinate origin at the leading edge with $x = 0$ and $y = 0$ on the solid surface. The fluid is slowed down with a layer that is close to the solid surface, which is called boundary layer. The

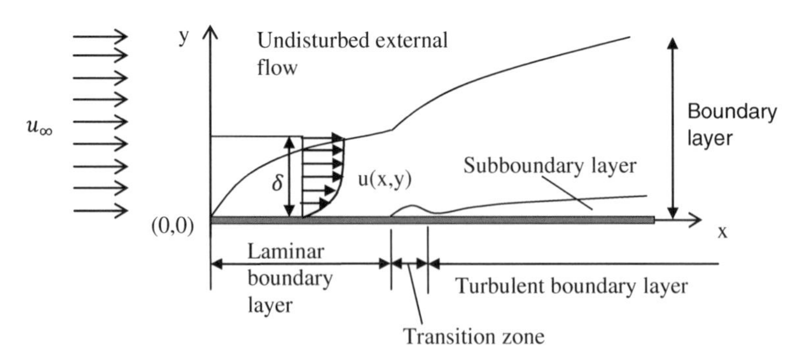

Fig. 2.3 Boundary layer concept

thickness of the boundary layer $\delta(x)$ is taken as the distance above the solid surface where the velocity reaches 0.99 u_∞. The boundary layer grows with the distance from the leading edge until it reaches a constant thickness. Then, we call it fully developed boundary layer.

The flow within a boundary layer also changes gradually from laminar to turbulent. And it is characterized with boundary layer Reynolds number Re_x based on u_∞ and x

$$\mathrm{Re}_x = \frac{\rho u_\infty x}{\mu} \tag{2.75}$$

The velocity profile within the laminar boundary layer is simplified as

$$\frac{u}{u_\infty} = \frac{2y}{\delta} - \frac{y^2}{\delta^2} \tag{2.76}$$

The skin drag is resulted from the shear stress on the solid surface (wetted area). For the real bluff bodies, flow separation occurs and it results in another drag effect called form drag or pressure drag. Overall, the drag is

$$F_D = C_D\left(\frac{1}{2}\rho u_\infty^2\right)A \tag{2.77}$$

where C_D is the drag coefficient based on the reference area, A.

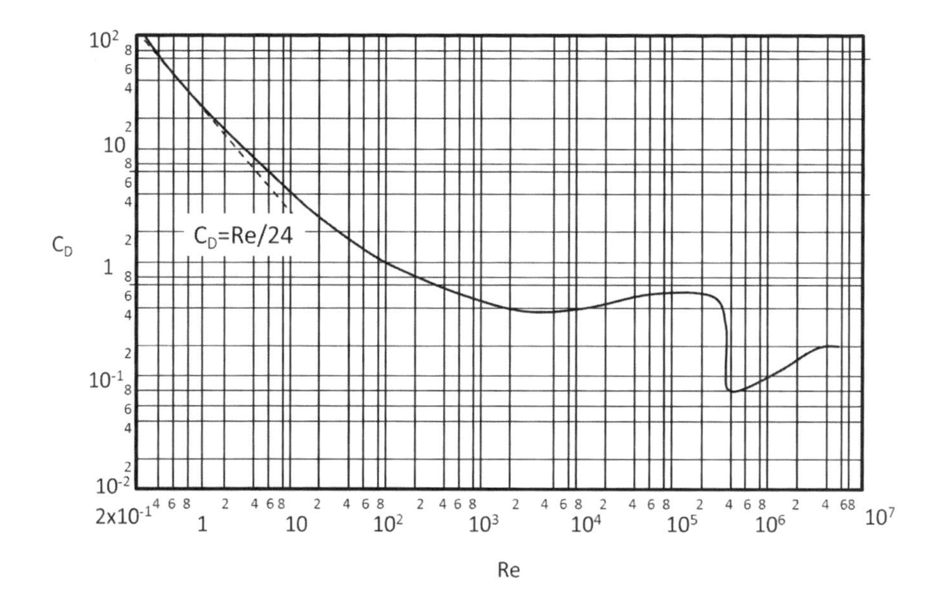

Fig. 2.4 Drag coefficient versus Reynolds number for a sphere (*Data source* [11])

The drag coefficients are geometry dependent and mostly determined experimentally. The drag coefficients of spheres at different Reynolds number are shown Fig. 2.4. More complicated cases can be found in most fluid dynamics books.

2.3 Gas–Liquid Interfacial Behavior

2.3.1 Solubility and Henry's Law

According to International Union of Pure and Applied Chemistry (IUPAC), solubility is the analytical composition of a saturated solution expressed as a proportion of a designated solute in a designated solvent. The most widely used solvent is liquid, which can be a pure substance or a mixture. The extent of the solubility of a substance in a specific solvent is measured as the saturation concentration, where adding more solute does not increase the concentration of the solution and begin to precipitate the excess amount of solute. Solubility may be stated in units of concentration (C_i, kg/m^3), mol fraction (x_i, mol/mol) and other units. The solubility of a substance depends on many physical and chemical properties of the solute and the solvent such as temperature, pressure, and the pH.

Consider a process shown in Fig. 2.5, where gas i is mixed with another insoluble gas, and gas i is soluble. The gas molecules will enter the liquid phase and become part of the liquid mixture. Given enough time, the system reaches equilibrium state.

The Henry's law governs the equilibrium state,

$$P_i = Hx_i \tag{2.78}$$

where P_i = partial pressure of gas i in the gas phase above the liquid, x_i = the equilibrium concentration of gas i in the liquid phase, and H = Henry's law constant with a unit that is determined by that of P_i/x_i. When the partial pressure is in

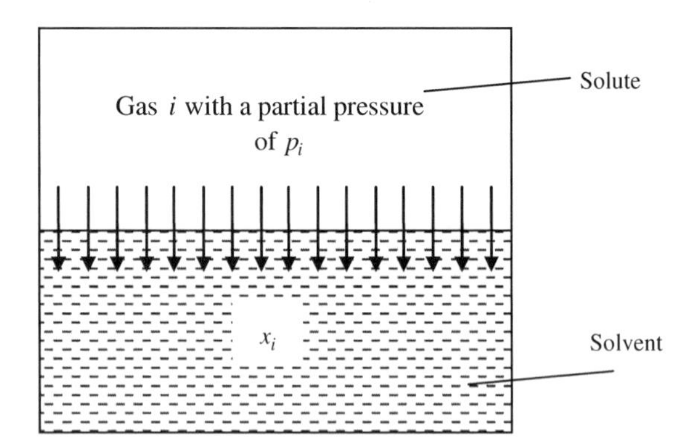

Fig. 2.5 Absorption of gas i into liquid and Henry's Law

Pa and x_i in mol/mol, for example, the corresponding unit of H is [Pa/(mol in liquid/mol of liquid)]. Common units of pressure are Pascal (Pa), atm, bar, and mmHg; those of solute concentration can be g/g, mol/mol, mol/m^3, mol/L, and so on. Users have to excuse unit conversion as needed.

In many engineering design practices, mole fraction of the air pollutant in the gas phase (y_i) is used to quantify the concentration of a gas in a gas mixture; it is a direct description of mass transfer between the two phases. In this case, the partial pressure of the target gas is calculated using Eq. (2.41), $P_i = y_i P$ and the Henry's law equation becomes

$$Py_i = Hx_i \tag{2.79}$$

A comprehensive compilation of Henry' law constants was made by Sander [15]. Based on the data therein, Henry's law constants for some typical gases in water at 25 °C or 298.15 K are summarized in Table 2.3.

Example 2.7: Gas solubility

Estimate the solubility of CO_2 in water in gram of CO_2 per kilogram of water under standard condition (25 °C, 1 atm).

Solution

The Henry's law constant for CO_2 in water at 20 °C and 1 atm is 29.41 atm/(mol of gas per kg of water) (Table 2.3). Because air is an ideal gas and CO_2 takes 0.0314 % of the volume in air, the partial pressure of CO_2 is 0.0314 % of the standard atmospheric pressure:

$$P_{CO_2} = 0.0314\,\% \times 1\,\text{atm} = 0.000314\,\text{atm}$$

Substituting H and P_{CO_2} values into Eq. (2.78) gives

$$x_{CO_2} = \frac{P_{CO_2}}{\text{H}} = \frac{0.000314\,\text{atm}}{29.41\left(\frac{\text{atm}}{\text{mol of CO}_2\text{ per kg of water}}\right)} = 1.068 \times 10^{-5}\,\text{mol of CO}_2/\text{kg of water}$$

The next step is to convert the unit into gram of CO_2 per g of water with the molar weight of CO_2 being 44 g/mole.

$$x_{CO_2} = \frac{1.068 \times 10^{-5}\text{mol CO}_2 \times \frac{44\,\text{g}}{\text{mol}}}{\text{kg of water}} = 4.698 \times 10^{-4}\,\frac{\text{g of CO}_2}{\text{kg of water}}$$

Liquid temperature affects the solubility of a gas and the consequent value of Henry's law constant. Usually, the higher temperature, the lower solubility, and the greater Henry's law constant. The Henry's law constants of some typical air pollutants at different temperatures are listed in Table 2.4. More solubility data can be found in Perry's Chemical Engineers' Handbooks or similar publications.

Table 2.3 Henry's Law constants for air emission related gases in water at T = 298 K

Gas		H at 298 K with the unit of			
		$\frac{\text{atm}}{\left(\frac{\text{mole of gas}}{\text{liter of water}}\right)}$	$\frac{\text{atm}}{\left(\frac{\text{mole of gas}}{\text{kg of water}}\right)}$	$\frac{\text{atm}}{\left(\frac{\text{mole of gas}}{\text{mole of water}}\right)}$	$\frac{\text{Pa}}{\left(\frac{\text{mole of gas}}{\text{mole of water}}\right)}$
Oxygen	O_2	769.23	769.23	4.27×10^4	4.33×10^8
Ozone	O_3	83.33	83.33	4.63×10^3	4.69×10^7
Ammonia	NH_3	0.02	0.02	0.93	9.38×10^3
Nitrogen monoxide	NO	526.32	526.32	2.92×10^4	2.96×10^8
Nitrogen dioxide	NO_2	83.33	83.33	4.63×10^3	4.69×10^7
Hydrogen sulfide	H_2S	10.00	10.00	555.56	5.63×10^6
Sulfur dioxide	SO_2	0.83	0.83	46.30	4.69×10^5
Mercury	Hg	10.75	10.75	597.37	6.05×10^6
Methane	CH_4	714.29	714.29	3.97×10^4	4.02×10^8
Carbon monoxide	CO	1010.10	1010.10	5.61×10^4	5.69×10^8
Carbon dioxide	CO_2	29.41	29.41	1633.99	1.66×10^7

Table 2.4 Henry's law constants for gases in water at different temperatures ($H = p_i/x_i$, atm/(mol gas/mol water))

Gas	0 °C	10 °C	20 °C	30 °C	40 °C	50 °C
He	129,000	126,000	125,000	124,000	121,000	115,000
H_2	57,900	63,600	68,300	72,900	75,100	76,500
N_2	52,900	66,800	80,400	92,400	104,000	113,000
CO	35,200	44,200	53,600	62,000	69,600	76,100
O_2	25,500	32,700	40,100	47,500	53,500	58,800
CH_4	22,400	29,700	37,600	44,900	52,000	57,700
C_2H_6	12,600	18,900	26,300	34,200	42,300	50,000
C_2H_4	5,520	7,680	10,200	12,700	–	–
CO_2	728	1,040	1,420	1,860	2,330	2,830
H_2S	268	367	483	609	745	884

The Henry's law indicates that the equilibrium mole fraction of a gas in liquid is proportional to the partial pressure of the gas above the liquid regardless of the total pressure. Generally, this linear relationship (Henry's law) is sufficiently accurate for pollutant gases at low partial pressures (see Fig. 2.6). This equilibrium state will be broken by change of the amount of the target gas in either gas or liquid phase. An increase in the gas phase concentration results in further absorption into the liquid; too much dissolved gas in the liquid phase results in desorption and a mass transfer from liquid to gas phase.

However, the Henry's law may not be valid when the partial pressure of a gas is too high. Fortunately, in most air emission control engineering problems, the partial pressures of gaseous pollutants of concern are low. Therefore, the Henry's law can be used to estimate the absorption rate with a reasonable accuracy.

2.3.2 Raoult's Law for Ideal Solution

For the gas–liquid system depicted in Fig. 2.5, molecules of the liquid phase also becomes part of the gas phase and form a vapor by evaporation. When the vapor

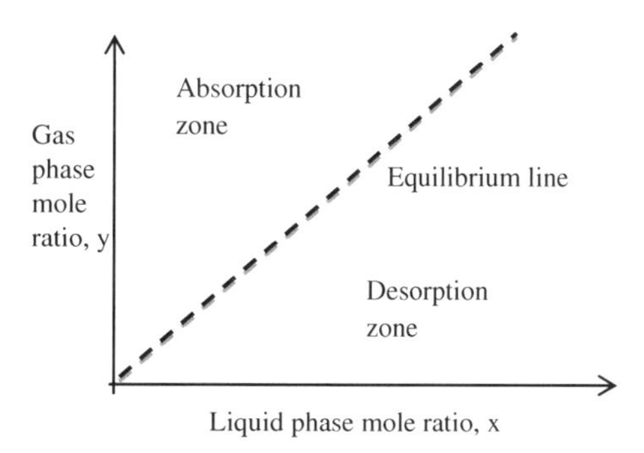

Fig. 2.6 Henry's law line

formed above the liquid-gas interface reaches its dynamic equilibrium, the rate at which liquid evaporates is equal to the rate that the gas is condensing back to liquid phase. This is called the vapor pressure. All liquids have vapor pressure, and the vapor pressure is constant regardless of absolute amount of the liquid substance.

The vapor pressure of the solution will generally decrease as solute dissolves in the liquid phase. As additional solute molecules fill the gaps between the solvent molecules and take up space, less of the liquid molecules will be on the surface and less will be able to break free to join the vapor.

Rather than explaining the Raoult's law based on sophisticated concept of entropy [8], let us explain it in a simple visual way. Consider a sealed container with only one species originally in liquid form (e.g., water). Eventually, we can get the saturated vapor at equilibrium which is sustained as the number of molecules leaving the liquid surface is equal to the number of molecules condensing back to the surface. Later on, we have added so much solute (e.g., salt into water) that there are less water molecules on the surface, because some spaces are taken by the solute salt molecules (Fig. 2.7). Since vapor is formed by the number of solvent (water) molecules that have enough energy to escape from the surface, with less solvent molecules on the surface, the vapor pressure will drop. However, reduced solvent molecules does not affect the ability of vaporized molecules to condense (or stick to the surface) because the vaporized molecules can be attached to both solvent and solute molecules. They are deemed to be able to attract each other, otherwise there would have been no solution in the first place. When the system reaches new equilibrium again, the vapor pressure is lowered.

The Raoult's law is mathematically described as

$$P_v = x_i P_v^0 \tag{2.80}$$

where x_i is the solute mole fraction in the solution, the unit of x_i is mol/mol and P_v^0 is the original vapor pressure. A solution that is governed by the Raoult's law is called an ideal solution.The Raoult's law only applies under ideal conditions in an ideal solution. It works fairly well for the solvent in dilute solutions, which we often

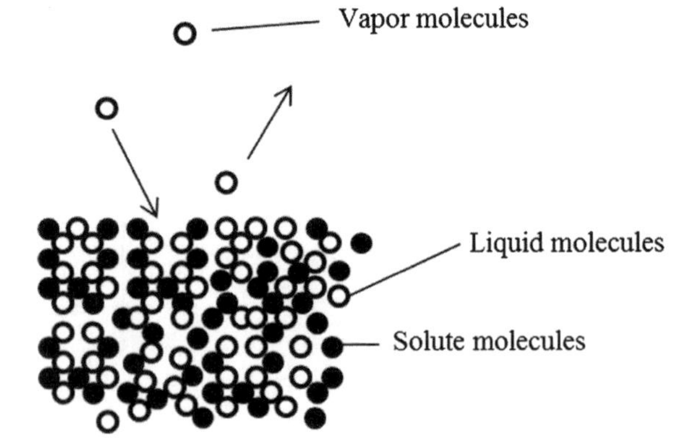

Fig. 2.7 Visualization of Rauolt's law

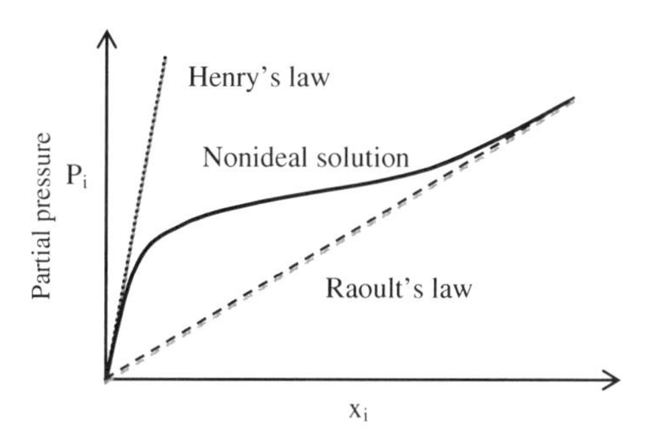

Fig. 2.8 Vapor pressure of a nonideal solution

encounter in air emission studies. In practice, there is no such thing as an ideal solution, it is unlikely for two different types of molecules behave as if they were the same: the forces of attraction between solvent and solute are exactly the same as between the original solvent molecules. However, very dilute solutions obey the Raoult's law to a reasonable approximation.

2.3.3 A Real Gas–Liquid System

For a real gas–liquid system, however, the vapor pressure decreases much faster than that calculated using Eq. (2.80) for extremely dilute solutions. As depicted in Fig. 2.8, the vapor pressure versus mole fraction of solvent in solution curve of a nonideal solution should follow Henry's law at low concentrations and Raoult's law at high concentrations.

2.3.4 Interfacial Mass Transfer

Lewis and Whitman (1924) two-film theory may be used to visualize the gas–liquid interfacial mass transfer. It is assumed that the gas and liquid phases are in turbulent contact with each other, and there is an interface area that separates these two phases. As shown in Fig. 2.9, near the interface, there is a small portion called film exists including a small portion (film) of the gas and another portion of liquid on either side of the interface. Beyond the films, fluids are assumed to be perfectly mixed with uniform concentrations. Mass transfer takes place in these two films. Fluids in these films are assumed to flow in a laminar or streamline motion. Therefore, molecular motion occurs by diffusion, which can be mathematically described. Concentration differences are negligible except in the films in the vicinity of the interface. Both films offer resistance to overall mass transfer. And the interface is at equilibrium described by Henry's law and it offers no resistance to mass transfer.

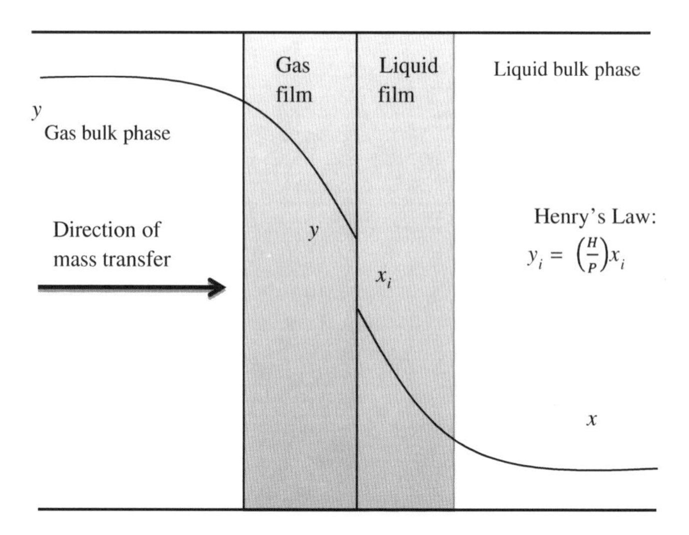

Fig. 2.9 Gas–liquid interfacial mass transfer two-film theory

In summary, according to the two-film theory, gas molecules are dissolved in the liquid phase by the following five steps.

1. Molecules migrate from the bulk-gas phase to the laminar gas film.
2. They penetrate through the gas film by diffusion.
3. These molecules cross the gas–liquid interface by diffusion.
4. Diffuse through the liquid film.
5. Finally, they mix into the bulk liquid phase.

Since it is assumed in this theory that both gas and liquid bulk phases are completely mixed, the interface is at equilibrium with respect to gas molecules transferring through the interface. And, it implies that all mass transfer resistance is resulted from molecular diffusion through the gas and liquid films. With this background introduction, we can derive the mathematical expressions that follow.

The mass transfer per unit interface area is quantified by

$$n' = k_y(y - y_i) \quad \text{gas film} \tag{2.81}$$

$$n' = k_x(x_i - x) \quad \text{liquid film} \tag{2.82}$$

In this equation, n' = mole transfer rate of gas (mol/m^2 s), k_y = gas phase mass transfer coefficient (mol/m^2 s), k_x = liquid phase mass transfer coefficient (mol/m^2 s) and x, y = mole fractions in the bulk liquid and gas phases respectively.

The mass transfer coefficients, k_x and k_y, are determined experimentally. However, it is impractical to determine x_i and y_i at the gas–liquid interface. The mass transfer equation can be described with the bulk phase parameters.

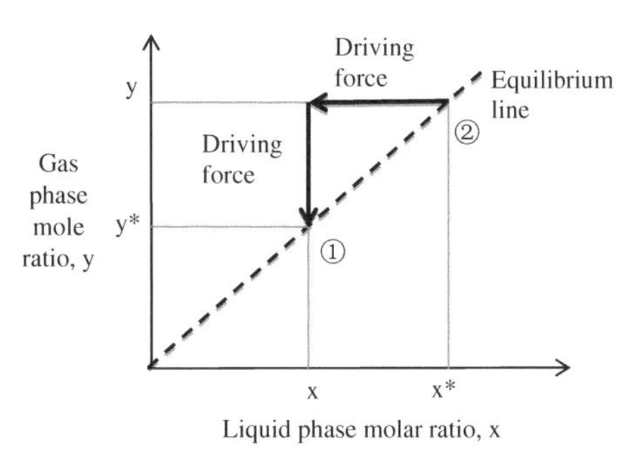

Fig. 2.10 Interfacial mass transfer diving force

$$n' = K_y(y - y^*) \quad \text{gas phase} \tag{2.83}$$

$$n' = K_x(x^* - x) \quad \text{liquid phase} \tag{2.84}$$

where K_x, K_y = Overall mass transfer coefficients for the liquid and gas phases respectively (mol/m^2 s). x^*, y^* = Hypothetical mole fraction corresponding to y, x in the bulk fluids, and they are determined by the Henry's law as

$$y = Hx^* \tag{2.85}$$

$$y^* = Hx \tag{2.86}$$

The difference between the hypothetical mole fraction and the corresponding actual mole fraction results in the driving force illustrated in Fig. 2.10.

If we consider gas phase only in Eq. (2.83) above, the total mass transfer rate (mole/m^2 s) is described as

$$n' = K_y(y - y^*) = K_y[(y - y_i) + (y_i - y^*)] \tag{2.87}$$

Dividing both sides by K_y gives,

$$\frac{n'}{K_y} = (y - y_i) + H(x_i - x) \tag{2.88}$$

The single-phase mass transfer Eqs. (2.81) and (2.84) above also give

$$\frac{n'}{k_y} = y - y_i \quad \text{and} \quad \frac{n'}{k_x} = x_i - x \tag{2.89}$$

Substitute Eq. (2.89) into (2.88) and dividing both sides with n', one can get

$$\frac{1}{K_y} = \frac{1}{k_y} + \frac{H}{k_x} \tag{2.90}$$

This equation indicates that the overall mass transfer coefficient (K_y) can be calculated from the single phase mass transfer coefficients (k_y and k_x). They are usually determined experimentally.

To understand Eq. (2.90), the mass transfer can be expressed in terms of driving potential (see Fig. 5.4) and the corresponding resistance

$$\frac{y - y^*}{R} = \frac{y - y^*}{R_x + R_y} \tag{2.91}$$

where R, R_x and R_y are overall resistance, liquid phase resistance, and gas phase resistance to mass transfer, respectively,

$$R = \frac{1}{K_y}, \quad R_y = \frac{1}{k_y}, \quad \text{and} \quad R_x = \frac{H}{k_x} \tag{2.92}$$

When the value of H is small, there is a great solubility of the target gas in the liquid phase; it indicates a low liquid phase film resistance and that the gas phase resistance is dominating (or gas phase mas transfer is the bottleneck). In this case, only a small amount of water is needed to hold certain amount of gases. In an engineering application, this is preferred in favor of low liquid consumption. On the other hand, when the value of Henry's law constant is large, it means a low solubility of the target gas in the liquid. Then, the mass transfer resistance is primarily attributed to the liquid phase. A large amount of liquid is required to hold the gas.

2.4 Practice Problems

1. Estimate the mass of a single molecule of hydrogen, CO_2 and ethane (C_2H_6).
2. Calculate the root-mean-square speed and kinetic energy of 1 mol of the following gases: He, N_2, O_2, Cl_2, and CH_4 at 300 K.
3. Which gas has the highest root-mean-square speed: Helium (He), Neon (Ne), Argon (Ar), Krypton (Kr), and Xenon (Xe)?
4. Name the two gases of the following: N_2, O_2, CO, CO_2, CH_4, and SO_2 that have the same kinetic energy and root-mean-square speed at 293 K.
5. Which gas has a lowest root-mean-square speed: Fluorine (F_2), Chlorine (Cl_2), SO_2, Krypton (Kr), and Nitrogen dioxide (NO_2)?
6. Explain why nitrogen (N_2) and carbon monoxide (CO) have the same molar kinetic energy at 293 K.
7. Compute the kinetic energy of one cubic meter of air, and CO_2 under standard condition.
8. At what temperature will the root-mean-square speed of CO_2 be the same as that of CH_4 at 300 K?
9. Which has the greater root-mean-square speed at 25 °C, NH_3 (g) or HCl (g)?

10. Which molecule has the highest root-mean-square speed (a) nitrogen molecule (N_2) at 100 °C, (b) Oxygen molecule (O_2) at 200 °C, (c) Chlorine molecule under STP conditions?
11. Assuming spherical oxygen molecules with a radius of 1.65×10^{-10} m, find the mean free path of oxygen molecules at sea level at 293 K.
12. If a gas after fuel combustion at 1000 K contains, by volume, 79 % N_2, 6 % O_2, 7 % CO_2, and 8 % H_2O, what is its molar weight? Can you approximate the molar weight of this flue gas with that of standard dry air?
13. What should be the pressure of N_2 (g) above the solution to increase the solubility of the N_2 (g) to a value of 100.0 mL N_2 per L?
14. What is the mole concentration of O_2 in a saturated solution when the O_2 pressure is 0.21 atm at 25 °C?
15. 15 mL of CO_2 (g) dissolves in 1.0L of water at 25 °C temperature and 1.0 atm pressure. What will be the mole concentration of CO_2 (g) in the saturated solution at 25 °C When the CO_2 pressure is 0.25 atm?
16. A truck tire at the gage pressure of 220 kPa and the temperature of 27 °C contains 12 L of air. Assume that the temperature drops to -40 °C. Determine the gage pressure in the tire.
17. At 27 °C and 1 atm the viscosity and density of air are 1.81×10^{-5} Pa.s and 1.2 kg/m^3, respectively. Determine the diameter and mean free path of an air molecule.
18. At 27 °C and 1 atm the viscosity and density of helium (He) are 1.9×10^{-5} Pa and 0.18 kg/m^3, respectively. Determine the diameter of helium molecule and diffusion coefficient of this gas.
19. At 20 °C and 1 atm the viscosity and density of methane are 2.01×10^{-5} Pa.s and 0.717 kg/m^3, respectively. Determine the diameter of methane molecule, mean free path, and diffusion coefficient.
20. Estimate the following maximum solubility of the air compounds CO_2, O_2 and CH_4 in water under standard conditions. Assume the gas phase is pure dry air.
21. The Henry's law constant of SO_2 at 20 °C in a SO_2-water system is 0.014×10^5 atm/mol fraction in water.
    What is the Henry's law constant with a unit of (Pa/mol fraction in water)? At 30 °C, H = 0.016×10^5 atm/mol fraction in water, does it imply a higher or lower SO_2 solubility in water?
22. Under certain condition, air density is 1.1 kg/m^3. Under the same condition, estimate the density of a gas mixture containing, by volume, 95 % CH_4 and 5 % CO? How many moles of CH_4 in a volume of 0.2 m^3 of this mixture gas?
23. Determine the volume occupied by 1 mol of ideal gas at 15 °C and 1 atm.
24. Find density of dry air at STP conditions (T = 25 °C and P = 1 atm). Assume that dry air has 79 % N_2, and 21.0 % O_2. Given molar weights of N_2 and O_2 are 28 and 32 g/mole, respectively.

References and Further Readings

1. Allen MD, Raabe OG (1985) Slip correction measurements of spherical solid aerosol particles in an improved millikan apparatus. Aerosol Sci Technol 4(3):269–286
2. Bird RB, Stewart WE, Lightfoot EN (1960) Transport phenomena. John Wiley and Sons Inc, New York
3. Cooper CD, Alley FC (2002) Air pollution control—a design approach, 3rd edn. Waveland Press, Inc., Long Grove, IL, USA
4. Crane Company (1988) Flow of fluids through valves, fittings, and pipe. Technical Paper No. 410 (TP 410). P A-5
5. CRC (2013) CRC handbook of chemistry and physics. In: Lide DR (ed), (94th edn), CRC Press, Inc., Boca Raton, Florida, USA
6. Einstein A (1905) On the kinetic molecular theory of thermal movements of particles suspended in a quiescent fluid An. Physik 17:549–560. (English translation: Investigation on the Theory of Brownian Movement, Furth R (ed), Dover, New York, 1956)
7. Flagan R, Seinfeld John H (2012) Fundamentals of air pollution engineering. Dover Publications Inc., New York, USA
8. Guggenheim EA (1937) The theoretical basis of Raoult's law. Trans Faraday Soc 33:151–156
9. Hibbeler RC (2010) Engineering mechanics: dynamics (12th Edn), Prentice Hall, Pearson Publishing Company, NJ, London
10. John HS, Richard CF, Auzmann W (1966) Kinetic theory of gases, W. A. Benjamin Inc., New York
11. Lapple CE, Shepherd CB (1940) Calculation of particle trajectories. Ind Eng Chem 32 (5):605–617
12. Montgomery RB (1947) Viscosity and thermal conductivity of air and diffusivity of water vapor in air. J Meteorol 4:193–196
13. National Research Council (1929) International critical tables, vol III. McGraw-Hill, New York, NY, USA
14. Otto E, Fissan H, Park SH, Lee KW (1999) The log-normal size distribution theory of Brownian aerosol coagulation for the entire particle size range: Part II. J Aerosol Sci 30 (1):16–34
15. Sander R (1999) Compilation of henry's law constants for inorganic and organic species of potential importance in environmental chemistry. http://www.mpch-mainz.mpg.de/~sander/res/henry.html
16. Starzak ME (2010) Energy and enthropy. Chapter 13 Maxwell and boltzmann distribution. Springer, Berlin p 199

Chapter 3
Basics of Gas Combustion

Combustion or burning is a complex sequence of chemical reactions between a fuel and oxygen. The main purpose of combustion is to produce energy and, occasionally, light. In a combustion reaction, the reactants are fuels and an oxidizer, and the products include various air pollutants and carbon dioxide. Meanwhile heat is generated for other engineering applications such as boiling water for steam production, engine powering, and space heating by raising the temperature of the air.

Topics to be covered in this chapter include the properties of air as a combustion oxidizer, combustion stoichiometry, thermodynamics of a combustion system, flame temperature calculation, and combustion chemical equilibrium. After reading this chapter, a reader is expected to be able to

- Understand the basics of gas fuel combustion;
- Explain how air pollutants are formed during fossil fuel combustion with air;
- Theoretically predict what and how much air pollutants can be formed by simple combustion of fuel with known chemical formulae;
- Theoretically calculate the temperature of combustion and energy production;
- Be ready to understand topics related to combustion of other fuels in Part 2.

3.1 Air–Fuel Ratio

Air–fuel ratio (A/F) is the amount of air to the amount of fuel in the combustion mixture. The amount can be quantified by mole or mass. If the air-to-fuel ratio is mole based, the value will be different from the one based on mass. In this book, we use only mole ratio to explain the basic concept, although mass ratio is important too. The mole air fuel ratio is determined by the mole numbers of the components of the mixture as

$$(A/F) = \frac{n_a}{n_f} = \frac{(\sum n_i)_a}{(\sum n_i)_f} \tag{3.1}$$

Z. Tan, *Air Pollution and Greenhouse Gases*, Green Energy and Technology,
DOI 10.1007/978-981-287-212-8_3

where n is mole amount, and the subscripts a and f stand for air and fuel, respectively.

The air–fuel ratio of a combustion mixture determines the combustion chemistry, and consequently the air emissions produced from the combustion process. There are three possible outcomes of the combustion process: stoichiometric, fuel lean, or fuel rich combustion.

- *Stoichiometric combustion* When the air and fuel are mixed at such a ratio that both the fuel and oxygen in the air are consumed completely. This ratio is called theoretical air–fuel ratio, and the combustion is considered as ideal combustion, or stoichiometric combustion.
- *Fuel rich combustion* When the oxygen in the air is not sufficient to burn the fuel completely, the combustion is referred to as fuel rich combustion. And, there will still be fuel left in the combustion product.
- *Fuel lean combustion* When there is more oxygen than needed to burn the fuel completely, the combustion is called fuel lean combustion, and there will be oxygen left in the combustion product.

An important term that is related to air–fuel ratio is equivalence ratio (ϕ). It is defined as the ratio of the theoretical air–fuel ratio to the actual air–fuel ratio of the mixture. Mathematically,

$$\phi = \frac{(A/F)_s}{(A/F)_{\text{mix}}} \tag{3.2}$$

where the subscript s stands for stoichiometric. The advantage of using equivalence ratio over air-to-fuel ratio is that the former is independent on the units being used for the computation of the air–fuel ratio.

The equivalence ratio is more commonly used than air fuel ratio as an indicator to show whether the mixture is stoichiometric, fuel lean, or fuel rich. When ϕ is greater than one there is always excess fuel in the mixture than what is needed for a stoichiometric combustion. ϕ less than one represents a deficiency of fuel in the mixture.

$$\begin{cases} \phi < 1 & \text{Fuel lean} \\ \phi = 1 & \text{Stoichiometric} \\ \phi > 1 & \text{Fuel rich} \end{cases} \tag{3.3}$$

Example 3.1: Air fuel ratio and equivalence ratio

Consider a mixture of one mole of ethane (C_2H_6) and 15 mol of air. Determine

(a) the air-to-fuel ratio of this mixture
(b) the equivalence ratio of this mixture if the stoichiometric air to fuel ratio based on volume is 16.66.

Solution

Note that one mole of air is simplified as a mixture of 0.21 mol of O_2 and 0.79 mol of N_2; the molar weight of air is 28.82 g/mole.

(a) Assuming the mixture is in the same container and reached steady state without chemical reactions. The A/F based on volume is simply

$$(A/F)_{\text{mix}} = \frac{n_a}{n_f} = 15$$

(b) The equivalence ratio is determined as

$$\phi = \frac{(A/F)_s}{(A/F)_{\text{mix}}} = \frac{16.66}{15} = 1.11$$

It is a fuel rich mixture.

3.2 Combustion Stoichiometry

The combustion is stoichiometric when all the atoms in the fuel are burned into their corresponding oxides, *for instance,* all the carbon to CO_2 and all hydrogen to H_2O. Consider a hydrocarbon fuel with α carbon and β hydrogen atoms in each molecule, the stoichiometric combustion process of a hydrocarbon fuel with a molecular formula $C_\alpha H_\beta$ can be described as

$$C_\alpha H_\beta + aO_2 \rightarrow bCO_2 + cH_2O \tag{3.4}$$

where the values of a, b and c depend on α and β.

A simple mass balance leads to the answers of a, b and c in terms of α and β (see Example 3.2), and Eq. (3.4) becomes

$$C_\alpha H_\beta + \left(\alpha + \frac{\beta}{4}\right)O_2 \rightarrow \alpha CO_2 + \frac{\beta}{2}H_2O \tag{3.5}$$

It indicates that in order to achieve the above stoichiometric combustion, for each hydrocarbon molecule, $(\alpha + \beta/4)$ oxygen molecules are required to convert the carbon and hydrogen into carbon dioxide and water, respectively.

In general, all liquid and gaseous fossil fuels are mixtures of multiple components. For ease of engineering analysis, however, they can be simplified as an average formula $C_\alpha H_\beta$, where α and β stand for the numbers of carbon and hydrogen atoms in the fuel, respectively. For a natural gas, its formula is simplified as CH_4, and for a typical liquid fuel, e.g., gasoline, $\alpha = 8$ and $\beta = 18$.

Example 3.2: Stoichiometric combustion
Determine the overall chemical equation for the stoichiometric combustion of propane (C_3H_8) with oxygen as oxidant.

Solution
Start with a general chemical reaction equation as

$$C_3H_8 + aO_2 \rightarrow bCO_2 + cH_2O$$

with a, b, and c to be determined.

According to the conservation of mass, the amount of atoms cannot be created or destroyed. Therefore, the coefficients of a, b and c in the above reaction formula can be computed from the atom balances of carbon, hydrogen, and oxygen:

$$\begin{aligned} \text{Carbon balance:} &\quad b = 3 \\ \text{Hydrogen balance:} &\quad 2c = 8 \\ \text{Oxygenbalance:} &\quad 2b + c = 2a \end{aligned}$$

Solving the above three equations gives $a = 5$, $b = 3$ and $c = 4$. Thus the above stoichiometry is

$$C_3H_8 + 5O_2 \rightarrow 3CO_2 + 4H_2O$$

3.2.1 Stoichiometric Combustion with Dry Air at Low Temperature

In most engineering applications, air instead of pure oxygen is used as an oxidizer for fossil fuel combustion. In other words, oxygen for combustion is obtained from the air for the sake of low cost. Our analysis herein further simplifies the ambient air as a mixture of 21 % oxygen and 79 % nitrogen molecules by volume. In this case, air is defined as

$$1\text{Air} = 0.21O_2 + 0.79N_2 \tag{3.6}$$

This simplified definition is reasonable in that in one mole of air there is about 0.21 mol of oxygen and 0.79 mol of nitrogen. And the corresponding density of air is 1.21 kg/m^3 and the molar weight of air is 28.82 g/mol under standard condition. All these values are close to the measured ones.

For combustion at low temperatures, it is common to assume that the molecular nitrogen in air is inert. Then the general combustion reaction of a hydrocarbon $C_\alpha H_\beta$ with air is

$$C_\alpha H_\beta + a'(0.21O_2 + 0.79N_2) \rightarrow bCO_2 + cH_2O + dN_2 \quad (3.7)$$

It is tedious to carry the decimals when (Air) is referred to as one compound in a chemical reaction formula. Therefore, in most combustion formula, we see the formula of air being defined by normalizing it against 1 mol of O_2. And Eq. (3.7) becomes

$$C_\alpha H_\beta + a(O_2 + 3.76N_2) \rightarrow bCO_2 + cH_2O + dN_2 \quad (3.8)$$

We have to be careful here that $(O_2 + 3.76N_2)$ is not (Air) but 4.76(Air). The coefficient of 3.76 comes from the mole ratio of N_2 to O_2 in a dry air, which is 0.79/0.21 = 3.76. And the coefficient 4.76 is from $1/0.21 \approx 4.76$.

Then we can determine a, b, c, and d. Similar to the approach in Example 3.2, the mass balances of atoms give:

$$\begin{aligned} \text{Carbon balance}: &\quad \alpha = b \\ \text{Hydrogen balance}: &\quad \beta = 2c \\ \text{Oxygenbalance}: &\quad 2(0.21)a = 2b + c \\ \text{Nitrogen balance}: &\quad 2(0.79)a = 2d \end{aligned}$$

One can easily solve the above four equations for the coefficients:

$$a = (\alpha + \beta/4)/0.21, b = \alpha, c = \beta/2, \text{ and } d = 3.76(\alpha + \beta/4)$$

Thus the stoichiometry of a general hydrocarbon $C_\alpha H_\beta$ burned with air can be described as

$$C_\alpha H_\beta + \left(\alpha + \frac{\beta}{4}\right)(O_2 + 3.76N_2) \rightarrow \alpha CO_2 + \frac{\beta}{2}H_2O + 3.76\left(\alpha + \frac{\beta}{4}\right)N_2 \quad (3.9)$$

The stoichiometric air–fuel ratio is also determined as $(A/F)_s = 4.76\left(\alpha + \frac{\beta}{4}\right)$.

Example 3.3: Stoichiometric combustion of propane

Determine the stoichiometric combustion equation for propane (C_3H_8) mixture with dry air.

Solution

For propane C_3H_8, $\alpha = 3$ and $\beta = 8$, comparing with Eq. (3.9),

We can get

$$C_3H_8 + \left(3 + \frac{8}{4}\right)(O_2 + 3.76N_2) \rightarrow 3CO_2 + \frac{8}{2}H_2O + 3.76\left(3 + \frac{8}{4}\right)N_2$$

The corresponding stoichiometry is

$$C_3H_8+5(O_2+3.76N_2) \rightarrow 3CO_2 + 4H_2O + 18.8N_2 \quad (3.10)$$

3.2.2 Fuel Lean Combustion

When there is excess air in a combustion process, the combustion is called fuel lean combustion and the corresponding mixture is called fuel lean mixture. At low combustion temperatures the nitrogen in the air and the extra oxygen appear in the products. For the fuel with a formula $C_\alpha H_\beta$ the reaction formula in general is

$$C_\alpha H_\beta+a(O_2+3.76N_2) \rightarrow bCO_2+cH_2O + dN_2+eO_2 \quad (3.11)$$

The coefficients b and c can be determined from the carbon balance and hydrogen balance, respectively ($b = \alpha; c = \beta/2$).

From the definition of the equivalence ratio above, one can get

$$(A/F)_{\text{mix}}= \frac{(A/F)_s}{\phi} = \frac{4.76}{\phi}\left(\alpha + \frac{\beta}{4}\right) \quad (3.12)$$

Now the fuel lean combustion reaction equation becomes,

$$C_\alpha H_\beta+\frac{4.76}{\phi}\left(\alpha + \frac{\beta}{4}\right)(0.21O_2+0.79N_2) \rightarrow \alpha CO_2+\frac{\beta}{2}H_2O + dN_2+eO_2 \quad (3.13)$$

where ϕ is less than one for a fuel lean mixture because of the excess air. The mole amount of nitrogen (d) and the extra oxygen in the excess air (e) can be determined by the atom balances of O and N:

$$\frac{4.76 \times 0.79}{\phi}\left(\alpha + \frac{\beta}{4}\right) = d \rightarrow d = \frac{3.76}{\phi}\left(\alpha + \frac{\beta}{4}\right)$$

$$e = \left(\frac{1}{\phi} - 1\right)\left(\alpha + \frac{\beta}{4}\right)$$

Therefore, the fuel-lean reaction formula for the combustion of $C_\alpha H_\beta$ perfectly mixed with excess air is,

$$\begin{aligned} &C_\alpha H_\beta+\frac{4.76}{\phi}\left(\alpha+\frac{\beta}{4}\right)(0.21O_2+0.79N_2) \\ &\rightarrow \alpha CO_2+\frac{\beta}{2}H_2O + \frac{3.76}{\phi}\left(\alpha+\frac{\beta}{4}\right)N_2+\left(\frac{1}{\phi}-1\right)\left(\alpha+\frac{\beta}{4}\right)O_2 \end{aligned} \quad (3.14)$$

Most of the time, it is normalized against O_2, and Eq. (3.14) becomes

$$\begin{aligned} & C_\alpha H_\beta + \frac{1}{\phi}\left(\alpha+\frac{\beta}{4}\right)(O_2+3.76N_2) \\ & \rightarrow \alpha\, CO_2 + \frac{\beta}{2}H_2O + \frac{3.76}{\phi}\left(\alpha+\frac{\beta}{4}\right)N_2 + \left(\frac{1}{\phi}-1\right)\left(\alpha+\frac{\beta}{4}\right)O_2 \end{aligned} \tag{3.15}$$

Example 3.4: Fuel lean combustion of octane

Considering a reaction of C_8H_{18} with 10 % excess air by volume, what is the equivalence ratio of the mixture?

Solution

The stoichiometric combustion of C_8H_{18} with dry air $(0.21O_2+0.79N_2)$ is

$$C_8H_{18}+59.52(0.21O_2+0.79N_2) \rightarrow 8CO_2 + 9H_2O + 47N_2$$

or

$$C_8H_{18}+12.5(O_2+3.76N_2) \rightarrow 8CO_2 + 9H_2O + 47N_2$$

The combustion with 10 % excess is then

$$C_8H_{18}+1.10 \times 12.5(O_2+3.76N_2) \rightarrow 8CO_2 + 9H_2O + 47N_2$$

where 1.10 indicates 10 % excess air.

Oxygen balance is

$$16+9 + 2a = 1.1(12.5)(2) \rightarrow a = 1.25$$

Similarly, nitrogen balance gives

$$b = 1.1(59.52)(0.79) = 51.7$$

Therefore,

$$\phi = \frac{(A/F)_s}{(A/F)_{mix}} = \frac{12.5/1}{1.1 \times 12.5/1} = 0.91$$

The equivalence ratio can simply be determined by definition

$$\phi = \frac{(A/F)_s}{(A/F)_{mix}} = \frac{1}{1.1} = 0.91$$

3.2.3 Fuel Rich Combustion with Dry Air at Low Temperatures

A mixture of fuel with less than stoichiometric air is fuel rich mixture, and the corresponding combustion is called fuel rich combustion. Since oxygen is insufficient to oxidize all the C and H in the fuel to CO_2 and H_2O, there may be, for example CO and H_2 in the products. The combustion reaction formula for hydrocarbon fuel with dry air can be derived using the similar approach above.

$$C_\alpha H_\beta + \frac{1}{\phi}\left(\alpha + \frac{\beta}{4}\right)(O_2 + 3.76N_2) \rightarrow [aCO_2 + bCO] + [cH_2O + eH_2] + dN_2 \quad (3.16)$$

where for fuel rich mixture with insufficient air, $1/\phi$ is less than one. Again from the atom balances for carbon, hydrogen, oxygen, and nitrogen, one can get

$$\text{Carbon balance:} \quad \alpha = a + b \quad (1)$$

$$\text{Hydrogen balance:} \quad \beta = 2c + 2e \quad (2)$$

$$\text{Oxygen balance:} \quad \frac{2}{\phi}\left(\alpha + \frac{\beta}{4}\right) = 2a + b + c \quad (3)$$

$$\text{Nitrogen balance:} \quad \frac{4.76}{\phi}\left(\alpha + \frac{\beta}{4}\right) \times 0.79 = d \quad (4)$$

Only d can be determined from the nitrogen balance. The other four (a, b, c and e) cannot be determined because there are four unknowns in three equations. To solve the problem, one more equation is needed. This equation can be derived from chemical equilibrium (see Sect. 3.3.2). For now, we can only write the chemical equations as follows with x and y to be determined from other known factors.

$$C_\alpha H_\beta + \frac{1}{\phi}\left(\alpha + \frac{\beta}{4}\right)(O_2 + 3.76N_2)$$
$$\rightarrow [xCO_2 + (1 - x)CO] + [yH_2O + (1 - y)H_2] + \frac{3.76}{\phi}\left(\alpha + \frac{\beta}{4}\right)N_2 \quad (3.17)$$

The preceding analyses were based on a common assumption that air and fuel are perfectly mixed. However, it is very challenging in engineering practices to achieve perfect mixing in the entire combustion device. Stoichiometric, fuel lean and fuel rich combustion all take place at different spots in the combustion device. As a result, the actual combustion formula is much more complicated than that in Eq. (3.17). For instance, if we consider only oxygen as one more extra product, the reaction becomes

$$C_\alpha H_\beta + \frac{1}{\phi}\left(\alpha + \frac{\beta}{4}\right)(O_2 + 3.76N_2)$$
$$\rightarrow [xCO_2 + (1-x)CO] + [yH_2O + (1-y)H_2] + \frac{3.76}{\phi}\left(\alpha + \frac{\beta}{4}\right)N_2 \quad (3.18)$$
$$+ \left[\frac{1}{\phi}\left(\alpha + \frac{\beta}{4}\right) - \frac{1+x+y}{2}\right]O_2 + \cdots$$

3.2.4 Complex Fossil Fuel Combustion Stoichiometry

As mentioned above, it is unusual to have pure hydrocarbon fuel. There are always atoms other than hydrogen and carbon in fossil fuels. For example, sulfur and oxygen are typical elements in fossil fuels, coal, oil or gas. There are also other additives in oil and gas. As a result, the combustion stoichiometry of these fuels becomes complicated. For a general fossil fuel the corresponding stoichiometric combustion can be described as, only for example,

$$\begin{aligned} &C_\alpha H_\beta O_o Cl_q S_s N_n + a(0.21O_2 + 0.79N_2) \\ &\quad \rightarrow bCO_2 + cH_2O + dHCl + eN_2 + fSO_2 \end{aligned} \quad (3.19)$$

Then the atom balances give

$$\text{Carbon balance}: \quad \alpha = b \quad (1)$$

$$\text{Hydrogen balance}: \quad \beta = 2c + d \quad (2)$$

$$\text{Chloride (Cl) balance}: \quad q = d \quad (3)$$

$$\text{Sulfur balance}: \quad s = f \quad (4)$$

$$\text{Oxygen balance}: \quad o + 2(0.21)a = 2b + c + 2f \quad (5)$$

$$\text{Nitrogen balance}: \quad n + 2(0.79)a = 2e \quad (6)$$

Solving the above six equations, the unknown coefficients, a through f, can be determined. The chemical reaction formula for the stoichiometric combustion of the complex fuel becomes,

$$C_\alpha H_\beta O_o Cl_q S_s N_n + 4.76\left(\alpha + \frac{\beta}{4} - \frac{q}{4} + s - \frac{o}{2}\right)(0.21O_2 + 0.79N_2)$$
$$\rightarrow \alpha CO_2 + \left(\frac{\beta - q}{2}\right)H_2O + qHCl + \left[\frac{n}{2} + 3.76\left(\alpha + \frac{\beta}{4} - \frac{q}{4} + s - \frac{o}{2}\right)\right]N_2 + sSO_2$$
$$(3.20)$$

Similarly, for fuel lean combustion with an equivalence ratio of ϕ, one can derive the combustion reaction formula from the atom balances and get

$$\begin{aligned} C_{\alpha}H_{\beta}O_{o}Cl_{q}S_{s}N_{n} + \frac{4.76}{\phi}\left(\alpha + \frac{\beta}{4} - \frac{q}{4} + s - \frac{o}{2}\right)(0.21O_2 + 0.79N_2) \\ \rightarrow \alpha CO_2 + \left(\frac{\beta - q}{2}\right)H_2O + qHCl + sSO_2 \\ + \left[\frac{n}{2} + \frac{3.76}{\phi}\left(\alpha + \frac{\beta}{4} - \frac{q}{4} + s - \frac{o}{2}\right)\right]N_2 \\ + \left[\left(\frac{1}{\phi} - 1\right)\left(\alpha + \frac{\beta}{4} - \frac{q}{4} + s - \frac{o}{2}\right)\right]O_2 \end{aligned} \tag{3.21}$$

Again, for fuel rich combustion, by mass balance of the atoms, there are more unknowns than the number of equations. It requires the knowledge of chemical kinetics and chemical equilibrium to acquire more equations needed.

3.3 Chemical Kinetics and Chemical Equilibrium

3.3.1 Chemical Kinetics

Chemical kinetics, also known as reaction kinetics, aims at the rate of a one-way chemical reaction. Now consider the reaction between A and B producing C and D.

$$aA + bB \rightarrow cC + dD$$

The reaction rate of A is a function of the concentrations of the reactants.

$$r_A = \frac{d[A]}{dt} = -k[A]^a[B]^b \tag{3.22}$$

where r_A is the reaction rate in (mole/$m^3 \cdot$ s), k = rate constant of the reaction; the unit of rate constant depending on coefficients a and b are s^{-1}, $m^3/(\text{mole.s})$, and $m^6/(\text{mole}^2\text{.s})$ for reactions of first, second, and third order, respectively. $[i]$ = mole concentration in (mole/m^3) and it is exchangeable with c_i in this book. The negative sign indicates that the reactants are consumed in the reaction. This negative sign signifies the reaction rate to be a positive value.

The rate constant of reaction (k) defines how fast an irreversible elementary reaction takes place. Some reactions like oxidation of iron in natural conditions may take years to complete, but others like combustion of CH_4 completes in less than a second.

k is not a real constant, because it is a function of the combustion condition, especially the temperature. The rate constant of reaction can be determined using a modified Arrhenius equation

$$k = AT^B \exp\left(-\frac{E_A}{RT}\right) \tag{3.23}$$

When $B = 0$, Eq. (3.23) becomes the Arrhenius equation

$$k = A \exp\left(-\frac{E_A}{RT}\right) \tag{3.24}$$

where A and B are the rate coefficients, E_A = Activation energy (J/mole) and R = Universal gas constant. The values of coefficients A, B, and E for H–O reactions are tabulated in Table 3.1 [1]. A more comprehensive table of rate constants for chemical reactions in combustion is given by Westley [18].

A global hydrogen–oxygen reaction actually proceeds via the multiple elementary reactions, collectively known as reaction mechanisms. These mechanisms

Table 3.1 Rate coefficients for H–O reactions

Reaction	A $\left[(\text{cm}^3/\text{mole})^{n-1}\right]^n$	B	E_A (J/mole)	Temperature Range (K)
$H + O_2 \rightarrow OH + O$	1.2×10^{17}	−0.91	69.1	300–2,500
$OH + O \rightarrow O_2 + H$	1.8×10^{13}	0	0	300–2,500
$O + H_2 \rightarrow OH + H$	1.5×10^{7}	2.0	31.6	300–2,500
$OH + H_2 \rightarrow H_2O + H$	1.5×10^{8}	1.6	13.8	300–2,500
$H + H_2O \rightarrow OH + H_2$	4.6×10^{8}	1.6	77.7	300–2,500
$O + H_2O \rightarrow OH + OH$	1.5×10^{10}	1.14	72.2	300–2,500
$H + H + M \rightarrow H_2 + M$				
M = Ar (low P)	6.4×10^{17}	−1.0	0	300–5,000
$M = H_2$ (low P)	0.7×10^{16}	−0.6	0	100–5,000
$H_2 + M \rightarrow H + H + M$				
M = Ar (low P)	2.2×10^{14}	0	402	2,500–8,000
$M + H_2$ (low P)	8.8×10^{14}	0	402	2,500–8,000
$H + OH + M \rightarrow H_2O + M$				
$M = H_2O$ (low P)	1.4	−2.0	0	1,000–3,000
$H_2O + M \rightarrow H + OH + M$				
$M = H_2O$ (low P)	1.6×10^{17}	0	478	2,000–5,000
$O + O + M \rightarrow O_2 + M$				
M = Ar (low P)	1.0×10^{17}	−1.0	0	300–5,000
$O_2 + M \rightarrow O + O + M$				
M = Ar (low P)	1.2×10^{14}	0	451	2,000–10,000

include, but are not limited to, those listed in Table 3.1, where M can be any species present in the combustion system that acts as a collision partner. Species such as H, O, OH and HO_2 are called radicals. Radicals play important roles in combustion. They are highly reactive and short-lived.

Example 3.5: Reaction rate

Determine the reaction rate at low pressure for the mechanism reaction

$$\mathrm{H} + \mathrm{H} + \mathrm{M} \rightarrow \mathrm{H_2} + \mathrm{M}$$

at temperature of 3,000 k, where M is H_2 as a function of [H].

Solution

According to Eq. (3.22), the chemical reaction rate for the reaction

$$H + H + \mathrm{M} \rightarrow \mathrm{M} + \mathrm{H_2}$$

is described in terms of the formation rate of H_2

$$\frac{\mathrm{d}[\mathrm{H_2}]}{\mathrm{d}t} = k[\mathrm{H}]^2$$

where the reaction rate constant k can be described using Eq. (3.25)

$$k = AT^B \exp\left(-\frac{E_A}{RT}\right) \tag{3.25}$$

Therefore, the chemical reaction rate is described as

$$\frac{\mathrm{d}[\mathrm{H_2}]}{\mathrm{d}t} = AT^B \exp\left(-\frac{E_A}{RT}\right)[\mathrm{H}]^2$$

There is no negative sign because H_2 is produced rather than consumed in this one-way reaction.

From Table 3.1, we can get

$$A = 0.7 \times 10^{16}, B = -0.6, \text{ and } E_A = 0$$

Thus the rate of reaction at T = 3,000 K is

$$\frac{\mathrm{d}[\mathrm{H_2}]}{\mathrm{d}t} = 0.7 \times 10^{16} \times 3{,}000^{-0.6} \exp(-0)[\mathrm{H}]^2 = 5.74 \times 10^{19}[\mathrm{H}]^2$$

The solution indicates that molecular hydrogen is produced extremely fast at high temperature.

3.3.2 Chemical Equilibrium

Now consider a general reversible reaction $aA + bB \leftrightarrow cC + dD$ that starts with a mixture of A and B without C and D. A and B are thus identified as the reactants and $C + D$ are the products. At any moment $t > 0$, once C and D present, they also react and produce A and B reversing the process. At microscopic level, reactions proceed both ways, but at macro level, the concentrations of A and B decrease and those of C and D increase until they all remain stable, or reached chemical equilibrium state. The equilibrium state is a dynamic one in which reactant and product concentrations remain constant, not because the reaction stops but because the rates of the forward and reverse reactions are equal.

Chemical reactions are dynamic in the sense that at any instant changes continue in a system on a microscopic scale, but everything appears to be constant from a macro scale point of view. This is called an equilibrium state. A combustion system has a tendency to reach an equilibrium state. When the combustion conditions such as pressure and temperature change, extra time is required for the system to reestablish the new equilibrium state.

For the chemical reaction

$$aA + bB \underset{k_b}{\overset{k_f}{\leftrightarrow}} cC + dD \quad (3.26)$$

where k_f, k_b are the rate constants for the forward and backward/reverse reactions, respectively.

By considering both forward and backward reactions, the net consumption rate of species A becomes

$$r_A = -k_f[A]^a[B]^b + k_b[C]^c[D]^d \quad (3.27)$$

When the reaction is at equilibrium, the reaction rate is zero $(r_A = 0)$. Solving this equation one can get the equilibrium constant based on concentration, which is the ratio of the reaction constants

$$K_C = \frac{k_f}{k_b} = \frac{[C]^c[D]^d}{[A]^a[B]^b} \quad (3.28)$$

The unit of K_C depends on the unit of concentration and the mole difference $\Delta n = (c + d) - (a + b)$; it is $[\,]^{(c+d)-(a+b)}$.

Example 3.6: Chemical equilibrium constant
At certain temperature, the equilibrium constant is 4.0 for the reaction. All compounds are in gas phase

$$CO + 3H_2 \leftrightarrow CH_4 + H_2O \tag{R1}$$

What is the equilibrium constant expression and value for the following reaction?

$$CH_4 + H_2O \leftrightarrow O + 3H_2 \tag{R2}$$

Solution

According to Eq. (3.28), the equilibrium constant for Reaction R_1 is

$$K_{C1} = \frac{[CH_4][H_2O]}{[CO][H_2]^3} = 4$$

Similarly, for R_2

$$K_{C2} = \frac{[CO][H_2]^3}{[CH_4][H_2O]} = \frac{1}{K_{C1}} = \frac{1}{4} = 0.25$$

Chemical equilibrium can also be described in terms of partial pressures of the gases when all the products are considered as ideal gases. The partial pressure-based equilibrium constant equation is described as

$$K_P = \frac{P_C^c P_D^d}{P_A^a P_B^b} \tag{3.29}$$

According to Dalton's Law, Eq. (2.40), $P_i = y_i P$, then Eq. (3.29) becomes

$$K_P = \frac{y_C^c y_D^d}{y_A^a y_B^b} (P)^{c+d-a-b} \tag{3.30}$$

With mole fractions $y_i = n_i/n$ into this equation becomes

$$K_P = \frac{n_C^c n_D^d}{n_A^a n_B^b} \left(\frac{P}{n}\right)^{c+d-a-b} \tag{3.31}$$

where n_i is the mole amount of the ith gas and n is the total mole amount of all the gases in the reaction system. And $n \geq (n_A + n_B + n_C + n_D)$ as there may be other gases present in the system.

Considering the ideal gas law, $PV = nRT$,

$$K_P = \frac{(n_C/V)^c (n_D/V)^d}{(n_A/V)^a (n_B/V)^b} (RT)^{c+d-a-b} = \frac{[C]^c [D]^d}{[A]^a [B]^b} (RT)^{c+d-a-b} \tag{3.32}$$

This equation describes the conversion between K_P and K_C as

$$K_P = K_c(RT)^{c+d-a-b} = K_c(RT)^{\Delta n} \tag{3.33}$$

where Δn is a general expression of the number of coefficient difference and $\Delta n = (c+d) - (a+b)$ for the equilibrium reaction $aA + bB \leftrightarrow cC + dD$.

Example 3.7: SOx chemical equilibrium

Consider a mixture of 64 kg of SO_2 and 32 kg of O_2 mixed with 56 kg of inert N_2 in a sealed reactor. After ignition the system eventually reaches chemical equilibrium at temperature T = 1,000 K and pressure $P = 2$ atm. Assume nitrogen does not participate in the chemical reactions. If the equilibrium constant (K_P) at 1,000 K is 1.8, determine the corresponding chemical mole fractions in the equilibrium products.

Solution

The mole amount of the mixture before reaction can be determined as follows:

$$n_{SO_2} = \frac{64{,}000\,\text{g}}{64\,\text{g/mole}} = 1{,}000\,\text{mole}$$

$$n_{O_2} = \frac{32{,}000\,\text{g}}{32\,\text{g/mole}} = 1{,}000\,\text{mole}$$

$$n_{N_2} = \frac{56{,}000\,\text{g}}{28\,\text{g/mole}} = 2{,}000\,\text{mole}$$

After ignition the chemical reaction most likely proceed as follows:

$$SO_2 + \frac{1}{2}O_2 \leftrightarrow SO_3$$

Since N_2 does not participate in the reaction as an inert gas, it works only as a dilution gas throughout the process. When the final products reach equilibrium, we assume there is s mole of SO_3. Then we can determine the mole amount of other gases before combustion and after reaching equilibrium by mass balance as follows.

Mole amount	n_{SO_2}	n_{O_2}	n_{SO_3}	n_{N_2}	Total n
Prior to combustion	1,000	1,000	0	2,000	4,000
At equilibrium after reaction	1,000 − s	1,000 − 0.5 s	s	2,000	4,000 − 0.5 s

The partial pressure-based equilibrium constant can be calculated using Eq. (3.31)

$$K_P = \frac{n_{SO_3}}{n_{SO_2} n_{O_2}^{0.5}} \left(\frac{P}{n}\right)^{1-1-0.5} = \frac{s}{(1{,}000 - s)(1{,}000 - 0.5s)^{0.5}} \left(\frac{2}{4{,}000 - 0.5s}\right)^{-0.5}$$

At 1,000 K, $K_P = 1.8$, by iteration we can solve this equation and get

$$s = 515$$

So the final mole amount of the species are determined as follows.

	SO_2	O_2	SO_3	N_2	Total
n_i(mole)	485	742.5	515	2,000	3742.5

Since chemical equilibrium constant depends on gas temperature, the temperature of the combustion product mixture has to be known in order to determine the equilibrium constant. When this information is known, an empirical formula known as *van't Hoff's* equation [7] can be used for the computation of the partial pressure based equilibrium constant K_P:

$$\frac{\mathrm{d}}{\mathrm{d}T}(\ln K_P) = \frac{\Delta \mathrm{H}_R}{RT^2} \tag{3.34}$$

where $\Delta \mathrm{H}_R$ = the enthalpy of reaction (J/mole) and it generally depends on the temperature.

Assuming there is no phase change, $\Delta \mathrm{H}_R$ can be calculated using Eq. (3.53).

When ΔH_R is assumed constant over a narrow range of temperature, the integration of Eq. (3.34) gives a form that is similar to the Arrhenius equation for a narrow range of temperature

$$K_P = K_{P0} \exp\left(\frac{\Delta \mathrm{H}_R}{RT}\right) \tag{3.35}$$

where K_{P0} can be determined using a reference temperature such as $T_0 = 298K$. From this analysis it can also be seen that the components in the combustion is dependent on the combustion temperature [5]. Therefore, a general form of the chemical equilibrium constant is

$$K_P = A \exp\left(-\frac{B}{T}\right) \tag{3.36}$$

For example, the chemical equilibrium constant of the reaction $\frac{1}{2}N_2 + \frac{1}{2}O_2 \leftrightarrow NO$ is

Table 3.2 Equilibrium constants based on partial pressure for chemical reactions

T (K)	$\ln(K_P)$			
	$\frac{1}{2}O_2 + \frac{1}{2}N_2 \leftrightarrow NO$	$CO_2 + H_2 \leftrightarrow CO + H_2O$	$CO_2 \leftrightarrow CO + \frac{1}{2}O_2$	$H_2O \leftrightarrow H_2 + \frac{1}{2} O_2$
298	−35.052	−11.554	−103.762	−92.208
500	−20.295	−4.9252	−57.616	−52.691
1,000	−9.388	−0.366	−23.529	−23.163
1,200	−7.569	0.3108	−17.871	−18.182
1,400	−6.27	0.767	−13.842	−14.609
1,600	−5.294	1.091	−10.83	−11.921
1,800	−4.536	1.328	−8.497	−9.826
2,000	−3.931	1.51	−6.635	−8.145
2,200	−3.433	1.648	−5.12	−6.768
2,400	−3.019	1.759	−3.86	−5.619
2,600	−2.671	1.847	−2.801	−4.648
2,800	−2.372	1.918	−1.894	−3.812
3,000	−2.114	1.976	−1.111	−3.086
3,200	−1.888	2.022	−0.429	−2.451
3,400	−1.69	2.061	0.169	−1.891
3,600	−1.513		0.701	−1.392
3,800	−1.356		1.176	−0.945
4,000	−1.216		1.599	−0.542

$$K_{P,\mathrm{NO}} = 4.71\exp\left(-\frac{10{,}900}{T}\right) \quad \text{for} \quad \frac{1}{2}\mathrm{N}_2 + \frac{1}{2}\mathrm{O}_2 \leftrightarrow \mathrm{NO} \tag{3.37}$$

Then the chemical equilibrium constants at different temperatures can be calculated. Table 3.2 shows some examples obtained at $P = 1\,\mathrm{atm}$.

3.3.3 *Chemical Equilibrium in Gaseous Combustion Products*

Now consider a general combustion reaction that converts oxygen and fuel (O and F) into a mixture of different product gases, *A*, *B*, … *Z*. After chemical reactions there are only reaction products left in the system.

$$\underbrace{oO + fF + \cdots}_{\substack{\text{Reactant}\\ \text{mixutre}}} \rightarrow \underbrace{aA + bB + cC + dD + \cdots + yY + zZ}_{\substack{\text{Reaction}\\ \text{products}}} \tag{3.38}$$

At a macroscopic scale, it seems that the combustion process is finished, but at a microscopic scale, there may be chemical equilibrium established among different products in the form of, say,

$$A \leftrightarrow B + \frac{1}{2}C \quad \text{and} \quad 2C + D \leftrightarrow 3X + \frac{1}{2}Y$$

These chemical equilibrium reactions determine the actual quantities of the products, numerically, $a, b, c \ldots z$. With the known equilibrium constants of these reactions, we can derive the amount of the gas products.

Let us use an example to explain how it works for a fuel rich combustion.

Example 3.8: Fuel rich combustion and chemical equilibrium

Combustion products from the reactant mixture of 1,000 mol of CO_2, 500 mol of O_2 and 500 mol of N_2 consist of CO_2, CO, O_2, N_2 and NO at 3,000 K and 1 atm. Determine the equilibrium composition of the combustion product.

Solution

Since the reactant mixture and the product compounds are already known, the combustion stoichiometry can be described as

$$CO_2 + \frac{1}{2}O_2 + \frac{1}{2}N_2 \rightarrow aCO + bNO + cCO_2 + dO_2 + eN_2$$

with a, b, c, d, e as coefficients to be determined.

From the mass balances for C, O, and N we can set up three equations:

$$\text{Carbon balance}: \quad a + c + 1 \tag{1}$$

$$\text{Oxygen balance}: \quad a + b + 2c + 2d = 3 \tag{2}$$

$$\text{Nitrogen balance}: \quad b + 2e = 1 \tag{3}$$

We need two more equations, which can be obtained from two equilibrium reactions:

$$CO_2 \leftrightarrow CO + {}^1\!/_2 O_2 \qquad (R_1)$$

$${}^1\!/_2 O_2 + {}^1\!/_2 N_2 \leftrightarrow NO \qquad (R_2)$$

The corresponding chemical equilibrium constants can be found in Table 3.2, which gives,

$$K_{P1}\,(3{,}000\,\text{K}) = \exp(-1.111) = 0.3273$$
$$K_{P2}\,(3{,}000\,\text{K}) = \exp(-2.114) = 0.1222$$

In the combustion products, for a mole of CO, there are b mole of NO, c mole of CO_2, d mole of O_2 and e mole of N_2, therefore, the mole fraction of the species in the combustion products can be determined as

$$y_{CO} = \frac{a}{a+b+c+d+e}, y_{NO} = \frac{b}{a+b+c+d+e}, y_{CO_2} = \frac{c}{a+b+c+d+e}$$

$$y_{O_2} = \frac{d}{a+b+c+d+e}, y_{N_2} = \frac{e}{a+b+c+d+e}$$

From the equilibrium chemical reaction $CO_2 \leftrightarrow CO + {}^1\!/_2 O_2$, the chemical equilibrium constant at 1 atm is

$$K_{P1} = \frac{y_{co} \cdot y_{O_2}^{1/2}}{y_{CO_2}} = \frac{ad^{1/2}}{c}(a+b+c+d+e)^{-1/2} = 0.3273 \qquad (4)$$

Similarly, for the second equilibrium reaction ½ O_2 + ½ $N_2 \leftrightarrow NO$, its equilibrium constant

$$K_{P2} = \frac{y_{NO}}{y_{O_2}^{1/2} y_{N_2}^{1/2}} = \frac{b}{d^{1/2} e^{1/2}} = 0.1222 \qquad (5)$$

From Eqs. (1)–(3), we can get

$$c = 1 - a$$
$$d = 1/2(1 + a - b)$$
$$e = 1/2(1 - b)$$

Put them into (4) and (5), we can get

$$\frac{a}{1-a}\left(\frac{1+a-b}{4+a}\right)^{1/2} = 0.3273 \qquad (4')$$

$$\frac{2b}{[(1+a-b)(1-b)]^{1/2}} = 0.1222 \qquad (5')$$

Solving Eqs. (4′) and (5′) we get

$$a = 0.3745 \quad b = 0.0675 \quad c = 0.6255 \quad d = 0.6535 \quad e = 0.4663$$

Then the mole fraction of each species is determined as follows:

$$y_{CO} = 0.17, y_{NO} = 0.03, y_{CO2} = 0.29, y_{O2} = 0.30, y_{N2} = 0.21.$$

3.3.4 The Pseudo-Steady-State Approximation

Many combustion reactions involve free radicals, which are very reactive intermediate species. They are consumed rapidly and present at very low concentration, but are critical to the combustion reactions. It is important to understand how they affect the chemical equilibrium and overall rate of constant.

Consider an overall reaction $A \rightarrow B + C$, which proceeds through the free radical of $A*$. Molecules of A first collide with other molecule (M) that does not change, to produce free radicals $A*$, followed by decomposition of A^* to $B + C$. These two step reactions can be described as follows:

$$A + M \underset{k_{b1}}{\overset{k_{f1}}{\leftrightarrow}} A^* + M \quad (1)$$

$$A^* \xrightarrow{k_2} B + C \quad (2)$$

Note that the first reaction is reversible and the second is one way. The rate of reaction before reaching equilibrium in terms of A^* for the first reversible reaction is

$$-\frac{d[A]}{dt} = \frac{d[A^*]}{dt} = -k_{f1}[A][M] + k_{b1}[A^*][M] \quad (3.39)$$

And for the second step, A^* is consumed only and the corresponding rate is

$$\frac{d[A^*]}{dt} = -k_2[A^*] \quad (3.40)$$

The overall consumption rate of $[A^*]$ considering both reactions is

$$\frac{d[A^*]}{dt} = -k_{f1}[A][M] + k_{b1}[A^*][M] - k_2[A^*] \quad (3.41)$$

Since $A*$ is so reactive that it is consumed almost immediately after its formation; this is referred to as the pseudo-steady-state approximation [7]. Thereby, mathematically, the left hand side of the above equation is zero, and it becomes

$$k_{f1}[A][M] - k_{b1}[A^*][M] + k_2[A^*] = 0 \quad (3.42)$$

This approximation allows us to determine the concentration of $A*$

$$[A^*] = \frac{k_{f1}[A][M]}{k_{b1}[M] - k_2} \tag{3.43}$$

Substituting $[A^*]$ into Eq. (3.39) leads to

$$\frac{d[A]}{dt} = \frac{k_{f1}k_2[A][M]}{k_{b1}[M] - k_2} \tag{3.44}$$

Now, the rate of reaction constant k for the overall reaction $A \xrightarrow{k} B + C$ is determined by

$$\frac{d[A]}{dt} = -k[A]$$

And it gives the overall reaction rate constant as,

$$k = \frac{k_{f1}k_2[M]}{k_2 - k_{b1}[M]} \tag{3.45}$$

3.4 Thermodynamics of Combustion System

Temperatures are assumed known in the preceding analyses. In case the temperature of the combustion product mixture is unknown yet, it has to be determined from the combination of chemical equilibrium and thermodynamics. Related topics are briefly reviewed as follows for the readers without related training background.

3.4.1 First Law of Thermodynamics

The first law of thermodynamics is about the conservation of energy; the total energy of the thermodynamic system and its surrounding environment is conserved. Physically, the first law states that, for a closed system, the change in the internal energy is equal to the sum of the amount of heat energy supplied to the system and the work done by the system on the surroundings. The first law of a combustion system is mathematically described as

$$\Delta U = Q - W \tag{3.46}$$

where ΔU = increase in the internal energy of the system (J), Q = heat added to the system (J), and W = work done by the system on the surroundings (J).

Since combustion takes place in a relatively short period of time, it allows us to conduct a simplified analysis without complex integration, considering a combustion reaction starting with reactants at state 1 (pressure P_1 and temperature T_1). After a constant pressure combustion the products are at state 2 (P_2, T_2). The system is at constant pressure ($P_1 = P_2 = P$), therefore, work is done by the system on the surroundings, where $W = P(V_2 - V_1)$. Applying first law of thermodynamics,

$$U_2 - U_1 = Q - P(V_2 - V_1) \tag{3.47}$$

which leads to the heat produced by the combustion process as

$$Q = (U_2 + PV_2) - (U_1 + PV_1) \tag{3.48}$$

Note that the enthalpy of a system at certain status is defined as

$$H = U + PV \tag{3.49}$$

where H = the total enthalpy of the system (J), U = the internal energy of the system (J), and V = the volume of the system.

Therefore, the heat added to the system is the total enthalpy difference

$$Q = H_2 - H_1 \tag{3.50}$$

The total enthalpy at status j = 1 or 2 is the summarization of the enthalpy of all the compounds,

$$H_j = \sum n_i h_i(T_j) \tag{3.51}$$

Then Eq. (3.50) becomes

$$Q = \sum_P n_i h_i(T_2) - \sum_R n_i h_i(T_1) \tag{3.52}$$

where n = mole amount of the component in fuel–air mixture or the product, $h(T)$ = enthalpy of a component at temperature T in J/mole. Subscripts i, P and R stand for the ith component, product, and reactant, respectively.

The combustion system can be defined as exothermic, isothermal or endothermic based on the heat of reaction as follows:

$$\begin{cases} Q < 0 & \text{exothermic reaction} \\ Q = 0 & \text{isothermal reaction} \\ Q > 0 & \text{endothermic reaction} \end{cases}$$

When the final temperature of the products is the same as the initial temperature of the reactants $T_1 = T_2 = T$, Eq. (3.52) can be used to define enthalpy of reaction (ΔH_R), which is defined as the heat per unit fuel (J/mole) released from a combustion. Mathematically, it is described as

$$\Delta H_R(T) = \frac{Q}{n_f} = \frac{1}{n_f}\left[\sum_P p_i h_i(T) - \sum_R r_i h_i(T)\right] \tag{3.53}$$

Again, the unit of enthalpy of reaction is J/(mole of fuel). It is temperature dependent. For the reaction of $aA + bB \rightarrow cC + dD$ that takes place under an isothermal steady state, its enthalpy of reaction is the enthalpy difference between the products and the reactants.

3.4.2 Enthalpy Scale for Reacting System

For a reacting system the working fluid changes molecularly from reactants to products while undergoing a combustion process. The enthalpy of a substance at condition of (P, T) is $h(P, T)$. The enthalpy of a substance at standard condition is defined as that at 1 atm and 298 K taking the symbol of h(STP). So,

$$h(P,T) = h(STP) + [h(P,T) - h(STP)].$$

The enthalpy of every element in its natural state at the standard condition is zero. The enthalpy of all other substances at STP is simply the heat of formation of the substance, since it is formed from its corresponding elements. For example,

$$\tfrac{1}{2}O_2(g) + H_2(g) \rightarrow H_2O(l)$$

Recall that at standard condition (1 atm and 298 K), $q = h_{H_2O(l)} - \tfrac{1}{2}h_{O_2(g)} - h_{H_2(g)} \equiv h^o_{f,H_2O(l)}$, i.e., $h_{H_2O(l)} = h^o_{f,H_2O(l)}$. The enthalpy of the ith component in a mixture is

$$\begin{aligned} h(P,T) &= h(STP) + [h(P,T) - h(STP)] \\ &= h^o_{f,i} + [h_i(P,T) - h_i(STP)] \end{aligned} \tag{3.54}$$

where $h^o_{f,i}$ is the heat of formation of substance i, and the term in the bracket is sensible enthalpy. Therefore, the enthalpy of the ith component in a mixture is

$$h_i(T) = h^o_{f,i} + \int_{298\,K}^{T} C_{p,i}\, dT \tag{3.55}$$

where $C_{p,i}$ is a function of temperature, it can be computed using

$$C_{p,i}(T) = a_i + b_i T \tag{3.56}$$

In this equation, the coefficients a_i, and b_i, are available in Table A.4, the approximate thermodynamic data for species of combustion interest [2, 7, 9, 16]. The integration of Eq. (3.55) then leads to

$$h_i(T) = h^o_{f,i}(\mathrm{T}) + \mathrm{a_i}(\mathrm{T} - \mathrm{T_{298K}}) + \frac{1}{2}\mathrm{b_i}(\mathrm{T}^2 - \mathrm{T}^2_{298K}) \tag{3.57}$$

The enthalpy of formation, h^o_f, of a compound is defined as the change of enthalpy that accompanies the formation of 1 mol of compound in its standard state from its constituent elements in their standard states. This definition indicates that the products and the reactants are at the standard state. The standard state is referred to as atmospheric pressure (101.325 Pa) and a temperature of 298 K.

Values for standard heat of formation for different species are tabulated in Table A.4 (selected from Burcat [2]). More can be found in Perry's Chemical Engineers' handbook or similar publications. Table A.4 shows the enthalpy of formation of an element in its natural state at the standard condition is zero. Enthalpy of formation can also be determined from their heating values that will be introduced in Sect. 3.4.3.

3.4.3 Heating Values

When a fuel undergoes combustion with oxygen, energy is released as heat. The maximum amount of heat is released when the combustion is stoichiometric, where all the hydrogen and carbon contained in the fuel is converted to CO_2 and H_2O. This maximum energy from 1 mol of fuel is called the heat of combustion or the heating value. Depending on the state of water in the combustion products, the heat of combustion for fuels is expressed as the higher heating value (HHV) or lower heating value (LHV). HHV is used when the water in the products is in the liquid state ($h_{\mathrm{H_2O}} = h_l$). LHV is used when the water in the products is in the vapor state ($h_{\mathrm{H_2O}} = h_g$), and the energy required to vaporize the water is not considered as heat of combustion.

The conversion between mole based LHV and HHV is thus

$$\mathrm{LHV} = \mathrm{HHV} - \frac{n_{\mathrm{H_2O}}}{n_{\mathrm{fuel}}} h_{fg} \tag{3.58}$$

where h_{fg} is the latent heat of vaporation of water (J/mole), and $n_{\mathrm{H_2O}}/n_{\mathrm{fuel}}$ gives the mole amount of water produced by 1 mol of fuel.

Since a typical flue gas or engine exhaust is at temperature greater than 100 °C, where water exists as vapor, LHVs are more relevant than HHVs to fuel combustion analysis. The heating values of typical fuels are available in the literature, for example, Fundamentals of Combustion Processes by McAllister et al. [12].

Table A.5 shows the HHV and LHV of some typical fuels.

The LHV can be related to the enthalpy of formation of a fuel. Enthalpy of formation is available at certain temperature, typically T = 298 K.

$$\mathrm{LHV}(298K) = \frac{1}{n_f}\left[\sum_R \left(n_i h^o_{f,i}\right) - \sum_P \left(n_i h^o_{f,i}\right)\right] = \Delta \mathrm{H}_R(298K) \tag{3.59}$$

where the enthalpy of formation of the fuel is used for the calculation of the total enthalpy of the reactants.

Example 3.9: Heating values

Consider stoichiometric combustion of methane with pure oxygen proceeds as $CH_4 + 2O_2 \rightarrow CO_2 + 2H_2O$. The mass-based HHV of methane is 55.65 MJ/kg at 298 K and 1 atm, estimate (a) LHV of CH_4, (b) enthalpy of formation of methane.

Solution

The conversion from HHV (MJ/kg) to HHV (MJ/kmole) is needed by considering the molar weight of the fuel given based on mass, and it can be converted by

$$\begin{aligned}\mathrm{HHV(MJ/kmole)} &= \mathrm{HHV(MJ/kg)} \times \mathrm{M_{fuel}} \\ &= 55.65\,\mathrm{(MJ/kg)} \times 16\,\mathrm{(kg/kmole)} \\ &= 890.4\,\mathrm{MJ/kmole\ (at\ 298K)}\end{aligned}$$

(a) The LHV can be determined by considering the latent heat of vaporization of water at 298 K, which is $h_{fg} = 43.92 \mathrm{MJ/kmole}$.

$$\begin{aligned}\mathrm{LHV} &= \mathrm{HHV} - \frac{n_{H_2O}}{n_{\mathrm{fuel}}} h_{fg} \\ &= 890.4 - 2 \times 43.92 \quad \mathrm{(MJ/kmole)} \\ &= 802.6 \quad \mathrm{(MJ/kmole)} \\ &= 802.6 \quad \mathrm{(MJ/kmole)\ or\ 50.1\,MJ/kg}\end{aligned} \tag{3.60}$$

(b) The combustion $CH_4 + 2O_2 \rightarrow CO_2 + 2H_2O$ gives

$$\mathrm{LHV} = \left(n_{CO_2} h^o_{f,CO_2} + n_{H_2O} h^o_{f,H_2O}\right) - \left(n_{CH_4} h^o_{f,CH_4} + n_{O_2} h^o_{f,O_2}\right) \tag{3.61}$$

The enthalpy of formation values can be found in Table A.4

Species	h_f^o (298 K) (J/mol)	Δh_f^o (298 K) (MJ/kmol)
O_2	0	0
CO_2	−394.088	−394.1
H_2O	−242.174	−242.2

Then we have

$$802.6 = \left(h^o_{f,CH_4}\right) - [(-394.1) + 2 \times (-242.2)]$$
$$h^o_{f,\mathrm{CH_4}} = 802.6 - 394.1 - 2 \times 242.2 = 75.9\,\mathrm{MJ/kmole}$$

The answer is close to the value of 75 MJ/kmole in Table A.4.

3.5 Adiabatic Flame Temperature

The flame temperature affects the formation of the air pollutants, especially NO_x. Adiabatic flame temperature (AFT) is employed to quantify the flame temperature. Depending on how the process is completed, constant volume AFT and constant pressure AFT are used accordingly. The constant volume AFT is the temperature of the products results from a complete combustion process that takes place without any work done on the surroundings because the volume of the combustion system does not change. The constant pressure AFT is the temperature of products of a complete combustion process that takes place without any heat transfer to the surroundings, but there is work done on the surroundings.

First of all, for either of the definitions, the combustion system is adiabatic (i.e., $Q = 0$). Therefore, the first law of thermodynamics becomes

$$\Delta U = W$$

For this adiabatic system it is clear that the constant volume AFT is the highest temperature that can be achieved for known mixture of fuel and oxidant, because work done to the surrounding environment and/or any incomplete combustion would lower the temperature of the products. Therefore, the constant pressure flame temperature is lower than that of constant volume process because some of the energy is utilized to change the volume of the system.

3.5.1 Constant Pressure Adiabatic Flame Temperature

Considering an adiabatic system with constant pressure complete combustion, the first law of thermodynamics gives

$$Q = \sum_P n_i h_i(T_a) - \sum_R n_i h_i(T_R) \tag{3.62}$$

where T_a is the constant pressure AFT. The temperature of the products of a complete combustion process can be determined for a given combustion reaction where n_i's are known for both the mixture of reactants and the combustion products.

The above equation gives Q = 0 for adiabatic flame and

$$\sum_R n_i\left[h^o_{f,i} + (h_i(T_R) - h_i(298K))\right] = \sum_P n_i\left[h^o_{f,i} + (h_i(T_a) - h_i(298K))\right] \tag{3.63}$$

Considering Eq. (3.57), the above equation becomes

$$\begin{aligned}&\sum_R n_i\left[\left(a_i(T_R - 298K) + \frac{b_i}{2}(T_R^2 - (298K)^2\right) + h^o_{f,i}\right]\\&= \sum_P n_i\left[\left(a_i(T_a - 298K) + \frac{b_i}{2}(T_a^2 - (298K)^2\right) + h^o_{f,i}\right]\end{aligned} \tag{3.64}$$

The values of coefficients a_i and b_i and the heat of formation, $h^o_{f,i}$ in the above equation can be found in Table A.4, and the adiabatic temperature T_a can be determined mathematically. As a common simplification, the temperature of the air–fuel mixture before combustion is assumed standard temperature, $T_R = 298$ K. In this case, Eq. (3.64) above is simplified as

$$\sum_R n_i h^o_{f,i} = \sum_P n_i\left[\left(a_i(T_a - 298K) + \frac{b_i}{2}(T_a^2 - (298K)^2\right) + h^o_{f,i}\right] \tag{3.65}$$

Example 3.10: Adiabatic Flame Temperature
CH_4 is pre-mixed with air at 298 K at an equivalence ratio of 1.0 and the combustion is complete. Determine the constant pressure AFT.

Solution
First of all, set up stoichiometric combustion reaction equation using the methods introduced in Sect. 3.4,

$$CH_4 + 2(O_2 + 3.76N_2) \rightarrow CO_2 + 2H_2O + 7.52N_2$$

For an adiabatic constant pressure system, with $T_R = T_0 = 298$ K, left hand side (LHS) of Eq. (3.65) is

$$\text{LHS} = h^o_{f,\text{CH}_4} + 2h^o_{f,\text{O}_2} + 7.52h^o_{f,\text{N}_2} \tag{1}$$

The right-hand side (RHS) of the equation is simplified as

$$\begin{aligned}\text{RHS} = &\left[\left(a_{\text{CO}_2}(T_a - 298K) + \frac{b_{\text{CO}_2}}{2}(T_a^2 - (298K)^2\right) + h^o_{f,\text{CO}_2}\right]\\ &+ 2\left[\left(a_{\text{H}_2\text{O}}(T_a - 298K) + \frac{b_{\text{H}_2\text{O}}}{2}(T_a^2 - (298K)^2\right) + h^o_{f,\text{H}_2\text{O}}\right]\\ &+ 7.52\left[\left(a_{\text{N}_2}(T_a - 298K) + \frac{b_{\text{N}_2}}{2}(T_a^2 - (298K)^2\right) + h^o_{f,\text{N}_2}\right]\end{aligned} \tag{2}$$

From Table A.4, we can get

Species	$h^o_{f,i}$ (J/mol)	$C_p(T)$ [J/mol K]	
		a_i	b_i
CO_2	−394,088	44.3191	0.0073
H_2O	−242,174	32.4766	0.00862
N_2	0	29.2313	0.00307
O_2	0	30.5041	0.00349
CH_4	−74,980	44.2539	0.02273

With these values

$$\begin{aligned}\text{LHS} =& -74{,}980(\text{J/mol})\\ \text{RHS} =& \left[\left(44.3191(T_a - 298) + \frac{0.0073}{2}(T_a^2 - (298\text{K})^2\right) - 394.088\right]\\ &+ 2\left[\left(32.4766(T_a - 298\text{K}) + \frac{0.00862}{2}(T_a^2 - (298\text{K})^2\right) - 242.174\right]\\ &+ 2(3.76)\left[\left(29.2313(T_a - 298\text{K}) + \frac{0.00307}{2}(T_a^2 - (298\text{K})^2\right) + 0\right]\end{aligned}$$

Equating LHS and RHS we have

$$0.02387\,T_a^2 + 330.26\,T_a - 903994 = 0$$

Solving this equation we can get, $T_a = 2341$ K; Another root $T_a = -16.177$ K is ignored because it is physically wrong. So the estimated AFT is

$$T_a = 2{,}341\ \text{K}$$

Note that, with this high temperature, it is no longer reasonable to assume that nitrogen does not react with oxygen. We can proceed with the calculation for better accuracy by considering the knowledge to be introduced in Sect. 7.7.1.

3.5.2 Constant Volume Adiabatic Flame Temperature

For an adiabatic constant volume combustion process the work done is zero to the surroundings. The first law of thermodynamics gives

$$Q = \sum_P n_i U_i(T_a) - \sum_R n_i U_i(T_R) = 0 \tag{3.66}$$

From $h = U + PV = U + RT$, one can get $U = h - RT$. And the above equation becomes

$$\sum_P n_i (h_i(T_a) - RT_a) = \sum_R n_i (h_i(T_R) - RT_R) \tag{3.67}$$

or

$$\sum_P n_i \left[h^o_{f,i} + (h_i(T_a) - h_i(T_0) - RT_a\right] = \sum_R n_i \left[h^o_{f,i} + (h_i(T_R) - h_i(T_0)) - RT_R\right] \tag{3.68}$$

Reorganizing the formula leads to

$$\begin{aligned}\sum_P n_i h_i(T_a) = &\left[\sum_R n_i (h_i(T_R) - h_i(T_0)\right] - \left[\sum_P n_i h^o_{f,i} - \sum_R n_i h^o_{f,i}\right] \\ &+ \left[\sum_P n_i h_i(T_0)\right] + \left[\sum_p n_i RT_a - \sum_R n_i RT_R\right]\end{aligned} \tag{3.69}$$

There is an extra term in the last bracket compared to the formula for constant pressure AFT. This term is positive because the flame temperature is always greater than that of the reactants. Therefore, mathematically it shows that the constant volume AFT is greater than the constant pressure one. The AFT is lower for constant pressure process since there is work done on the surroundings.

3.6 Practice Problems

1. Determine the stoichiometric air/fuel mass ratio and product gas composition for combustion of heptane (C_7H_{16}) in dry air.
2. Determine the stoichiometric air required for the combustion of butane in kg of air per kg of fuel.

3. Propane is burned in dry air at an equivalence ratio of 0.85. Determine,
   a. air/fuel mass ratio
   b. product gas composition
4. Ethane is burned with 50 % excess air. Determine
   a. air/fuel molar ratio
   b. equivalence ratio of the mixture
5. Octane (C_8H_{18}) is burned with 25 % excess air. Determine
   a. stoichiometric and actual air/fuel ratio, both on a mole and mass basis
   b. equivalence ratio of the mixture
6. Ethylene (C_2H_4) is burned in air at an equivalence ratio of 0.75. Determine constant pressure AFT. Both fuel and air are initially at temperature of 298 K.
7. Acetylene (C_2H_2) is burned in air at an equivalence ratio of 0.91. Determine constant pressure AFT by assuming that both fuel and air are initially at temperature of 298 K.
8. Determine which is a better fuel methanol (CH_3OH) or methane by comparing their constant pressure AFTs. Assume that fuel and air are initially at 298 K and equivalence ratio is 1 in both cases. Given that enthalpy of formation of liquid methanol at 298 K is –239,000 J/mole.
9. A fuel oil contains 88 % carbon and 12 % hydrogen by weight. Its HHV is 46.4×10^6 J/kg. Determine the effective chemical formula and enthalpy of formation for this fuel.
10. Determine the stoichiometric air to fuel ratio for a fuel containing 50 % by weight octane (C_8H_{18}), 25 % by weight methanol (CH_3OH), and 25% by weight ethanol (C_2H_5OH), in kg of air per kg of fuel.
11. Consider a fuel having formula C_5H_5N is burned completely to CO_2, H_2O, and NO. Assume that there is no formation of thermal NO. Calculate the stoichiometric amount of air required in kg of air per kg of fuel and only NO in products.
12. Calculate the stoichiometric amount of air required (in kg of air per kg of fuel) when one mole of $C_5H_{10}S$ is burned completely and SO_2 in products.
13. Consider the reaction $N_2 + O_2 \leftrightarrow 2NO$. Calculate the equilibrium concentration of NO (in ppmv) for a flue gas. The flue gas composition is 79 % N_2, 6 % O_2, 7 % CO_2 and 8 % H_2O at temperature of 1,500 K and pressure of 1 atm. Given $K_p = 1.1 \times 10^{-5}$.
14. NO and NO_2 formation in flames can be represented by the following two reactions:

$$N_2 + O_2 \leftrightarrow 2NO \qquad (R_1)$$

$$NO + 1/2\, O_2 \leftrightarrow NO_2 \qquad (R_2)$$

Calculate the equilibrium concentrations of NO and NO_2 in ppmv for air that is held at 1,500 K long enough to reach equilibrium. Given K_p for R_1 and R_2 are 1.1×10^{-5} and 1.1×10^{-2}, respectively.

15. Consider a fuel having formula $C_{10}H_{20}$ NS is burned completely to CO_2, H_2O, NO, and SO_2. Assume that there is no formation of thermal NO. Calculate the stoichiometric amount of air required in kg of air per kg of fuel.
16. Calculate the equilibrium concentration of sulfur trioxide in ppm when 500 mL of sulfur dioxide is mixed with 999.500 mL of air. Assume T = 1000K and P = 2 atm.
17. Calculate the equilibrium concentrations of NO and NO_2 in ppmv from the following two reactions:

$$N_2 + O_2 \leftrightarrow 2NO \qquad (R1)$$

$$NO + 1/2\ O_2 \leftrightarrow NO_2 \qquad (R2)$$

The reactions take place in air at temperature of 2,200 K. Given K_p for R1 and R2 are 3.5×10^{-3} and 2.6×10^{-3}, respectively.

18. A car is driven 10,000 miles/year on gasoline. Assume that its milage is 22 miles per gallon (MPG). The gasoline weighs 6.0 pounds/gallon and contains 85 % carbon by weight. Determine the amount of carbon dioxide emitted per year.
19. Consider constant pressure complete combustion of stoichiometric liquid butane (C_4H_{10})-air mixture initially at 298 K and 1 atm. Estimate the AFT.
20. Calculate the stoichiometric air required for the complete combustion of $C_{11}H_{22}N$ to CO_2, H_2O and NO (assuming no thermal NO is formed) in kg of air per kg of fuel.
21. Under certain conditions, a thermochemical conversion (TCC) process produced a liquid fuel that contains 80 % by mass carbon and 20 % by mass hydrogen. This fuel is then burned in dry air at an equivalence ratio of 1.0. initial temperature of the air–fuel mixture is at T_0 = 298 K and 1 atm.

    (a) Determine the formula of this fuel
    (b) Determine the AFT assuming complete combustion.

22. Consider combustion of a mixture of 1 mol of CH_4 mixed with air at stoichiometric condition. The flame temperature is measured to be 2,000 K and the pressure is 1 atm. Assume that only the following gases, CO_2, CO, H_2O, H_2, O_2 and N_2, present in product. Determine the mole amount of CO_2 produced through this combustion process, when only the following two equilibrium reactions are considered:

$$CO_2 + H_2 \leftrightarrow CO + H_2O$$

$$CO_2 \leftrightarrow CO + 1/2\ O_2$$

References and Further Readings

1. Baulch DL, Bowman CT, Cobos CJ, Cox RA, Just T, Kerr JA, Pilling MJ, Stocker D, Troe J, Tsang W, Walker RW, Warnatz J (2005) Evaluated kinetic data for combustion modeling. J Phys Chem Ref Data 34:757
2. Burcat A (1984) Thermochemical data for combustion. In: Gardiner WC (ed) Combustion chemistry. Springer, New York
3. Cooper CD, Alley FC (2002) Air pollution control—a design spproach, 3rd edn. Waveland Press, Inc., IL
4. CRC (2013) CRC handbook of chemistry and physics, 94th edn. Lide DR (ed), CRC Press, Inc., Boca Raton
5. Denbigh K (1981) The principles of chemical equilibrium, 4th edn. Cambridge University Press, London
6. Dryer FL, Glassman I (1973) High temperature oxidation of CO and methane. In: 14th 369 Symposium on international on combustion, The combustion Institution, Pittsburg, pp 987–1001
7. Flagan RC, Seinfeld JH (2012) Fundamentals of air pollution engineering. Dover Press, New York
8. Heinsohn R, Kabel R (1999) Sources and control of air pollution. Prentice Hall, Upper Saddle River, p 36
9. Kee RJ, Rupley FM, Miller JA (1987) The CHEMKIN thermodynamic data base. SANDIA report SAND87-8215. Sandia National Laboratories, Livermore
10. Leib T, Pereira C (2007) Reaction kinetics, Sect. 7. In: Green D, Perry R (eds) Perry's chemical engineers' handbook, 8th edn. McGraw-Hill, New York, pp 7-5–7-7
11. Li S, Williams FA (1999) NO_x formation in two-stage methane-air flames. Combust Flame 118:399–414
12. McAllister S, Chen J-Y, Fernandez-Pello AC (2011) Fundamentals of combustion processes. Springer, New York
13. NIST (2005) NIST Chemistry Web Book, NIST Standard Reference Database Number 69
14. Palmer HB, Beers JM (1974) Combustion technology. Academic Press, New York
15. Sonibare JA, Akeredolu FA (2004) A theoretical prediction of non-methane gaseous emissions from natural gas combustion. Energy Policy 32:1653–1665
16. Stull DR, Prophet H (eds) (1971). JANAF Thermochemical tables. U.S. Department of Commerce, Washington DC, and addenda
17. Turns SR (2000) An introduction to combustion: concepts and application, 2nd edn. McGraw-Hill, New York
18. Westley F (1980) Table of recommended rate constants for chemical reactions occuring in combustion, issued in April 1980 by national standard reference data system. Document No. C13.48.67. http://www.nist.gov/data/nsrds/NSRDS-NBS67.pdf. Accessed 12 June 2014

Chapter 4
Properties of Aerosol Particles

As a counterpart of Chap. 2, this chapter covers the basic properties and dynamics of aerosol particles. An aerosol is a mixture of solid particles and/or liquid droplets suspended in a gas. The gas phase can be air or other gases. In air pollution studies, we can also call it particulate matter. In this book, particulate matter (PM) is interchangeable with *aerosol* without examining their fine differences. The particles can be either solid aerosol particles or liquid droplets with little deformation or evaporation.

This chapter starts with classic particle dynamics followed by basic terms that are widely used in air emission engineering including aerodynamic diameter, equivalent diameters, Stokes number, Stokes law, adhesion and reentrainment of particles, and diffusion of particles in the air. At the end, particle size statistics and dynamics are introduced.

4.1 Particle Motion

It is important to analyze the behavior of particles in various force fields, which guides the design and operation of many particulate air pollution control devices. In most of the cases, the particle is subject of at least two forces acting along opposite directions, one is parallel and another opposite to the direction of motion. The latter is the resistance of the surrounding gas to the particle in motion.

4.1.1 Particle Reynolds Number

In particle dynamics, the particle Reynolds number follows similar definition except that characteristic length is the particle diameter, d_p, and that the velocity in Eq. (2.69) is replaced with the magnitude of the velocity (speed) of gas with respect to that of the particle, $|u - v|$.

Z. Tan, *Air Pollution and Greenhouse Gases*, Green Energy and Technology,
DOI 10.1007/978-981-287-212-8_4

$$Re_p = \frac{\rho_g d_p |u - v|}{\mu} \tag{4.1}$$

In this equation, the density is the gas density and subscript g is added in order to differentiate it with the density of solid particles. With a coordinate fixed on the moving particle, $|u - v|$ should always be a positive quantity.

The particle Reynolds number can be used to estimate the flow condition around the particles. There are four regime for particle dynamics as follows.

$$\begin{cases} \text{Stokes regime :} & Re_p < 1 \\ \text{Transient regime :} & 1 < Re_p < 5 \\ \text{Turbulent regime :} & 5 < Re_p < 1000 \\ \text{Newton's regime :} & Re_p > 1000 \end{cases}$$

For Stokes regime, the flow around the particle is laminar and the frictional force is dominant over the inertia force. In the transient regime, the flow around the particle starts to develop turbulence, both inertial and frictional forces are important. In the turbulent regime, the flow around the particle is turbulent and drag decreases with Re_p. The flow around the particle becomes highly turbulent when $Re_p > 1000$ and the drag is considered constant, and the inertial force is dominant.

4.1.2 Stokes' Law

Stokes' law is derived from the Navier-Stokes equations, which are nearly insolvable due to the nonlinear partial differential equations. Stokes solved these equations with the following assumptions:

(1) The inertial forces are negligible compared with the viscous forces.
(2) There are no walls or other particles nearby. Because of the small size of airborne particles, the fraction of the particles near a wall is negligible.
(3) The particle is a rigid sphere. This means that there is a limitation of Stokes law for liquid droplets or soft particles.
(4) The fluid is incompressible. When the fluid is a gas, say air, the assumption does not imply that air is incompressible, but that it does not compress near the surface of the particle. This is equivalent to assuming that the relative velocity between the gas and the particle is much less than the speed of sound, and it is valid for airborne particles.
(5) There is no slipping between the fluid and particles i.e., the fluid velocity at the particle's surface is zero. When the relative velocity is not zero, a correction is needed.
(6) The motion of the particle is constant.

With the first assumption, the drag force on the moving particle can now be described by

$$F_D = 3\pi\mu d_p|u - v| \tag{4.2}$$

where u is the velocity of air and v is that of the particle.

In Stokes regime, the resistance experienced by a moving particle in a gas is described by the above equation. In practice, this equation is not perfect. When particle $Re_p = 1.0$, the error is 12 %. The error can be reduced to 5 % at a $Re_p = 0.3$.

The same drag force can also be calculated using Eq. (4.3),

$$F_D = C_D\left(\frac{1}{2}\rho u_\infty^2\right)A \tag{4.3}$$

which was derived from Newton's law. Substituting the cross-section area (or rigorously, the frontal area) $A = \pi d_p^2/4$ into Eq. (4.3) leads to

$$F_D = C_D\left[\frac{1}{2}\rho_g(u - v)^2\right]\left(\frac{1}{4}\pi d_p^2\right) \tag{4.4}$$

Comparing Eqs. (4.2) and (4.4), we can get the drag coefficient in the Stokes regime,

$$C_D = \frac{24\mu}{\rho_g d_p|u - v|} = \frac{24}{Re_p} \tag{4.5}$$

This is the equation for the dashed straight-line portion at the up-left corner of Fig. 2.4. Note that when we first introduced Fig. 2.4 and Eq. (4.3), we assumed the solid phase (the sphere) is static and only air moves around the sphere. It would be easy to understand the concepts in this section by assuming air is quiescent and only the aerosol particle is moving. Indeed $|u - v|$ represents the relative motion of particle with respect the air.

Overall, for spherical particles, the following relationships can be used to estimate drag coefficient.

$$C_D = \begin{cases} \frac{24}{Re_p} & Re_p \leq 1 \\ \frac{Re_p}{24}\left(1 + 0.15Re_p^{0.687}\right) & 1 < Re_p \leq 1000 \\ 0.44 & Re_p > 1000 \end{cases} \tag{4.6}$$

4.1.3 Dynamic Shape Factor

In the Stokes' analysis above, the particles were assumed rigid spheres. But in reality, most of the particles are nonspherical. Being cubic, cylindrical, crystal, or irregular. The shape of a particle affects its aerodynamic behavior by influencing its drag resistance. Therefore, a correction factor called the dynamic shape factor is necessary to correct the Stokes' law.

The dynamic shape factor, taking symbol S_f herein, is defined as the ratio of the actual drag force of the nonspherical particle to the drag force of a sphere having the same volume and velocity as the nonspherical particle. The dynamic shape factor S_f is then given by

$$S_f = \frac{F_D}{3\pi\mu|u - v|d_e} \tag{4.7}$$

which gives the drag on a nonspherical particle in Stokes regime as

$$F_D = 3\pi\mu|u - v|d_e S_f \tag{4.8}$$

where d_e is the equivalent volume diameter. It is the diameter of a sphere having the same volume as that of the nonspherical particle. Note that the shape factor is 1 for spherical particles. Most of the dynamic shape factors are greater than 1.0.

Dynamic shape factors are usually determined experimentally by measuring the settling velocity of geometric models in liquids. For irregular particles, settling velocities were measured indirectly using the elutriation devices [11]. An elutriation device separates particles based on their size, shape and density.

4.1.4 The Knudsen Number and Cunningham Correction Factor

An important assumption of the Stokes' law is that there is no slipping between the gas and the aerosol particles. It is also referred to as continuum flow. However, when the particle is getting smaller and smaller, approaching the mean free path of the gas molecules, this assumption of continuum transport is no longer valid. The dimensionless parameter that defines the nature of the aerosol is the Knudsen number (Kn), which is the ratio of gas mean free path to particle radius.

$$Kn = 2\lambda/d_p \tag{4.9}$$

When $Kn \ll 1$, the particle diameter is much greater than the mean free path of the gas, and the particle is in the continuum regime. This applies to the preceding analyses. On the other hand, when $Kn \gg 1$, the particle size is much smaller than the gas mean free path, and its behavior is like a gas molecule. The particle size between these two extremes defines the transition regime.

When the system is in noncontinuum regime, *Cunningham correction factor* (C_c), is used to correct the drag force.

$$F_D = \frac{3\pi\mu d_p |u - v|}{C_c} \tag{4.10}$$

The Cunningham correction factor for aerosol particles can be determined using the following equation recommended by Allen and Raabe [2]

$$C_c = 1 + Kn\left[1.142 + 0.558\exp\left(-\frac{0.999}{Kn}\right)\right] \tag{4.11}$$

where λ is the gas mean free path that was introduced in Sect. 2.1.7.

There are several alternative equations for the Cunningham correction factor, differing only in the numerical factors. For example, Whitby et al. [28] (cited by Otto et al. [19]) used a formula as

$$C_c = 1 + 1.392Kn^{1.0783} \tag{4.12}$$

And Flagan and Seinfeld [9] give a more complex one as follows.

$$C_c = \begin{cases} 1 + 1.257Kn \approx 1.0 & Kn < 0.001 \\ 1 + Kn\left[1.257 + 0.40\exp\left(-\frac{1.10}{Kn}\right)\right] & 0.001 < Kn < 100 \\ 1 + 1.657Kn & Kn > 100 \end{cases} \tag{4.13}$$

Equation (4.11) is used in this book unless otherwise specified. However, readers are suggested to choose the equations in accordance to their specific applications.

Both Knudsen number and Cunningham correction factor are dimensionless parameters. Since the theoretical value of C_c is always greater than 1, the drag force experienced with slipping effect considered is always smaller than the value calculated with nonslipping assumption.

For a small particle with irregular shape, both slip factor and shape factor are supposed to be considered for particle dynamics analysis. However, the task is so complicated that it outvalues the outcome. In air emission engineering, it is well acceptable to use the approximate factor calculated for the equivalent volume sphere for most irregular particles. The slip factor for randomly oriented fibers ($L/d < 20$) is 0–12 % greater than that for the equivalent volume sphere. Drag coefficients of different particle shapes are available in the literature (e.g., [12, 15]).

4.2 Rectilinear Particle Motion

Steady rectilinear particle motion is the simplest yet important type of particle motion in particle dynamics. It is the foundation for the mechanisms of particle separation from air stream, namely particulate air pollution control.

4.2.1 Particle Acceleration

Consider a particle with constant mass m that is released in quiescent air with an initial velocity of zero. Newton's second law of motion must hold at any instant $t > 0$.

$$\sum \vec{F} = m\frac{\mathrm{d}\vec{v}(t)}{\mathrm{d}t} \tag{4.14}$$

where $\vec{v}(t)$ is the particle velocity in the static air at time t, and the mass of the particle is considered as a constant when there is no evaporation or growth. In this case only two forces, a constant force of gravity and a drag force, act on the falling particles. The drag force depends on the particle velocity at any instant, ignoring additional acceleration of the surrounding air.

At any instant, the drag force is given by Stokes' law. Taking the positive direction downward, the above vector equation can be described using magnitudes

$$\sum F = mg - F_D \Rightarrow mg - \frac{3\pi\mu d_p v(t)}{C_c} = m\frac{\mathrm{d}v(t)}{\mathrm{d}t} \tag{4.15}$$

Note that in this equation we ignored the bouyant force. This is valid for typical condtions, when the aerosol particle density is much great than that of the air.

Rearranging the above equation and integration with the initial condition of $v = 0$ at $t = 0$ leads to

$$\left(\frac{3\pi\mu\, d_p}{mC_c}\right)\int_0^t \mathrm{d}t = -\int_0^{v(t)} \frac{\mathrm{d}v(t)}{\left[v(t) - \left(\frac{mgC_c}{3\pi\mu d_p}\right)\right]} \tag{4.16}$$

Integrating both sides and replacing m with $\frac{1}{6}\pi\rho d_p^3$ leads to

$$-\frac{t}{\left(\frac{\rho_p d_p^2 C_c}{18\mu}\right)} = \ln\left(\frac{\frac{\rho_p d^2 g C_c}{18\mu} - v(t)}{\frac{\rho_p d^2 g C_c}{18\mu}}\right) \tag{4.17}$$

If we define a constant

$$\tau = \frac{\rho_p d_p^2 C_c}{18\mu} \tag{4.18}$$

then we get the settling velocity of the particle, $v(t)$, at any time, t.

$$v(t) = g\tau\left[1 - \exp\left(-\frac{t}{\tau}\right)\right] \tag{4.19}$$

This equation shows that when t approaches infinity, the settling speed of the aerosol particle approaches a constant, which is the maximum speed that the particle can reach. In aerosol dynamics, this maximum speed is called terminal setting velocity, denoted by v_{TS}.

$$v_{\mathrm{TS}} = v(t \to \infty) = g\tau = \frac{\rho_p d_p^2 g C_c}{18\mu} \tag{4.20}$$

Then Eq. (4.19) can be simplified as

$$\frac{v(t)}{v_{\mathrm{TS}}} = 1 - \exp\left(-\frac{t}{\tau}\right) \tag{4.21}$$

This equation gives the speed $v(t)$ of a particle at any time t after it is released in still air in a gravitational field.

Figure 4.1 is produced using Eq. (4.21) above. It shows that the particle reaches 95 % of its terminal settling speed when t = 3τ. However, from a practical point of view, within ±5 % error the particle speed reaches v_{TS} when t is 3τ and after that it remains constant.

As shown in Table 4.1, particles having aerodynamic diameters less than or equal to10 μm, reach their terminal speed in less than 1 ms. Even a 100 μm particle reaches its terminal speed in less than 0.1 s. Therefore, it is acceptable to assume that a particle reaches its terminal speed instantly with a negligible error.

Since we can use 95 % of the maximum settling speed to represent its terminal settling speed, the terminal settling speed equation can be simplified as

$$v_{\mathrm{TS}} = \frac{\rho_p d_p^2 g C_c}{18\mu} \tag{4.22}$$

Note that for a particle with irregular shape the shape factor has to be taken into consideration. The analysis is identical to that for spherical particles. The terminal settling velocity becomes

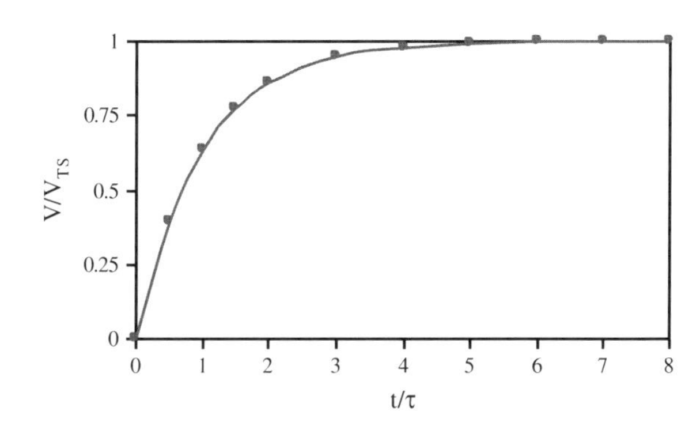

Fig. 4.1 Dimensionless speed versus dimensionless time for an accelerating particle

Table 4.1 Time required for particles of standard density to reach their terminal velocity at standard conditions

Particle diameter, d_p (μm)	95 % of its maximum settling speed, $0.95v_{TS}$ (m/s)	Time to reach $0.95v_{TS}$ $t = 3\tau$
0.01	0.0000001	0.0000000
0.1	0.0000008	0.0000003
1	0.0000331	0.0000107
10	0.0028980	0.0009329
100	0.2863625	0.0921817

$$v_{TS} = \frac{\rho_p d_e^2 g C_c}{18\mu S_f} \tag{4.23}$$

Example 4.1: Terminal settling time
Consider a spherical glass particle with a diameter of 30 μm and a density of 2,500 kg/m^3 is released from rest in still air. How long it will take to reach its terminal velocity?

Solution
From $d_p = 30\,\mu\text{m}$, ρ_p = 2,500 kg/m^3, and $\mu = 1.81 \times 10^{-5}$ Pa s, we can get

$$Kn = 2\lambda/d_p = 2 \times \frac{0.066\,\mu\text{m}}{30\,\mu\text{m}} = 0.0044$$

Since $0.001 < Kn < 100$, the Cummingham correction factor is calculated using

$$\begin{aligned} C_c &= 1 + Kn\left[1.142 + 0.558\exp\left(-\frac{0.999}{Kn}\right)\right] \\ &= 1 + 0.0044\left[1.142 + 0.558\exp\left(-\frac{0.999}{Kn}\right)\right] = 1.055 \end{aligned}$$

The time it takes for the particle to reach its terminal speed is,

$$t = 3\tau = 3\left(\frac{\rho_p d_p^2 C_c}{18\mu}\right) = 3\left(\frac{2500 \times \left(30 \times 10^{-6}\right)^2 1.055}{18 \times 1.81 \times 10^{-5}}\right) = 0.0073\,\text{s}$$

Example 4.2: Sneezing droplet settling
Scientific research results indicated the total average size distribution of the droplets by coughing was 0.58–5.42 μm, and 82 % of droplet nuclei were centered in 0.74–2.12 μm. A spherical droplet with a diameter of 5 μm and density of 1,000 kg/m^3 is discharged from the mouth horizontally in still air.

If the mouth is about 1.5 m above the floor, how long would it take for the 5-μm droplet to settle down to the floor.

Solution Given $d_p = 5$ μm, $\rho_p = 1{,}000$ kg/m^3, and $\mu = 1.81 \times 10^{-5}$ Pa s,we can get,

$$Kn = 2\lambda/d_p = 2 \times \frac{0.066\ \mu\text{m}}{5\ \mu\text{m}} = 0.0264$$

$$C_c = 1 + Kn\left[1.142 + 0.558\exp\left(-\frac{0.999}{Kn}\right)\right] = 1.033$$

$$t = 3\tau = 3\left(\frac{\rho_p d_p^2 C_c}{18\mu}\right) = 3\left(\frac{1000 \times (5 \times 10^{-6})^2 \times 1.0333}{18 \times 1.81 \times 10^{-5}}\right) = 0.00024\ \text{s}$$

$$v_{\text{TS}} = \frac{\rho_p d^2 g C_c}{18\mu} = \frac{1000 \times (5 \times 10^{-6})^2 \times 9.81 \times 1.033}{18 \times 1.81 \times 10^{-5}} = 0.000753\ \text{m/s}$$

$$\Delta t = \frac{H}{v_{\text{TS}}} = \frac{1.5\ \text{m}}{0.000753\ \text{m/s}} = 1992\ \text{s} = 33\ \text{min}$$

It could be much slower for a smaller particle.

4.2.2 Settling at High Reynolds Numbers

The particle motion is in Newton's regime for $Re > 1.0$ and the corresponding settling speed in a gravitational field can be determined by equating the drag force and the force of gravity, which gives

$$v_{\text{TS}} = \frac{4\rho_p d_p g}{3C_D \rho_g} \tag{4.24}$$

where the drag coefficient C_D was introduced before. However, it is not straightforward to find v_{TS} using this equation. Because C_D depends on the particle Reynolds number. It cannot be determined without knowing v_{TS}. One way to solve this problem is an iterative solution obtained by substituting an initial guess of v_{TS} into the above equation and by trying different values of v_{TS} until the solution converges with a desired accuracy.

4.2.3 Aerodynamic Diameter

Aerodynamic diameter (d_a) is an equivalent diameter that finds many applications in aerosol characterization and particulate emission control. It is defined as the diameter of a spherical particle with a standard density of 1,000 kg/m^3 that has the same settling velocity as the real particle, when both of them present in the same gravitational field.

Figure 4.2 illustrates the aerodynamic diameter of an irregular particle. Imagine two particles, one is spherical with a standard density (ρ_o) and a diameter (d_a) and another is nonspherical, are released in the same calm air. They have the same aerodynamic diameter if they fall with the same settling velocity. In another word, the aerodynamic diameter of the nonspherical particle is d_a.

For the same gravitational settling velocity, it can be written in terms of aerodynamic diameter and equivalent geometric diameter as

$$v_{\mathrm{TS}} = \frac{\rho_p d_e^2 g C_c}{18\mu S_f} = \frac{\rho_0 d_a^2 g C_c}{18\mu} \tag{4.25}$$

where ρ_0 is the standard particle density, 1,000 kg/m^3, which is the same as water at normal condition.

The aerodynamic diameter of a particle is the key particle property for evaluating or comparing the performances of different types of particulate air cleaners, as to be introduced later in this book. Many particulate matter emission control devices, such as cyclones and filters, separate particles from the gas/air stream aerodynamically.

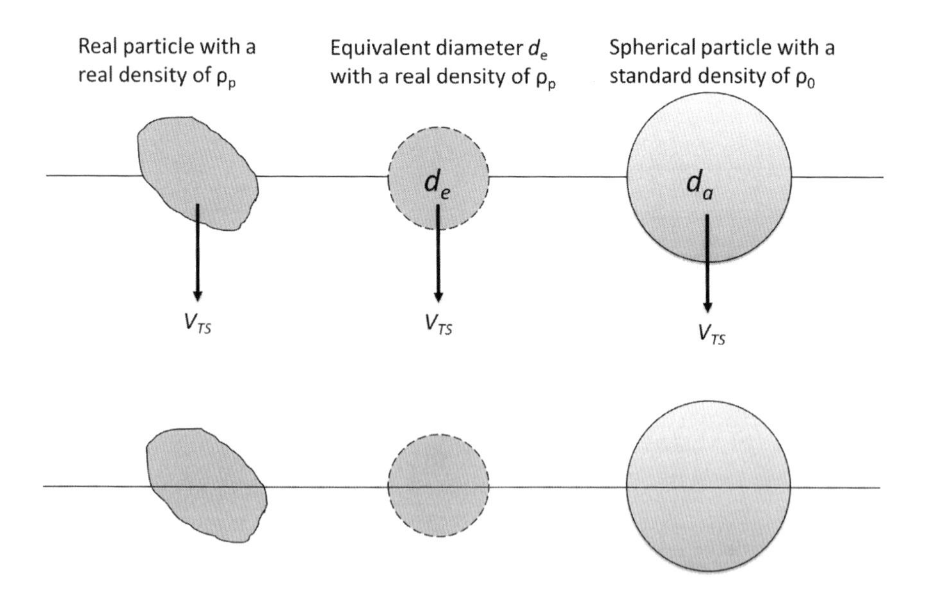

Fig. 4.2 An irregular particle and its equivalent volume diameter and aerodynamic diameter

Consequently, it is not necessary to know the physical size, shape or density of a particle if its aerodynamic diameter is determined.

From the above equation, one can get the formula for aerodynamic diameter,

$$d_a = d_e \left(\frac{\rho_p}{\rho_0 S_f} \right)^{1/2} \tag{4.26}$$

Example 4.3: Aerodynamic diameter

Estimate the aerodynamic diameter of a spherical steel particle with a geometric equivalent diameter d_e = 10 μm and ρ_p = 8,000 kg/m³.

Solution

Since the particle is spherical, its shape factor, $S_f = 1$

$$d_a = d_e \left(\frac{\rho_p}{\rho_0 S_f} \right)^{1/2} = 10 \times 10^{-6} \left(\frac{8000}{1000 \times 1} \right)^{1/2} = 2.83 \times 10^{-5}\,\text{m} = 28.3\,\mu\text{m}$$

For a particle with this great density, its aerodynamic particle diameter is much greater than its geometric equivalent diameter.

4.2.4 Curvilinear Motion of Aerosol Particles

Curvilinear motion is a motion when a particle follows a curved path. A classic example of curvilinear motion is the projectile of a particle with a horizontal initial velocity in the still air. It is more complicated for a particle-air mixture that flows around an obstacle. Very small particles with negligible inertia tend to follow the gas while large and heavy particles tend to continue in a straight line due to the great inertia.

The inertia of a particle in curvilinear motion is characterized by the Stokes number (*Stk*), like the Reynolds number in fluid mechanics for the characterization of a fluid flow. The Stokes number is defined as the ratio of the stopping distance of a particle to a characteristic dimension of the obstacle.

$$Stk = \frac{\tau u_0}{d_c} = \frac{\rho_p d_p^2 C_c u_0}{18 \mu d_c} \tag{4.27}$$

where the characteristic dimension d_c in the above equation can be defined differently according to applications. And the definition of Stokes number may be application specific. u_0 is the undisturbed air speed. In standard air, a particle with $Stk \gg 1.0$ will continue in a straight line as the fluid turns around the obstacle. But for a particle with $Stk \ll 1$, it will follow the fluid streamlines closely.

Example 4.4: Stokes number

Estimate the Stokes number of a 1 μm spherical particle with a density of 8,000 kg/m^3 in an air flowing at 1 m/s normal to a cylinder of diameter 10 cm, assuming standard conditions.

Solution

Given d_e = 1 μm, u_o = 1 m/s, d_c = 10 cm, ρ_p = 8000 kg/m^3, we get

$$Kn = 2\lambda/d_p = 2\lambda/d_e = 2 \times 0.066/1 = 0.132$$

$$C_c = 1 + Kn\left[1.142 + 0.558\exp\left(-\frac{0.999}{Kn}\right)\right] = 1.166$$

$$Stk = \frac{\rho_p d_p^2 C_c u_0}{18\mu d_c} = \frac{8000 \times \left(1 \times 10^{-6}\right)^2 \times 1.166 \times 1}{18 \times 1.81 \times 10^{-5} \times 0.1} = 2.86 \times 10^{-4}$$

Note that the geometric diameter is used in calculating the Knudsen number because Kn is a geometric ratio rather than an aerodynamic property.

The small Stokes number indicates that this particle follows the air under standard conditions and it is difficult to separate the particle from the air simply by inertia. However, it is feasible under other conditions, when the particle viscosity is reduce and the particle Cunningham correction factor is enhanced by great air mean free path. In-depth analysis will be introduced in Chap. 13.

4.2.5 *Diffusion of Aerosol Particles*

Diffusion of gas borne particles takes place when there is a gradient of particle concentration in the space. It results in a net transport of particles from a region of higher concentration to that of lower concentration. The transport flux of the gas borne particles is defined by Fick's first law of diffusion. In the absence of external forces, Fick's law is the same as that for gases,

$$J = -D_p \frac{\mathrm{d}n}{\mathrm{d}x} \tag{4.28}$$

where J = flux of particles, expressed in terms of the number of particles per unit area per unit time (#/s•m^2), D_p = diffusivity of the particles in the gas (m^2/s), and $\mathrm{d}n/\mathrm{d}x$ = gradient in number concentration of particles (1/m).

The diffusion coefficient for an aerosol particle is given by the Stokes-Einstein equation [12].

$$D_p = \frac{kTC_c}{3\pi\mu d_p} \tag{4.29}$$

The particle diffusion coefficient has a unit of m^2/s. For large particles when $C_c = 1$, the particle diffusion coefficient is inversely proportional to the particle size. For smaller particles when the slip effect becomes significant, D_p approaches a value that is inversely proportional to d_p^2, which is very similar to the diffusion coefficient for gas molecules. This is because the behavior of finer particles approaches that of gas molecules.

Example 4.5: Particle diffusion coefficients
Calculate the diffusion coefficients for 100 nm and 10 μm particles at standard conditions.

Solution
We know that $k = 1.3806 \times 10^{-23}$kg• m^2/(s^2 K), $T = 288$ K; $\lambda = 66$ nm.

Case I For particles with $d_p = 100$ nm

$$Kn = 2\lambda/d_p = 2\lambda/d_e = 2 \times 66/100 = 1.32$$

$$C_c = 1 + Kn\left[1.257 + 0.40\exp\left(-\frac{1.10}{Kn}\right)\right] \quad 0.001 < Kn < 100$$

$$C_c = 1 + 1.32\left[1.257 + 0.40\exp\left(-\frac{1.10}{1.32}\right)\right] \approx 2.66$$

The particle diffusion coefficient $D_p = \frac{kTC_c}{3\pi\mu d_p}$

$$= \frac{1.3806 \times 10^{-23} \times 288 \times 2.66}{3\pi \times 1.81 \times 10^{-5} \times 100 \times 10^{-9}}$$

$$= 6.21 \times 10^{-10} m^2/s$$

Case II For particles with $d_p = 10$ μm, repeat the calculation and we get

$$C_c = 1.0166$$

$$D_p = \frac{kTC_c}{3\pi\mu d_p} = \frac{1.3806 \times 10^{-23} \times 288 \times 1.0166}{3\pi \times 1.81 \times 10^{-5} \times 10 \times 10^{-6}} = 2.33 \times 10^{-17} m^2/s$$

4.2.6 Particle Deposition on Surface by Diffusion

As introduced above, most of the fine particles adhere when they impact on a surface. In this case, the particle concentration in the space where the surface is can be assumed to be zero and there is a concentration gradient established in the region near the surface. This results in a continuous diffusion of particles to the surface and a gradual decrease in particle concentration in the gas.

Consider a plane vertical surface that is immersed in a large space filled with gas and particles. It can be assumed that the gas velocity near the surface is zero. Then the rate at which particles are removed from the space, by deposition onto the surface can be determined following the analysis below. Let x be the horizontal distance from the surface. Then the particle concentration, $n(x, t)$, in the space at x at any time t, must satisfy Fick's second law of diffusion.

$$\frac{\mathrm{d}n}{\mathrm{d}t} = -D_p \frac{\mathrm{d}^2 n}{\mathrm{d}x^2} \quad \left\{ \begin{array}{lll} n(x,0) = n_0 & \text{for} & x > 0 \\ n(0,t) = 0 & \text{for} & t > 0 \end{array} \right\} \tag{4.30}$$

It is assumed that the initial particle concentration in the entire space is n_0. The boundary condition of $n(0,t) = 0$ is based on the assumption that all particles will adhere on the surface once they come into contact. The general solution of this equation is

$$n(x,t) = \frac{n_0}{\sqrt{\pi D_p t}} \int_0^x \exp\left(-\frac{z^2}{4D_p t}\right) \mathrm{d}z \tag{4.31}$$

where z is the dummy variable for integration.

Then the concentration gradient at the surface, $\mathrm{d}n/\mathrm{d}x$ at $x = 0$, is

$$\left.\frac{\mathrm{d}n}{\mathrm{d}x}\right|_{x=0} = \frac{\mathrm{d}}{\mathrm{d}x}\left[\frac{n_0}{\sqrt{\pi D_p t}} \int_0^x \exp\left(-\frac{z^2}{4D_p t}\right)\mathrm{d}z\right]\Bigg|_{x=0} = \frac{n_0}{\sqrt{\pi D_p t}} \tag{4.32}$$

The rate of deposition of particles onto a unit area of surface at any time, t, is then described as

$$J = -D_p \frac{\mathrm{d}n}{\mathrm{d}x}\Big|_{x=0} = -n_0 \sqrt{\frac{D_p}{\pi t}} \tag{4.33}$$

Integrating over time in this equation from 0 to t, the cumulative number of particles, $N(t)$, deposited per unit area of surface is

$$N(t) = \int_{0}^{t} n_0 \sqrt{\frac{D_p}{\pi z}} \mathrm{d}z = 2n_0 \sqrt{\frac{D_p t}{\pi}} \tag{4.34}$$

where $N(t)$ has a unit of (#/m^2).

Again, z is the dummy variable for integration. In engineering practice, there is always limited space for containers, and therefore, this equation can only be used to set the upper limit of losses to the walls of the container.

The preceding analyses were based on the properties of a single particle without considering the particle size distribution. The same principles may be applied to a group of particles with a uniform size well dispersed in air, which is referred to as monodisperse aerosol.

4.3 Particle-Surface Interaction

The particle is subjected to adhesion force between the particle and the solid surface and pull-off force. When the adhesion force is greater than or equal to the pull-off force, the particle adhere to the surface, otherwise, the particle leaves the surface. The former is called attachment and the latter reentrainment. The adhesion force is primarily the van der Waals force between two rigid spheres [3] and the pull-off force can be quite different including, but are not limited to, drag, shear, and impaction from other aerosol particles.

Unlike a gas molecule, most aerosol particles, especially the submicron sized ones, attach firmly to a surface they come in contact. When the particle comes in contact with another particle, both particles adhere to each other, and the process is referred to as coagulation. Various particle separation devices such as fiber filters are designed to separate particles from the gas stream by taking advantage of the adhesion. It is especially important for submicron particles because the adhesive forces on submicron particles exceed other common forces by orders of magnitude.

Closely related to adhesion of particles are the processes of resuspension and reentrainment. They are important for the buildup and removal of particles on surfaces. For example, for fugitive dust emissions caused by vehicles on paved and unpaved roads.

Wang et al. [27] investigated the resuspension of micron sized particles (0.4–10 μm) from a flat surface simulating a ventilation duct. The results in Fig. 4.3 show that the resuspension rate depends on the time, air speed, and the particle size. In-depth analyses suggest that more particles slide and roll off rather than being lifted into the air flow.

Not all airborne particles adhere to the surfaces in contact. It depends on the properties of both the particle and the surface, and the nature of the impact between the particle and the surface. A large solid particle impacting a hard surface at high

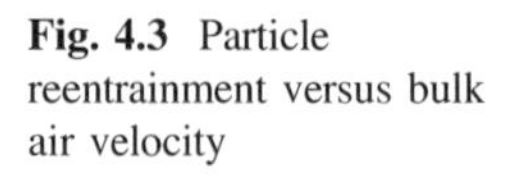

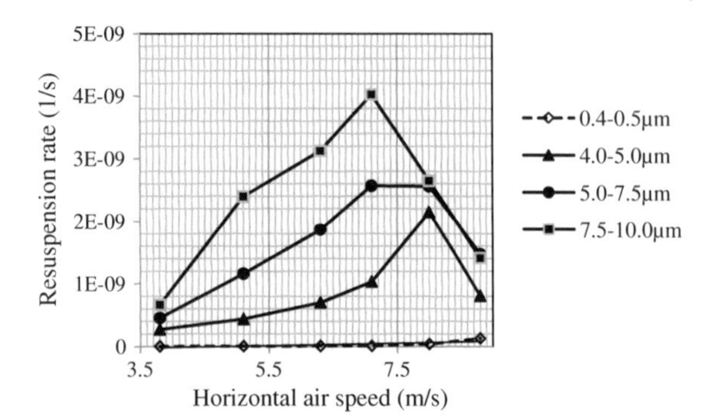

Fig. 4.3 Particle reentrainment versus bulk air velocity

velocity can rebound from the surface. When the speed of the particles is great enough, part of its kinetic energy is dissipated in the deformation process during the particle-surface impact, and part is converted elastically to kinetic energy of rebound. If the rebound energy exceeds the energy required to overcome the adhesive forces, the particle will bounce away from the surface rather than adhering to it.

Particle bounce has been studied for solid particle from impactor and fibrous filters. Overall, the bounce is likely to take place for the large particles of hard materials traveling at a great speed. In addition, the roughness of the surface plays an important role. Bouncing does not occur for droplets of liquid or easily deformed materials. The coating of surfaces with oil improves particle adhesion but reduces the bounce. On the other hand, when particles are are small enough, they also may rebound because of the high thermal speed. This is called thermal rebound (see Chap. 14).

4.4 Particle Coagulation

Particle size distribution in the air is constantly changing over time, primarily because of coagulation. When particles collide on each other by certain mechanisms, they may attach to each other by the van de Waals force and form larger ones. This phenomenon is referred to as particle coagulation or agglomeration or coalescence. The mechanisms for particle coagulation may include, but are not limited to, Brownian motion, collision, electrostatics, gravity, and gas phase turbulence.

Many models have been developed for aerosol particle coagulation and they are available in the literature. Most of them are based on the classic Smoluchowski [23] equation (Cited by Geng et al. [10]), by ignoring evaporation and condensation.

$$\frac{\partial}{\partial t}f(v,t) = \frac{1}{2}\int_0^V K(u, v-u)f(u,t)f(v-u,t)\mathrm{d}u - f(v,t)\int_0^\infty K(u,v)f(u,t)\mathrm{d}u \quad (4.35)$$

where $K(u, v)$ specifies the collision rate between particles of volume u and v. f is the size distribution density function.

It is assumed in conventional aerosol analysis that particles attach to each other upon collision, therefore, the collision rate can also be considered as the coagulation rate.

Because of the dynamic change of the size distribution, the Smoluchowski [23] equation can only be solved numerically. However, analytical solutions can be obtained for simple cases with assumptions.

4.4.1 Monodisperse Aerosol Coagulation

The analysis is simplified if we only consider monodispere coagulation, where the particle diameters are within a narrow range. Smoluchowski [23] developed a model for Brownian coagulation of monodisperse particles in the continuum regime. It is applicable to particles with sizes in a narrow range, say $1 < d_{pA}/d_{pB} < 1.25$. The corresponding Brownian monodisperse coagulation efficient for the continuum regime ($Kn \ll 1$) and free molecule regime ($Kn \gg 1$) are simplified as

$$K = \begin{cases} 8kT3\mu & \text{continuum regime,} \quad Kn \ll 1 \\ 9.8\sqrt{kT\rho_p d_p} & \text{free molecule regime,} \quad Kn \gg 1 \end{cases} \quad (4.36)$$

where k is Boltzmann's constant and μ is the dynamic viscosity of the carrier gas. Otto et al. [19] introduced an analytical model for the regime in between, but its form is complicated.

Integration of Smoluchowski equation using the initial condition of $N(t=0) = N_0$ leads to the solution for the total particle concentration as a function of time and the coagulation coefficient above.

$$N(t) = \frac{2N_0}{2 + KN_0 t} \quad (4.37)$$

where N_0 is the initial particle concentration (#/m^3).

By defining the characteristic dimensionless time

$$\tau = \frac{2}{KN_0} \quad (4.38)$$

Equation (4.37) becomes,

$$N(t) = \frac{N_0}{1 + t/\tau} \tag{4.39}$$

Since $N = N_0/2$ when $t = \tau$, τ is also referred to as the half-value time, which is the time it takes for the particle number concentration to drop to half of its initial value.

Example 4.6: Half value time of monodisperse aerosol
Calculate the half-value time of spherical monodisperse aerosol with an initial concentration of 10^3 particles/cm^3 and an initial particle diameter of $d_p = 1$ nm. Assume standard condition and only Brownian coagulation is considered.

Solution
The Knudsen number of 1 nm particle is

$$Kn(1\text{nm}) = 2 \times 0.066/0.001 = 132 > 100$$

It is within the free molecule region. Therefore,

$$K = 9.8\sqrt{\frac{kT}{\rho_p d_p}} = 9.8\sqrt{\frac{(1.38 \times 10^{-23}) \times 298}{1{,}000 \times (1 \times 10^{-9})}} = 6.28 \times 10^{-7}$$

Then the half-value time is

$$\tau = \frac{2}{KN_0} = \frac{2}{6.28 \times 10^{-7} \times 10^3 \left(\frac{1}{\text{cm}^3}\right) \times 10^6 \left(\frac{\text{cm}^3}{\text{m}^3}\right)} = 0.003 \text{ s}$$

As seen from the result, particles at 1 nm do not have a great life time. While this is practically inaccurate, it does qualitatively show that aerosol particle diameter is a dynamic parameter.

4.4.2 Polydisperse Coagulation

Coagulation coefficients for different mechanisms have been summarized by Geng et al. [10]. Brownian motion is the dominating mechanisms for the coagulation of very fine particle under normal atmospheric condition; for this reason, our analysis is focused on Brownian coagulation.

One of the widely used Brownian coagulation coefficients is the Fuchs (1964) equation for binary collision.

$$K(a,b) = \frac{4\pi(a+b)(D_A + D_B)}{\frac{a+b}{a+b+\delta_{AB}} + \frac{4(D_A+D_B)}{(a+b)\bar{c}_{AB}}} \quad (4.40)$$

where a and b are the radii of the particles of concerns, $\bar{c}_{AB}$ is the average particle thermal velocity, δ_{AB} is the jump distance, and D_i is the diffusivity of the particle of size $i\,(=A \text{ or } B)$. If we use particle diameters, d_{pA} and d_{pB}, instead of the radii, Eq. (4.40) becomes

$$K\left(d_{pA}, d_{pB}\right) = \frac{2\pi\left(d_{pA} + d_{pB}\right)(D_A + D_B)}{\frac{d_{pA}+d_{pB}}{d_{pA}+d_{pB}+2\delta_{AB}} + \frac{8(D_A+D_B)}{\left(d_{pA}+d_{pB}\right)\bar{c}_{AB}}} \quad (4.41)$$

The particle diffusion coefficient is calculated using

$$D_i = \frac{kTC_c}{3\pi\mu d_{pi}} \quad (4.42)$$

and the mean thermal velocity are calculated using

$$\bar{c}_{AB} = \left(\bar{c}_A^2 + \bar{c}_B^2\right)^{1/2} \quad (4.43)$$

where the mean thermal velocity of the particle can be calculated using the equation for gas molecules (Eq. 2.7) by replacing the mass of a molecule with the mass of the particle.

$$\bar{c}_i = \sqrt{\frac{8kT}{\pi m_i}} \quad (4.44)$$

Similar to Eq. (4.43), the jump distance, δ_{AB}, is determined as

$$\delta_{AB} = \left(\delta_A^2 + \delta_B^2\right)^{1/2} \quad (4.45)$$

where the individual jump distance is calculated with

$$\delta_i = \frac{\left(d_{pi} + l_i\right)^3 - \left(d_{pi}^2 + l_i^2\right)^{1.5}}{3d_{pi}l_i} - d_{pi} \quad (4.46)$$

$$l_i = \frac{8D_i}{\pi\bar{c}_i} \quad (4.47)$$

It is obviously a very tedious work to accurately predict the particle size change by polydisperse coagulation. Readers are referred to the literature for more complicated models for polydisperse coagulation.

4.5 Aerosol Particle Size Distribution

Although monodisperse particles are used in describing particles with a narrow size distribution in nanoparticle manufacturing, threat agents and bioaerosols, most engineers deal with polydisperse aerosol. Polydisperse aerosol is a group of particles with different sizes suspended in the air. In typical urban atmosphere, particle concentration can reach as high as 10^7–10^8/cm^3; their diameters can range from a few nanometers to around 100 μm [22].

A great amount of literature shows that the size distribution of most polydisperse aerosol particles is lognormal. To understand lognormal distribution, it is necessary to revisit what we learned about normal distribution. Normal distribution is also called Gaussian distribution. For a variate x with a mean of $\bar{x}$ and standard deviation of σ, the normal probability fractional distribution function in the domain $-\infty < x < \infty$ is described as

$$f(x) = \frac{1}{\sigma\sqrt{2\pi}}\exp\left[-\left(\frac{x-\bar{x}}{\sigma\sqrt{2}}\right)^2\right] \tag{4.48}$$

where $f(x) \leq 1$. The corresponding cumulative distribution, which describes the probability of the variable below certain value, is

$$\emptyset(u) = f(x < u) = \frac{1}{2}\left[1 + \text{erf}\left(\frac{u-\bar{u}}{\sigma_u\sqrt{2}}\right)\right] \tag{4.49}$$

where the error function is used to characterize the measurement errors. It is described as

$$\text{erf}(y) = \frac{2}{\sqrt{\pi}}\int_0^y \exp(-z^2)\text{d}z \tag{4.50}$$

An error function $\text{erf}(y)$ is symmetric about its origin. For a quantity that is lognormally distributed, its logarithm is governed by normal distribution by replacing x with $\log x$, $\bar{x}$ with $\log\bar{x}$ and σ with $\log\sigma$.

$$f(\log x) = \frac{1}{\log\sigma\sqrt{2\pi}}\exp\left[-\left(\frac{\log x - \log\bar{x}}{\sqrt{2}\ln\sigma}\right)^2\right] \tag{4.51}$$

For its specific applications to aerosol particles, the probability of the particle number fraction can be described as

$$f(\log d_p) = \frac{1}{\log\sigma\sqrt{2\pi}}\exp\left[-\left(\frac{\log d_p - \log\bar{d}_{pg}}{\sqrt{2}\ln\sigma}\right)^2\right] \tag{4.52}$$

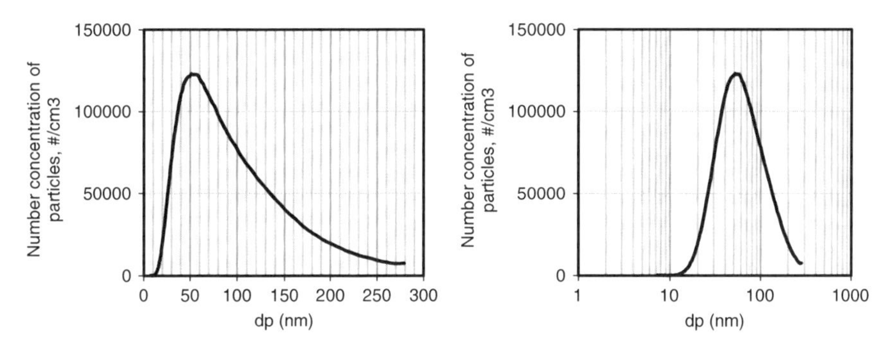

Fig. 4.4 Aerosol size distribution (*left* linear scale x-axis, *right* logarithm x-axis)

Since $f(\log d_p) = d_p \times f(d_p)$ [9], the equation can be transformed into

$$f(d_p) = \frac{1}{\sqrt{2\pi}d_p\log\sigma}\exp\left[-\left(\frac{\log d_p - \log\overline{d}_{pg}}{\sqrt{2}\log\sigma}\right)^2\right] \tag{4.53}$$

Figure 4.4 shows how the curves look like when the particle number concentrations are presented in linear scale x-axis and logarithm scale *x*-axis.

The corresponding cumulative distribution is then similar to Eq. (4.49)

$$\emptyset(d_p) = \frac{1}{2}\left[1 + \text{erf}\left(\frac{\log d_p - \log\overline{d}_{pg}}{\sqrt{2}\log\sigma}\right)\right] \tag{4.54}$$

This equation can be used to explain the physical meanings of $\overline{d}_{pg}$ and σ. When $d_p = \overline{d}_{pg}$, $\emptyset(d_p) = 0.5$ and it indicates that 50 % of the particles are less than $\overline{d}_{pg}$. Therefore, $\overline{d}_{pg}$ is the *median* diameter of the particles. The role of σ can be better understood letting $d_p = \sigma\overline{d}_{pg}$, which gives $\emptyset(d_p) = 0.84$. This means that 84 % of the particles are smaller than $(\sigma\overline{d}_{pg})$. So σ is the geometric standard deviation. Similarly, we can get that 95 % of the particles are smaller than $(2\sigma\overline{d}_{pg})$ and 99.5 % are smaller than $(3\sigma\overline{d}_{pg})$.

In aerosol technology, we usually are also interested in the absolute number of particles besides the probability function in Eq. (4.54). Multiplying both sides of Eq. (4.54) with the total number of the particles gives us a practical equation as

$$F(d_p) = N\emptyset(d_p) = N\left[1 + \text{erf}\left(\frac{\log d_p - \log\overline{d}_{pg}}{\sqrt{2}\log\sigma}\right)\right] \tag{4.55}$$

It describes the total number below particles of size d_p in a population. Equation (4.55) can also be extended to describing the surface and volume distributions, by replacing $n(d_p)$ with $\pi d_p^2 \cdot n(d_p)$ and $\left(\pi d_p^3/6\right) \cdot n(d_p)$ for surface and

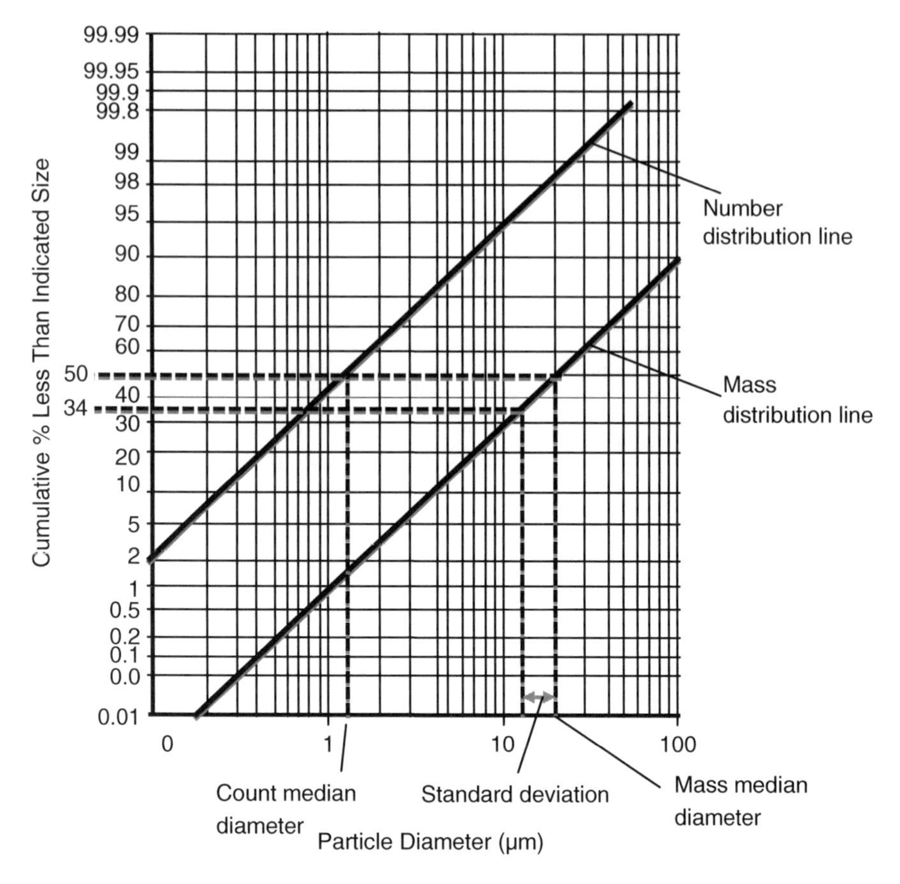

Fig. 4.5 Cumulative lognormal distributions for the same aerosol sample

volume, respectively. We can see that both surface and volume also follow log-normal distributions with the same standard deviation but with different median diameters. They are related to the number geometric mean diameter as follows.

$$\log\overline{d}_{pgA} = \log\overline{d}_{pg} + 2(\log\sigma)^2 \tag{4.56}$$

$$\log\overline{d}_{pgV} = \log\overline{d}_{pg} + 3(\log\sigma)^2 \tag{4.57}$$

where $\overline{d}_{pgA}$ and $\overline{d}_{pgV}$ are the geometric surface and volume median diameters. Respectively.

A log-probabiltiy paper based on Eq. (4.55) has been widely used to aerosol particle size distribution analysis. If we plot the number, surface, and volume cumulative distributions in the same log probability paper, they are parallel straight lines, as seen in Fig. 4.5. We can determine the median size of particles (based on number or mass) and the geometric standard deviation using a chart like this.

In an engineering practice, log probability paper is often used in engineering practice to determine the aerosol particle size distribution. One kind of log probability paper is given in Fig. A.1 as an example. Note that the x-axis is for the probability and the y-axis is for the particle diameter.

4.6 Practice Problems

1. Calculate the aerodynamic diameter for an iron oxide spherical particle with geometric equivalent diameter, d_p = 0.21 μm and density, ρ_p = 5,200 kg/m^3.
2. Calculate the aerodynamic diameter for a sand particle with geometric equivalent diameter, d_p = 0.3 μm and density, ρ_p = 3,500 kg/m^3.
3. A spherical grain of concrete dust is falling down to the floor through standard calm air. The particle geometric diameter is 2 μm and the particle density is 2,500 kg/m^3. Determine

   (a) the aerodynamic particle diameter of the particle
   (b) how long it will take for the particle to reach its terminal velocity
   (c) terminal settling velocity of the particle
   (d) flow condition around this particle, laminar or turbulent
   (e) how long it will take this particle will take to fall down a distance of 1 m?

   Given: standard room air density = 1.21 kg/m^3, viscosity = 1.81×10^{-5} Pa s. Air mean free path = 0.066 μm.
4. A spherical particle with a diameter of 2 μm and a density of 5,200 kg/m^3 is released from rest in still air. How long it will take to reach its terminal velocity?
5. A dust grain with a diameter of 5 μm and a density of 5,000 kg/m^3 is released from rest in still air. Calculate the terminal settling velocity and drag force exerted on this particle.
6. A particle with a diameter of 0.08 μm and a density of 2,000 kg/m^3 is released from rest in still air. What is the terminal velocity and drag force if particle is released from rest in still air?
7. Calculate the diffusion coefficient of a 2-μm particle in air at 25 °C and 1 atm.
8. Calculate the diffusion coefficient of a smoke particle having 0.05 μm diameter in air at 45 °C and 1 atm.
9. Calculate the Cunningham correction factor for the following particles at 373 K and 1 atm. (a) 0.055 μm, (b) 0.55 μm, and (c) 5.5 μm.
10. A dust particle having 0.75 μm diameter escapes through a filter of vacuum cleaner at a height 1 m above the floor. Assume that the particle is spherical and its density is 1,200 kg/m^3. How long it will take to settle on the floor?
11. Two particles having diameters 1 and 10 μm, respectively, are released from a height of 25 m above the ground. Assume that the particles are spherical with 1,200 kg/m^3 of density and ambient air is still. How long will they take to settle on the ground?

12. Starting from Eq. (4.55) derive the fractional distribution equation the surface and volume distributions to show that both surface and volume also follow lognormal distributions with the same standard deviation but with different median diameters.
13. A 1-μm cubic particle discharged from a 20-m tall stack travels with a horizontal wind at a constant speed of 10 m/s. Assuming its density is 1,500 kg/m^3, estimate

    (a) its volume *equivalent* diameter,
    (b) its aerodynamic diameter, and
    (c) the horizontal distance it travels before it falls to the ground.

14. Calculate the half-value time of spherical monodisperse aerosol with initial concentration of 10^6 particles/cm^3 and initial particle diameter of $dp = 100$ μm. Assume standard condition and only Brownian coagulation is considered.
15. The figure below shows the cumulative mass distribution of a sample taken from coal fly ash. Assume the particles are lognormally distributed. Find the geometric standard deviation, and mass median diameter using the figure below.

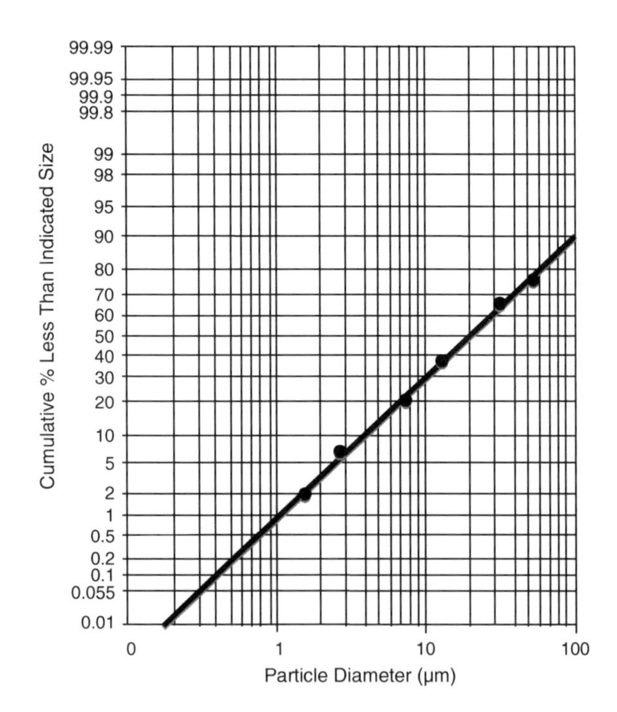

References and Further Readings

1. Anastasio C, Martin S (2001) Atmospheric nanoparticles. In: Banfield JF, Navrotsky A (eds) Nanoparticles and the environment. reviews in mineralogy and geochemistry, vol. 44. Mineralogical Society of America, Chantilly, pp 293–349
2. Allen MD, Raabe OG (1985) Slip correction measurements of spherical solid aerosol particles in an improved Millikan apparatus. Aerosol Sci Technol 4(3):269–286
3. Bradley R (1932) The cohesive force between solid surfaces and the surface energy of solids, London, Edinburgh, and Dublin. Phil Mag J Sci Taylor & Francis, 13:853–862
4. Castellano P, Ferrantel R, Curinil R, Canepari S (2009) An overview of the characterization of occupational exposure to nanoaerosols in workplaces. J Phys: Conf Ser 170:012009
5. Cohen B, Xiong J, Fang C, Li W (1998) Deposition of charged particles on lung airways. Health Phys 74:554–560
6. Dahneke B (1971) The capture of aerosol particles by surfaces. J Colloid Interface Sci 37:342–353
7. Derjaguin BV, Muller VM, Toporov YP (1974) Effect of contact deformations on the adhesion of particles. J Colloid Interface Sci 53:314–326
8. Ferreira AJ, Cemlyn-Jones J, Cordeiro C (2013) Nanoparticles, nanotechnology and pulmonary nanotoxicology. Pneumologia 19:28–37
9. Flagan RC, Seinfeld JH (2012) Fundamentals of air pollution engineering. Dover Press, New York, USA
10. Geng J, Park H, Sajo E (2013) Simulation of aerosol coagulation and deposition under multiple flow regimes with arbitrary computational precision. Aerosol Sci Technol 47:530–542
11. Hettler EN, Gulliver JS, Kayhanian M (2011) An elutriation device to measure particle settling velocity in urban runoff. Sci Total Environ 409(24):5444–5453
12. Hinds W (1999) Aerosol technology: properties, behavior, and measurement of airborne particles, 2nd edn. Wiley-Interscience, New York
13. Jaques P, Kim C (2000) Measurement of total lung deposition of inhaled ultrafine particles in healthy men and women. Inhal Toxicol 12:715–731
14. Johnson L, Kendall K, Roberts A (1971) Surface energy and the contact of elastic solids. Proc R Soc Lond A.324:301–313
15. Kulkarni P, Baron PA, and Willeke K (eds). (2011) Aerosol measurement: principles, techniques, and applications. Wiley, Hoboken
16. Marlow W, Brock J (1975) Calculations of bipolar charging of aerosols. J Colloid Interface Sci 51:23–31
17. Maugis D (2000) Contact, adhesion and rupture of elastic solids. Springer, Berlin
18. Mouret G, Chazelet S, Thomas D, Bemer D (2011) Discussion about the thermal rebound of nanoparticles. Sep Purif Technol 78:125–131
19. Otto EH, Fissan H, Park SH, Lee KW (1999) The log-normal size distribution theory of Brownian aerosol coagulation for the entire particle size range: Part II. J Aerosol Sci 30 (1):16–34
20. Phares D, Rhoads K, Wexler A (2002) Performance of a single-ultrafine-particle mass spectrometer. Aerosol Sci Technol 36:583–592
21. Pui DYH, Fruin S, McMurry P (1988) Unipolar diffusion charging of ultrafine aerosols. Aerosol Sci Technol 8(2):173–187
22. Seinfeld J, Pandis S (2006) Atmospheric chemistry and physics 2nd edn, chapter 8 properties of the atmospheric aerosol, pp 350–395
23. Smoluchowski M (1917) Versuch einer mathematischen Theorie der Koagulationskinetik kolloider L¨osungen. Z Phys Chem XCII(2):129–168
24. Stahlmecke B, Wagener S, Asbach C, Kaminski H, Fissan H, Kuhlbusch T (2009) Investigation of airborne nanopowder agglomerate stability in an orifice under various differential pressure conditions. J Nanopart Res 11:1625–1635

25. Tan Z, Wexler A (2007) Fine particle counting with aerodynamic particle focusing and corona charging. Atmos Environ 41:5271–5279
26. Tsai C, White D, Rodriguez H, Munoz C, Huang C, Tsai C, Barry C, Ellenbecker M (2012) Exposure assessment and engineering control strategies for airborne nanoparticles: an application to emissions from nanocomposite compounding processes. J Nanopart Res 14:989
27. Wang S, Zhao B, Zhou B, Tan Z (2012) An experimental study on short-time particle resuspension from inner surfaces of straight ventilation ducts. Build Environ 53 (2012):119–127
28. Whitby ER, McMurry PH, Shankar U, Binkowski FS (1991) Modal aerosol dynamics modeling. *EPA report*, 600/3-91/020

Chapter 5
Principles for Gas Separation

With a good understanding of the properties of gaseous and particulate air pollutants, this chapter starts with basic principles for the separation of unwanted gases from the air. It is divided into two major sections as adsorption and absorption. Topics covered in this chapter include adsorbate and adsorbent, adsorption affinity, adsorption isotherm, adsorption wave, absorption, absorption equilibrium, and chemical assisted absorption. Condensation is not covered because it applies mainly to gases with very high concentration, which do not happen often in air emission control.

5.1 Adsorption

5.1.1 General Consideration

Adsorption is a process by which gas molecules are attracted to the surfaces of a solid or liquid and consequently separated from the main gas stream. The adsorbing material is called adsorbent, and the gas molecules to be adsorbed is called adsorbate.

Adsorption can be classified as physisorption and chemisorption. The differences between these two types of adsorptions are summarized in Table 5.1. The former is caused by van der Waals forces and the latter involves chemical reactions between the adsorbent and the adsorbate. Thereby the mechanisms and models for chemical adsorption are more complicated than physical adsorption.

Physioadsorption is a surface phenomenon where the adsorbate will stay on the surface of the adsorbent. The attractive force normal to the surface tends to grab adjacent molecules of adsorbate. For example, activated carbon has a tendency to adsorb volatile organic compounds (VOCs) from an air stream. Chemisorption is driven by a chemical reaction that takes place at the adsorbate-adsorbent interface. As a result, a new chemical species is generated at the interface.

Both physisorption and chemisorption may occur simultaneously at the same adsorption interface and physical adsorption can continue after the chemical adsorption layer is completed. However, the bonding strength of chemisorption is

Z. Tan, *Air Pollution and Greenhouse Gases*, Green Energy and Technology,
DOI 10.1007/978-981-287-212-8_5

Table 5.1 Physisorption and chemisorption

	Physical adsorption	Chemical adsorption
Driving force	van der Waals force between molecules. No change of properties of either the adsorbent or the adsorbate	Chemical reactions between the adsorbent and adsorbate. Forming new adsorption products
Heat release rate	Low, about 20 kJ/g mol	High due to chemical reaction, 20–400 kJ/g mol
Reversibility	Can be easily reversed by reducing the pressure at the temperature at which the adsorption took place	Difficult to reverse. Requires very high temperature or positive ion bombardment to remove the adsorbate
Equilibrium pressure	Physical adsorption of a gas is related to liquefaction or condensation, it only occurs at pressures and temperatures close to those required for liquefaction. Low-pressure adsorptions take place mainly in fine porous adsorbents by capillary effect	Chemisorption can take place at much lower pressures and much higher temperatures than physical adsorption
Thickness of reaction layers	A physical adsorption layer at equilibrium can be several molecules thick	A chemisorption layer can only be one molecule thick, because the newly formed compound layer prevents the further reaction of the adsorbent and the adsorbate

stronger than that of physisorption. A description of the chemisorption bond requires a detailed understanding of molecules outside surfaces and the electronic structure of atoms. The author of this book does not intend to extend the scope of this book to interfacial chemistry; readers interested in this topic are directed to specialized books devoted to chemisorption.

Common physioadsorbents include

- activated carbon,
- silica gel,
- activated alumina, and
- aluminosilicates (molecular sieves).

Activated carbon is a char-like material with a great surface area. Silica gel is a hard, granular, and porous material made by precipitation from sodium silicate solutions treated with an acid. Activated alumina is an aluminum oxide activated at high temperature and used primarily for moisture adsorption. Aluminosilicates are made of porous synthetic zeolites and are used primarily in gas separation processes.

Activated carbon is the most common adsorbent for air emission control. Activated carbon is made by the carbonization of carbon rich coal or biomass (wood, fruit pits, or coconut shells) followed by activation with hot air or steam. It is produced by a two-step process. First, pyrolysis of raw material with a high carbon source such as coal, wood and nutshells results in charred highly carbonaceous solid residue. Then activation of the charred residue by oxidation forms pores and passages

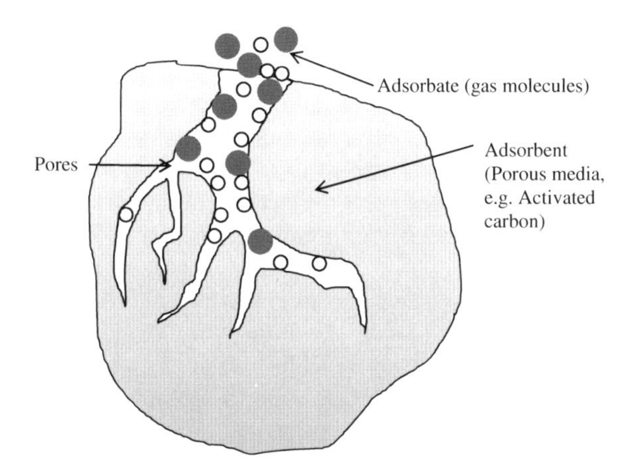

Fig. 5.1 Schematic diagram of pores in adsorbent

with high surface area and different polarities for different air pollutants. Chemical treatment is also necessary to further improve its performance. The final properties of activated carbon depend on the treatment of the activated carbon (e.g., [7]).

Activated carbons can be made in many forms, depending on the need: pellet and granular forms are common for adsorption beds; fibrous structures are common for air filters, for example, activated carbon cloth can be used for removal of water vapor [3].

Most engineered adsorbents are characterized with many pores and consequent large surface areas and low bulk density. If an adsorbent is porous, then its actual surface area is much greater than external area.

Figure 5.1 illustrates how adsorbate molecules are adsorbed on the surfaces of the pores in an adsorbent. The bulk sizes of activated carbon particles can be up to a few millimeters in form of granules or in micrometers for powder.

For most adsorbents, the sizes of the pores are sufficiently small and the corresponding surface area is enormous. The pore diameter of a molecular sieve can be 30 or 40 Angstrom (Å) (1 Å = 0.1 nm). Pore sizes in activated carbon range from 10–10,000 Å in diameter. Those larger than 1,000 Å are called macropores and the pores smaller than 1,000 Å are micropores. The surface areas of the pores can be 500–1,500 m^2/g of activated carbon [1, 7]. Large amount of internal pores in an adsorbent greatly reduces its bulk density. Bulk densities of dry granular activated carbon are 220–500 kg/m^3, and those of powders are 340–740 kg/m^3. The real density of the granular activated carbon itself may be about 2,000 kg/m^3.

5.1.2 Adsorption Affinity

Adsorbent affinity is the attraction between the adsorbent and the adsorbate and it is critical to successful adsorption. Affinities can be designed by adjusting the pore diameters of the adsorbent. For example, silica gel is used for adsorbing water vapor from the air because it has an affinity for water. In air pollution control, activated carbon is an ideal adsorbent for many VOCs because it has an affinity for hydrocarbon gases. Both the surface area of the adsorbent and the affinity of the adsorbent for the adsorbate affect adsorption efficiency. Adsorption efficiency increases with the increase in surface area of adsorbent. Bulk sizes of the porous adsorbent particles do not affect maximum adsorption capabilities, but they affect the time to achieve equilibrium significantly.

Affinities can also be designed by adjusting the pore diameters of the adsorbent. For example, a molecular sieve with a pore diameter of 30 Å can adsorb light gases, such as NH_3 and H_2O. However, when the diameter of the pores is increased to 40 Å, the sorbent is more effective on larger molecules, such as CO_2 and SO_2. Further increasing the pore diameters can enable the adsorption of large organic molecules, such as benzene, phenol, and toluene.

5.1.3 Adsorption Isotherm

Physical and thermodynamic properties of different gases are particularly useful for engineers who design adsorption systems, and analysis of isotherms can yield important data. Adsorption equilibrium is reached when a stable relationship exists between the concentration of the species in air and the amount of adsorbate adsorbed per unit mass of adsorbent. The adsorption equilibrium is a function of temperature.

An adsorption isotherm is a relationship of equilibrium adsorbent capability versus the adsorbate concentration at a given temperature. Note that equilibrium absorption is rarely achieved in real engineering operations. Factors that can reduce the adsorption capacity include moisture in the air, heat waves, or residue moisture in the adsorbent. This actual adsorption capacity is sometimes called working adsorption capacity.

$$M_{eq} = f(C) \tag{5.1}$$

M_{eq} is the adsorbate to adsorbent mass ratio at equilibrium (kg adsorbate/kg adsorbent), while C is adsorbate concentration in gas phase (kg/m^3, ppmv, etc.)

There are several models for this relationship described in Eq. (5.1), of which Langmiur and Freundlich isotherms are widely used [11]. Both are introduced below.

5.1.3.1 Langmuir Isotherm

Irving Langmuir was awarded the Nobel Prize in 1932 for his investigations concerning surface chemistry. Langmuir model is an empirical model and it was developed in the 1910s based on the following assumptions:

- The adsorbed layer is made up of a single layer of molecules.
- Adsorption is a reversible process.
- The process is dynamic: the adsorbed molecules do not move on the surface of the adsorbent, but they may reenter the air stream.
- The enthalpy of adsorption is the same for all molecules independently of how many have been adsorbed.

At equilibrium the number of molecules being adsorbed equals to the number of molecules leaving the adsorbed state, i.e., *Rate of adsorption = Rate of desorption*

$$k_1 C\left(M_{\max} - M_{\text{eq}}\right) = k_2 M_{\text{eq}} \tag{5.2}$$

where C = the gas concentration (kg/m^3); M_{eq} = equilibrium loading capacity with a unit of kg/kg, $M_{\max}$ = the maximum loading potential of adsorbate that can be loaded to per mass of adsorbent with same unit as M_{eq}. k_1 and k_2 are mass transfer coefficients, and their unit changes with that of $M_{\max}$ and M_{eq}.

Solving Eq. (5.2) leads to the description of $M_{\text{eq}}/M_{\max}$, which stands for the ratio of equilibrium loading to maximum loading potential.

$$\frac{M_{\text{eq}}}{M_{\max}} = \frac{k_1 C}{k_1 C + k_2} \tag{5.3}$$

With a new constant $K_L = k_1/k_2$, Eq. (5.3) becomes

$$M_{\text{eq}} = \frac{M_{\max} K_L C}{1 + K_L C} \tag{5.4}$$

$M_{\max}$ is a constant for a fixed design with fixed amount of known adsorbent; the mass transfer coefficients can also be treated as constants for certain operating conditions. Then K_L and $M_{\max}$ are both constants in the Langmuir model; they are to be determined experimentally. To make sure $K_L C$ is dimensionless, the unit of K_L is the same as that of $1/C$.

The Langmuir isotherm can be rearranged as

$$\frac{1}{M_{\text{eq}}} = \frac{1}{M_{\max}} + \left(\frac{1}{K_L M_{\max}}\right)\frac{1}{C} \tag{5.5}$$

It shows that $1/M_{\text{eq}}$ versus $1/C$ is a linear relationship.

In a typical experiment, one can continuously monitor the up- and down-stream adsorbate concentrations. When they are approaching each other it means that an equilibrium state is reached. Then the amount of adsorbate can be determined by measuring the mass of the adsorbent before and after the experiment.

We can plot the experimental data with $1/C$ as x-axis and $1/M_{\text{eq}}$ as y-axis. By linear regression, the slope of the straight line is $1/K_L M_{\max}$ and the intercept is $1/M_{\max}$.

Table 5.2 Langmuir isotherm constants measured at 25 °C by linear regression

$M_{eq} = \frac{M_{max} K_L C}{1+K_L C}$

Adsorbate	AC type	M_{max} (g/g)	K_L (1/ppmv)
Acetone	A	0.401	0.00067
	B	0.389	0.00057
	D	0.330	0.000520
	E	0.391	0.000491
Benzene	A	0.389	0.00492
	C	0.393	0.00157
	D	0.124	0.00379
Carbon tetrachloride	A	0.959	0.00206
	D	0.677	0.00177
Chloroform	D	0.631	0.000685
Dichloromethane	A	0.298	0.000716
Diethylamine	A	0.461	0.00139
Ethanol	C	0.353	0.00145
	D	0.333	0.00133
Hexane	A	0.337	0.00924
Isopropanol	A	0.456	0.00185
Methylacetate	A	0.304	0.00170
Methylchloroform	A	0.666	0.00443
Methanol	B	0.394	0.000349
	C	0.334	0.000446
	D	0.334	0.000317
	E	0.396	0.00027
Nitrobenzene	D	0.518	0.140
o-xylene	D	0.378	0.0238
Toluene	B	0.447	0.00686
	C	0.384	0.00873
	D	0.681	0.00371
	E	0.441	0.00629
Trichloroethylene	B	0.703	0.00265
	C	0.650	0.00331
	D	0.625	0.00242
	E	0.708	0.00204

Some of the Langmuir constants for coconut-based activated carbon are available in Table 5.2 [14]. The properties of the coconut-based activated carbon materials are listed in Table 5.3. The constants depend on the adsorbate–adsorbent combination.

Table 5.3 Coconut-based activated carbon reported [14]

AC type	Density (g/cm^3)	Surface area (m^2/g)	Pore volume (cm^3/g)
A	0.38–0.44	1,500–1,625	0.9–1
B	0.44	1,270	0.7
C	0.41	1,090	0.94
D	0.45	1,098	0.57
E	0.43	1,240	0.65

5.1.3.2 Freundlich Isotherm

Freundlich isotherm is also the empirical model and is described as

$$M_{\mathrm{eq}} = K_F C^{\frac{1}{n}} \tag{5.6}$$

where K_F and n are constants for a specific adsorbate–adsorbent system at certain condition. A logarithm conversion on both side transforms this equation into a linear one as

$$\log_{10} M_{\mathrm{eq}} = \log_{10} K_F + \frac{1}{n} \log_{10} C \tag{5.7}$$

Again this linear relationship allows the coefficients to be determined by linear regression of a few experimental data points. $M_{\mathrm{eq}} = K_F$ when $C = 1$, and the slope of the straight line is $1/n$.

Although both Langmuir and Freundlich models are empirical, there are some differences. The Langmuir isotherm is a model with some theoretical analyses and assumptions. It assumes reversible adsorption and desorption of the adsorbate molecules. The Freundlich isotherm is an empirical model without assumption. In general, the Langmuir isotherm works well for typical single component and high adsorbate concentration. The Freundlich isotherm can be used for mixtures of compounds and it agrees well with experimental data.

Freundlich is more relevant to air emission studies where air pollutants are diluted. With $a = \log_{10} K_F$ and $b = 1/n$, Eq. (5.7) becomes

$$\log_{10} M_{\mathrm{eq}} = a + b \log_{10} C \tag{5.8}$$

For a greater accuracy, Yaws et al. [16] refined the model with one more term

$$\log_{10} M_{\mathrm{eq,g/100\,g}} = a + b \log_{10} C_{\mathrm{ppmv}} + d \left(\log_{10} C_{\mathrm{ppmv}}\right)^2 \tag{5.9}$$

The values of a, b, and d are pollutant specific. Experimental data obtained using 243 VOCs adsorbed using activated carbon are available in the literature [16]. Some of them are listed in Table 5.4 for training purpose only in this book. Users are reminded that the units have to match on both sides of the above equation. In order

Table 5.4 The values of coefficients a, b, and d for Eq. (5.9) with $M_{eq,g/100\,g}$ in (g adsorbate/100 g adsorbent) and C_{ppmv} in ppmv

Formula	Name	a	b	d
$CBrCl_3$	Bromotrichloromethane	1.39842	0.23228	−0.02184
$CBrF_3$	Bromotrifluoromethane	−1.46247	0.58361	−0.01044
CBr_2F_2	Dibromodifluoromethane	0.82076	0.30701	−0.01384
CBr_3F	Tribromofluoromethane	−1.43748	0.55503	−0.00450
CCl_2F_2	Dichlorodifluoromethane	−0.07350	0.40145	−0.01404
CCl_2O	Phosgene	−0.64469	0.60428	−0.02986
CCl_3F	Trichlorofluoromethane	0.17307	0.40715	−0.01915
CCl_3NO_2	Chloropicrin	1.26745	0.20841	−0.01288
CCl_4	Carbon tetrachloride	1.07481	0.28186	−0.02273
$CHBr_3$	Tribromomethane	1.73184	0.19948	−0.02246
$CHCl_3$	Chloroform	0.67102	0.36148	−0.02288
CHN	Hydrogen cyanide	−4.39245	1.08948	−0.00740
CH_2BrCl	Bromochloromethane	0.61399	0.41353	−0.02531
CH_2BrF	Bromofluoromethane	0.45483	0.36332	−0.01606
CH_2Br_2	Dibromomethane	1.08376	0.37211	−0.03238
CH_2Cl_2	Dichloromethane	−0.07043	0.49210	−0.02276
CH_2I_2	Diiodomethane	1.94756	0.14984	−0.01947
CH_2O	Formaldehyde	−2.48524	0.69123	−0.00375
CH_2O_2	Formic acid	−1.77731	1.09503	−0.06354
CH_3Br	Methyl bromide	−1.23835	0.78564	−0.05521
CH_3Cl	Methyl chloride	−1.91871	0.62053	−0.00549
CH_3Cl_3Si	Methyl trichlorosilane	1.07198	0.24275	−0.01911
CH_3I	Methyl iodide	0.73997	0.32985	−0.01330
CH_3NO	Formamide	1.30981	0.25274	–
CH_3NO_2	Nitromethane	−0.32847	0.70602	−0.05111
CH_4	Methane	−4.31008	0.77883	−0.00628
CH_4Cl_2Si	Methyl dichlorosilane	0.73271	0.29305	−0.01822
CH_4O	Methanol	−1.96739	0.82107	−0.01393
CH_4S	Methyl mercaptan	−1.12288	0.60573	−0.02094
CH_5N	Methylamine	−1.93548	0.64710	−0.01057
CN_4O_8	Tetranitromethane	1.49047	0.18181	−0.01894
CO	Carbon monoxide	−5.18782	0.90121	−0.01358
COS	Carbonyl sulfide	−1.42882	0.51061	0.00028
CO_2	Carbon Dioxide	−3.65224	0.80180	−0.00328
CS_2	Carbon disulfide	−0.18899	0.47093	−0.01481
$C_2Br_2F_4$	1,2-Dibromotetrafluoroethane	0.90388	0.25693	−0.00974
C_2ClF_5	Chloropentafluoroethane	0.08264	0.34756	−0.01343
$C_2Cl_3F_3$	1,1,2-Trichlorotrifluoroethane	1.27368	0.18656	−0.01231

(continued)

Table 5.4 (continued)

Formula	Name	a	b	d
C_2Cl_4	Tetrachloroethylene	1.40596	0.20802	−0.02097
$C_2Cl_4F_2$	1,1,2,2-Tetrachlorodifluoroethane	1.37307	0.17625	−0.01465
$C_2HBrClF_3$	Halothane	0.92405	0.31204	−0.02004
C_2HCl_3	Trichloroethylene	1.02411	0.29929	−0.02539
C_2HCl_3O	Dichloroacetyl chloride	1.23647	0.26219	−0.02596
C_2HCl_3O	Trichloroacetaldehyde	1.17362	0.26971	−0.02513
C_2HCl_5	Pentachloroethane	1.64566	0.13515	−0.01572
$C_2HF_3O_2$	Trifluoroacetic acid	−0.12577	0.59373	−0.03445
C_2H_2	Acetylene	−2.24177	0.82454	−0.03390
$C_2H_2Br_4$	1,1,2,2-Tetrabromoethane	-	-	-
$C_2H_2Cl_2$	1,1-Dichloroethylene	0.48740	0.33282	−0.01622
$C_2H_2Cl_2$	cis-1,2-Dichloroethylene	0.47567	0.39061	−0.02554
$C_2H_2Cl_2$	trans-1,2-Dichloroethylene	0.47567	0.39061	−0.02554
$C_2H_2Cl_2O_2$	Dichloroacetic acid	1.69237	0.09630	–
$C_2H_2Cl_4$	1,1,1,2-Tetrachloroethane	1.44097	0.19166	−0.01995
$C_2H_2Cl_4$	1,1,2,2-Tetrachloroethane	1.52322	0.17848	−0.02019
C_2H_3Cl	Vinyl chloride	−0.98889	0.66564	−0.04320
C_2H_3ClO	Acetyl chloride	0.03627	0.45526	−0.02093
$C_2H_3ClO_2$	Methyl Chloroformate	0.41186	0.42776	−0.02776
$C_2H_3Cl_3$	1,1,1-Trichloroethane	0.97331	0.28737	−0.02277
$C_2H_3Cl_3$	1,1,2-Trichloroethane	1.17163	0.27791	−0.02746
C_2H_3N	Acetonitrile	−0.79666	0.63512	−0.02598
C_2H_3NO	Methyl isocyanate	−1.07579	0.85881	−0.06876
C_2H_4	Ethylene	−2.27102	0.61731	−0.01467
$C_2H_4Br_2$	1,1-Dibromoethane	1.37260	0.25671	−0.02516
$C_2H_4Br_2$	1,2-Dibromoethane	1.44231	0.25500	−0.02666
$C_2H_4Cl_2$	1,1-Dichloroethane	0.54485	0.36091	−0.02192
$C_2H_4Cl_2$	1,2-Dichloroethane	0.55343	0.37072	−0.02161
$C_2H_4Cl_2O$	Bis(chloromethyl)ether	0.95599	0.33784	−0.03200
$C_2H_4F_2$	1,2-Difluoroethane	−3.97902	2.51862	−0.31617
C_2H_4O	Acetaldehyde	−1.17047	0.62766	−0.02475
C_2H_4O	Ethylene oxide	−2.42379	0.94878	−0.04062
$C_2H_4O_2$	Acetic acid	−0.05553	0.68410	−0.06071
$C_2H_4O_2$	Methyl formate	−0.99586	0.61693	−0.01847
C_2H_4S	Thiacyclopropane	0.02258	0.45520	−0.02154
C_2H_5Br	Bromoethane	0.31783	0.43549	−0.03072
C_2H_5Cl	Ethyl chloride	−0.50828	0.50364	−0.02179
C_2H_5ClO	2-Chloroethanol	0.74164	0.46933	−0.05158
C_2H_5I	Ethyl iodide	1.00356	0.32123	−0.02405
C_2H_5N	Ethyleneimine	−1.16912	0.91238	−0.07400

(continued)

Table 5.4 (continued)

Formula	Name	a	b	d
C_2H_5NO	N-Methylformamide	1.23333	0.21723	–
$C_2H_5NO_2$	Nitroethane	0.44968	0.49708	−0.04612
C_2H_6	Ethane	−2.40393	0.68107	−0.01925
C_2H_6O	Ethanol	−0.51153	0.67525	−0.04473
C_2H_6OS	Dimethyl sulfoxide	1.24042	0.31302	−0.04768

Source [16]. More data can be found in Table A.6

to use these values, the unit of $M_{\text{eq},\text{g}/100\,\text{g}}$ must be (g adsorbate/100 g adsorbent) and C_{ppmv} in ppmv.

For many engineering processes, the last term $\text{d}\left(\log_{10}C_{\text{ppmv}}\right)^2$ is negligible, and a simplified formula can be used for estimation with a reasonable accuracy. For adsorbate concentrations lower than 50 ppmv, the error is less than 5 %.

Example 5.1: VOC adsorption using AC

In an automobile assembling shop, the concentration of n-butanol ($C_4H_{10}O$) in the room air is 5 ppmv. The density of n-butanol is 3.06 kg/m^3 under standard room air conditions. A carbon filter bed is used for air cleaning, and the airflow rate is 0.1 m^3/s through the filter. Determine

a. the adsorption capacity of the activated carbon filter,
b. the total carbon mass needed for the bed, assuming the working adsorption capacity is 40 % of the maximum potential and the bed service life is one year.

Solution

a. From Table 5.4, we have the adsorption constants of $C_4H_{10}O$:

$$a = 0.89881;\; b = 0.32534;\; d = -0.03648$$

Then Eq. (5.9) leads to

$$\begin{aligned}\log_{10}M_{\text{eq},\text{g}/100\,\text{g}} &= a + b\log_{10}C_{\text{ppmv}} + d\left(\log_{10}C_{\text{ppmv}}\right)^2\\ &= 0.89881 + 0.32534\log_{10}(5) - 0.03648[\log_{10}(5)]^2 = 1.1084\end{aligned}$$

$$M_{\text{eq},\text{g}/100\,\text{g}} = 10^{1.1084} = 12.8\,(\text{gram of } C_4H_{10}O \text{ per 100 gram of carbon})$$

b. The total amount of n-butanol passing through the carbon bed in 1 year is

$$\begin{aligned}m = QC\rho_g t &= 0.1\,\frac{\text{m}^3\text{air}}{\text{s}} \times 0.000005\,\frac{\text{m}^3\text{n}-\text{bu}}{\text{m}^3\text{air}} \times 3.05\,\frac{\text{kg}}{\text{m}^3\text{n}-\text{bu}} \times (365 \times 24 \times 3600)\,\text{s}\\ &= 48.1\,\text{kg}\end{aligned}$$

The actual carbon mass needed is

$$m_{\text{carbon}} = \frac{m}{M_{\text{eq.g/100 g}}} \times \frac{100}{40} = \frac{48.1 \text{ kg of } C_4H_{10}O}{0.128 \text{ kg of } C_4H_{10}O \text{ per kg of carbon}} \times \frac{100}{40}$$
$$= 939.5 \text{ kg of carbon}$$

5.1.4 Adsorption Wave

An adsorption wave is used to determine the kinetics of adsorption for an adsorbent column. Consider a gas stream passing through a column packed with adsorbent. The concentration of the adsorbate in the gas before cleaning is denoted as C_0. The adsorption process does not take place uniformly throughout the bed. When the polluted air stream passes through it, three different zones are developed as shown in Fig. 5.2. The dark area near the inlet of the column is the saturated adsorbent because it adsorbs most of the pollutant gas initially and becomes saturated. Immediately downstream is the area where adsorption is active. The remaining portion of the bed adsorbs little pollutant gas and it is considered fresh.

If we zoom in to the entire active zone, the concentration of the adsorbate at the entrance of the active zone can be assumed C_0 because it is adjacent to the saturation zone where there is no loss of adsorbate. In the active adsorption zone, the adsorbate concentration is essentially reduced from C_0 to zero in an *S*-shape. As more adsorbate enters the adsorption bed, the saturation zone grows longer and longer and it looks as if a wave is propagating within the column. The wave could be steep or quite flat, depending on various factors including the adsorption capacity of the adsorbent, the flow rate and the retention time of the gas stream.

The entire adsorbent bed is nearly saturated as the wave approaches the exit of the column. At this instant, the adsorbent column loses its function and the concentration of the adsorbate at the exit increases. This is referred to as the breakthrough point, or breakpoint. The breakthrough point can also be defined as the ratio of the outlet to inlet concentrations depending on the application and the

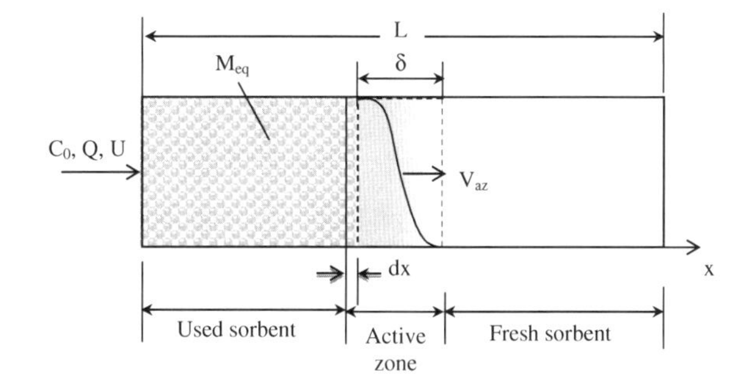

Fig. 5.2 Adsorption wave in a stationary adsorption column

emission limits set by emission standards. For many toxic chemicals, a measurable concentration at the exit, even if it is less than 1 % of the inlet concentration, can be defined as the breakthrough point.

The kinetics of adsorption wave can be analyzed as follows. Consider an adsorption column of length L with an adsorption wave width of δ. The concentration of the adsorbate in the gas stream is C_0 (in mass of adsorbate/volume of gas), and the load of the adsorbate in the saturated adsorbent is M_{eq} (in mass of adsorbate per mass of adsorbent). Within the active adsorption zone, the concentration C of adsorbate in the gas varies from C_0 to zero and the load of adsorbate in the adsorbent C_x varies from M_{eq} to zero. While gas moves through the column at a bulk face velocity of U, the saturation zone keeps growing as fresh adsorbent adsorbs the incoming adsorbate, resulting in propagation of active adsorption zone at a speed V_{az}. In practice, $U \gg V_{az}$, therefore, the relative speed of the gas stream with respect to the active adsorption zone can be assumed as U.

During the process described above, the rate of adsorbate entering adsorbent column is equal to the rate of adsorbate adsorbed by the adsorbent if we ignore other mechanisms for gas separation or leaking. That is,

$$QC_0 = M_{eq}(\rho_b A V_{az}) \tag{5.10}$$

where Q = volumetric flow rate of the gas (m^3/s), C_0 = the incoming adsorbate concentration (kg/m^3), ρ_b = bulk density of the adsorbent (kg/m^3), A = cross-sectional area of the adsorption column (m^2), and M_{eq} = equilibrium loading with a unit of kg/kg.

The bulk density of the adsorbent depends on the packing density of the column and it is not the physical density of the adsorbent material itself. Usually the bulk density is much less than the real density of the material and it depends on the packing density.

The left-hand side of Eq. (5.10) stands for the mass flow rate (in kg/s) of the adsorbate into the column, and the term $(\rho_b A V_{az})$ gives the mass rate of adsorbent used (kg/s). In an engineering practice where the specifications of the adsorption column and operation conditions are known, one can get the values of ρ_s, A, Q and C_0. M_{eq} is required in order to solve Eq. (5.10) for the wave speed V_{az}, and M_{eq} is usually determined experimentally and described using adsorption isotherm.

A simple manipulation of Eq. (5.10) leads to the expression of adsorption wave propagation speed,

$$V_{az} = \frac{QC_0}{M_{eq}(\rho_b A)} \tag{5.11}$$

The mathematical description of the relationship between M_{eq} and C_0 depends on the model introduced above or measured data. When using data in Table 5.4, we must pay attention to the units of C_0 and M_{eq}. In Eqs. (5.10) and (5.11) they are different from those of $M_{eq,g/100\,g}$ and C_{ppmv} listed in Table 5.4. Correct unit conversion is necessary.

$$M_{\mathrm{eq}} = 0.01 M_{\mathrm{eq,g/100\,g}} \tag{5.12}$$

$$C_0 = 10^{-6} \rho_g C_{\mathrm{ppmv}} \tag{5.13}$$

where the factor of 10^{-6} converts adsorbate concentration from ppmv to $\left(\mathrm{m^3\ adsorbate\ per\ m^3\ of\ gas}\right)$. ρ_g is not the density of the entire gas phase, but rather that of the adsorbate ($\mathrm{kg/m^3}$).

Example 5.2: Adsorption wave propagation speed calculation
Same as the parameters given in Example 5.1 above, in an automobile assembling shop, the concentration of n-butanol ($C_4H_{10}O$) in the room air is 5 ppmv. The density of n-butanol is 3.06 $\mathrm{kg/m^3}$ under standard condition. A carbon filter bed is used for air cleaning, and the airflow rate is 0.1 $\mathrm{m^3/s}$ through the bed. The activated carbon bed is manufactured in such a way that its bulk density of the activated carbon is 400 $\mathrm{kg/m^3}$. The cross-sectional area of the bed is 2 $\mathrm{m^2}$. Estimate the propagation speed of the adsorption wave.

Solution
From Example 5.1, we obtained

$$M_{\mathrm{eq,g/100\,g}} = 12.8\,(\text{gram of } C_4H_{10}O \text{ per 100 gram of A.C.})$$

Unit conversion gives

$$M_{\mathrm{eq}} = 0.01 M_{\mathrm{eq,g/100\,g}} = 0.128(\text{kg VOC/kg AC})$$
$$C_0 = 10^{-6} \rho_g C_{\mathrm{ppmv}} = 10^{-6}(3.06) \times 5 = 1.53 \times 10^{-5}(\text{kg VOC/m}^3)$$

Then we can get the adsorption wave propagation speed as

$$\begin{aligned} V_{\mathrm{az}} &= \frac{Q C_0}{M_{\mathrm{eq}} \rho_b A} \\ &= \frac{0.1(\mathrm{m^3/s}) \times 1.53 \times 10^{-5}\left(\text{kg VOC/m}^3\right)}{0.0128(\text{kg VOC/kg AC}) \times \left(400\ \text{kg AC/m}^3\right) \times 2(\mathrm{m^2})} \\ &= 0.01494 \times 10^{-6}(\mathrm{m/s}) = 0.0538(\mathrm{mm/h}) \end{aligned}$$

5.1.5 Breakthrough Time

With the known wave propagation speed, we can easily predict the operating time of a fresh column.

$$t_x = \frac{L - \delta}{V_{\mathrm{az}}} \tag{5.14}$$

where L is the length/height of the column and δ is the width of the adsorption wave (see Fig. 5.2). This is also called the breakthrough time. The lifetime of an adsorbent column bed before the breakthrough point can be estimated using this equation, when regeneration or replacement of adsorbent column is necessary in order to maintain an effective adsorption process.

Although δ can determined by the analysis of gas-solid mass transfer, no single model applies to all gas-solid systems. The exact wave width depends on the packing density, the activated carbon properties and incoming adsorbate concentration. There have been several models developed. One widely used model for breakthrough time t_x is the modified Wheeler equation [8, 14], which is for single adsorbate.

$$t_x = \frac{M_{\mathrm{eq}}}{QC_0}\left[W - \frac{\rho_b Q}{K_x}\ln\left(\frac{C_0}{C_x} - 1\right)\right] \tag{5.15}$$

where t_x = break through time (s), M_{eq} = adsorption capacity (kg/kg), C_0 = inlet concentration (kg/m^3), W = weight of the adsorbent (kg), ρ_b = bulk density of the packed adsorbent (kg/m^3), Q = gas phase volumetric flow rate (m^3/s), K_x = mass transfer rate coefficient (1/s), C_x = exit concentration (kg/m^3). This exit concentration is determined based on local emission standards or air quality requirement.

Note that the last term is related to the breakthrough efficiency, which is the column adsorption efficiency when breakthrough occurs.

$$\eta_x = 1 - \frac{C_x}{C_0} \tag{5.16}$$

Then the breakthrough time Eq. (5.15) can be described in terms of breakthrough efficiency as

$$t_x = \frac{M_{\mathrm{eq}}}{QC_0}\left[W - \frac{\rho_b Q}{K_x}\ln\left(\frac{\eta_x}{1 - \eta_x}\right)\right] \tag{5.17}$$

Example 5.3: Adsorption column breakthrough time

Continuing from Example 5.2 above, we obtained $M_{\mathrm{eq}} = 0.128$(kg VOC/kg AC), $C_0 = 1.53 \times 10^{-5}$(kg VOC/m^3), $Q = 0.1$(m^3/s), $\rho_b = 400$ kg AC/m^3. Now, the mass transfer coefficient of the adsorbent is $K_x = 20\,\mathrm{s}^{-1}$. The cross-sectional area of the bed is given as 0.1 m^2. The designed adsorption efficiency is $\eta_x = 0.9$. That is, the breakthrough occurs when the outlet n-butanol concentration reaches 10 % of the inlet concentration. Determine the length of the bed if it is to be replaced every 2 months (assuming 60 days).

Solution

First we convert the unit of breakthrough time from months to seconds.

$$t_x = 60 \times 24 \times 3600\,\text{s} = 5{,}184{,}000\,\text{s}$$

From Eq. (5.17) we can get the weight of activated carbon in the filter as

$$\begin{aligned} W &= \frac{QC_0 t_x}{M_{\text{eq}}} + \frac{\rho_b Q}{K_x} \ln\left(\frac{\eta_x}{1-\eta_x}\right) \\ &= \frac{0.1(\text{m}^3/\text{s}) \times (1.53 \times 10^{-5}\,\text{kg/m}^3) \times \left(5.184 \times 10^6\,\text{s}\right)}{0.128(\text{kg/kg})} \\ &\quad + \frac{\left(400\,\text{kg/m}^3\right) \times 0.1(\text{m}^3/\text{s})}{20\,\text{s}^{-1}} \ln\left(\frac{0.9}{0.1}\right) \\ &= 61.965(\text{kg}) + 4.39(\text{kg}) = 66.3\,\text{kg} \end{aligned}$$

So, the required length of the filter bed is

$$L = \frac{W}{A\rho_b} = \frac{66.3\,\text{kg}}{0.1\,\text{m}^2 \times 400\,\text{kg/m}^3} = 1.66\,\text{m}$$

Breakthrough curves for binary mixtures on a solid adsorbent are more complicated than those for pure compounds. Mixed compounds compete for the same adsorption sites, weakly adsorbed compound may also be replaced with more strongly adsorbed compound. As a result, there may be more than one adsorption wave propagating in the adsorption column, the stronger one moves slower than the weaker one. More in-depth analysis can be found in the literature (e.g. [14]).

With the development of numerous similar theories for adsorption phenomena, few of them generally agreed with a wide range of experimental data. The measurements have shown that increasing the partial pressure of the adsorbate at a given temperature results in increase in adsorption capacity of the adsorbent, while an increase of the temperature of the adsorbate in the air at a certain partial pressure results in decrease in adsorption capacity. Therefore, heating can regenerate many saturated adsorbent. In addition, gases or vapors with heavier molecules can be more effectively adsorbed than the lighter ones.

5.1.6 Regeneration of the Adsorbent

When breakthrough occurs, adsorbent in an adsorption column can be regenerated instead of being disposed of. Regeneration of an adsorbent is a process that drives the adsorbate out of the saturated adsorbent, which can be referred to as desorption. It is done by passing the regeneration fluid, which often is steam or hot air, through the saturated adsorbent column.

As a result, there is a desorption wave in the column but it propagates in the opposite direction of the adsorption wave. The propagation speed of desorption wave can be determined by similar analysis. Readers are referred to the literature for in-depth analysis.

5.2 Absorption

Absorption is a volumetric process where gases penetrate into the structure of the solid or liquid, most likely by diffusion. Similar to the classification of adsorption, absorption can also be classified into physioabsorption and chemioabsorption. A physioabsorption involves negligible chemical reactions, for example solution of oxygen in water. The physical absorption process involves a mass transfer by means of molecular and turbulent diffusion governed by Henry's law (see Sect. 2.3.1). A chemioabsorption process results in new substances from the chemical reactions. For example, when a flue gas passes through a water spray, the droplets will capture the gaseous SO_2 molecules by absorption. Captured SO_2 molecules dissolve in the body of a water droplet and produces H_2SO_3.

An absorption-based process is also commonly referred to as wet scrubbing. However, one has to be careful that wet scrubbing of particles is based on the principles of impaction and coagulation, whereas wet scrubbing of gaseous pollutants is based on absorption. In this section, we explain the basics using counter flow absorption tower.

5.2.1 *Counter Flow Absorption Tower*

Consider a single-stage counter flow gas-liquid system as shown in Fig. 5.3. The gas enters from the bottom and exits at the top; the liquid enters from the top and

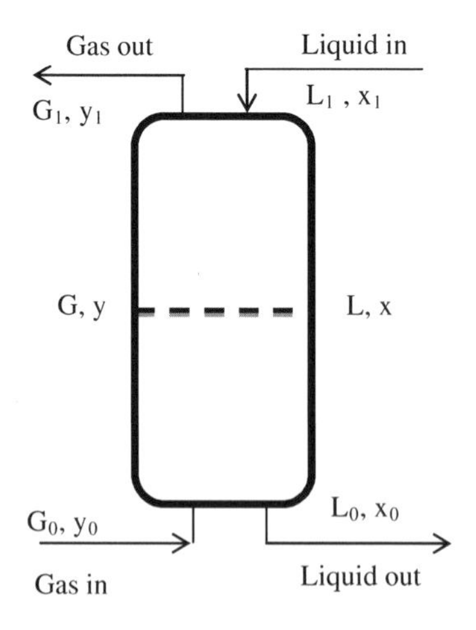

Fig. 5.3 Schematic diagram of a counter-flow absorption tower

exits at the bottom. At any cross section within the absorption tower, all the molecules removed from the gas phase enter the liquid phase, then the mass balance at any cross section within the tower gives

$$\mathrm{d}(Gy) = \mathrm{d}(Lx) \tag{5.18}$$

where L and G are the mole flow rates of the liquid and the gas, respectively (mole/s). y and x are mole fractions of the target pollutant in gas and liquid, respectively. Note that both the gas phase and liquid phase mole flow rates (G and L) are variables within the tower, because gas molecules leave the gas phase enter the liquid phase.

Integration of Eq. (5.18) from any point to top (1) leads to

$$yG - y_1G_1 = xL - x_1L_1 \tag{5.19}$$

The gas phase contains the adsorbate molecules and the inert carrier gas (mostly air), and their mole fractions are, respectively, y and $1 - y$. The liquid phase also contains inert liquid and the pollutants absorbed from the gas phase, and their mole fractions are, x and $1 - x$, respectively. The liquid phase without contamination is called solute free liquid.

Denote the mole flow rates of the carrier gas and solute fee liquid as $\bar{G}$ and $\bar{L}$, respectively, which remain constant throughout the process.

$$G = \frac{\bar{G}}{1-y} \qquad L = \frac{\bar{L}}{1-x} \tag{5.20}$$

In reality, there may be some evaporation of liquid phase, and extra liquid is added as make up liquid to keep the mass balance, but we have to ignore this in the following analysis. With Eq. (5.20), Eq. (5.19) can be rewritten in terms of $\bar{G}$ and $\bar{L}$ as

$$\left(\frac{y}{1-y}\right)\bar{G} + \left(\frac{x_1}{1-x_1}\right)\bar{L} = \left(\frac{x}{1-x}\right)\bar{L} + \left(\frac{y_1}{1-y_1}\right)\bar{G} \tag{5.21}$$

Applying this equation to the entire tower from bottom (0) to top (1) leads to:

$$\left(\frac{y_0}{1-y_0} - \frac{y_1}{1-y_1}\right)\bar{G} = \left(\frac{x_0}{1-x_0} - \frac{x_1}{1-x_1}\right)\bar{L} \tag{5.22}$$

If we define the mole ratio of target gas to that of the corresponding fluid as

$$X = \frac{x}{1-x} \qquad Y = \frac{y}{1-y} \tag{5.23}$$

then the mass balance Eqs. (5.21) and (5.22) above can be simplified as

$$(Y - Y_1)\bar{G} = (X - X_1)\bar{L} \tag{5.24}$$

$$(Y_0 - Y_1)\bar{G} = (X_0 - X_1)\bar{L} \tag{5.25}$$

where Y and X are the mole ratio of pollutant to the corresponding fluid with unit of (mole of pollutant per mole of solute free fluid). They are different from y and x. y and x must be less than one but Y and X can be any values, although most often they are also less than one.

Equation (5.25) can be rewritten as

$$\frac{\bar{L}}{\bar{G}} = \frac{(Y_0 - Y_1)}{(X_0 - X_1)} \tag{5.26}$$

The term on the left-hand side is called liquid to gas mole flow rate ratio. In a design process, the mole ratio of the pollutant in gas phase, Y_0, the carrier gas flow rate $\bar{G}$ are likely known from the source of the air emission. The mole ratio of the pollutant in liquid phase at the inlet, X_1, is usually provided by the solvent supplier. The two unknown parameters are the mole ratio of pollutant in the liquid phase X_0 and that is determined by the mole flow rate of the solute-free liquid ($\bar{L}$).

Example 5.4: Liquid to gas ratio in absorption tower

A mixture of air and H_2S is forced to pass through a single-stage counter flow water absorption scrubber. The inlet mole fraction of H_2S in air is 50 ppmv and outlet being 10 ppmv. Outlet H_2S in water is 20 ppmv. The total pure air flow rate into the scrubber is 80 mol/s. What is the pure water flow rate into the scrubber?

Solution

From the problem description, we can get the following known parameters. For the gas phase, the incoming mole fraction of H_2S in the air is $y_0 = 50 \times 10^{-6}$ (mole H_2S/mole air) and the exiting mole fraction of H_2S in the air is $y_1 = 10 \times 10^{-6}$ (mole H_2S/mole air). Then we can get the corresponding mole ratios as

$$\text{Bottom/gas:} Y_0 = \frac{y_0}{1 - y_0} = \frac{50 \times 10^{-6}}{1 - 50 \times 10^{-6}} \approx 50 \times 10^{-6}$$

$$\text{Top/gas:} Y_1 = \frac{y_1}{1 - y_1} = \frac{10 \times 10^{-6}}{1 - 10 \times 10^{-6}} \approx 10 \times 10^{-6}$$

Similarly for the liquid phase, the mole fraction of H_2S in the incoming liquid phase is $x_1 = 0$ (mole H_2S/mole water) and exiting pollutant mole fraction in the liquid is $x_0 = 20 \times 10^{-6}$ (mole H_2S/mole water). The corresponding mole ratios are

$$X_0 = \frac{x_0}{1 - x_0} = \frac{20 \times 10^{-6}}{1 - 20 \times 10^{-6}} \approx 20 \times 10^{-6}$$

$$X_1 = \frac{x_1}{1 - x_1} = 0$$

Substitute the mole ratios and the carrier gas mole flow rate of $\bar{G} = 80$ (mole/s) into the mass balance Eq. (5.26), and we have:

$$\frac{\bar{L}}{\bar{G}} = \frac{(Y_0 - Y_1)}{(X_0 - X_1)} \rightarrow \frac{\bar{L}}{80} = \frac{(50 - 10)}{(20 - 0)} = 2$$

So the solute-free water flow rate is 160 mol/s. Assuming water molar weight of 18 g/mole, we can calculate the water mass flow rate and it is 2.88 kg/s.

This example also shows quantitatively that for the cases with very low mole fractions of y and x, $1 - y \approx 1$ and $1 - x \approx 1$, then $X = x$ and $Y = y$. Then the mass balance Eq. (5.22) becomes

$$\frac{\bar{L}}{\bar{G}} = \frac{y_0 - y_1}{x_0 - x_1} \quad (\text{for } x \ll 1, y \ll 1) \tag{5.27}$$

5.2.2 Absorption Equilibrium Line and Operating Line

Equation (5.24) can be rewritten as

$$Y = \frac{\bar{L}}{\bar{G}}(X - X_1) + Y_1 \tag{5.28}$$

This equation implies that the gas phase mole ratio Y is a linear function of liquid phase mole ratio X if $\frac{\bar{L}}{\bar{G}}$, X_1, and Y_1 are constant. Therefore, it describes the relationship between the gas phase mole ratio and liquid phase mole ratio at any elevation in an operating tower. If we plot a Y versus X in a figure, it is thereby called the absorption operating line.

5.2.2.1 Absorption Equilibrium Line

A special operating line is *equilibrium line.* If the absorption tower operates in such a manner that the gas-liquid system at any elevation reaches equilibrium, then for any gas mole ratio Y, there is a corresponding equilibrium liquid mole ratio X*. And Y can be related to X* according to the Henry's law. The corresponding linear function defines the absorption equilibrium line.

The equilibrium line can be determined as follows. Equation (5.23) gives

$$x = X/(1+X) \text{ and } y = Y/(1+Y)$$

Substituting them into the Henry's law Eq. (2.76)

$$Py = Hx$$

We can describe the Henry's Law in terms of the mole ratios (X and Y) instead of the mole fractions (x and y),

$$\frac{Y}{1+Y} = \frac{H}{P}\left(\frac{X}{1+X}\right) \tag{5.29}$$

Manipulation of the equation gives the mole ratio in liquid phase as

$$Y = \frac{H}{P}\left(\frac{1+Y}{1+X}\right)X \tag{5.30}$$

For most air emission control by absorption, the concentration of the target gas in air is usually very low, which means $X \ll 1$; $Y \ll 1$. With this critical assumption, which is justified in most applications, Eq. (5.30) can be simplified as

$$Y = \frac{H}{P}X \tag{5.31}$$

In order to differentiate the equilibrium line from the operating line, we put forward two more terms, hypothetical or equilibrium mole ratio: Y* and X*. They are defined as

$$Y^* = \frac{H}{P}X \quad Y = \frac{H}{P}X^* \tag{5.32}$$

A line can be drawn from this equation on X-Y axes. Since this equation is derived from the equilibrium assumption, it is called equilibrium line (see Fig. 5.4). The slope of this line is

$$m = \frac{H(1+Y)}{P(1+X)} \approx \frac{H}{P} \tag{5.33}$$

5.2.2.2 Absorption Operating Line

In operating an absorption tower in engineering practice, the equilibrium between gas and liquid phase cannot be established because of the short residence time when they encounter each other. Therefore, for a given gas mole ratio Y, the corresponding liquid mole ratio X should be always less than the equilibrium mole ratio.

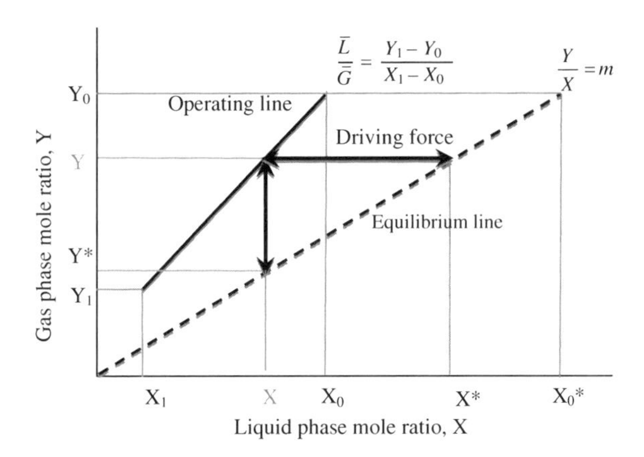

Fig. 5.4 Absorption equilibrium line and operating line

$$Y = \frac{\bar{L}}{\bar{G}}(X - X_1) + Y_1 \tag{5.34}$$

In Fig. 5.4, we also can see the driving force of absorption is the gas phase mole ratio difference Y-Y* or liquid phase mole ratio difference X*-X. Note that (X_1, Y_1) are usually fixed in Fig. 5.4.

5.2.2.3 Absorption Minimum Operating Line

Depending on the engineering design, the driving force of the absorption can be changed by adjusting the operating line. The slope of each line is the corresponding liquid to gas ratio $\frac{\bar{L}}{\bar{G}}$. As seen in Fig. 5.5, there is a minimum operating line where equilibrium is reached at the bottom of the tower, and it corresponds to the lowest operating line right above the equilibrium line in Fig. 5.5. A practical term is the minimum liquid to gas

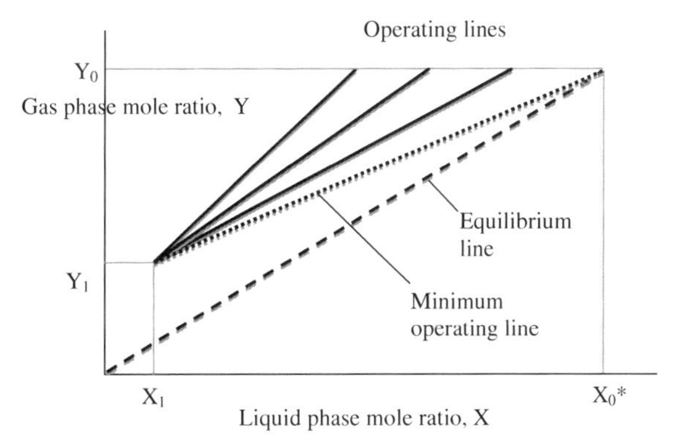

Fig. 5.5 Absorption equilibrium line and different operating lines

ratio defined by the minimum operating line. Once it is determined, the actual operating liquid-to-gas ratio is set higher than this minimum ratio.

The minimum liquid-to-gas flow ratio can therefore be determined using the above Eq. (5.26) based on the equilibrium value of X_0^*,

$$\left(\frac{\bar{L}}{\bar{G}}\right)_{\min} = \frac{Y_0 - Y_1}{X_0^* - X_1} \tag{5.35}$$

$$X_0^* = \frac{x_0^*}{1 - x_0^*} \tag{5.36}$$

and X_0^* would be the maximum possible mole ratio of the target gas in the liquid phase if it were allowed to come to equilibrium with the gas entering the tower in the gas phase. x_0^* can be taken from the equilibrium line with the inlet gas inlet y_0 or determined using Eq. (5.37).

$$x_0^* = Py_0/H \tag{5.37}$$

Example 5.5: Absorption minimum operating line

A mixture of air and H_2S is force to pass through a single-stage counter flow water absorption scrubber. The inlet mole fraction of H_2S in air is 50 ppmv. The total gas flow rate into the scrubber is 80 mol/s and the pure water flow rate into the scrubber is 10 mol/s. Assuming that the gas-water system is at equilibrium state at 30 °C, and the atmospheric pressure is 101,325 Pa. Find the mole fraction of H_2S in gas phase at the exit. Assume that the system within the tower is air-H_2S for gas and water-H_2S for liquid phases.

Solution

Since the air pollutant concentration in the air is very low and the consequent concentration in the liquid phase should also be very low, the simplified mass balance equation can be used to solve this problem. Substitute $\bar{L}$ = 10 mol/s, $\bar{G}$ = 80 mol/s, x_1 = 0 mol H_2S/mole water, y_0 = 0.00005 mol H_2S/mole air into the simplified mass balance equation (5.27) gives

$$y_1 = 5 \times 10^{-5} - \frac{10}{80}(x_0 - 0)$$

The inlet mole fraction of H_2S in the air can be determined from the assumption of equilibrium state. Applying Henry's law to the equilibrium state to the bottom of the system gives

$$Py_0 = Hx_0$$

where the Henry's Law constant can be found from Table 2.4 as H = 609 × 1. 1 × 10^5 Pa/mole fraction in water at 30 °C and P = 1.013 × 10^5 Pa with y_0 = 50 × 10^{-6}, x_0 can be determined as

$$x_0 = 8.32 \times 10^{-8} \quad \text{(mole of H}_2\text{S/mole of water)}$$

Then the outlet H_2S mole fraction in the air is

$$y_1 \approx 50 \text{ ppmv} = y_0$$

This result indicates that the wet scrubber that operates at equilibrium governed by Henry's law is basically useless in absorbing the air pollutant into the liquid phase. This is calculated based on the assumption of equilibrium at the bottom of the system. This actually makes sense because that is what equilibrium implies—no net mass transfer between gas and liquid phases. Practically speaking, this system works along the minimum operating line, but it is not that effective in gas absorption. In practice, the operating line must be different from this line.

Example 5.6: Absorption operating line
A packed bed wet scrubber is designed to remove high concentration SO_2 from the exhaust of a sulfuric acid plant. It is expected to achieve a removal efficiency of 95 %. The incoming SO_2 concentration is 10 %. Pure water is used as an absorbent and the solute-free liquid to gas ratio is 1.5 times the minimum ratio. Assume that the system operates at an average temperature of 30 °C and 1 atm. The equilibrium data for SO_2 in air and water at this temperature are as follows [18]

Partial pressure p_{SO_2} (mmHg)	0.6	1.7	4.7	8.1	11.8	19.7	36	52	79
SO_2 concentration in water c_{SO_2} (g SO_2/100 g water)	0.02	0.05	0.1	0.15	0.2	0.3	0.5	0.7	1

Plot the equilibrium line, the minimum operating line and the operating line in the same figure.

Solution
To determine the equilibrium line Y versus X, we need to determine the mole fraction changes in liquid and gas phases in the scrubber.

Step 1: Determine the equilibrium line by the mole fraction in gas by Dalton's law, Eq. (2.40) and liquid phases, respectively

$$y = y_{SO_2} = \frac{P_{SO_2}}{P}$$

$$x = x_{SO_2} = \frac{\frac{c_{SO_2}}{64 \text{ g/mole}}}{\frac{c_{SO_2}}{64 \text{ g/mole}} + \frac{100 \text{ g}}{18 \text{ g/mole}}}$$

Table 5.5 SO_2 solubility data at 30 °C and 1 atm

Partial pressure	Solubility	SO_2 mole fraction in gas	Mole fraction in liquid	Mole ratio	Mole ratio
p_{SO_2} (mmHg)	c_{SO_2} (g SO_2/100 g water)	$y = \frac{p_{SO_2}}{760\ \text{mmHg}}$	x_{SO_2}	$Y = \frac{y}{1-y}$	$X = \frac{x}{1-x}$
0.6	0.02	0.000789	0.00005625	0.00079	0.000056
1.7	0.05	0.00224	0.00014	0.00224	0.00014
4.7	0.1	0.00618	0.000281	0.00622	0.000281
8.1	0.15	0.01066	0.000422	0.01077	0.000422
11.8	0.2	0.0155	0.000562	0.01577	0.000563
19.7	0.3	0.0259	0.000843	0.0266	0.000844
36	0.5	0.04737	0.00140	0.0497	0.00141
52	0.7	0.06841	0.001965	0.07345	0.00197
79	1	**0.104**	**0.0028**	0.1160	0.00281

where 64 and 18 are the molar weights of SO_2 and water, respectively, with the unit of g/mole. Using these equations, we can calculate the SO_2 mole fractions in gas phase and liquid phase (Table 5.5).

And the corresponding the equilibrium line by linear regress is

$$Y = 41.69X - 0.006 \quad \left(R^2 = 0.994\right) \tag{1}$$

The equilibrium line Y versus X is plotted in Fig. 5.6.

We also can get the linear regression of y versus x as

$$y = 37.747x - 0.0043 \quad \left(R^2 = 0.998\right) \tag{2}$$

The minimum water flow rate can be determined using Eq. (5.21) with the inlet SO_2 mol fraction $y_0 = 10\ \% = 0.10$, and exit SO_2 mol fraction $y_1 = 0.10 \times (1–95\ \%) = 0.005$. In the pure water entering the top of the tower, $x_1 = 0$, and at the bottom of tower, the SO_2 mol fraction in the air is $y_0 = 0.01$ and

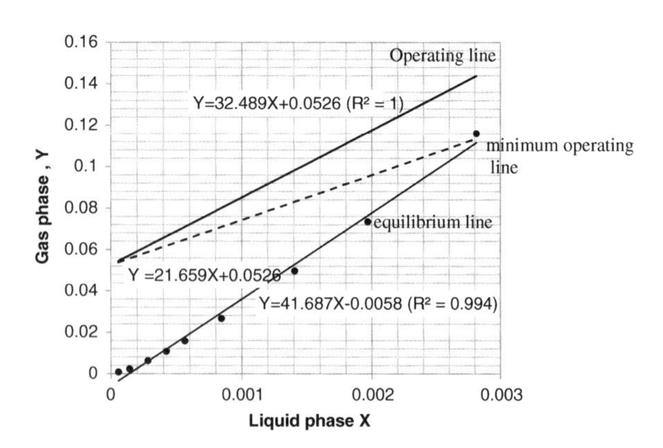

Fig. 5.6 SO_2-equilibrum-line and operation line

the corresponding equilibrium mole fraction in the water exiting the bottom of the tower is determined from the equilibrium line, Eq. (2)

$$x_0^* = 0.0028$$

Then the corresponding minimum liquid-to-gas ratio is determined by substituting these values into Eq. (5.35), together with $x_1 = 0$, $y_1 = 0.05$,

$$\left(\frac{\bar{L}}{\bar{G}}\right)_{min} = \frac{\frac{y_0}{1-y_0} - \frac{y_1}{1-y_1}}{\frac{x_0^*}{1-x_0^*} - \frac{x_1}{1-x_1}} = 21.67$$

Then the minimum operating line can be determined using Eq. (5.21)

$$\left(\frac{y}{1-y}\right) + \left(\frac{x_1}{1-x_1}\right)\left(\frac{\bar{L}}{\bar{G}}\right)_{min} = \left(\frac{x}{1-x}\right)\left(\frac{\bar{L}}{\bar{G}}\right)_{min} + \left(\frac{y_1}{1-y_1}\right)$$

In Fig. 5.6, the minimum operating line is defined as

$$Y = 21.66X + 0.053 \tag{3}$$

The actual operating liquid to gas ratio is 1.5 times of this liquid flow rate

$$\frac{\bar{L}}{\bar{G}} = 1.5 \times \left(\frac{\bar{L}}{\bar{G}}\right)_{min} = 32.49$$

Then we can determine the actual operating line using Eq. (5.21) again, with $x_1 = 0$, and $y_1 = 0.05$, and $\left(\frac{\bar{L}}{\bar{G}}\right)_{min} = 32.49$

$$\left(\frac{y}{1-y}\right) + \left(\frac{x_1}{1-x_1}\right)\left(\frac{\bar{L}}{\bar{G}}\right) = \left(\frac{x}{1-x}\right)\left(\frac{\bar{L}}{\bar{G}}\right) + \left(\frac{y_1}{1-y_1}\right)$$

Which leads to, in Fig. 5.6,

$$Y = 32.49 + 0.053 \tag{4}$$

All the three lines defined by Eqs. (1), (3) and (4) are shown in Fig. 5.6.

5.2.3 Height of the Packed Absorption Tower

Another engineering interest is the height of the tower, where the minimum liquid flow rate is known. Referring to the same schematic diagram in Fig. 5.3, and the

Table 5.6 Packing data (Adapted from Bhatia [2])

Packing type	Material	Nominal size (in.)	Area per volume (m²/m³)
Berl saddle	Ceramic and porcelain	¼	899
		½	509
		1	259
		2	105
Intalox saddle	Plastic	1	207
		2	108
		3	89
	Ceramic	¼	984
		½	623
		1	256
		2	118
Raschin ring	Ceramic and porcelain	½	374
		1	190
		1½	118
		2	92
		3	62
Pall ring	Metal	5/8 × 0.018 thick	341
		1½ × 0.03 thick	128
Tellerette		1	180
		2	125
		3	98

two-film model (see Chap. 2), the mass transfer in the liquid phase within the elemental height, dz, in the tower is

$$d(Lx) = k_x(x_i - x)dA_i \quad (5.38)$$

where dA_i is the gas–liquid interfacial surface area over the height of dz, and it depends on the packing material property, a (m²/m³ of tower), which is the surface area per bulk volume in the tower.

Typical packing materials and the corresponding surface areas are summarized in Table 5.6 [2]. The packing materials are designed to maximize the gas-liquid interfacial area with low resistance to the flow. Nonetheless, these values are only for guide purpose only. Users are strongly recommended to get updated data from packing suppliers if needed.

Consider a tower with a cross-sectional area A_c filled with packing materials characterized with a (m²/m³). The interfacial surface area over the height of dz is

$$dA_i = aA_c dz \tag{5.39}$$

where A_c is the cross section area (m^2) of the tower that is normal to the flow direction. The liquid mole flow rate L (mole/s) is generally a variable and it is $L = \bar{L}/(1-x)$. Introducing this variable into Eq. (5.38) leads to

$$d\left(\frac{x}{1-x}\right) = \frac{k_x}{\bar{L}}(x_i - x)aA_c dz \tag{5.40}$$

It gives

$$dz = \frac{\bar{L}}{k_x a A_c (x_i - x)} d\left(\frac{x}{1-x}\right) = \frac{\bar{L}}{k_x a A_c (x_i - x)} \frac{dx}{(1-x)^2} \tag{5.41}$$

Manipulation of the equation gives

$$dz = \left(\frac{\bar{L}}{k_x a A_c}\right) \frac{dx}{(x_i - x)(1-x)} \tag{5.42}$$

Integration of the above equation leads to the total height of the tower

$$Z = \int_{x_1}^{x_0} \left(\frac{\bar{L}}{k_x a A_c}\right) \frac{dx}{(x_i - x)(1-x)} \tag{5.43}$$

Rigorously speaking, $\bar{L}/(k_x a A_c)$ is a variable in that the single phase mass transfer coefficients (k_x) increases as the liquid flows downward. For the simplicity of calculation, we can use the mean values of $\bar{L}/(k_x a A_c)$ and it can be treated as a constant. Then

$$Z = \frac{\bar{L}}{k_x a A_c} \int_{x_1}^{x_0} \frac{dx}{(x_i - x)(1-x)} \tag{5.44}$$

When the total mass transfer coefficient K_x is available, we can use the equilibrium mole fraction x^*, for calculation

$$Z = \frac{\bar{L}}{K_x a A_c} \int_{x_1}^{x_0} \frac{dx}{(x^* - x)(1-x)} \tag{5.45}$$

By similar analysis for the gas phase, we also can get

$$Z = \frac{\bar{G}}{k_y a A_c} \int_{y_1}^{y_0} \frac{dy}{(y - y_i)(1-y)} \tag{5.46}$$

$$Z = \frac{\bar{G}}{K_y a A_c} \int_{y_1}^{y_0} \frac{dy}{(y - y^*)(1 - y)} \tag{5.47}$$

Let us continue with our analysis by liquid phase. In design practice, the terms in front of the integration sign in Eq. (5.44) is often referred to as height of transfer unit (HTU) calculated using liquid phase mass transfer coefficient. For example, the HTU based on liquid phase mass transfer is

$$\text{HTU}_x = \frac{\bar{L}}{k_x a A_c} \tag{5.48}$$

The corresponding term by integration in Eq. (5.44), is called number of transfer unit (NTU),

$$\text{NTU}_x = \int_{x_1}^{x_0} \frac{dx}{(x_i - x)(1 - x)} \tag{5.49}$$

By the same approach we can get the HTU_y and NTU_y for gas phase. They all can be described using the overall mass transfer coefficients K_x, K_y as well. Nonetheless, the packed tower height is

$$H = \text{HTU} \times \text{NTU} \tag{5.50}$$

5.2.3.1 Packed Tower Diameter and Flooding Velocity

Body diameter is another important parameter of a packed tower. It is mainly limited by the gas velocity at which liquid droplets become entrained in the exiting gas stream.

$$D = \left(\frac{4Q}{\pi \bar{u}_g} \right)^{1/2} \tag{5.51}$$

where the diameter D is in m, Q is the volumetric gas flow rate in (m^3/s) and $\bar{u}_g$ is the mean gas face speed in m/s. When the gas flow rate reaches a point that the liquid is held in the void spaces between the packing materials, the corresponding gas-to-liquid ratio is termed as **loading point**. A further increase in gas flow rate (or gas velocity) will prevent the liquid from moving downward causing the liquid to fill up the void spaces in the packing. As a result, the gas–liquid interface surface area drops substantially and thereby the absorption efficiency decreases dramatically. And, the pressure drop increases greatly too. This condition is referred to as flooding, and the corresponding gas velocity is called flooding velocity. As a typical engineering practice, the diameter of a packed tower should enable the operation at 50–75 % of the flooding velocity.

A common and relatively simple procedure for estimating flooding velocity and minimum column diameter is to use a generalized flooding and pressure drop correlation. One version of the flooding and pressure drop relationship for a packed tower is the Sherwood Chart [12]. Readers are referred to literature for in-depth understanding on this topic in engineering design of a packed bed wet scrubber.

5.2.4 Chemical Absorption

In many engineering applications, the resistance in the liquid phase mass transfer is reduced by converting the dissolved gas into other materials. This reduction in the solute concentration in the liquid allows more gases to be absorbed at a much lower consumption of the liquid absorbent. For example, base solvents are used for the capture of acidic gases. The most common acid gases include sulfur dioxide (SO_2), hydrogen chloride (HCl), and hydrogen fluoride (HF). Nitric oxides and carbon dioxide formed in most combustion processes are also mildly acidic. Common alkalis include lime, soda ash, and sodium hydroxide. Sodium hydroxide is usually fed in solution. One classic example is de-SO_2 by spray of limestone or sodium hydroxide. The alkali requirements are usually calculated based on the quantities of acidic gases captured and the molar ratios necessary for the corresponding chemical reactions.

Consider the packed-bed wet scrubber again and assume that the fresh liquid solvent contains little dissolved gas of concern. At steady state, the mass transfer rate (in mole/s) within the liquid phase is

$$N = k_x aV(x_i - x_{ss}) \tag{5.52}$$

where V the total reactor volume (m^3) and aV together is the interfacial contact area for mass transfer (m^2); x and x_{ss} are the mole fraction of the target gas at gas–liquid interface and that in the bulk liquid at steady state, respectively.

The target gas transferred into the liquid is either physically stored in the liquid or consumed by chemical reactions:

$$N = \bar{L}\left(\frac{x_{ss}}{1 - x_{ss}}\right) + kx_{ss} \tag{5.53}$$

Again $\bar{L}$ is the mole flow rate (in mole/s) of the solute free liquid entering the tower. k (mole/s) is the chemical reaction coefficient corresponding to x_{ss}. When k is available based on the mass concentration c_{ss} rather than x_{ss}, a conversion between units is needed. The first term on the right-hand side of Eq. (5.53) stands for amount of physically dissolved gas; the last term for chemically absorbed gas.

Combine Eq. (5.52) and (5.53), we have:

$$N = k_x aV(x_i - x_{ss}) = \bar{L}\left(\frac{x_{ss}}{1 - x_{ss}}\right) + kx_{ss} \tag{5.54}$$

For cases where $x_{ss} \ll 1$, $1 - x_{ss} \approx 1$, then

$$k_y aV(y_i - y_{ss}) = \bar{L}y_{ss} + ky_{ss} \tag{5.55}$$

Solving this equation we can get the steady state concentration of the target gas in the liquid phase

$$\frac{x_{ss}}{x_i} = \frac{k_x aV}{k_x aV + \bar{L} + k} \tag{5.56}$$

This equation indicates that there are three factors that affect the steady state absorption ratio, which is defined as x_{ss}/x_i, and they are $k_x aV, \bar{L}$ and k. They stand for the effects of interfacial mass transfer, liquid flow rate, and kinetic rate of chemical reaction, respectively. Practically, it is challenging to determine the mole fraction at the interface, x_i, although it could be estimated by extensive theoretical analysis.

5.2.4.1 Enhanced Absorption Factor, e

A more practical approach to this problem is to employ an enhanced absorption factor, e. It is defined as the ratio of extra amount of target gases absorbed into the liquid by chemical absorption to that by physical absorptions.

$$e = \frac{x'}{x} \tag{5.57}$$

where x' stands for the extra absorption resulted from chemical absorption. The theoretical enhanced absorption factor could be very high, but the actual value depends on the design and operation of the tower. Then, with chemical absorption considered, Eq. (5.19) becomes

$$yG - y_1 G_1 = (x + x')L - x_1 L_1 = x(1 + e)L - x_1 L_1 \tag{5.58}$$

It indicates that the amount of liquid flow rate is decreased by a factor of $(1 + e)$. Then all results obtained by the analysis for physical absorption can be applied to chemical absorption, by multiply $\bar{L}$ with a factor of $(1 + e)$. For example, Eq. (5.21) leads to

$$\frac{y}{1-y} - \frac{y_1}{1-y_1} = \left(\frac{x}{1-x} - \frac{x_1}{1-x_1}\right)\left(\frac{\bar{L}}{\bar{G}}\right)(1+\mathrm{e}) \tag{5.59}$$

This equation also indicates that the amount of liquid to gas ratio is reduced by enhanced chemical absorption.

5.3 Practice Problems

1. The concentration of acetone (C_3H_6O) in a machine shop is 3 ppmv. The density of acetone vapor is 2.01 kg/m^3 under standard conditions. A carbon filter bed is used for air cleaning. The airflow rate through the filter is 0.05 m^3/s. Determine

   a. The adsorption capacity of the activated carbon filter using Yaws data (1995).
   b. The total amount of carbon needed for the bed, assuming the working adsorption capacity is 75 % of the saturated adsorption and the bed service life is 1 year.

2. In a machine shop, the concentration of acetone (C_3H_6O) in the room air is 0.00001 kg/m^3. An activated carbon filter is used in an air recirculation system to remove the acetone. The bulk density of the activated carbon is 400 kg/m$^{3.}$ The airflow rate of the recirculation system is 20 l/s. The cross-sectional area of the bed is 1.5 m^2. Determine

   a. The speed of the adsorption wave
   b. The length of the bed if the bed is to be replaced every 30 days.

3. A carbon adsorption unit is designed to control the emission from an air stream having 250 ppmv of toluene flowing through the bed at a rate of 1,500 m^3/min. The process works 24 h/day at 27 °C temperature and 1 atm pressure. Assume the working capacity of the carbon is 0.15 kg toluene/kg of carbon. Determine the amount of carbon needed in all beds of a 4-bed adsorber if two beds are online for 4 h and then regenerated while the other two working.
4. A carbon adsorption unit is designed to control the emission from an air stream having 250 ppmv of toluene is flowing through the bed at a rate of 1,500 m^3/min. Calculate the cross sectional area of a bed if maximum allowable superficial velocity through the bed is 20 m/min.
5. A carbon adsorption unit is designed to control the emission from an air stream having 2,500 ppmv of ethylene glycol vapor is flowing through the bed at a rate of 1,500 m^3/min at a temperature of 27 °C and pressure of 1 atm. Calculate the working adsorption capacity of the carbon. Assume that the working adsorption capacity of the bed is 50 % of the adsorption capacity, and that the bed is filled

with 200 kg of activated carbon. The expected breakthrough efficiency is 90 %. Estimate its breakthrough time. Assume $K_x = 20\ s^{-1}$ and packing density of 400 kg/m^3.

6. An activated carbon column is used to remove tribromomethane (2.89 g/cm^3) from the air at standard condition. The mass flow rate of air is 2.5 kg/s and the concentration of the pollutant in the inlet air stream is 0.0025 kg/m^3. The bulk density of the activated carbon is 500 kg/m^3, and the mass transfer coefficient of the adsorbent $K_x = 25\ s^{-1}$. The adsorption bed is 2 m deep and 1.5 m^2 in cross-sectional area. The designed adsorption efficiency is $\eta_x = 0.95$. That is, the breakthrough point of the bed is considered to be when the outlet concentration reaches 5 % of the inlet concentration. Estimate the life time of this column.
7. A packed bed is designed to remove SO_2 by pure water absorption from a sulfuric acid plant. The incoming SO_2 concentration is 10 %. The water flow rate is 1.5 times the minimum water flow rate and the inert gas flow rate $\bar{G} = 500$ kg/hr. The tower operates at an average temperature of 30 °C and 1 atm. The equilibrium data for SO_2 in air and water at this temperature.

Partial pressure p_{SO_2} (mmHg)	Solubility c_{SO_2} (g SO_2/100 g water)
0.6	0.02
1.7	0.05
4.7	0.10
8.1	0.15
11.8	0.20
19.7	0.30
36.0	0.50
52.0	0.70
79.0	1.00

If the corresponding mass transfer coefficients are

$$k_y a = 0.6634 L^{0.82}$$

$$k_x a = 0.09944 L^{0.25} G^{0.70}$$

where L and G are the liquid and gas flow rates in kg/(m^{2}h)

(a) determine minimum water flow rate,
(b) the equilibrium line,
(c) the operating line is the same figure,
(d) the gas phase mass transfer coefficient at the gas inlet,
(e) the gas phase mass transfer coefficient at the gas outlet,
(f) the liquid phase mass transfer coefficient at the liquid inlet,
(g) the liquid phase mass transfer coefficient at the liquid outlet.

References and Further Readings

1. Albright L (2008) Albright's chemical engineering handbook. CRC Press, USA
2. Bhatia MV (1977) Packed tower and absorption design. In: Cheremisinoff PN, Young RA (eds) Air pollution control and design handbook. Marcel Dekker, New York
3. Cal M, Rood M, Larson S (1997) Gas phase adsorption of volatile organic compounds and water vapor on activated carbon cloth. Energy Fuels 11:311–315
4. Fair JR (1969) Sorption processes for gas separation. Chem Eng 76(15):90–110
5. Fair JR, Steinmeyer MA, Penney WR, Crocker BB (2008) Gas absorption and gas-liquid system design. In: Green DW, Maloney JO (eds) Perry's chemical engineers' handbook, 8th edn. McGraw-Hill, New York (Section 14)
6. Green DW, Maloney JO (eds) (2008) Perry's chemical engineers' handbook, 8th edn. McGraw-Hill, New York
7. Hsi H-C, Rood M, Rostam-Abadi M, Chang Y-M (2013) Effects of sulfur, nitric acid, and thermal treatments on the properties and mercury adsorption of activated carbons from bituminous coals. Aerosol Air Qual Res 13:730–738
8. Jonas LA, Rehrmann JA (1972) The kinetics of adsorption of organo-phosphorus vapors from air mistures by activated carbons. Carbon 10:657–663
9. Lewis W, Whitman W (1924) Principles of gas absorption. Ind Eng Chem 16(12):1215–1220
10. Mantell CL (1951) Adsorption, 2nd edn. New York, McGraw-Hill
11. Reed BE, Matsumoto MR (1993) Modeling cadmium adsorption by activated carbon using the Langmuir and Freundlich isotherm expressions. Sep Sci Technol 28:2179–2195
12. Sherwood TK, Shipley GH, Holloway FA (1938) Ind Eng Chem 30:765–769
13. Turk A (1978) Adsorption. In: Stern A (ed) Air pollution, 3rd edn, vol IV. Academic Press, New York, pp 329–363
14. Vahdat N (1997) Theoretical study of the performance of activated carbon in the presence of binary vapor mixtures. Carbon 35:1545–1557
15. Wark K, Warner CF, Davis WT (1998) Air pollution. Addison–Wesley Longman, Berkeley, pp 220–277
16. Yaws CL, Bu L, Nijhawan S (1995) Determining VOC adsorption capacity. Pollut Eng 27 (2):34–37. http://www.pollutionengineering.com/
17. Young DM, Crowell AD (1962) Physical adsorption of gases. Butterworth, London
18. Flagan RC, Seinfeld JH (2012) Fundamentals of air pollution engineering. Dover Press, New York, USA

Chapter 6
Separation of Particles from a Gas

As the counterpart of Chap. 5, this chapter covers the basics behind various particulate emission control devices. It starts with a general introduction to particle separation efficiency followed by systematic introduction to the principles for gravity settling chambers, centrifugal separators (cyclones), electrostatic precipitators, and filters.

6.1 General Consideration

Particle separation is a critical step in many energy and environmental engineering applications. In addition to the reduction of particulate emission before a flue gas is discharged to atmosphere, particle separation is an important step for alternative fuel development. For example, in order to clean gaseous fuel from gasification, heavy gas molecules have to be removed from the stream using membranes. However, this cannot be achieved without the removal of particulate matter from the gas stream, otherwise the membrane will lose its function by clogging.

In general, a particle can be separated from its carrier gas by gravitational settling, cyclonic separation, filtration, wet scrubbing, thermal force separation, and electrical separators like electrostatic precipitators. The thermal force separators work in principle, but it is not effective in handling large volume of gas flow or for large particles [15], therefore, it will not be introduced in this book.

The performance of a particle separation device can be quantified by the following three main parameters: pressure drop, capability, and most importantly, efficiency.

6.1.1 Particle Separation Efficiency

The efficiency of a particle separator can be described by grade efficiency curve, which gives the separation efficiency as function of particle size. It can also be quantified by total efficiency.

Z. Tan, *Air Pollution and Greenhouse Gases*, Green Energy and Technology,
DOI 10.1007/978-981-287-212-8_6

Consider a particle separation device, which could be any of the devices to be introduced shortly. N_i and N_o are the numbers of particles with size d_p before and after the device, respectively. The total amount of particles collected by the device is $N_c = N_i - N_o$. Then the separation efficiency for particles with a size d_p is defined as the

$$\eta(d_p) = \frac{N_c}{N_i} = 1 - \frac{N_o}{N_i}. \tag{6.1}$$

The performance of a particle separation device can also be described with a penetration efficiency (P).

$$P(d_p) = \frac{N_i}{N_o}. \tag{6.2}$$

Obviously, the relationship between P and η is defined as

$$\eta = 1 - P \tag{6.3}$$

$\eta(d_p)$ represents the efficiency for the particles having same diameter, d_p. It is also referred to as grade collection efficiency.

An important parameter in fractional efficiency is the so-called "cut size", d_{50}, which is the particle size for which the separation efficiency is 50 %.

In air pollutant control, we deal with polydisperse particles. It leads to another term called total efficiency. The total efficiency by considering all the particles is

$$\eta = \frac{\int_0^\infty N_o(d_p)dd_p}{\int_0^\infty N_i(d_p)dd_p} = \frac{\int_0^\infty \eta_p(d_p)N_i(d_p)dd_p}{\int_0^\infty N_i(d_p)dd_p} \tag{6.4}$$

When the particle size distribution is measured using discrete data, it can be estimated by

$$\eta = \frac{\sum \eta(d_{pi})N_i(d_{pi})}{\sum N_i(d_{pi})} \tag{6.5}$$

If the particle density m is assumed to be the same before and after the device and all the particles with the same size have the same mass, we can replace number N with mass m; Eq. (6.4) becomes

$$\eta = \frac{\int_0^\infty \mathrm{m}_{po}(d_p)d(d_p)}{\int_0^\infty \mathrm{m}_{pi}(d_p)d(d_p)} = \frac{\int_0^\infty \eta(d_p)\mathrm{m}_{pi}(d_p)d(d_p)}{\int_0^\infty \mathrm{m}_{pi}(d_p)d(d_p)} \tag{6.6}$$

Similarly, when the particle size distribution is measured using discrete data, Eq. (6.5) can be rewritten as

$$\eta = \frac{\sum_i \eta(d_{pi}) m_{pi}(d_{pi})}{\sum_i m_{pi}(d_{pi})} \tag{6.7}$$

With the fractional efficiency curve of a device determined, the total efficiency for polydisperse particles can be calculated using above equation. In the following analyses, only fractional efficiency will be introduced to avoid duplication of work.

6.1.2 Particle Separation Efficiency of Multiple Devices

In engineering applications, usually more than one unit is employed in order to achieve high efficiency or to handle a great amount of air flow. The former is achieved by connecting more than one unit in serial and the latter in parallel. Consider k identical devices arranged in serial, the number of particles entering the ith unit is the same as that penetrating through the $(i - 1)$th unit. Then the penetration through all the k units is

$$P = \Pi P_i = P_1 P_2 \ldots P_k \tag{6.8}$$

And the corresponding efficiency is

$$\eta = 1 - P = 1 - P_1 P_2 \ldots P_k \tag{6.9}$$

In this analysis, we actually made a critical assumption that the particle separation efficiencies of the identical units are the same. In reality, it is actually invalid because the particle separation efficiency of a unit depends on the incoming particle concentration, which keeps decreasing when the units are connected in serial. Therefore, careful interpretation of this equation should be executed.

Example 6.1: General particle separation efficiency
A filter has an efficiency of 85 %. What is the total efficiency if two of them working in serial.

Solution
With a single filter efficiency of 85 %, the corresponding penetration efficiency is 15 %. For two filters in serial, the total efficiency is thereby

$$\eta = 1 - P_1 P_2 = 1 - 0.15 \times 0.15 = 0.9775$$

The total filtration efficiency is 97.75 %.

6.2 Gravity Settling Chambers

A gravity settling chamber was mainly used for the separation of large particles from the air stream. In practice, a gravity settling chamber is only effective for particles with a diameter of 50 μm or larger. As seen in Fig. 6.1, it classifies the particles by gravitational force. A consequential drawback of these devices is the large footprint. Depending on the gas velocity, a chamber can be designed to operate at laminar flow or turbulent flow.

6.2.1 Laminar Flow Model

Consider a gravity settling chamber with dimensions of height H, length, L and width W. A particle having diameter d_p enters the chamber as shown in Fig. 6.2. Under laminar flow condition, the trajectory of any particle should be a straight line. As it moves from left to the right together with the airflow at a speed of U, it settles at a terminal speed of v_{TS}. The concept of terminal settling speed was introduced above in Sect. 4.2.1.

If the particle barely touches the lower right edge of the chamber when it enters the chamber at a height H_c, then all the particles of the same size entering the chamber above H_c will penetrate through the chamber; those below H_c will be

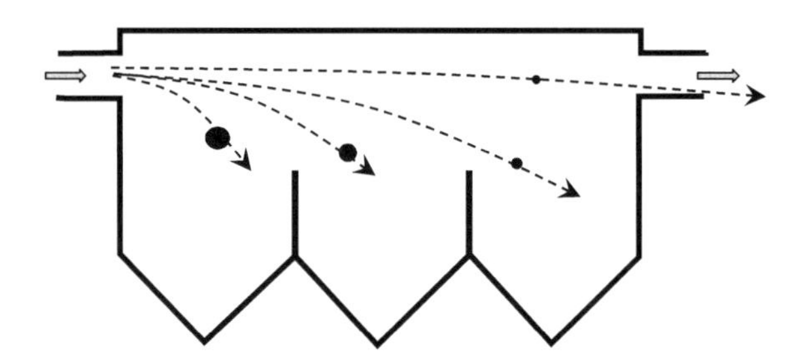

Fig. 6.1 A schematic diagram of a gravity settling chamber

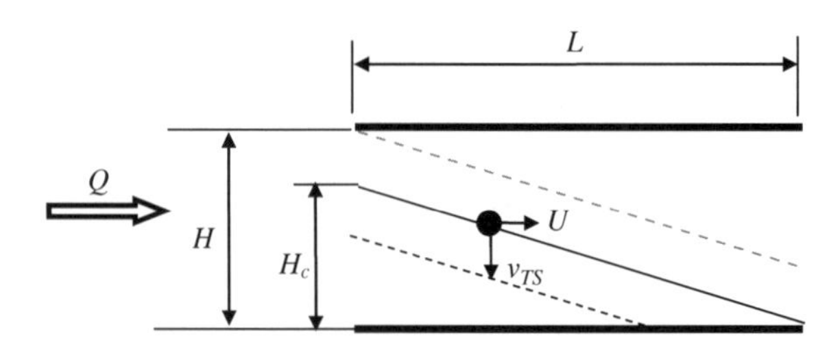

Fig. 6.2 A laminar flow gravity settling model

collected on the surface of the lower plate. This height H_c can be referred to as the critical height. Then the fractional efficiency of the chamber for this group of particles (d_p) is

$$\eta(d_p) = \frac{H_c}{H} \tag{6.10}$$

The time for a particle entering the chamber at the critical height, H_c, falls down through a vertical distance H_c while traveling a horizontal distance L, is

$$t = \frac{L}{U} = \frac{H_c}{v_{\mathrm{TS}}} \tag{6.11}$$

It leads to

$$H_c = \frac{v_{\mathrm{TS}} L}{U} \tag{6.12}$$

Substitute Eq. (6.12) into the efficiency Eq. (6.10) above, and we have

$$\eta(d_p) = \frac{v_{\mathrm{TS}} L}{UH} \tag{6.13}$$

where the gas incoming speed can be determined from the flow rate of the air passing through the chamber,

$$U = \frac{Q}{WH} \tag{6.14}$$

where W is the width of the chamber. The terminal settling speed of a spherical particle falling in a gravitational field is

$$v_{\mathrm{TS}} = \frac{\rho_p d_p^2 g C_c}{18\,\mu} \tag{6.15}$$

Therefore, the fractional particle separation efficiency can be described as

$$\eta(d_p) = \frac{\rho_p g d_p^2 C_c}{18\,\mu} \frac{LW}{Q} \tag{6.16}$$

6.2.2 Turbulent Flow Model

For turbulent flow model, we assume that the particles and the air are completely mixed at any cross section that is normal to the direction of airflow. Consider an element with an infinitesimal length dx. The concentration of particles within which is assumed to be uniform and the particles settle down at a terminal velocity of v_{TS}.

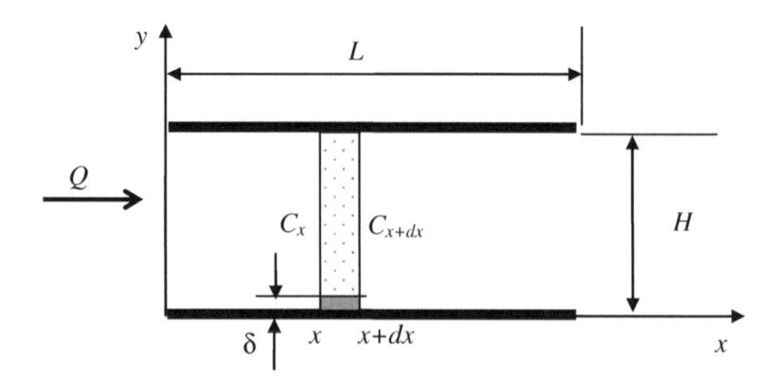

Fig. 6.3 A turbulent flow model of gravity settling chamber

During an infinitesimal period of time dt, the particles at the bottom of the chamber within a distance $\delta = v_{\mathrm{TS}}\,\mathrm{d}t$ above the lower collecting plate are considered collected since only these particles can reach the surface (shaded area in the Fig. 6.3).

Then the amount of particles entering the element defined by dx equals to the total depositing on the bottom surface and that penetrating through the element, i.e.,

$$C_x UWH = Cv_{\mathrm{TS}}W(\mathrm{d}x) + C_{x+\mathrm{d}x}UWH \tag{6.17}$$

The ratio of the amount of settled particles to the total amount of particles that enter the element defined by dx equals to the shaded area over the total elemental area, which gives

$$\frac{\mathrm{d}C}{C} = -\frac{v_{\mathrm{TS}}}{H}\mathrm{d}t \tag{6.18}$$

The negative sign indicates the decreasing particle number concentration along x-direction.

Substitute the elemental residence time, $\mathrm{d}t = \mathrm{d}x/U$, into the above equation, and we have,

$$\frac{\mathrm{d}C}{C} = -\frac{v_{\mathrm{TS}}}{HU}\mathrm{d}x \tag{6.19}$$

Integration of both sides leads to

$$\int_{C_i}^{C_o}\frac{\mathrm{d}C}{C} = -\int_0^L\frac{v_{\mathrm{TS}}}{HU}\mathrm{d}x \tag{6.20}$$

We can get the penetration efficiency of the particles through the chamber

$$P = \exp\left(-\frac{v_{\mathrm{TS}}L}{HU}\right) \tag{6.21}$$

The corresponding particle separation efficiency is thereby

$$\eta(d_p) = 1 - \exp\left(-\frac{v_{TS}L}{HU}\right) = 1 - \exp\left(-\frac{v_{TS}LW}{Q}\right) \tag{6.22}$$

It is important to note that in the analysis above, we assumed that a particle is collected and stays on the collection surface once it reaches there. This is actually more applicable to a sticky particle than a hard bumpy one. Particle bouncing and resuspension, also referred to as re-entrainment, introduced in Chap. 4, can significantly reduce the particle separation efficiency. Unfortunately, there is very limited knowledge about particle re-entrainment in particle separation due to its extreme complexity. terminal precipitating velocity,Therefore, the analytical formulae above, as those to come for other technologies, can only be used for guidance only.

Example 6.2: Gravity settling chamber efficiency
Consider a gravity settling chamber that is 1-m wide (W = 1 m) and 1-m high (H = 1 m). Air flow rate is 1 m^3/s (=3,600 m^3/h) and assume laminar flow within the chamber. Estimate its separation efficiency versus aerodynamic diameter under standard ambient condition.

Solution
Using Eq. (6.16), we can calculate the fraction efficiency for different particle size as follows:

d_p (μm)	$\eta(d_p) = \frac{\rho_p g d_p^2 C_c}{18\mu}\frac{LW}{Q}$ (%)
10	0.03
100	3
150	7
200	12
250	19
350	37
500	75
575	100

The fractional efficiency of a gravitational settling chamber is so low that it can no longer meet more and more stringent emission control requirements. As a result, there has been a sharp decline in the use of gravity settling chamber, although there are still a few of them in commercial use. However, similar analysis applies to electrostatic precipitator and, to a certain degree, to cyclone, which are introduced as follows.

6.3 Electrostatic Precipitation

The model analysis of electrostatic precipitation is very similar to that for the gravity settling chamber as discussed in Sect. 6.2 except that the driving force is now not gravitational but electrical. And the electrical field is arranged horizontal rather than vertical. Consider a flow through a pair of vertical plates H apart from each other with length L and depth b into the paper.

By replacing the gravitational settling velocity in Eq. (6.16) above with the terminal precipitating speed V_E, we can get the equation for laminar condition as follows:

$$\eta(d_p) = \frac{V_E L}{UH} = \frac{V_E L b}{UHb} = \frac{V_E A}{Q} \quad \text{(Laminar)} \tag{6.23}$$

where A is the area of one plate collecting particles.

Following the similar analysis for gravitational setting chamber, we can also get the efficiency for complete mixing condition.

$$\eta(d_p) = 1 - \exp\left(-\frac{V_E A}{Q}\right) \quad \text{(Turbulent)} \tag{6.24}$$

where the terminal precipitating speed, V_E, in the electrical force field can be determined by equating the electrical force on the particles and the drag force,

$$qE = \frac{3\pi V_E d_p \mu}{C_c} \tag{6.25}$$

where q is the charge carried by the particles (columns) and E is the electric field intensity (V/m). This equation leads to

$$V_E = \frac{qEC_c}{3\pi d_p \mu} \tag{6.26}$$

Equation (6.26) shows that the precipitation speed of a particle depends on the charge carried by the particle, q, and the strength of the electrical field, E. They are determined as follows.

6.3.1 The Electric Field Intensity

The intensity of an electric field, E, is determined by the electrode geometry and the voltage difference that is applied between the electrodes.

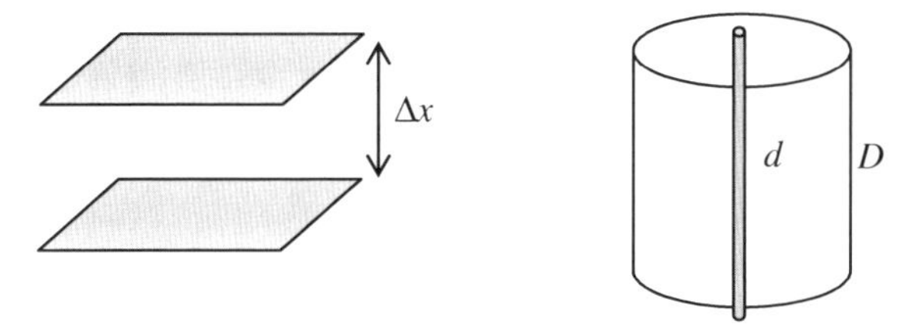

Fig. 6.4 Two typical electrode configurations

$$E = -\nabla V \tag{6.27}$$

where E is the electric field intensity in (V/m), V is the voltage. The exact form of the electric field depends on the configuration of the electrodes.

Two typical electrode configurations are shown in Fig. 6.4. One is parallel plates and another wire-tube. The difference in voltage between the two electrodes is V. Then the electrical field intensity between the two parallel plates is uniform and it is

$$E = \frac{V}{\Delta x} \tag{6.28}$$

However, for the wire-tube type, the electrical field intensity is a function of radial position, r,

$$E(r) = \frac{V}{r\ln(D/d)} \tag{6.29}$$

where d = diameter of the wire, D = diameter of the tube, and r = radial position, and $d/2 < r < D/2$

In reality, the electric field created by the electrode system may also be affected by the presence of electrons, ions, and other charged particles in the gas stream. This alters the electric field strength especially near the collection electrode.

6.3.2 Particle Charging

The success of ESP operation depends primarily on the charging of the particles. There are many ways to charge airborne particles, but only corona discharge can generate sufficient amount of ions for industrial electrostatic precipitators. Corona discharge is accomplished by applying high voltages in the order of kV on the discharge electrodes and grounding the collector plates. When the electric field intensity is greater than the electric breakdown intensity (typically about 30 kV/cm for ambient air), ions such as N^{2+} and O^{2+} and electrons, e^-, are produced at the electrode.

The particles can be charged by the ions generated by a corona discharger. There are two distinctive charging mechanisms, one being diffusive charging and another field charging. For either one, there is a saturation of charging because a particle can carry only certain amount of ions. With more and more ions charged on the particle, they also create another electric field preventing more ions from coming closer to the particle.

Ions charged to a particle can be positive, negative, or both. Depending on the polarity of the ions, the charging process is defined as unipolar or bipolar charging. Unipolar charging is much more effective than bipolar charging and thereby widely employed in industrial ESPs. Although there is not much difference between the effectiveness of positive and negative charging processes, positive charging will generate ozone, which is considered as secondary air pollutant. Therefore, negative charging is preferred and widely used.

6.3.2.1 Diffusive Charging

Airborne ions share the thermal energy of the gas molecules and obey the same law of kinetics theory. Diffusion of the ions in the air may result in collisions between the ions and the airborne particles, and thereby the attachment between the particles and the ions. This process is referred to as diffusive charging.

Consider an ion that is approaching a particle already being charged with n ions. The potential energy of the air with a distance r away from the particle is

$$P = \frac{K_E n e^2}{r} \tag{6.30}$$

where $K_E = 9 \times 10^9$ Nm^2/C^2 is a force constant.

According to White (1951), the spatial distribution of the concentration of airborne ions in a potential field is

$$N_i(r) = N_{i0}\exp\left(-\frac{P}{kT}\right) \tag{6.31}$$

where N_{i0} = ion concentration in the charging zone, k = the Boltzmann constant, T = absolute temperature in K, and P = potential energy in J.

Substitute Eq. (6.30) into (6.31), ion concentration near the particle at a radial distance of r becomes

$$N_i(r) = N_{i0}\exp\left(-\frac{K_E n e^2}{r}\frac{1}{kT}\right) \tag{6.32}$$

At the surface of the particle, where $r = d_p/2$, Eq. (6.32) leads to,

$$N_i(d_p/2) = N_{i0}\exp\left(-\frac{2K_E ne^2}{d_p kT}\right) \tag{6.33}$$

The number flux of ions can be determined by assuming that all the ions are captured when they strike the particle.

At $r = d_p/2$, the number of ions loss from the air to the surface of the particle is then described as

$$\frac{dn}{dt} = \frac{N_{i0}\bar{c}_i \pi d_p^2}{4}\exp\left(-\frac{2K_E ne^2}{d_p kT}\right) \tag{6.34}$$

where n is the number of ions moved from air to the surface of the particle, $\bar{c}_i$ = mean thermal speed of the ions and $(\pi d_p^2/4)$ stands for the surface area of the particle. Integration of the above equation leads the number of ions charged on the particles at time t

$$n(t) = \frac{d_p kT}{2e^2 K_E}\ln\left(1 + \frac{d_p K_E \bar{c}_i \pi e^2 N_{i0}}{2kT}t\right) \tag{6.35}$$

where $\bar{c}_i$ is the mean thermal speed of ions. Under normal condition, the mean thermal speed of ion $\bar{c}_i$ is about 239 m/s [41]. k is Boltzmann constant (1.38×10^{-23} J/K), $K_E = 9 \times 10^9$ Nm^2/C^2, and N_{i0} is ion concentration.

6.3.2.2 Field Charging

With the presence of electric field, ions are forced to move along the direction of the electric field. Leading to a high rate of collision between the ions and the particles. This is referred to as field charging mechanism. The number of ions charged to a particle by field charging depends on the properties of the particle, its size, and the intensity of the electric field, E:

$$n(t) = \frac{Ed_p^2}{4eK_E}\left(\frac{3\varepsilon_r}{2+\varepsilon_r}\right)\left(\frac{t}{t+\tau}\right) \tag{6.36}$$

where E = intensity of the electric field with a typical value of 10^6 V/m, ε_r = relative permittivity or dielectric constant of the particle with respect to vacuum, and $\varepsilon_r = \varepsilon/\varepsilon_0$; $\varepsilon_0 = 8.854 \times 10^{-12}$ C/Vm is the permittivity of a vacuum. The permittivity of typical particles can be found in handbooks. τ is the charging constant and it varies with the field condition.

$$\tau = \frac{1}{\pi N_{i0} K_E e Z_e} \tag{6.37}$$

where Z_e = mobility of the ions, and the average value of the mobility of air ions is about 1.5×10^{-4} m^2/V s. A typical charging constant is $\tau = 0.01 - 0.1$ s using the above equation.

The number of ions that is eventually charged to a particle depends on charging time, the concentration of ions in the charging zone, and the electric mobility of these ions, which determines the moving speed of the ions in response to the electric field E. In a typical industrial application, the particle residence time t is of the order of 10 s. Therefore, $t \gg \tau$ and we can assume maximum field charging, which is from Eq. (6.36).

$$n_{\max} = \frac{Ed_p^2}{4eK_E}\left(\frac{3\varepsilon_r}{2+\varepsilon_r}\right) \tag{6.38}$$

6.3.2.3 Combined Charging

Combined charging takes into consideration both charging mechanisms. The total number of ions charged to a particle is calculated using

$$n(t) = \frac{d_p kT}{2e^2 K_E}\ln\left(1 + \frac{d_p K_E \bar{c}_i \pi e^2 N_{i0}}{2kT}t\right) + \frac{Ed_p^2}{4eK_E}\left(\frac{3\varepsilon_r}{2+\varepsilon_r}\right)\left(\frac{t}{t+\tau}\right) \tag{6.39}$$

On the right-hand side of the equation, the first term is for the effect of diffusive charging and the last for field charging. The relative importance of diffusive and field charging depends on the size of the particles to be charged. Since both diffusive and field charging are a function of d_p^2, which indicates that it is not the aerodynamic diameter but rather the actual surface area that affects the charging effect.

Figure 6.5 shows the relative importance of diffusive and field charging; calculation was based on the following parameters:

$$\begin{array}{lll} K_E = 9\times10^9\,\mathrm{Nm^2/C^2} & Z_e = 1.50\times10^4\,\mathrm{m^2/Vs} & N_{io} = 5.00\times10^{14}\,\mathrm{ion/m^3} \\ e = 1.60\times10^{-19}\,\mathrm{C} & \bar{c}_i = 240\,\mathrm{m/s} & k = 1.38\times10^{-23}\,\mathrm{J/K} \\ T = 293\,\mathrm{K} & E = 1.0\times10^6\,\mathrm{V/m} & t = 0.1\,\mathrm{s},\quad \varepsilon_r = 1 \end{array}$$

It shows that diffusion charging is an important mechanism for particles smaller than 200 nm in diameter, whereas field charging dominates for particles larger than 2 μm.

In order to implement the charging mechanisms, diffusive, field, or both, there have to be enough ions generated. Among the ion generation technologies, corona discharge has been believed to be the most effective in producing sufficient ions. It has been well known that ozone and aerosol particles, especially smaller ones, are

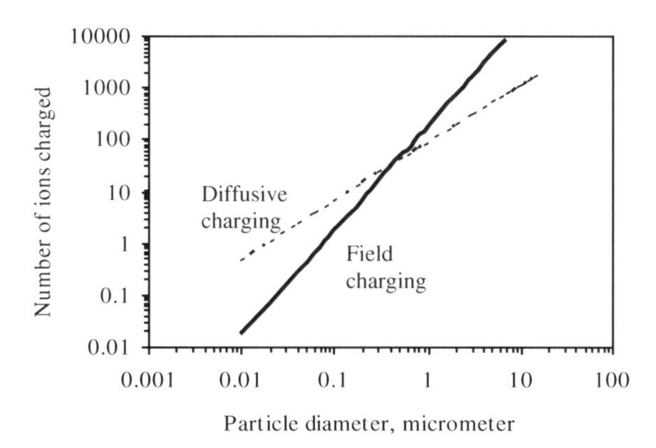

Fig. 6.5 Comparison of field and diffusive charging of particles

produced by corona discharge, and they are considered as secondary air pollutants. More information about nanoparticle generation in corona charger can be found in Chap. 14.

6.4 Cyclone

A cyclone separates particles from the air by "centrifugal force". As depicted in Fig. 6.6, particles and air enter the device from the side. When the air changes direction following the curvature of the device, particles tend to remain the same flow direction due to inertia. As a result, the particles move along radial direction with respect to the air. This is commonly referred to as centrifugal force, which is actually conceptually incorrect.

Figure 6.6 shows the top view of a cyclone body, the height and width of the rectangular inlet are H and W, respectively, and radii of the outer and inner tubes are

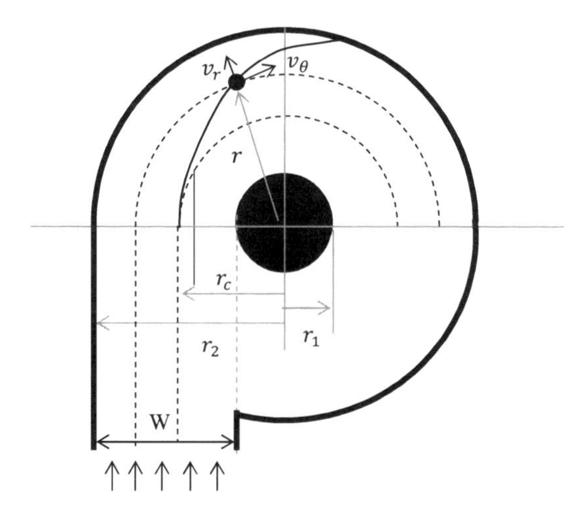

Fig. 6.6 Particle motion in rotating fluids

r_2 and r_1, respectively. Assume there is no leakage and the gas is incompressible, the flow rate Q is constant from the inlet through the annular chamber and the outlet.

6.4.1 Cyclone Fractional Efficiency

Assume the airflow within the cyclone is laminar. There is a critical radius, r_c, where a particle to be located at this position eventually reaches the inner surface of the outer body of the cyclone. Then, similar to the analysis for gravity settling chamber and electrostatic precipitator, the fractional particle separation efficiency is

$$\eta(d_p) = \frac{r_2 - r_c}{r_2 - r_1} = \frac{r_2 - r_c}{W} \tag{6.40}$$

At steady state, the average tangential velocity of the airflow is

$$\bar{u}_g = \frac{Q}{WH} = \frac{Q}{(r_2 - r_1)H} \tag{6.41}$$

With the laminar flow assumption, the airflow does not mix along the radial direction. The radial component of velocity of the particles can be derived from Newton's Second Law,

$$m_p \frac{\mathrm{d}v_r}{\mathrm{d}t} = F_C - F_D = 0 \tag{6.42}$$

where v_r is the radial speed of the particle. Assume a spherical particle with a density ρ_p and a diameter of d_p, the centrifugal force, F_C, and the drag force F_D exerted on the particle at any position r are, respectively

$$F_C = m_p \frac{v_\theta^2}{r} = \frac{\rho_p \pi d_p^3}{6} \frac{v_\theta^2}{r} \tag{6.43}$$

$$F_D = 8\,\pi \mu d_p v_r \tag{6.44}$$

In this equation, we ignored the Cunningham coefficient because cyclones are used primarily for separating particles with large sizes. In the Stokes region, $F_C = F_D$, and it leads to

$$v_r = \frac{\rho_p d_p^2 v_\theta^2}{18 \mu r} \tag{6.45}$$

It is commonly assumed that the particle follows air stream along tangential direction, that is, $v_\theta = u_\theta$. However, the exact description of air tangential speed u_θ depends on the researcher.

6.4.1.1 Crawford Model

Crawford [6] derived a formula by applying fluid dynamics to the air phase, which leads to

$$v_\theta = u_\theta = \frac{Q}{H\ln(r_2/r_1)}\frac{1}{r} \tag{6.46}$$

Substituting Eq. (6.46) into (6.45) leads to

$$v_r = \frac{\rho_p d_p^2}{18\mu}\left[\frac{Q}{H\ln(r_2/r_1)}\right]^2\frac{1}{r^3} \tag{6.47}$$

Over an infinitesimal period of time, dt, the particle moves outward along radial direction a distance of d$r = v_r$dt and an arc length along the tangential direction, rd$\theta = v_\theta$dt. For the same dt

$$\mathrm{d}t = \frac{\mathrm{d}r}{v_r} = \frac{r\mathrm{d}\theta}{v_\theta} \tag{6.48}$$

It leads to

$$\frac{\mathrm{d}r}{r\mathrm{d}\theta} = \frac{v_r}{v_\theta} \tag{6.49}$$

Substituting Eqs. (6.46) and (6.45) into (6.49) leads to

$$\frac{r\mathrm{d}r}{\mathrm{d}\theta} = \frac{\rho_p d_p^2}{18\mu}\frac{Q}{H\ln(r_2/r_1)} \tag{6.50}$$

where the right-hand side of this equation is constant for fixed particle size and cyclone configuration. For a particle entering the cyclone at $r = r_c$ when $\theta = 0$, its radial position in the cyclone is defined by

$$r^2 - r_c^2 = \left[\frac{\rho_p d_p^2}{9\mu}\frac{Q}{H\ln(r_2/r_1)}\right]\theta \tag{6.51}$$

When the particle reaches the collecting wall, $r = r_2$, and the corresponding angle θ_2 is determined by

$$r_2^2 - r_c^2 = \left[\frac{\rho_p d_p^2}{9\mu}\frac{Q}{H\ln(r_2/r_1)}\right]\theta_2 \tag{6.52}$$

Substituting r_c back into Eq. (6.40) leads to the description of the fractional efficiency as

$$\eta(d_p) = \frac{r_2 - \left[r_2^2 - \frac{\rho_p d_p^2 Q \theta_2}{9\mu H \ln(r_2/r_1)}\right]^{1/2}}{r_2 - r_1} \tag{6.53}$$

It seems like we have now a mathematical description but it requires a certain calibration because θ_2 is unknown. In addition, it is unlikely that there is a laminar flow in an industrial scale cyclone. Alternatively, we can use the empirical model developed by Lapple [19] that follows.

6.4.1.2 Lapple Model

A semi-empirical model was developed by Lapple [19]. The average radial speed is described in terms of the migration time. It can be considered as the average terminal speed in the centrifugal field. On average, before the particle reaches the inner surface of the cyclone outer wall, corresponding to $r = r_2$, the average terminal speed is

$$v_r = \frac{r_2 - r_c}{t} \tag{6.54}$$

where t is the corresponding residence time. Combination of Eqs. (6.45) and (6.54) leads to

$$r_2 - r_c = \frac{\rho_p d_p^2 v_0^2}{18\mu r} t \tag{6.55}$$

Substituting Eq. (6.55) into (6.40) leads to the fractional efficiency of the particles with diameter d_p

$$\eta(d_p) = \frac{r_2 - r_c}{W} = \frac{\rho_p d_p^2 v_0^2}{18\mu r W} t \tag{6.56}$$

The residence time depends on the engineering design of the cyclone. Many models of commercial cyclones have been developed since the end of the nineteenth century. Typical cyclones are classified into four basic categories, as depicted in Fig. 6.7, based on airflow direction:

(1) reverse flow with a tangential inlet (involute),
(2) reverse flow with a guide vane inlet (vane-axial),
(3) uniflow with a tangential inlet, and
(4) uniflow with guide vanes.

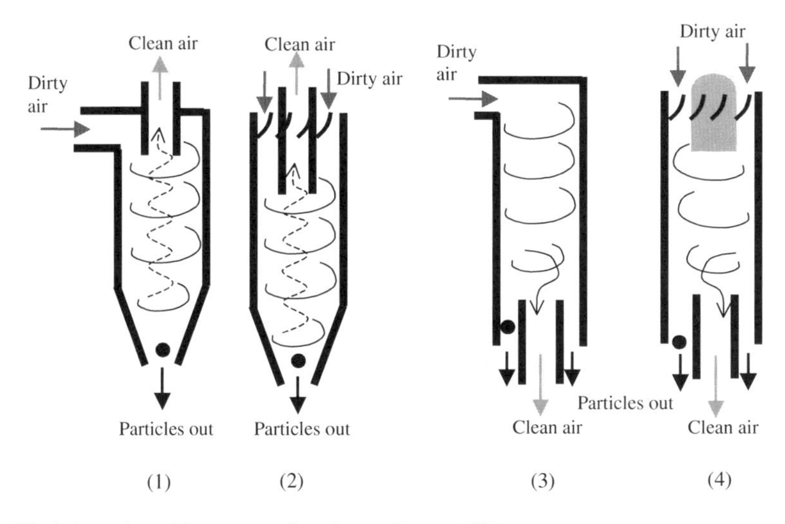

Fig. 6.7 Schematics of four types of cyclones (*Source* [42])

Our analysis continues with the mostly widely used cyclone model, involute cyclone. An involute cyclone is a conventional reverse flow cyclone that has been well standardized and commercialized. It can be classified into three types based on the flow rate capacity and efficiency: conventional cyclone, high efficiency cyclone, and high throughput cyclone.

For the widely used, so-called Lapple cyclones [19], design parameters are given in Fig. 6.8 and Table 6.1 [21] with a typical gas inlet velocity in the range of 15–30 m/s [4]. The values in Table 6.1 are for guidance only. Actual design may vary.

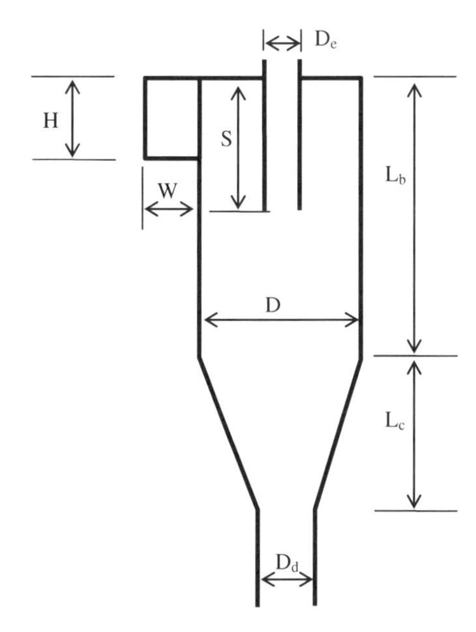

Fig. 6.8 Schematic diagram of a Lapple cyclone

Table 6.1 Design parameters for a Lapple cyclone

	High efficiency	Conventional	High throughput
Height of inlet, H/D	0.5–0.44	0.5	0.75–0.8
Width of inlet, W/D	0.2–0.021	0.25	0.375–0.35
Diameter of gas exit, D_e/D	0.4–0.5	0.5	0.75
Length of vortex finder, S/D	0.5	0.625–0.6	1.5–1.7
Length of body, L_B/D	1.4	1.75	1.7
Cone length, L_C/D	2.5	2	2.5–2
Diameter of dust outlet	0.375–0.4	0.25–0.4	0.375–0.4

The residence time of an involute cyclone is estimated by

$$t = \frac{\text{Circumference of the vortex} \times \text{Number of turns of the air}}{\text{Superficial gas speed}} \tag{6.57}$$

where the number of turns the gas flow makes before turning upward to the vortex finder (N_e) is

$$N_e = \frac{L_B + 0.5L_C}{H} \tag{6.58}$$

where L_B is the length of cyclone main body, and L_C is the length of the cyclone lower cone. Then the residence time can be calculated using

$$t = \frac{2\pi r N_e}{u_0} \tag{6.59}$$

Equation (6.56) then becomes

$$\eta(d_p) = \left(\frac{\pi N_e \rho_p}{9\mu W \bar{u}_g}\right) d_p^2 v_0^2 \tag{6.60}$$

If the average tangential speeds are approximated as the same, $v_0 \approx u_0 \approx \bar{u}_g$, then Eq. (6.60) becomes

$$\eta(d_p) = \left(\frac{\pi \bar{u}_g N_e \rho_p}{9\mu W}\right) d_p^2 \tag{6.61}$$

Instead of using this equation for direct calculation of the particle fractional efficiency, Lapple [19] presented a semi-empirical equation by introducing the cut size into the analysis. The corresponding cut size is first determined using Eq. (6.61) by letting $\eta = 0.5$

$$d_{50} = \left(\frac{4.5 \mu W}{\pi N_e \bar{u}_g \rho_p} \right)^{1/2} \tag{6.62}$$

With the computed cut size above, the fractional efficiency of the cyclone is described as a function of particle size, d_p, and cut size, d_{50}:

$$\eta(d_p) = \frac{1}{1 + (d_{50}/d_p)^2} \tag{6.63}$$

Example 6.3: Cyclone efficiency
A conventional cyclone has a body diameter of 20 cm and other geometries are listed in the table as follows. It operates at an inlet volumetric flow rate of 360 m^3/h. Assume standard condition, plot its fractional efficiency curve versus particle aerodynamic diameter.

	Ratio
Height of inlet, H/D	0.5
Width of inlet, W/D	0.25
Length of body, L_B/D	1.75
Cone length, L_C/D	2

Solution
First, we calculate the dimension of the cyclone as follows:

	Dimension
Body diameter, D (m)	0.2
Height of inlet, H	0.1
Width of inlet, W	0.05
Diameter of gas exit, D_e	0.1
Length of body, L_B	0.35
Cone length, L_C	0.4

The number of turns from Eq. (6.58) is

$$N_e = \frac{L_B + 0.5 L_C}{H} = \frac{0.35 + 0.5 \times 0.4}{0.1} = 5.5$$

The inlet area of this cyclone is

$$A = HW = 0.005 (\text{m}^2)$$

The average inlet air speed is then

$$\bar{u}_g = \frac{Q}{A} = \frac{360\ \frac{\mathrm{m}^3}{\mathrm{h}} \times \frac{1\,\mathrm{h}}{3{,}600\,\mathrm{s}}}{0.005\,\mathrm{m}^2} = 20\,\mathrm{m/s}$$

Cut size is determined using Eq. (6.62):

$$d_{50} = \left[\frac{9\mu W}{2\pi N_e \bar{u}_g (\rho_p - \rho_g)}\right]^{1/2} = \left[\frac{9 \times 1.81 \times 10^{-5} \times 0.05}{2\pi \times 5.5 \times 20(1000 - 1.21)}\right]^{1/2}$$
$$= 3.43 \times 10^{-6}\,\mathrm{m} \quad \text{or} \quad 3.43\,\mu\mathrm{m}$$

Then the fractional efficiency is determined by using Eq. (6.63):

$$\eta(d_p) = \frac{1}{1 + (d_{50}/d_p)^2}$$

The curve is shown in Fig. 6.9. This example shows that a typical cyclone works effectively for particles larger than a few micrometers.

Their separation efficiency is, however, limited to 90 % or so for a cyclone of reasonable size (diameters up to 1 m) with reasonable pressure drop, and the separation efficiency rapidly deteriorates for particles smaller than 10 μm.

In the example above, we did not use the vortex finder. Vortex finder is indirectly related to the performance of a cyclone. Agglomeration of particles at the inlet region is the result of stronger centrifugal forces on larger particles than on smaller ones, causing a "sweeping" effect. At the same time, particles may short-cut from the inlet to the gas outlet if the "vortex finder" does not penetrate deep enough into the cyclone.

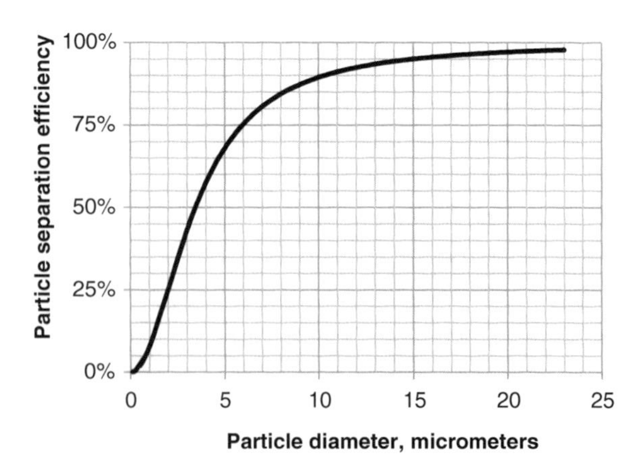

Fig. 6.9 Cyclone efficiency versus aerodynamic diameter

6.4.2 Pressure Drop of Cyclone

Pressure drop is the second important cyclone performance indicator, after collection efficiency. The pressure drop across a Lapple cyclone can be estimated by:

$$\Delta P = K\left(\frac{\rho_g \bar{u}_g^2}{2}\frac{HW}{D_e^2}\right) \tag{6.64}$$

which contains the design dimensions of H, W and D_e. For the coefficient K, a value in the range of 16–18 is suggested, with $K = 16$ as a recommended value [4].

A cyclone is capable of reducing dust concentrations in a gas stream from several g/m^3 to below 0.1 g/m^3. Cyclones can also be applied for removing water from oil at oil fields or solids from water. They are considered effective as low-cost preseparators for gas cleanup purposes.

6.4.3 Other Cyclone Models

Many analytical or semi-empirical models have been developed to predict the collection efficiencies of reverse flow cyclones under laminar or complete mixing assumptions. In these theoretical analyses, dimensionless geometric parameters are frequently defined. Dirgo and Leith [9] have summarized the models that were developed prior to 1985. It was stated that Lapple's cut-size theory based on time flight approach was widely cited in North American literature, while Barth's theory based on static particle approach was more often referred to in Europe. Both theories are based on laminar flow assumption. The well-known Leith–Licht model [22] developed in the 1970s was based on the assumption that flow was turbulent and uncollected particles were completely and uniformly mixed. Barth's theory and Leith–Licht theory were closest to Dirgo and Leith's experimental results [9] obtained from a Stairmand high efficiency cyclone.

Several uniflow cyclone models have been published too. Most researchers assume that increasing the separation length favors the solids separation efficiency, as the residence time increases allowing more particles to migrate to the wall of the cyclone. Summer et al. [32] reported an optimum separation length of around 1 cyclone diameter. Gauthier et al. [12] found that the optimum length increased with the inlet air velocity. These assumptions were validated by experimental results of large particle separation in a small uniflow cyclone with a tangential inlet. The diameter of the cyclone was 50 mm and the particles had a mean diameter of 29 μm. These models, however, might not apply for separation of fine particles in cyclones handling high airflow rates, where high turbulence exists.

Ogawa et al. [24] analyzed the separation mechanism of fine solid particles for a uniflow cyclone and demonstrated that particle cut size could be smaller than 5 μm. However, this theory is also based on small scale models. Their outer diameters range from 30 to 99 mm and the lengths range from 90 to 279 mm. Furthermore, uniflow cyclones of commercial size suitable for large volume of air cleaning, especially those applicable for dusty airspaces, also need to be examined. Tan [33] derived a model for uniflow cyclone with tangential inlet and concentric exit.

Overall, cyclones are characterized with their simple structure, low cost, small footprint, and large capacity. Theoretically, high flow rate leads to higher particle separation efficiency. On the other hand, the most important contras of cyclones are high pressure drop, and relatively low efficiency for fine particles as demonstrated in Example 6.3. For reverse flow cyclones, the layer of collected particles may come in contact with the flow field of the gas, leading to re-entrainment. Most critical is the position near the bottom outlet for the collected dust, where the downward swirl turns upward into the inner vortex toward the gas outlet. At that point strong re-entrainment of collected particles may occur, which most certainly will leave the cyclone with the gas.

6.5 Filtration

Filtration is a process where particles are separated from a fluid using porous media called filter. Filtration is widely used for particle–fluid separation. Our focus here is air filtration, although similar principles apply to particle–liquid separation. For aerosol particles to be captured by a filter, they are first transported from the air to filter medium surface, and then collide with the surface. Particles cannot be captured by the filter unless they reach the surface of the filter media. However, successful transport does not ensure the particles being captured.

According to conventional particle dynamics, the collision between the particles and the filter surface may result in

- some particles adhere to the surface and they are considered removed from the air,
- other particles rebound from the surface and remain airborne.

The resultant filtration efficiency is thereby,

$$\eta = \eta_{\text{ts}} \times \eta_{\text{ad}} \tag{6.65}$$

where η_{ts} is the transport efficiency and η_{ad} is the adhesion efficiency. In classic filtration models, it is assumed that a particle is permanently removed from the gas stream once it reaches the filtration medium surface, i.e.,

$$\eta_{\text{ad}} \equiv 1 \tag{6.66}$$

Thereby, the filtration efficiency is the same as transport efficiency.

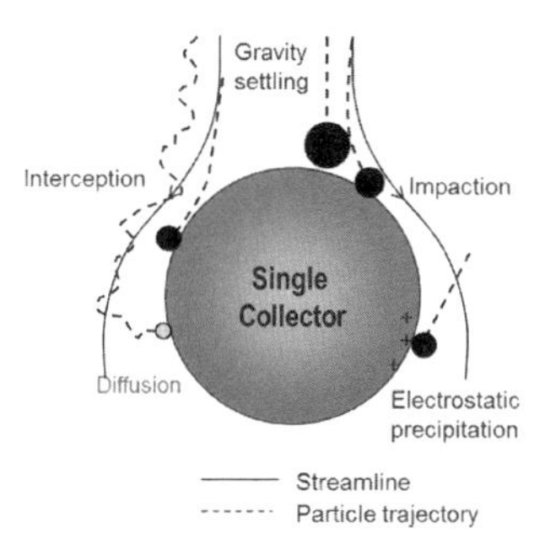

Fig. 6.10 Five particle transport mechanisms in filtration

$$\eta = \eta_{\text{ts}} \tag{6.67}$$

There are five basic particle *transport* mechanisms in filtration: interception, inertial impaction, diffusion, gravity settling, and electrostatic attraction. They are illustrated in Fig. 6.10. They are also applicable to particle filtration in liquid media and even particle feeding in biology [7]. While the method of analysis may be different for each specific case, the mechanisms are the same.

- *Interception*: Interception happens when a particle follows a gas streamline that comes within one particle radius to the surface of the filter media. The particle is captured because of its finite size. When interception dominates, it is assumed that the particles follow the streamlines perfectly, and they do not depart from the streamlines of the gas phase. This characteristic is unique and different from the other four mechanisms.
- *Inertial impaction*: When there is a sudden change in the flow direction near the media, a particle may be captured because of its inertia. This is referred to as *inertial impaction*. During an inertial impaction, the particle to be captured crosses the gas streamlines and reaches the surface of the filter media.
- *Diffusion*: Airborne particles can be captured by a surface due to the Brownian motion. This is especially true for small particles and also for the particles near the filter media surface.
- *Electrostatic attraction*: Electrostatic attraction takes place when the particles and the filter are charged. It follows the principle of particle dynamics in an electric field. The electrostatic attraction can be extremely important of all the mechanisms but is equally difficult to quantify because the charge on the particles or on the filters are often unknown. In most analytical works, electrostatic attraction is neglected due to the shortage of information.
- *Gravitational settling*: Gravitational settling takes place when a particle falls onto the filter during its motion. It is an inertial separation process. It is effective only for large particles or in air moving at low speed.

Filtration theory is essentially concerned with the prediction of the particle collection efficiency and the pressure drop as the carrier gas passes through the filter. Theoretically, they can be accurately calculated if the gas flow within a filter could be mathematically described precisely. However, the random orientations of the filter media, especially the fibers within a filter, make it nearly impossible to achieve accurate mathematical solution. Instead, simplified filtration models are introduced as follows.

6.5.1 Single Fiber Filtration Efficiency

The simplest model consists of an isolated cylinder (simulating a single fiber) or a hole (for single pore) or a sphere (simulating a granule) in an otherwise undisturbed flow of fluid (gas or liquid). This may be justified for perfectly designed filters with excessively high porosity. Although the model based on isolated filter medium has been quite useful in illustrating the relative importance of different transport mechanisms, it is challenging to relate the rate of particle deposition on an isolated medium to the collection efficiency of a real filter that is filled with random fibers or granules.

A better approach, although still quite far from being realistic, is the model based on Kuwabara [18] cell model of forces experienced by randomly distributed parallel circular cylinders or spheres in a viscous flow at small Reynolds number. This cell model has been used by several researchers as a starting point for their filtration models, especially for fibrous filter models. Unlike the isolated cylinder or sphere models, these newer models address the influence of the surrounding filter media. However, the Kuwabara model and its modified versions still have limitations because they cannot be justified. Kirsh and Fuchs [17] developed a 3D model to represent a real filter more closely than earlier models; this model is too complicated to be solved easily. Therefore, we will not discuss it.

Yeh and Liu [38, 39] developed a model of flow over a staggered array of cylinders (Fig. 6.11) to investigate Kuwabara's model with slip effect taken into consideration. With this they quantified the effects of inertial impaction, diffusion, and interception. These models were also experimentally validated [38, 39]. And,

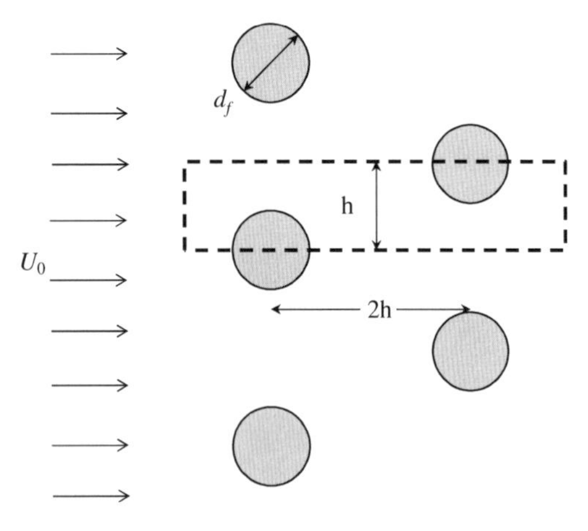

Fig. 6.11 Staggered array filter model

later on, the equations for diffusion and interception were further simplified by Lee and Liu [20]. Our discussion is focused on the staggered array models as follows.

In the staggered array model, air approaches the cylinders with a uniform speed of U_0. The diameter of the uniform cylinders is d_f. In the cylindrical coordinates defined by (r, θ) with origin at the center of the cylinder, the corresponding dimensionless stream line function is

$$\begin{aligned} \Psi &= \frac{\sin\theta}{2Y}\left[\frac{A}{r^*} + Br^* + 2r^*\ln r^* - \frac{\alpha}{2}(r^*)^3\right] \\ A &= \frac{\left(1-\frac{1}{2}\alpha\right)\left(1-Kn_f\right)+\alpha Kn_f}{1+Kn_f} \\ B &= \frac{(1-\alpha)\left(Kn_f-1\right)}{1+Kn_f} \end{aligned} \tag{6.68}$$

where α is called the solidity of the filter, it is the ratio of solid volume to the entire filter bulk volume; $r^* = 2r/d_f$ is a dimensionless distance from the center of the fiber; the hydrodynamic factor Y is described as

$$Y = -\frac{\ln\alpha}{2} - \frac{3}{4\left(1+Kn_f\right)} + \frac{\alpha}{1+Kn_f} + \frac{Kn_f(2\alpha-1)^2}{4\left(1+Kn_f\right)} - \frac{\alpha^2}{4} \tag{6.69}$$

In this equation, the Knudsen number is defined as the ratio of the mean free path of the gas molecules to the radius of the filter fiber. Similar to Eq. (4.8), it is expressed as

$$Kn_f = 2\lambda/d_f \tag{6.70}$$

It quantifies the slip effect in particle filtration; Kn_f is significant for $d_f < 2$ μm. The mean free path of air (λ) under standard conditions is about 0.066 μm. In this case, $2\lambda/d_f < 0.066$ % when $d_f > 2$ μm.

For cases with $\alpha \ll 1$ and $Kn_f \ll 1$, the streamline function becomes

$$\Psi = \frac{\sin\theta}{2Y}\left(\frac{1}{r^*} - r^* + 2r^*\ln r^*\right) \tag{6.71}$$

For a typical filtration process under normal conditions, $Kn_f \ll 1$, and the Kuwabara hydrodynamic factor in Eq. (6.69) can be simplified by letting $Kn_f \to 0$:

$$Y = -\frac{\ln\alpha}{2} - \frac{3}{4} + \alpha - \frac{\alpha^2}{4} \tag{6.72}$$

In certain special filtration processes such as those under low pressure or those with very fine filter diameter, Eq. (6.69) should be used for a great accuracy in calculation.

The single fiber filtration efficiency by inertial impaction per unit length of fiber due to inertial *impaction* is calculated as

$$\eta_{\mathrm{ip}} = \left(\frac{Stk_m}{2Y}\right) I \tag{6.73}$$

where I is described as

$$I = \begin{cases} \left(29.6 - 28\alpha^{0.62}\right)R^2 - 27.5R^{2.8} & \text{for } R < 0.4 \\ 2 & \text{for } R \geq 0.4 \end{cases} \tag{6.74}$$

There is no simple equation for I when $R \geq 0.4$. As an approximation, a value of $I = 2.0$ for $R \geq 0.4$ can be used.

In these equations, R is called the interception parameter and it is defined as ratio of particle diameter d_p to the filter fiber diameter d_f:

$$R = \frac{d_p}{d_f} \tag{6.75}$$

Stk_m is the modified Stokes number defined by the particle Stokes number with $d_c = d_f$ (Eq. (6.76)) divided by the hydrodynamic factor, Y

$$Stk = \frac{\rho_p d_p^2 C_c U_0}{18\mu d_f} \tag{6.76}$$

$$Stk_m = \frac{Stk}{2Y} \tag{6.77}$$

The single fiber efficiency due to diffusion per unit length of fiber is described as

$$\eta_{\mathrm{D}} = \frac{3.65(Pe_m)^{-\frac{2}{3}} + 0.624(Pe_m)^{-1}}{2Y} \tag{6.78}$$

where Pe_m is the modified Peclet number. Similar to the modification to Stokes number, the modified Peclet number is

$$Pe_m = \frac{Pe}{2Y} \tag{6.79}$$

and the Peclet number is

$$Pe = \frac{U_0 d_f}{D_p} \tag{6.80}$$

where U_0 is the air velocity approaching the filter fiber, and D_p is particle diffusion coefficient that can be calculated using Eq. (6.81).

$$D_p = \frac{kTC_c}{3\pi\mu d_p} \tag{6.81}$$

Peclet number shows the effect of convective transport over diffusive transport of particles.

Later on, Lee and Liu [20] further simplified the Eq. (6.78) for diffusion as,

$$\eta_{\text{D}} = 2.6\left(\frac{1-\alpha}{Y}\right)^{1/3} \quad Pe^{-2/3} \tag{6.82}$$

This simplified equation shows that $\eta_D \propto Pe^{-2/3}$ for a filter with a fixed solidity.

A general equation for single fiber filtration by *interception* can be described as (Lee and Liu [20], p. 152)

$$\eta_{\text{it}} = \frac{1+R}{2Y}\left[2\ln(1+R) - (1-\alpha) + \left(1-\frac{\alpha}{2}\right)(1+R)^{-2} - \frac{\alpha}{2}(1+R)^2\right] \tag{6.83}$$

Equation (6.83) is a complete expression based on Kuwabara flow fields with a wide range of R and α. Simpler forms are given by Lee and Liu [20], p. 152 with limitations. The respective simplified equations for the cases $R \ll 1$ or $\alpha \ll 1$ are as follows:

$$\eta_{\text{it}} = \frac{(1-\alpha)}{Y}\frac{R^2}{1+R} \quad \text{for } R \ll 1 \tag{6.84}$$

$$\eta_{\text{it}} = \frac{2(1+R)\ln(1+R) - (1+R) - (1+R)^{-1}}{2Y} \quad \text{for } \alpha \ll 1 \tag{6.85}$$

6.5.1.1 Electrostatic Attraction

In most filtration models, electrical attraction is ignored, not because of its little importance, but rather the complex in the quantification of its effect. Nonetheless, Brown [3] gave a review of the theory of particle separation by electrostatic attraction. He introduced the single fiber efficiency for a neutral fiber and a particle with charge q as

$$\eta_{\mathrm{E}} = 1.5\left[\left(\frac{\varepsilon_r - 1}{\varepsilon_r + 1}\right)\right] \frac{q^2}{12\pi\mu U_0 \varepsilon_0 d_p d_f^2} \tag{6.86}$$

where ε_r = relative permittivity of the fiber and ε_0 = permittivity of a vacuum. Note that this equation was validated with the experimental measurements using glass fiber filters.

As introduced in Sect. 6.2, gravitational settling is not effective for micron particles, and it works only for large particles. Similarly, it does not play an important role in filtration compared to other mechanisms. We will skip its analytical solution.

6.5.1.2 Total Efficiency of a Single Fiber

With the filtration efficiency of each mechanism determined, the total efficiency of a single fiber per *unit length* is

$$\eta_{\mathrm{sf}} = 1 - (1 - \eta_{\mathrm{it}})(1 - \eta_{\mathrm{ip}})(1 - \eta_{\mathrm{D}})(1 - \eta_{\mathrm{E}}) \tag{6.87}$$

In the computation of each component, the value should be less than or equal to 1.0.

Although each of the five filtration mechanisms plays a role in many cases, their relative importance is size dependent. As shown in Fig. 6.12, the contribution of diffusion drops with the increase of the particle diameter, whereas the effects of other mechanisms increase with particle diameter.

Example 6.4: Filtration efficiency based on different mechanisms
A fiber filter has a solidity of 3 %. The average diameter of the fiber is 5 μm. Estimate the single fiber efficiency based on, interception, impaction, and diffusion, respectively, when the face velocity is 0.05 m/s.

Solution
In this example, the following parameters are considered as constant

$$d_f = 5\,\mu\mathrm{m}, \quad \alpha = 0.03, \quad Kn_f = \frac{2\lambda}{d_f} = \frac{2(0.066)}{5} = 0.0264$$

$$Y = -\frac{\ln\alpha}{2} - \frac{3}{4} + \alpha - \frac{\alpha^2}{4} = 1.033, \quad \mu = 1.81 \times 10^{-5}\,\mathrm{Pa\,s}$$

The following variables can be calculated in an Excel sheet for different particle diameters

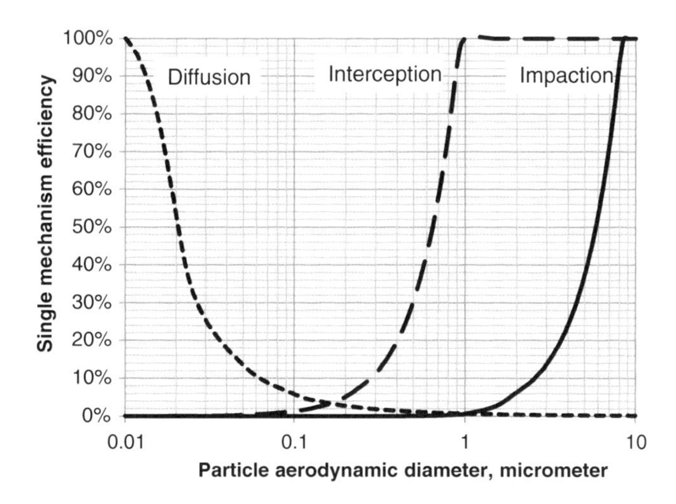

Fig. 6.12 Relative importance of different filtration mechanisms

$$R = \frac{d_p}{d_f}, \quad Stk_f = \frac{\rho_p d_p^2 C_c U_0}{18\mu d_f}, \quad Stk_m = \frac{Stk}{2Y}$$

$$I = \begin{cases} (29.6 - 28\alpha^{0.62})R^2 - 27.5R^{2.8} & \text{for } R < 0.4 \\ 2 & \text{for } R \geq 0.4 \end{cases}$$

$$D_p = \frac{kTC_c}{3\pi\mu d_p}, \quad Pe = \frac{U_0 d_f}{D_p}, \quad Pe_m = \frac{Pe}{2Y}$$

The single fiber filtration efficiency by inertial interception, impaction, and diffusion, *per unit length* of fiber is calculated using

$$\eta_{\mathrm{it}} = \frac{1+R}{2Y}\left[2\ln(1+R) - (1-\alpha) + \left(1 - \frac{\alpha}{2}\right)(1+R)^{-2} - \frac{\alpha}{2}(1+R)^2\right]$$

$$\eta_{\mathrm{ip}} = \frac{I}{2Y} Stk_m$$

$$\eta_{\mathrm{D}} = \frac{3.65(Pe_m)^{-\frac{2}{3}} + 0.624(Pe_m)^{-1}}{2Y}$$

The curves are plotted in Fig. 6.12. As indicated in the figure, when all other conditions are the same, diffusion is dominating for small particles. However, its effectiveness also drops at high air speed. It can be easily seen by repeating the example by changing face velocity with $U_0 = 0.1$ m/s.

As shown in Fig. 6.12, one filtration mechanism often predominates over the other for certain size group. In general, interception and impaction are negligible for small particles, but they become important for particles larger than 1 μm in diameter. Diffusion is the only important mechanism for particles below 0.2 μm, but drops quickly with the increase of particle diameter. The total efficiency is at the bottom for particles around 0.2 μm. This is because this group of particles is too large for diffusion and too small for impaction or interception to be effective.

6.5.2 Overall Fibrous Filtration Efficiency

Eventually we need to quantify the overall efficiency of an actual filter. "Overall filtration efficiency" is used here to avoid the confusion with the term "total efficiency" that has been used above to describe the total efficiency of a single fiber, Eq. (6.87). The overall filtration efficiency, η, can be derived from the total single fiber efficiency, η_{sf}, by the following simplified one-dimensional analysis.

As depicted in Fig. 6.13, analysis of overall filtration efficiency, consider a filter with a bulk thickness of L along x-direction, which is also the face velocity direction. The filter is filled with homogeneous fibers of diameter, d_f, and the length of the uniform single fiber is denoted as ds_f. Within an elemental thickness of dx, the solidity, α, from its definition is,

$$\alpha = \frac{\left(\pi d_f^2/4\right) \times ds_f}{A_c \times \mathrm{d}x} \tag{6.88}$$

where A_c is the bulk cross section area of the filter that is normal to the face velocity. It can be determined by the air flow rate

$$A_c = \frac{Q}{U_\infty} \tag{6.89}$$

where U_∞ is the bulk face speed, or the air speed approaching the filter. It is less than that approaching the fiber within the filter, U_0, because of the existence of solid fibers

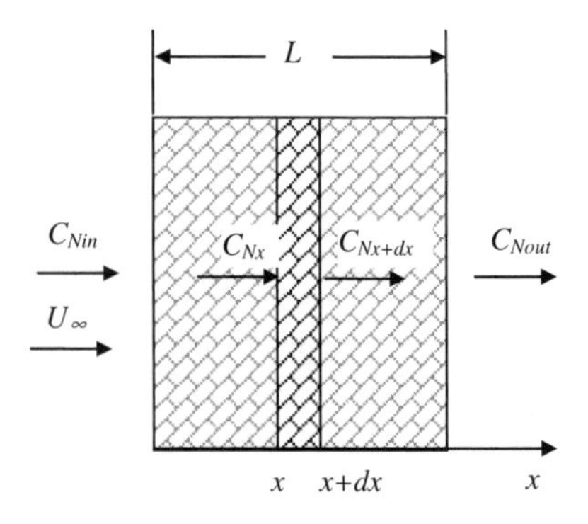

Fig. 6.13 Analysis of overall filtration efficiency

$$U_0 = \frac{U_\infty}{1-\alpha} = \frac{Q}{(1-\alpha)A_c} \tag{6.90}$$

The number concentration of particles lost per unit volume from the bulk air over the distance dx equals to that captured by the fiber with a single fiber efficiency η_{sf} corresponding to an approaching flow rate of $U_0 d_f \times ds_f$, where $d_f \times ds_f$ defines the cross section area of the fiber with the length of ds_f and the diameter of d_f.

$$Q \times dC_N(x) = -\eta_{\text{sf}} C_N(x) U_0 d_f \times ds_f \tag{6.91}$$

where C_N is the number concentration of particles (#/m^3), the single fiber efficiency η_{sf} is determined by Eq. (6.87).

The LHS stands for the decrease in particle number per unit time after air passes through the bulk filter thickness dx, whereas the RHS stands for the reason of this decrease calculated as if all these particles passed through a single fiber with a cross section area of $\left(d_f \times ds_f\right)$ with an approaching speed of U_0 and a single fiber efficiency of η_{sf}. Equations (6.88) and (6.91) give,

$$\begin{aligned} A_c U_\infty \times dC_N(x) &= -C_N(x)\eta_{\text{sf}} U_0 d_f \times \frac{4\alpha A_c \times \mathrm{d}x}{\pi d_f^2} \\ \frac{dC_N(x)}{C_N(x)} &= -\left(\frac{U_0}{U_\infty}\frac{\eta_{\text{sf}} 4\alpha}{\pi d_f}\right)\mathrm{d}x \end{aligned} \tag{6.92}$$

Consider Eq. (6.90), we have

$$\frac{dC_N(x)}{C_N(x)} = -\left[\frac{\eta_{\text{sf}} 4\alpha}{(1-\alpha)\pi d_f}\right]\mathrm{d}x \tag{6.93}$$

At any instant, all the parameters in the bracket on the right-hand side are constants for the particle size of d_p, which allows us to integrate from inlet to outlet of the filter:

$$\int_{C_{N\text{i}}}^{C_{N\text{o}}} \frac{dC_N(x)}{C_N(x)} = \int_0^L -\left[\frac{\eta_{\text{sf}} 4\alpha}{(1-\alpha)\pi d_f}\right]\mathrm{d}x \tag{6.94}$$

which gives the overall fractional penetration through the filter

$$P\left(d_p\right) = \frac{C_{N\text{o}}}{C_{N\text{i}}} = \exp\left[\frac{-\eta_{\text{sf}} 4\alpha L}{(1-\alpha)\pi d_f}\right] \tag{6.95}$$

Consequently, we can get the overall fractional filtration efficiency of the filter as

$$\eta(d_p) = 1 - \exp\left[\frac{-\eta_{sf}4\alpha L}{(1-\alpha)\pi d_f}\right] \tag{6.96}$$

Equation (6.96) shows that we can calculate the fractional efficiency of a filter with fixed specifications (α, d_f, L) as long as the single fiber efficiency is determined. This equation also shows that the overall filtration efficiency increases over time as the solidity increases too when particles take more void space in filter.

Example 6.5: Filtration total efficiency

A filter is made of fiberglass with a solidity of 3 %, and it is 2 mm thick. The average diameter of the fiber is 5 μm. When the face velocity is 0.05 m/s, estimate its overall fractional filtration efficiency as a function of particle aerodynamic diameter under standard conditions (consider only interception, impaction, and diffusion).

Solution

Following the approach in Example 6.4, we can get the same single mechanism filtration efficiency for interception, impaction, and diffusion.

Then the single fiber total filtration efficiency per unit length of fiber is calculated using Eq. (6.87)

$$\eta_{sf} = 1 - (1 - \eta_{it})(1 - \eta_{ip})(1 - \eta_D) \tag{6.97}$$

and the overall filtration efficiency of the filter itself is calculated using Eq. (6.96)

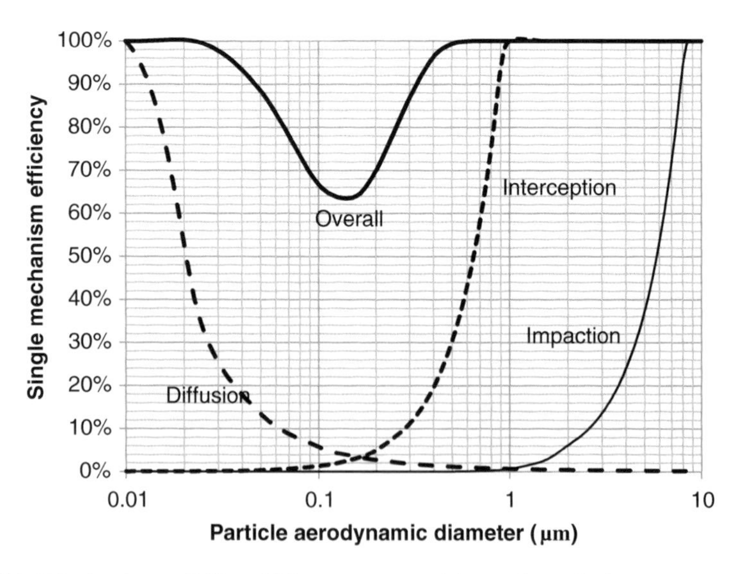

Fig. 6.14 Calculated overall filter efficiency versus particle aerodynamic diameter

$$\eta(d_p) = 1 - \exp\left[\frac{-\eta_{sf} 4\alpha L}{(1-\alpha)\pi d_f}\right] \tag{6.98}$$

The results are shown in Fig. 6.14.

Up to this point we assumed that filter fibers are oriented normal to the incoming air flow. It is clearly not the case in most engineering applications. Fiber orientation also affects the filtration efficiency and resistance to the air flow. The overall filtration efficiency is affected by the three-dimensional randomness of the fiber orientations too [31]. A filter with fibers randomly arranged in planes perpendicular to the approaching air velocity is more efficient than a filter with fibers randomly arranged in three dimensions.

The other assumptions in the preceding analysis of fibrous filtration are that all the aerosol particles are spherical and that the adhesion efficiency is 100 %. These simplifications do not introduce much of error because the likelihood of an aerosol particle adhering to a fibrous filter depends on not only the air flow velocity but also the particle–filter interfacial characteristics. Nonspherical particles are more likely to be captured than spherical ones. And, functions that correct for imperfect adhesion can be empirically derived for particular cases.

6.5.3 Fibrous Filter Pressure Drop

The pressure drop across a fiber filter is caused by the combined effect of each fiber resisting the flow of air past it. Davies [8] defined a dimensionless filter pressure coefficient as

$$C_{\Delta P} = \frac{\Delta P}{4\mu U_0 L / d_f^2} \tag{6.99}$$

With a known pressure drop coefficient, the pressure drop can be calculated as

$$\Delta P = C_{\Delta P} \frac{4\mu U_0 L}{d_f^2} \tag{6.100}$$

By dimensionless analysis and experimental correlation, Davies [8] obtained $C_{\Delta P}$ as a function of solidity in the range of 0.06–0.3 as follows,

$$C_{\Delta P} = 16\alpha^{1.5}(1 + 56\alpha^3) \tag{6.101}$$

Combination of Eqs. (6.100) and (6.101) leads to the total pressure drop over a bulk filter as

$$\Delta P = \frac{64\mu L U_0}{d_f^2}\alpha^{1.5}\left(1 + 56\alpha^3\right) \tag{6.102}$$

where the pressure drop ΔP is in Pascal. It is thus directly proportional to thickness of the filter L and inversely proportional to cross section area of the fiber d_f^2.

From the staggered array model, Yeh and Liu [38, 39] calculated the drag force over the fibers as

$$F_D = \frac{4\pi\mu U_0}{Y} \tag{6.103}$$

and the corresponding *pressure drop per unit thickness* of the filter is

$$\Delta P' = \frac{16\mu\alpha U_0}{Y d_f^2} \tag{6.104}$$

Then the total pressure drop over the entire array is determined as

$$\Delta P = L\Delta P' = \frac{16L\mu\alpha U_0}{Y d_f^2} \tag{6.105}$$

The corresponding pressure drop coefficient can then be determined by comparing Eq. (6.105) with (6.100).

$$C_{\Delta P} = \frac{4\alpha}{Y} \tag{6.106}$$

With Y in the denominator, Yeh's equation is complicated in form. In addition, as seen in Fig. 6.15, Yeh's model gives greater pressure drops than Davies model.

Figure 6.15 is produced for a filter with d_f = 10 μm, U_∞ = 0.2 m/s for α = 0.0–0.10 and L = 5, 10, 20 mm.

First of all, the pressure drop increases with filter solidity; the relationship based on these equations is not linear, so the pressure drop begins to dramatically increase as solidity increases. Increasing the thickness of the filter also causes the pressure drop to increase, but this increase is linear as the factor for thickness is not raised to any exponent and acts as a scalar quantity in the relationship (thus, this would also be true for velocity). Both of these responses make intuitive sense, as logically one would expect that the addition of filter material (whether by increasing solidity or thickness) would increase the pressure drop.

When comparing these two different models, it is clear that, for the same filter and operating conditions, Yeh's equations predict a greater pressure drop. Yeh's equations predict pressure drops about 1.5 times higher than Davies's. When solidity is 1 (though, this may be outside of the equation's domain), the result of

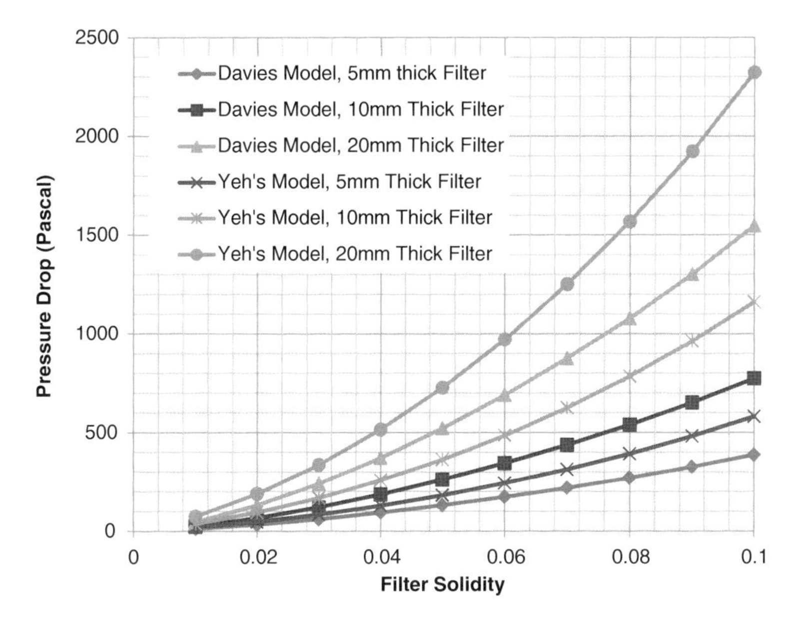

Fig. 6.15 Comparison between Yeh's and Davies models

$$Y = -\frac{\ln(\alpha)}{2} - \frac{3}{4} + \alpha - \frac{\alpha^2}{4}$$

evaluates to zero, and the pressure drop is infinite because the filter is completely blocked; a situation that Davies's model cannot predict.

Actually both models overestimate the pressure drop of a fibrous filter, but Davies's model is closer to experiments and therefore more widely used.

6.5.4 Particle Accumulation

A factor that is not considered so far in the preceding analysis is the effect of particle buildup on the fiber surface as the particle collection proceeds. Particles buildup on the fiber surface and it can be within the filter or on the surface of the filter, depending on the solidity of the filter and the particle sizes. The smaller the particle sizes, the deeper it goes into the bulk filter, and these particles are captured by internal fibers. This type of filter is also called internal filter. Larger particles are captured on the surface and these particles form a layer of dust, which are commonly called dust cake in industry. Since surface filters are designed to collect particles on the surface, they are usually as thin as a piece of fabric cloth. Surface filters can be reused after removal of the dust cake.

The buildup of particles inside and/or on a filter surface results in more or less structural changes in the filter and consequently alters the filtration efficiency and its resistance to flow. These particles deposited on the filter surface can themselves act as filter media. These factors should be taken into consideration in engineering design and practices. And it is discussed in Chap. 10, Sect. 10.2.2.

6.5.5 Granular Filtration

When the filter media is made of granules rather than fibers, the corresponding filters are called granular filters. It can be considered as an internal filter with respect to a pack of granules or surface filter with respect to each single layer. The earliest application of granular filtration was for water treatment. Layers or deep beds of solid granules (e.g., sands) have been used for a long time for precleaning drinking water. Recently, it has been tested for air emission control, especially for high-temperature applications. Low-cost granules such as sand, silicate, or alumina gravel can function very well at temperatures of as high as 450–500 °C. However, at higher temperatures, sintering of the granules and fine particles on granules may take place, leading to extreme filter clogging.

Granular filtration model was mainly based on the numerical analysis by Rajagopalan and Tien [30] predicting the trajectories of the particles moving around the filter granules under various conditions. By taking advantage of regression analysis, they developed the classic Rajagopalan–Tien model, referred to as RT-model hereby. Similar to the single fiber analysis, a single granule also captures aerosol particles by diffusion, interception, sedimentation efficiency, and electrostatic precipitation (ESP).

Starting from the single granule analysis, they derived the total filtration efficiency of a granular column filled with uniform granules. The correlation between the single granule efficiency and the overall packed bed efficiency is

$$\eta(d_p) = 1 - \exp\left[-1.5 f \alpha \eta_{\mathrm{sG}} \left(\frac{h}{d_G}\right)\right] \tag{6.107}$$

where f is an empirical fitting factor, representing the fraction of contacts between particles and collector granules; η_{sG} is the single granule efficiency; h is the height of the packed bed filled with granules; α is the filter solidity that depends on d_G, the diameter of the granules.

The internal pores of the individual granules are disregarded in the equations. The average bed solidity can be determined using the equations proposed by Pushnov [27].

$$\alpha = \begin{cases} 1 - \left[a\left(\frac{d_{\text{bed}}}{d_G}\right)^{-n} + b\right] & \text{for } \frac{d_{\text{bed}}}{d_G} > 2 \\ 1 - 12.6\left(\frac{d_{\text{bed}}}{d_G}\right)^{6.1} \exp\left(-\frac{3.6 d_{\text{bed}}}{d_G}\right) & \text{for } \frac{d_{\text{bed}}}{d_G} < 2 \end{cases} \tag{6.108}$$

where d_{bed} = body diameter of packed bed. The coefficients of a, b, and n are constants and dependent on the shape of the granules. For *spherical* granules,

$a = 1,\ b = 0.375$ and $n = 2$

By ignoring the ESP effect, the filtration efficiency of *a single granule* is described in terms of several dimensionless variables as

$$\eta_{\text{sG}} = 4A_s^{1/3}\left(\frac{D_p}{U_0 d_G}\right)^{2/3} + A_s N_{\text{vdw}}^{1/8}\left(\frac{d_p}{d_G}\right)^{15/8} + 3.38 \times 10^{-3} A_s Gr^{1.2}\left(\frac{d_p}{d_G}\right)^{-0.4} \tag{6.109}$$

where D_p = particle diffusion coefficient, U_0 = approaching air speed, d_G = granule diameter, d_p = particle diameter, Gr = gravity number, which is defined as the ratio of gravity settling speed to approaching air speed, N_{vdw} = van der Waals number, and A_s = filter porosity parameter. The first term on the right-hand side characterizes the diffusion effect, second term for van der walls effect, and the last term for gravitational effect.

With the definition of gravity settling speed defined in Eq. (6.110)

$$v_{\text{TS}} = \frac{\rho_p d_p^2 g C_c}{18\mu} \tag{6.110}$$

the gravity numberGr is described as

$$Gr = \frac{V_{\text{TS}}}{U_0} = \frac{\rho_p d_p^2 g C_c}{18\mu U_0} \tag{6.111}$$

The filter porosity parameter can be described using the solidity, α.

$$A_s = \frac{2 - 2\alpha^{\frac{5}{3}}}{2 - 3\alpha^{\frac{1}{3}} + 3\alpha^{\frac{5}{3}} - 2\alpha^2} \tag{6.112}$$

The van der Waals number, N_{vdw}, is described in terms of Hamaker constantH_{ij}

Table 6.2 Hamaker constants

Particle-media	Hamaker constant (J)	References
Glass beads–Air	5×10^{-19}	[1]
NaCl particles–Air	0.64×10^{-19}	[2]
Silica–Air	0.65×10^{-19}	[13]

$$N_{\mathrm{vdw}} = \frac{4H_{\mathrm{ij}}}{9\pi\mu d_p^2 U_0} \tag{6.113}$$

where H_{ij} = Hamaker constant that characterizes the interaction between aerosol particles and the granule. In general, the Hamaker constant is case-specific and it quantifies the interaction between the airborne particles and the fluids [2]. Three example constants are listed in Table 6.2. Much more of the Hamaker constants of inorganic materials can be found in the paper by Bergstrom [2].

Later on, Tufenkji and Elimelech [35] further improved the granular filtration model by including interception in the simulation of particle motion. The single granule collection efficiency by this new model is described as

$$\begin{aligned}\eta_{\mathrm{sG}} = {} & 2.4A_s^{\frac{1}{3}}Pe^{-0.715}N_{\mathrm{vdw}}^{0.052}\left(\frac{d_p}{d_G}\right)^{-0.081} + 0.55A_sN_A^{\frac{1}{8}}\left(\frac{d_p}{d_G}\right)^{\frac{15}{8}} \\ & + 0.22N_{\mathrm{vdw}}^{0.053}Gr^{1.11}\left(\frac{d_p}{d_G}\right)^{-0.24}\end{aligned} \tag{6.114}$$

where Pe is the particle Peclet number (Pe) and N_A is the attraction number that characterizes the effects of van der Walls attraction and fluid velocity on particle deposition due to interception.

$$N_A = \frac{H_{\mathrm{ij}}}{3\pi\mu d_p^2 U_0} \tag{6.115}$$

While this equation was validated using water filtration, Golshahi et al. [13] found that neither of the single fiber efficiencies is way below the experimental measurements for air filtration. Therefore, by correlating the experimental data for all cases tested, with the TE model Eq. (6.114) using the least square method, they proposed another equation for the single granule efficiency,

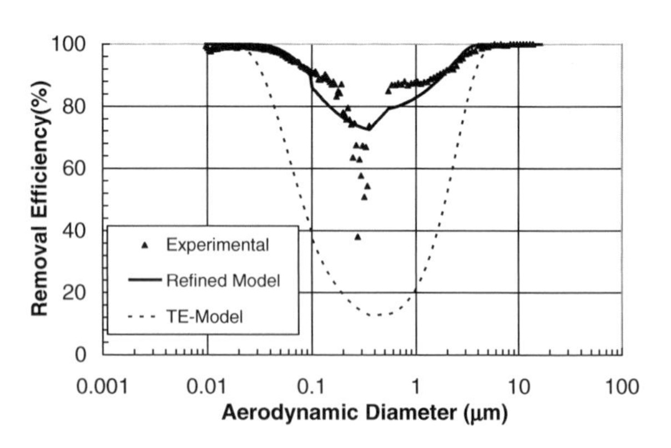

Fig. 6.16 Comparison of the models with the experiments

$$
\begin{aligned}
\eta_{\mathrm{sG}} = & 14A_s^{1/3}Pe^{-0.23}N_{\mathrm{vdw}}^{0.052}\left(\frac{d_p}{d_G}\right)^{0.34} + 0.55A_sN_A^{\frac{1}{8}}\left(\frac{d_p}{d_G}\right)^{1.675} \\
& + 0.22N_{\mathrm{vdw}}^{0.053}Gr^{1.11}\left(\frac{d_p}{d_G}\right)^{-0.24}
\end{aligned} \tag{6.116}
$$

Figure 6.16 shows a comparison between the TE model Eq. (6.114) and the refined model Eq. (6.116) with the experiments. Experiments were conducted using 2-mm glass beads in dry air with a bed thickness of 12.7 cm. The corresponding air flow rate was 65 L/min [13].

In granular filtration, particles are collected as they pass through a bed of granules by the same mechanisms that operate in a fibrous filter. In air cleaning, granular filtration is used primarily for sticky, corrosive, or high-temperature particles. Sometime, granules are driven to move for circulation, regeneration, and/or low resistance to air flow.

6.6 Practice Problems

1. The size distribution and collection efficiency of a particle separator as a function of particle size is shown in the table below. Estimate the cut size and the overall collection efficiency of the control device.

Particle size range (μm)	Mass fraction	Efficiency (%)
0–20	0.11	15
20–40	0.25	25
40–60	0.35	50
60–80	0.19	75
80–100	0.10	100

2. An industrial plant is using cyclone and electrostatic precipitator in serial for air emission control. A cyclone has an 80 % efficiency and it is followed by an electrostatic precipitator. The inlet air to the cyclone has a dust load of 150 g/m^3. What are the collection efficiency of the electrostatic precipitator and the allowable concentration of fly ash in the air that exits from the electrostatic precipitator, in order for the whole system to meet the total collection efficiency of 99 %?
3. Find the precipitation velocity of a 1 μm particle between two parallel plates with a potential difference of 1000 V. The distance between the two plates is

0.01 m and the particle has 100 element charges. The electrical mobility of the particle with a single charge is 1.1×10^{-9} m^2/Vs.

4. A 2 mm thick home furnace filter is made of fiberglass with 3 % solidity. The average diameter of the fiber is 20 μm. The filter is perpendicular to the face airflow of 1.5 m/s. Assume standard room conditions and neglect the electrostatic effect, determine the filter efficiency without circulation for particles of 5 μm in aerodynamic diameter.
5. A 98 % efficient electrostatic precipitator removes fly ash from combustion gases flowing at 10,000 m^3/min. Calculate the required collection area, for an effective precipitation velocity of 5.5 m/min.

Particle size range (μm)	Mass percent in size range
0–2	10
2–12	20
12–20	25
20–40	25
40–70	15
70–100	5

6. A wire-tube electrostatic precipitator has a wire diameter of 5 mm and a tube diameter of 50 cm. The potential between the wire and the cylinder is 5000 V. What is the electrical field intensity at 200 mm radial position?
7. An electrostatic precipitator has a total collection plate area of 5,500 m^2 and treats 8,000 m^3/min of air. If the effective drift velocity is 6 m/min, calculate the actual collecting efficiency.
8. Calculate the overall collection efficiency of a conventional Lapple cyclone with inlet height of 1 m and width of 0.5 m. Air enters the cyclone with a flow rate of 250 m^3/min. The density of particles flowing through the cyclone is 1,500 kg/m^3. The size distribution of the particles is shown in table below.
9. Calculate the cut size diameter of a conventional Lapple cyclone having inlet height of 1 m and width of 0.5 m. The standard air viscosity is 0.0072 kg/m-h. Air flows into the cyclone with a flow rate of 450 m^3/min. The density of particles flowing through the cyclone is 1,200 kg/m^3.
10. Calculate the pressure drop of a conventional Lapple cyclone having inlet height of 0.5 m, width of 0.25 m, and body inner diameter of 0.5 m. The standard air containing particles flow into the cyclone with a flow rate of 150 m^3/min. The density of particles flowing through the cyclone is 1,500 kg/m^3 and K = 16.
11. Calculate the following collection efficiencies of a single fiber per unit volume in a filter for particles of 8 μm in aerodynamic diameter under standard density and neglecting electrostatic effect. The filter bulk thickness is 5 mm, its solidity is 5 %. The average diameter of the fiber is 25 μm. The face velocity is 2 cm/s.

a. collection efficiency by interception
b. collection efficiency by inertial impaction
c. collection efficiency by diffusion
d. overall fiber collection efficiency

12. Calculate the pressure drop across the filter described in problem 11 above under standard condition.

References and Further Readings

1. Attard P, Schulz JC, Rutland MW (1998) Dynamic surface force measurement. I. van der Waals collisions. Rev Sci Instrum 69:3852–3866
2. Bergstrom L (1997) Hamaker constants of inorganic materials. Adv Colloids Interface Sci 70:125–169
3. Brown R (1993) Air filtration: an integrated approach to the theory and applications of fibrous filters. Pergamon, Oxford
4. Cooper CD, Alley FC (2002) Air pollution control—a design approach, 3rd edn. Waveland Press, Inc., Long Grove
5. Corbin RG (1987) A method for the location of sparks in electrostatic precipitators. Appl Acoust 22:297–317
6. Crawford M (1976) Air pollution control theory. McGraw-Hill Book Company, New York
7. Rubenstein DI, Koehl MAR (1977) The mechanisms of filter feeding: some theoretical considerations. Am Nat 111(981):981–994
8. Davies CN (ed) (1973) Air filtration. Academic Press, London
9. Dirgo J, Leith D (1985) Cyclone collection efficiency: comparison of experimental results with theoretical predictions. Aerosol Sci Technol 4:401–415
10. Engelbrecht HL (1981) Rapping systems for collecting surfaces in an electrostatic precipitator. Environ Int 6:297–305
11. Flagan R, Seinfeld JH (2012) Fundamentals of air pollution engineering. Dover Publications Inc., New York
12. Gauthier TA, Briens CL, Bergougnou MA, Galtier P (1990) Uniflow cyclone efficiency study. Powder Technol 62:217–225
13. Golshahi L, Abedi J, Tan Z (2009) Granular filtration for airborne particles—correlation between experiments and models. Can J Chem Eng 87(5):726–731
14. Goncalves JAS, Alonso DF, Costa MAM, Azzopardi BJ, Coury JR (2001) Evaluation of the models available for the prediction of pressure drop in venturi scrubbers. J Hazard Mater B 81:123–140
15. Hinds W (1998) Aerosol technology. Wiley & Sons, New York
16. Jiao J, Zheng Y (2007) A multi-region model for determining the cyclone efficiency. Sep Purif Technol 53:266–273
17. Kirsh AA, Fuchs NA (1968) Studies of fibrous filters—III: diffusionsal deposition of aerosol in fibrous filters. Ann Occup Hyg 11:299–304
18. Kuwabara S (1959) The forces experienced by randomly distributed parallel circular cylinders or spheres in viscous flow at small Reynolds numbers. J Phys Soc Japan 14(4):527–532 (in English)
19. Lapple CE (1951) Processes use many collector types. Chem Eng 58:144–151
20. Lee KW, Liu BYH (1982) Theoretical study of aerosol filtration by fibrous filters. Aerosol Sci Technol 1(2):147–161
21. Leith D, Mehta D (1972) Cyclone performance and design. Atmos Environ 7:529–549

22. Leith D, Licht W (1972) Collection efficiency of cyclone type particle collectors, a new theoretical approach. AIChE Symp Ser: Air-1971 68:196
23. Maloney JO, Wilcox AC (1950) Centrifugation bibliography, engineering bulletin, No. 25. University of Kansas Publications, Lawrence
24. Ogawa A, Sugiyama K, Nagasaki K (1993) Separation mechanism for fine solid particles in the uniflow type of the cyclone dust collector. In: Paper presented at Filtech Conference, Horsham, West Sussex, UK, pp. 627–640
25. Ozis F, Singh M, Devinny J, Sioutas C (2004) Removal of ultrafine and fine particulate matter from air by a granular bed filter. J Air Waste Manag Assoc 54:935–940
26. Perkins HC (1974) Air pollution, International student edn. McGraw-Hill, Kogakusha
27. Pushnov AS (2006) Calculation of average bed porosity. Chem Pet Eng 42:14–17
28. Qian F, Zhang J, Zhang M (2006) Effects of the prolonged vertical tube on the separation performance of a cyclone. J Hazard Mater 136:822–829
29. Quevedo J, Patel G, Pfeffer R, Dave R (2008) Agglomerates and granules of nanoparticles as filter media for submicron particles. Powder Technol 183:480–500
30. Rajagopalan R, Tien C (1976) Trajectory analysis of deep-bed filtration with the spherein-cell porous media model. AIChE J 22:523–533
31. Spielman L, Goren SL (1968) Model for predicting pressure drop and filtration efficiency in fibrous media. Environ Sci Technol 2:279–287
32. Summer RJ, Briens CL, Bergougnou MA (1987) Study of a novel uniflow cyclone design. Can J Chem Eng 65:470–475
33. Tan Z (2008) An analytical model for the fractional efficiency of a uniflow cyclone with a tangential inlet. Powder Technol 183(2):147–151
34. Tien C, Ramarao BV (2007) Granular filtration of aerosols and hydrosols, 2nd edn. Elsevier, Oxford
35. Tufenkji N, Elimelech M (2004) Correlation equation for predicting single collector efficiency in physicochemical filtration in saturated porous media. Environ Sci Technol 38:529–536
36. Wu MS, Lee K-C, Pfeffer R, Squires AM (2005) Granular-bed filtration assisted by filter cake formation 3. Penetration of filter cakes by a monodisperse aerosol. Powder Technol 155:62–73
37. Xiang R, Park SH, Lee KW (2001) Effects of cone dimension on cyclone performance. J Aerosol Sci 32:549–561
38. Yeh HC, Liu BYH (1974) Aerosol filtration by fibrous filters—I. Theoretical. J Aerosol Sci 5:191–204
39. Yeh HC, Liu BYH (1974) Aerosol filtration by fibrous filters—II. Experimental. J Aerosol Sci 5:205–217
40. Zhao B (2005) Development of a new method for evaluating cyclone efficiency. Chem Eng Process 44:447–451
41. Pui D, Fruin S, McMurry P (1988) Unipolar diffusion charging of ultrafine aerosols. Aerosol Sci Technol 8:173–187
42. Tan Z (2004) Mechanisms of particle separation in aerodynamic air cleaning, PhD dissertation, University of Illinois at Urbana-Champaign, Illinois, USA

Part II
Engineering Applications

Chapter 7
Combustion Process and Air Emission Formation

The combustion thermochemistry introduced in Chap. 3 applies to only simple cases. In engineering practice, fuels and oxidizers are seldom premixed to avoid explosion. Instead, they are often delivered separately into the combustion chamber and then mixed immediately prior to combustion. The combustion takes place in a flame rather than the entire combustion system. As a result, the physical processes as well as the fuel properties govern actual fuel combustion. Fluid mechanics, thermodynamics, and heat and mass transfer govern the fuel-oxidizer mixing, the combustion process, and air emissions.

In the process of combustion, the chemical energy of fuel is converted into thermal energy. There are a variety of fuels used in combustion for energy production in different engineering applications. The fuels can be solid, liquid, gaseous, or their mixtures. Each of them also has its own family. Most commonly used solid fuels include coal, charcoal, coke, and biomass; liquid ones include gasoline, diesel, and recently, bioethanol and biodiesel. Natural gas is the most widely used gaseous fuel which contains primarily methane (CH_4) and other trace gases.

Depending on the fuel properties the flue gases may contain high concentrations of the oxides of sulphur and nitrogen (SO_x, and NO_x), fine particulates, and trace elements like mercury. The actual combustion processes and their air emission formation mechanisms are introduced in this chapter.

7.1 Gaseous Fuel Flame

Gaseous fuels and oxidizers enter a stationary combustion device separately and combustion takes place in a diffusion flame (Fig. 7.1). The combustion sustains by the heat released from combustion. Mixing ratio is described using the overall equivalence ratio introduced in Sect. 3.2. However, this value is not uniform everywhere and the local equivalence ration varies from 0 to 1, representing zones of pure oxidizer or pure fuel. When the gaseous fuel is injected into the combustion chamber, there is a central core area containing pure fuel. The surrounding oxidizer and outer

Z. Tan, *Air Pollution and Greenhouse Gases*, Green Energy and Technology,
DOI 10.1007/978-981-287-212-8_7

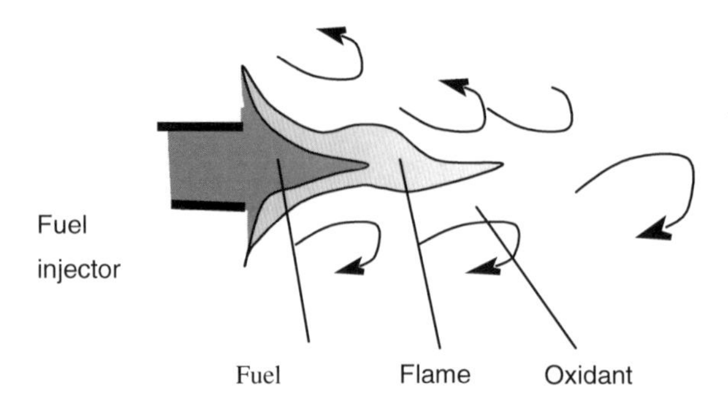

Fig. 7.1 Schematic diagram of conventional diffusion flame

part of the fuel core are mixed by diffusion forming a layer that enables combustion at an equivalence ratio of around 1, which defines the diffusion flame. Main combustion occurs in this diffusion flame. At the location that is closer to the injector, fuel rich combustion takes place resulting in soot particle formation.

Turbulent mixing is required to achieve high combustion efficiency and low air emissions of soot, CO, HC, and so on. As a result, more energy is needed to inject the fuel gas at the same flow rate. This energy consumption increases as the fifth power of the burner size [13]. Multiple small burners rather than a single large one can be employed in large facility.

7.2 Liquid Fuel Combustion

Liquid fuels are atomized and burned in the form of droplets. They can be burned in both stationary (such as a power plant) and mobile systems (like a truck engine). The combustion efficiency and the air emissions depend on the fuel type as well as the size and volatility of the fuel droplets.

The particle dynamics introduced in Chap. 4 can be used to describe the droplet dynamics in the combustion chamber. It has been found that the droplet drag coefficient is very close to that for a solid sphere of the same diameter [49], therefore, it is reasonable to assume that fine spray droplets would follow the carrier gas in a combustion chamber and the motion is in the Stokes region, i.e., $Re_{\mathrm{p}} < 1$ (see Sect. 4.1.2). The corresponding drag coefficient is then described as

$$C_{\mathrm{D}} = 24/Re_{\mathrm{p}} \tag{7.1}$$

At high Reynolds number, large droplets may break into smaller ones and these small droplets will have low Reynolds numbers again. In many practical combustion analyses, we can assume laminar flow near the droplet surface.

When the fuel droplets enter a combustion system, elevated surrounding temperature enables the evaporation of the droplets and combustion takes place between the oxidizer and the fuel vapors.

7.2.1 Droplet Vaporization

There have been many models for fuel droplet evaporation in spray combustion. Most of them describe fuel droplets as spherical; the only relative motion between the droplets and gas involves radial convection due to vaporization. A critical review was given by Sirignano [42] for droplet vaporization in a high-temperature environment. However, it would deviate too much from the scope of this book if we continued on the combustion theories. For this reason, only a simple analysis is introduced as follows.

The energy required to vaporize a fuel droplet can be calculated using

$$q' = c_p(T_v - T_0) + h_{fg} \tag{7.2}$$

where c_p is the specific heat of the fuel, T_v is the vaporization temperature for the liquid fuel (K), T_0 is the initial temperature when the droplets are sprayed into the combustion system, and h_{fg} is the latent heat of vaporization of the fuel (J/mol). The energy needed for the vaporization of the liquid droplets is taken from the surrounding gases by thermal radiative and conductive heat transfer, which can be found in many advanced heat and mass transfer books.

The droplet size decreases as the evaporation proceeds. Assuming vapor velocity profile is symmetric around the center of the droplet and the rate of vaporization is described as

$$\dot{m}_v = \rho_v u 4\pi r^2 = \rho_v u_s 4\pi r_s^2 \tag{7.3}$$

where $\dot{m}_v$ is the vaporization rate (kg/s); it is also the loss rate with respective to the liquid droplet. ρ_v is vapor density, u_s is the vapor speed leaving the surface and r_s is the radius of the droplet. $4\pi r_s^2$ is the surface area of the droplet sphere.

Assuming that only conductive heat transfer dominates and that the radial profiles of temperature and compositions are quasi-steady, the change of the droplet diameter can be related to the heat transfer from the surrounding to the droplet surface. Conservation of energy for the droplet leads to

$$\dot{m}_v c_p(T - T_v) + \dot{m}_v h_{fg} = 4\pi r^2 k \frac{dT}{dr} \tag{7.4}$$

where T is the temperature at r, k is the thermal conductivity of the liquid droplet at the surface. LHS is the energy for liquid evaporation and RHS is for the heat transfer by conduction.

The fuel vapor resulted from the evaporation is then transported by diffusion and convention from the surface of the droplet to the surrounding gas phase. The corresponding convective diffusion can be described as

$$\dot{m}_{\mathrm{v}}(1-f_{\mathrm{v}}) = -4\pi r^2 \rho_{\mathrm{v}} D \frac{\mathrm{d}f_{\mathrm{v}}}{\mathrm{d}r} \tag{7.5}$$

where f_{v} is the vapor mass fraction at the droplet surface. D is the diffusivity of the vapor in the surrounding gas phase.

Integration of Eqs. (7.4) and (7.5), respectively, from the droplet surface $r = r_{\mathrm{s}}$ to $r \to \infty$ leads to

$$\ln\left(\frac{T_{\mathrm{s}} - T_{\mathrm{v}} + h_{\mathrm{fg}}/c_{\mathrm{p}}}{T_{\infty} - T_{\mathrm{v}} + h_{\mathrm{fg}}/c_{p}}\right) = -\left(\frac{c_{\mathrm{p}}}{4\pi k}\right)\frac{\dot{m}_{\mathrm{v}}}{r_{\mathrm{s}}} \tag{7.6}$$

$$\ln\left(\frac{1-f_{\mathrm{v,s}}}{1-f_{\mathrm{v},\infty}}\right) = -\left(\frac{1}{4\pi\rho_{\mathrm{v}}D}\right)\frac{\dot{m}_{\mathrm{v}}}{r_{\mathrm{s}}} \tag{7.7}$$

where subscript s is the for the droplet surface. T_{s} is the surface temperature of the droplet. Equation (7.6) gives the evaporation rate as

$$\dot{m}_{\mathrm{v}} = \frac{4\pi r_{\mathrm{s}} k}{c_{\mathrm{p}}} \times \ln\left[\frac{c_{\mathrm{p}}(T_{\infty} - T_{\mathrm{v}}) + h_{\mathrm{fg}}}{c_{\mathrm{p}}(T_{\mathrm{s}} - T_{\mathrm{v}}) + h_{\mathrm{fg}}}\right]. \tag{7.8}$$

At thermodynamic equilibrium state, the surface temperature is assumed to be the same as the vaporization temperature, $T_{\mathrm{s}} \approx T_{\mathrm{v}}$, then Eq. (7.8) is simplified as

$$\dot{m}_{\mathrm{v}} = \frac{4\pi r_{\mathrm{s}} k}{c_{\mathrm{p}}} \ln\left[1 + \frac{c_{\mathrm{p}}(T_{\infty} - T_{\mathrm{s}})}{h_{\mathrm{fg}}}\right]. \tag{7.9}$$

Conservation of mass leads to the relationship between the mass of the droplet $m_{\mathrm{l}} = \rho_{\mathrm{l}}(4\pi/3)r_{\mathrm{s}}^3$ and the vaporation rate can is $\dot{m}_{\mathrm{v}} = -\mathrm{d}m_{\mathrm{l}}/\mathrm{d}t$

$$\dot{m}_{\mathrm{v}} = -\frac{\mathrm{d}m_{\mathrm{l}}}{\mathrm{d}t} = \frac{\mathrm{d}}{\mathrm{d}t}\left(\frac{4}{3}\pi\rho_{\mathrm{l}} r_{\mathrm{s}}^3\right) = -4\pi\rho_{\mathrm{l}} r_{\mathrm{s}}^2 \frac{\mathrm{d}r_{\mathrm{s}}}{\mathrm{d}t}. \tag{7.10}$$

Combination of Eqs. (7.9) and (7.10) leads to

$$-4\pi\rho_{\mathrm{l}} r_{\mathrm{s}}^2 \frac{\mathrm{d}r_{\mathrm{s}}}{\mathrm{d}t} = \frac{4\pi r_{\mathrm{s}} k}{c_{\mathrm{p}}} \times \ln\left[1 + \frac{c_{\mathrm{p}}(T_{\infty} - T_{\mathrm{s}})}{h_{\mathrm{fg}}}\right]$$

$$-r_{\mathrm{s}}\,\mathrm{d}r_{\mathrm{s}} = \left\{\frac{k}{\rho_{\mathrm{l}} c_{\mathrm{p}}} \ln\left[1 + \frac{c_{\mathrm{p}}(T_{\infty} - T_{\mathrm{s}})}{h_{\mathrm{fg}}}\right]\right\} \mathrm{d}t. \tag{7.11}$$

As a simplification, the term in { } can be assumed constant at quasi-steady state. Then the droplet diameter over time can be determined by integration from the initial droplet diameter $r_s = r_{s0}$ at $t = 0$ and r_s at any time t.

$$-\int_{r_{s0}}^{r_s} r_s \, dr_s = \int_0^t \left\{ \frac{k}{\rho_l c_p} \ln\left[1 + \frac{c_p(T_\infty - T_s)}{h_{fg}}\right] \right\} dt \tag{7.12}$$

$$\frac{1}{2}\left(r_{s0}^2 - r_s^2\right) = \frac{tk}{\rho_l c_p} \ln\left[1 + \frac{c_p(T_\infty - T_s)}{h_{fg}}\right] \tag{7.13}$$

Equation (7.13) can be used to estimate the time for a liquid fuel droplet to complete vaporization, or life time of the droplet. Substitute $r_s = 0$ into Eq. (7.13), we have

$$t_e = \frac{r_{s0}^2}{\frac{2k}{\rho_l c_p} \cdot \ln\left[1 + \frac{c_p(T_\infty - T_s)}{h_{fg}}\right]} \tag{7.14}$$

The lifetime of the droplets increases with their sizes, and it is in the order of millisecond [42]. The life time t_e is important to characterize the combustion efficiency of atomized liquid fuel droplets and the consequent air emissions. If t_e is longer than the residence time of the combustion system, the droplets will not be burned completely because of incomplete vaporization. As a result, liquid fuel droplets exist in the flue gas (or exhaust). They may be part of the particulate emissions. Meanwhile residual fuel evaporation continues but these extra vapors are not burned at low temperature; this result in extra volatile organic compound (VOC) emissions.

Since t_e is proportional to the square of initial droplet diameter, r_{s0}^2, it is imperative that fine spray droplets favor combustion efficiency and lower the air emissions. However, the upper limit of the droplet size depends on the design of nozzles employed.

7.2.2 Vapor Combustion

When the combustible vapor is mixed with the oxidizers at high temperature, combustion takes place in a thin flame surrounding the droplet. As depicted in Fig. 7.2, vapor concentration decreases from the droplet surface r_s to the flame r_f, and vapor is oxidized instantaneously in the thin flame. Oxidizer is transported to the flame from the surrounding environment by diffusion that is driven by the concentration gradient. Part of the heat produced by the combustion is transferred to the surface of the fuel droplet to sustain the droplet vaporization. Combustion in the zone between the flame and the droplet surface is highly fuel rich due to the great fuel vapor concentration, and the combustion products join the vapor as combustible gases. The combustion at $r > r_f$ is fuel lean and complete owing to the sufficient oxygen supply.

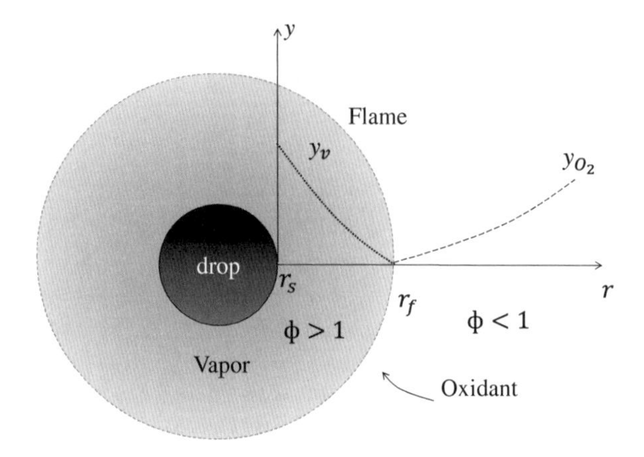

Fig. 7.2 Single droplet combustion model

7.3 Solid Fuel Combustion

7.3.1 Solid Fuels

Solid fuels are easy to transport, store, and produce. They have moderate ignition temperature. They are mainly used for combustion in stationary combustion processes; the combustion products are characterized with high ash content and low combustion efficiency. Coal is the most abundant solid fuel. It represents about 1/3 of the global primary energy production. The top five coal consumers are China, USA, India, Russia, and Germany.

In the combustion stoichiometry analysis (Chap. 3), the chemical formula of the fuel has to be determined. Properties of typical solid fuels are determined by proximate analysis and ultimate analysis.

- Proximate analysis of solid fuels determines moisture, volatile matter, ash, and fixed carbon (in coals and cokes) to rank the fuels by comparing the ratio of combustible to incombustible constituents.
- Ultimate analysis provides more information to include elemental analysis so the simplified chemical formula can be obtained for stoichiometry analysis.

ASTM International Standard [1] (MNL11271M, Proximate Analysis) specifies how to conduct the proximate analysis of coal. The result separates the products into four groups:

(1) moisture,
(2) volatile matter, consisting of gases and vapors driven off during pyrolysis,
(3) fixed carbon, the nonvolatile fraction of coal, and
(4) ash, the inorganic residue remaining after combustion.

Table 7.1 Examples of proximate analysis and ultimate analysis results

Coal rank	Proximate analysis (wt% as received)			
	Fixed carbon	Volatile matter	Moisture	Ash
Anthracite	81.1	7.7	4.5	6.0
Bituminous	54.9	35.6	5.3	4.2
Subbituminous	43.6	34.7	110.5	11.2
Lignite	27.8	24.9	36.9	10.4

Coal Rank	Ultimate Analysis (wt% moisture and ash free)					Net Heating Value (moisture and ash free) (MJ/kg)
	C	H	O	N	S	
Anthracite	91.8	3.6	2.5	1.4	0.7	36.2
Bituminous	82.8	5.1	10.1	1.4	0.6	36.1
Subbituminous	76.4	5.6	14.9	1.7	1.4	31.8
Lignite	71.0	4.3	23.2	1.1	0.4	26.7

Source Higman and van Burgt [27]

Another ASTM International Standard [2] (MNL11272M, Ultimate Analysis) describes how to conduct the ultimate analysis of coal and coke. It determines the carbon, hydrogen, sulfur, nitrogen, and ash in the material as a whole, and estimates the amount of oxygen by difference.

Examples of proximate analysis and ultimate analysis results are shown in Table 7.1.

Example 7.1: Fuel formula In an ultimate analysis of a bituminous coal, elemental analysis shows that the weight fractions of C, H, O, N, and S are, respectively, 82.8, 5.1, 10.1, 1.4, and 0.6. Determine its molecular formula.

Solution From the elemental analysis result, we have Table 7.2. Therefore, the molecular formula can be described as $C_{184}H_{136}O_{16.83}N_{2.67}S$

Note that the sulfur content is 0.6 %, and it is considered as low sulfur coal.

Table 7.2 Fuel formula calculation in Example 7.1

Element	C	H	O	N	S
Fraction (%)	82.8	5.1	10.1	1.4	0.6
Mass/mole	12	1	16	14	16
Mole weight	6.90	5.10	0.63	0.10	0.04
Normalized against S	184	136	16.83	2.67	1

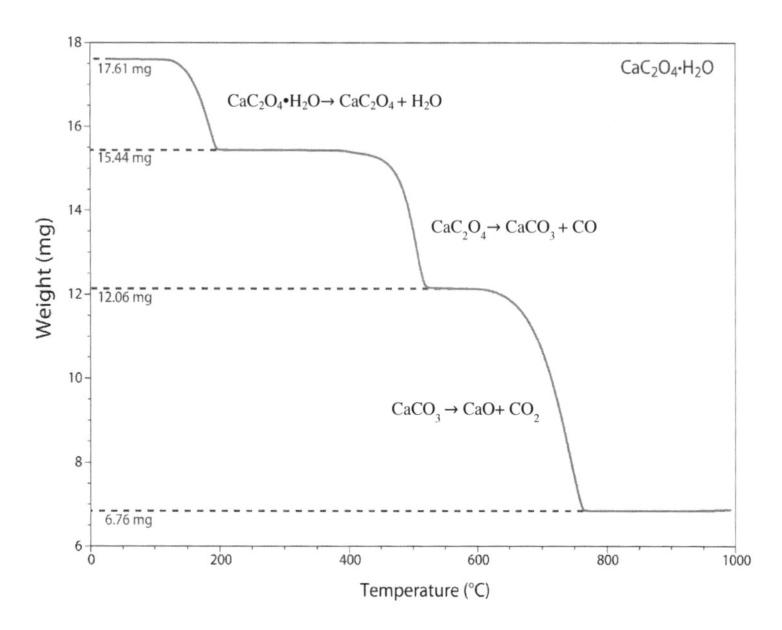

Fig. 7.3 An example of TGA analysis results (used with permission from David Harvey, DePauw University)

- Thermogravimetric analysis
  ASTM specified methods are time-consuming and require a significant amount of samples. An alternative method is thermogravimetric analysis (TGA), which requires a small sample size and gives results fast.

Figure 7.3 shows an example of TGA analysis. It was obtained for $CaC_2O_4 \cdot H_2O$ by heating a sample of 17.61 mg from room temperature to 1,000 °C at a rate of 20 °C/min [23]. Each change in mass results from the loss of a volatile products, and chemical compositions of the volatile products can be determined.

By using the values in Fig. 7.3, we can determine that the resultant gaseous products include, step by step from low to high temperature, 2.17 mg of water, 3.38 mg of carbon monoxide, and 5.30 mg of carbon dioxide. The final residue is CaO.

7.3.1.1 Coal Classification

ASTM (D388–12) Standard, *Classification of Coals by Rank*, defines the rank of coals quantitatively. The rank of a coal is determined based on its degree of metamorphism or alternation (or in a plain word, age). In the order of progressive alternation, coals can be classified into

- Peat
- Lignite coal

- Sub-bituminous coal
- Bituminous coal, and
- Anthracite coal

Peat is considered as the youngest coal in Europe and Asia, but it is considered as biomass in Canada. In Northern Ontario, the Ring of Fires, there is a great amount of peat covered in the forest. The majority of its weight (about 95 %) is water and it requires extensive energy to dry it before combustion, and the consequent thermal efficiency is the lowest.

By underground heating and pressurizing over thousands of years, peat becomes lignite. Lignite is also considered as an immature coal because of its high water content and low heating value. As time passes, lignite becomes darker and harder as sub-bituminous coal followed by bituminous and finally anthracite coals. This terminal rank represents the ultimate maturation of coals.

7.3.2 Solid Fuel Combustion

Solid fuels, coal or biomass, are burned after size reduction. Pulverized coal or biomass can be as small as few micrometers and most of them are smaller than 0.2 mm by mass. They are injected into a furnace where it is mixed with oxidants for combustion.

The combustion process of these solid particles is modeled by

- devolatilization or pyrolysis,
- volatile combustion, and
- char combustion.

7.3.2.1 Devolatilization

Devolatilization or pyrolysis is a complicated process that involves heat and mass transfer as well as chemical decompositions. In general, when a coal particle enters a combustion chamber where the gas is hot, the heating results in the release of volatiles from the pores of the coal particle. The greater the heating rate, the faster the volatiles release. Meanwhile, the size of the particle changes too. Typical heating rate in a pulverized coal combustor is about 10^4-10^5 K/s and the corresponding local temperature may be as high as 2,100 K [14].

The simplest model for this process is a single step model based on a global kinetics that applies to most solid fuels like coal, biomass, and plastic [48].

$$\text{Fuel} \xrightarrow{k_f} xV + (1-x)\text{Ch} \tag{7.15}$$

where V and Ch stand for volatiles and chars, respectively.

Two-step models take into consideration the heating rate and the competition between different volatiles.

$$\begin{aligned} &\text{Fuel} \xrightarrow{k_{f1}} x_1 V_1 + (1 - x_1)\text{Ch}_1 \\ &\text{Fuel} \xrightarrow{k_{f2}} x_2 V_2 + (1 - x_2)\text{Ch}_2 \end{aligned} \tag{7.16}$$

This simplified reaction formula indicates that the particle after devolatilization is decomposed into volatiles and chars, and there is nothing else. The corresponding reaction rate can be described using (modified) Arrhenius expressions

$$k_i = A_i \exp\left(-\frac{E_i}{RT}\right); \quad i = 1, 2 \tag{7.17}$$

With the 2-step model, the conversion rate of the fuel can be described as

$$-\frac{\mathrm{d}x_F}{\mathrm{d}t} = k_{f1} x_F + k_{f2} x_F \tag{7.18}$$

where x_F is the remaining mass fraction of the solid fuel. The 2-Step model parameters are summarized in Table 7.3.

The 1-step model is easy to use but it could not produce a good agreement over a broad temperature range. The 2-step model is more practical for its ease to use and reasonable kinetic agreement over a wide range of temperature; however, it is not based on physical mechanisms [14]. Other models for devolatilization can be found in literature, e.g., the report by Fletcher [14].

Table 7.3 Parameters for calculation of reaction rates of 2-step devolatilization

$k_{f1} = A_1 \exp\left(-\frac{E_1}{RT}\right)$	$k_{f2} = A_2 \exp\left(-\frac{E_2}{RT}\right)$	References
$A_1 = 2.0 \times 10^5$	$A_2 = 1.3 \times 10^7$	[30]
$E_1 = 25{,}000$	$E_2 = 40{,}000$	
$x_1 = 0.3$	$x_2 = 1.0$	
$A_1 = 3.7 \times 10^5$	$A_2 = 1.46 \times 10^{10}$	[46]
$E_1 = 17{,}600$	$E_2 = 60{,}000$	
$x_1 = 0.39$	$x_2 = 0.8$	
$A_1 = 3.7 \times 10^5$ (1/s)	$A_2 = 1.46 \times 10^{10}$ (1/s)	[43]
$E_1/R = 8{,}857$ K	$E_2/R = 30{,}200$ K	
x_1 = proximate analysis volatile matter	$x_2 = 0.8$	

7.4 Formation of VOCs and PAHs

Volatile organic compounds are a family of organic compounds that are volatile in nature. They are mainly lower (C_1–C_4) paraffin, olefins, aldehydes (e.g., formaldehyde), ketones (e.g., acetone) and aromatics (e.g., benzene, toluene, benzaldehyde, phenol). Sometimes, polycyclic aromatic hydrocarbon (PAH) (boiling point 218 °C) is also referred to as a VOC. The organic compounds that are not involved in the formation of smog such as methane, CO, and halogenated organics like 1,1,1-trichloroethane and CFCs are not considered as VOCs.

PAHs are produced during combustion when temperature is about 500–800 °C, and they are oxidized further above 800 °C. Therefore, PAHs mainly present in the low temperature zone of the flame or a combustion facility due to the poor fuel/oxygen mixing. Heavy duty diesel engines (mainly trucks) emit ~ 1,300 μg/km of lighter PAHs such as pyrene, fluoranthene, etc. while gasoline-fueled cars emit ~ 100 μg/km of more hazardous heavier PAHs such as benzo(a)pyrene and dibenz (a,h)anthracene [36].

7.5 Formation of CO and CO_2

7.5.1 Volatile Oxidation

The volatiles from liquid fuel vaporization or solid fuel devolatilization contain various combustible gas molecules. It is challenging to decide their exact formula if we were to use the combustion principles introduced in Chap. 3. Alternatively, we can use a simple step reaction model [48] for volatile combustion. Assuming oxygen is the oxidant,

$$\text{Volatiles} + O_2 \xrightarrow{k_{f1}} CO + H_2O \tag{7.19}$$

$$CO + \frac{1}{2}O_2 \xrightarrow{k_{f2}} CO_2. \tag{7.20}$$

The rate of combustion really depends on the rate of mixing.

Another key factor is the heat of combustion of the volatiles, which can be calculated or experimentally determined. With this information known, the reaction rate can be estimated using the combustion kinetics introduced in Chemical Kinetics and Chemical Equilibrium in Chap. 3.

7.5.2 Char Oxidation

Similar to the volatile oxidation analysis above, a simple model [13, p. 150] for the oxidation of char or other similar solid carbon-based fuels is

$$C_s + \frac{1}{2}O_2 \xrightarrow{k_{f1}} CO \tag{7.21}$$

$$C_s + CO_2 \xrightarrow{k_{f2}} CO \tag{7.22}$$

where C_s is in solid phase carbon. The oxidizers are not limited to O_2 and CO_2; other species such as OH, O, and H_2O can also be effective oxidizers.

The oxidation takes place at the surface of the solid char and the corresponding rate of reaction is governed by the diffusion of oxidizers to the surface of the char.

The apparent rate of char oxidation can be estimated using the equation given by Flagan and Seinfeld [13]

$$r = A\exp\left(-\frac{E}{RT}\right)\left(P_{O_2}^n\right) \qquad \left(\mathrm{kg/m^2 \cdot s}\right) \tag{7.23}$$

where $P_{O_2}^n$ is the partial pressure of the oxidier, with a unit of atm. A, E/R, and n are constants for specific solid fuels and they can be determined experimentally.

The formation of carbon monoxide and hydrogen carbon is driven by local fuel rich combustion in the flame and imperfect mixing between fuel and oxygen.

The level of CO in the exhaust or flue gas depends on the design of the combustion system. Oxidation of CO to CO_2 takes place in the luminous zone by the following mechanisms [26].

$$\begin{gathered} CO + OH \rightarrow CO_2 + H \\ k_{f1} = 4.4T^{1.5}\exp(372/T) \qquad \mathrm{m^3/mol \cdot s} \end{gathered} \tag{7.24}$$

$$\begin{gathered} H + CO_2; \rightarrow CO + O + H \\ k_{f2} \cong 0 \end{gathered} \tag{7.25}$$

$$\begin{gathered} CO + O_2 \rightarrow CO_2 + O \\ k_{f3} = 2.5 \times 10^6\exp(-24{,}060/T) \qquad \mathrm{m^3/mol \cdot s} \end{gathered} \tag{7.26}$$

$$\begin{gathered} CO + O + M \rightarrow CO_2 + M \\ k_{f4} = 53\exp(2{,}285/T) \qquad \mathrm{m^6/mol \cdot s} \end{gathered} \tag{7.27}$$

$$\begin{gathered} CO + HO_2 \rightarrow CO_2 + OH \\ k_{f5} = 1.5 \times 10^8\exp(-11{,}900/T) \qquad \mathrm{m^3/mol \cdot s.} \end{gathered} \tag{7.28}$$

The rates of formation of O, OH, and H are usually assumed to be at equilibrium and they can be estimated from the following reactions

$$\begin{gathered}\frac{1}{2}O_2 \leftrightarrow O \\ K_{P,O} = 3{,}030\exp(-30{,}790/T) \qquad \text{atm}^{-\frac{1}{2}}\end{gathered} \tag{7.29}$$

$$\begin{gathered}\frac{1}{2}H_2O \leftrightarrow \frac{1}{4}O_2 + H \\ K_{P,H} = 44{,}100\exp(-42{,}500/T) \qquad \text{atm}^{-3/4}\end{gathered} \tag{7.30}$$

$$\begin{gathered}\frac{1}{2}H_2O + \frac{1}{4}O_2 \leftrightarrow OH \\ K_{P,OH} = 166\exp(-19{,}680/T) \qquad \text{atm}^{1/4}.\end{gathered} \tag{7.31}$$

The overall reaction for the oxidation of CO is

$$CO + \frac{1}{2}O_2 \leftrightarrow CO_2 \tag{7.32}$$

It is very sensitive to temperature. The chemical equilibrium constants at different temperatures can be found in Table 3.2 above.

Dryer [10] gave an empirical equations for the consumption rate of CO and the rate of formation of CO_2 as follows

$$-\frac{d[CO]}{dt} = 10^{14.6\pm0.25}\exp\left(-\frac{40{,}000 \pm 1250}{RT}\right)[CO][H_2O]^{\frac{1}{2}}[O_2]^{\frac{1}{4}}\ (\text{mol/cm}^3\cdot\text{s}) \tag{7.33}$$

$$-\frac{d[CO_2]}{dt} = 10^{14.75\pm0.4}\exp\left(-\frac{43{,}000 \pm 2{,}200}{RT}\right)[CO][H_2O]^{\frac{1}{2}}[O_2]^{\frac{1}{4}}\ (\text{mol/cm}^3\cdot\text{s}) \tag{7.34}$$

7.6 Formation of SO_2 and SO_3

Nearly all fossil fuels contain sulfur atoms. Some of the sulfur in fuels is eventually oxidized to SO_2 and SO_3. Typical values for the sulphur content of various fuels are given in Table 7.4 [51]. Sulphur in coal is present in both organic and inorganic forms, the latter being pyretic sulphur (FeS_2) and sulphates (Na_2SO_4, $CaSO_4$, $FeSO_4$). Organic sulphur is present in the form of sulphides, mercaptanes, bisulphides, thiophenes, thiopyrones, etc. These organic compounds are also found in crude oils and gases.

Table 7.4 Typical values for sulphur content of fuels (wt%, dry)

Fuel		Sulphur content (wt%, dry)
Fossil fuels	Coal	0.2–5
	Oil	1–4
	Natural gas	0–10
	Light fuel oil	<0.5
	Heavy fuel oil	<5
	Peat	<1
	Petroleum coke	~5
Biomass	Wood	<0.1
	Straw	~0.2
	Bark	<2

The more sulfur in the fuel, the higher level of SO_2 emission. SO_2 is mainly formed when the sulfur elements are oxidized by O_2,

$$S + O_2 \rightarrow SO_2 \tag{7.35}$$

and SO_2 can be oxidized to SO_3.

Their concentrations at equilibrium can be determined by the overall reaction, using the knowledge we learned in Chap. 3.

$$\begin{aligned} &SO_2 + {}^1\!/\!{}_2O_2 \leftrightarrow SO_3 \\ &K_{P,SO_3} = 1.53 \times 10^{-5} \exp\left(\tfrac{11{,}760}{T}\right) \qquad \text{atm}^{-\frac{1}{2}} \end{aligned} \tag{7.36}$$

The van't Hoff equation (7.36) of the chemical equilibrium constant indicates that the concentration of SO_3 increases with the decreasing combustion temperature. The kinetics of SO_2 oxidation without catalytic effect can be described as follows [51].

$$SO_2 + O + M \rightarrow SO_3 + M \quad k_{f1} = 8.0 \times 10^4 \exp(-1{,}400/T)\ \text{m}^6/\text{mol}\cdot\text{s} \tag{7.37}$$

$$SO_3 + O + M \rightarrow SO_2 + O_2 + M \quad k_{f2} = 7.04 \times 10^4 \exp(785/T)\ \text{m}^6/\text{mol}\cdot\text{s} \tag{7.38}$$

$$SO_3 + H \rightarrow SO_2 + OH \quad k_{f3} = 1.5 \times 10^7\ \ \text{m}^3/\text{mol}\cdot\text{s} \tag{7.39}$$

Assuming constant-temperature and fuel-lean combustion, the net rate of SO_3 formation is thus described by Eq. (7.40)

$$r_{SO_3} = \frac{d[SO_3]}{dt} = k_{f1}[SO_2][O][M] - k_{f2}[SO_3][O][M] - k_{f3}[SO_3][H] \tag{7.40}$$

Sulfur mass conservation gives $[SO_2] + [SO_3] = [S] = \text{constant}$, or $[SO_2] = ([S] - [SO_3])$ and Eq. (7.40) becomes

$$r_{SO_3} = k_{f1}([S] - [SO_3])[O][M] - k_{f2}[SO_3][O][M] - k_{f3}[SO_3][H] \tag{7.41}$$

When the SO_3 concentration reaches steady state, corresponding to $r_{SO_3} = 0$, the equilibrium SO_3 concentration is described in terms of [O] and [H] as

$$[SO_3]_s = \frac{k_{f1}[O][M][S]}{(k_{f1} + k_{f2})[O][M] + k_{f3}[H]} \tag{7.42}$$

This equation can be used to estimate the steady-state SO_3 concentration as a fraction of the total SO_x concentration $[S] = [SO_2] + [SO_3]$. Mostly [O] and [H] are assumed constant.

The formation of sulphuric pollutants during solid fuel combustion can also be described by the following step reactions [51].

Fuel sulfur is first heated and devolatilized through,

$$\text{Fuel} - S(s) \rightarrow H_2S + COS + \cdots + \text{Char} - S(s) \tag{7.43}$$

Both solid and gases (vapors) are produced through this reaction. The solid phase char sulfur can be oxidized through the following reactions, where H_2S, SO_2 and COS are produced.

$$\text{Char} - S + O_2 \rightarrow SO_2 \tag{7.44}$$

$$\text{Char} - S + CO_2 \rightarrow COS \tag{7.45}$$

$$\text{Char} - S + H_2O \rightarrow H_2S. \tag{7.46}$$

The gas phases produced during the above three reactions can further react through

$$H_2S + 1\tfrac{1}{2}O_2 \rightarrow SO_2 + H_2O \tag{7.47}$$

$$H_2S + CO_2 \rightarrow COS + H_2O \tag{7.48}$$

$$H_2S + CO \rightarrow H_2 + COS \tag{7.49}$$

$$H_2S + COS \rightarrow CS_2 + H_2O \tag{7.50}$$

$$CS_2 \rightarrow C + 2/x\,S_x. \tag{7.51}$$

In the products, the concentrations of SO_2 and SO_3 at equilibrium can still be determined by the overall reaction in Eq. (7.36)

$$SO_2 + 1/2O_2 \leftrightarrow SO_3 \quad (7.52)$$

For coal combustion, sulfur in fuels is eventually oxidized to SO_2 and/or SO_3, with a small amount being bound to ashes. For combustion of liquid fuels in engine, SO_2 concentration in the engine exhaust depends on the type of fuel burned. Marine engine exhaust usually has the highest SO_2 emission, and it can be in a level that is comparable with coal fired power plants, (about 1 %) because marine diesel fuels are of very low quality and high sulfur content.

The chemical equilibrium constant in Eq. (7.36) indicates that the concentration of SO_3 increases with the decreasing combustion temperature. When the temperature is below 900 K, SO_3 would have been the dominant SO_x air pollutant at equilibrium. However, in a typical flue gas or engine exhaust, SO_2 is still dominating over SO_3. One of the reasons is that the conversion from SO_2 to SO_3 is too slow to reach an equilibrium state. In a typical engineering practice, about 3 % of SO_2 is converted into SO_3. Typical SO_2 concentration in a coal fired power plant flue gas stream is in the order of 1,000 ppm (0.1 %). It shall be much lower in modern gasoline or natural gas fired stationary combustion facilities because sulfur has been removed from the fuels.

The oxidation of SO_2 to SO_3 is usually slow unless the combustion temperature is above 1,100 °C or there are catalysts present to expedite the reaction. For example, sulphur in the liquid fuels supplied to automobiles is oxidized to SO_2 in the engine. But in the exhaust SO_3 is present due to the presence of various catalytic metals in the fuel, engine housing, and/or the filtration systems. For stationary combustion sources like a coal-fired power plant, there are also various metals in the fuel, and some of them may serve as catalysts for the oxidation of SO_2.

7.7 NO_x

Nitrogen oxides (NO_x) represent the following seven oxides of nitrogen [45].

- nitric oxide (NO)
- nitrogen dioxide (NO_2)
- nitrous oxide (N_2O)
- dinitrogen dioxide (N_2O_2)
- dinitrogen trioxide (N_2O_3)
- dinitrogen tetroxide (N_2O_4)
- dinitrogen pentoxide (N_2O_5)

However, NO_x are often referred only to NO and NO_2 by the environmental protection agencies in most jurisdictions. It is because these two gases are the major contributors to air pollution. Therefore, only NO and NO_2 are introduced in the following section.

7.7.1 Nitric Oxide

7.7.1.1 Thermal NO

For the analysis above using simple chemical stoichiometry (Chap. 3), it was assumed that nitrogen (N_2) is inert and does not react with oxygen. This is no longer true when the reaction temperature is high enough. As a result, NO is produced by the oxidation of N_2 in the air through the following chemical reaction,

$$\begin{gathered}\frac{1}{2}N_2 + \frac{1}{2}O_2 \leftrightarrow NO \\ K_{P,NO} = 4.71\exp\left(-\frac{10{,}900}{T}\right)\end{gathered} \tag{7.53}$$

As indicated by the formula for equilibrium constant ($K_{P,NO}$), oxidation of nitrogen proceeds slowly at low temperatures, but very fast at high temperatures. The formation of NO from molecular nitrogen was first proposed by Zeldovich [50], and it is thus referred to as the Zeldovich mechanism. NO formed by this mechanism is also called thermal NO.

Thermal NO formation was first introduced in the 1940s. The formation of thermal NO is very sensitive to the combustion temperature, and its formation is negligible when temperature is 1,000 °C or lower. On the other hand, the rate of NO formation is significant and increases exponentially when the temperature is over 1,400 °C [20]. The reactions in Table 7.5 can be used to describe the step reactions of thermal NO formation.

First of all, the concentration of oxygen atoms required for the initiation of reaction is strongly dependent on temperature because more oxygen atoms are available to the reaction at higher temperature. Secondly, this reaction produces nitrogen atoms, which is the bottleneck that limits the rate of NO formation, which is also sensitive to temperature.

Table 7.5 Thermal NO step reactions and rate constants

Reaction	Rate constant $(m^3/mol \cdot s)$	Equation
$N_2 + O \underset{k_{b1}}{\overset{k_{f1}}{\leftrightarrow}} NO + N$	$k_{f1} = 1.8 \times 10^8 \exp\left(-\frac{38{,}370}{T}\right)$	(7.54)
	$k_{b1} = 3.8 \times 10^7 \exp\left(-\frac{425}{T}\right)$	
$N + O_2 \underset{k_{b2}}{\overset{k_{f2}}{\leftrightarrow}} NO + O$	$k_{f2} = 1.8 \times 10^4 T \times \exp\left(-\frac{4{,}680}{T}\right)$	(7.55)
	$k_{b2} = 3.8 \times 10^3 T \times \exp\left(-\frac{-20{,}820}{T}\right)$	
$N + OH \underset{k_{b3}}{\overset{k_{f3}}{\leftrightarrow}} NO + H$	$k_{f3} = 7.1 \times 10^7 \times \exp\left(-\frac{450}{T}\right)$	(7.56)
	$k_{b3} = 1.7 \times 10^8 \times \exp\left(-\frac{24{,}560}{T}\right)$	

The net rate of NO formation under a constant pressure and temperature condition can be calculated by considering all three step reactions in Table 7.5.

$$\begin{aligned} r_{NO} = \frac{d[NO]}{dt} &= \{k_{1f}[N_2][O] - k_{1b}[NO][N]\} \\ &+ \{k_{2f}[O_2][N] - k_{2b}[NO][O]\} \\ &+ \{k_{3f}[N][OH] - k_{3b}[NO][H]\}. \end{aligned} \tag{7.57}$$

When all the reactions in Table 7.5 reach equilibrium state, the NO formation rate attributed to each of three step reactions can be described as, respectively,

$$\begin{aligned} r_1 &= k_{1f}[N_2]_e[O]_e = k_{1b}[NO]_e[N]_e \\ r_2 &= k_{2f}[O_2]_e[N]_e = k_{2b}[NO]_e[O]_e \\ r_3 &= k_{3f}[N]_e[OH]_e = k_{3b}[NO]_e[H]_e \end{aligned} \tag{7.58}$$

where the subscript e stands for equilibrium. For easy analysis, we also define two dimensionless concentrations as follows.

$$a = \frac{[NO]}{[NO]_e}; \quad b = \frac{[N]}{[N]_e} \tag{7.59}$$

Then the rate of NO formation before reaching equilibrium state Eq. (7.57) can be rewritten as

$$r_{NO} = r_1(1 - ab) + r_2(b - a) + r_3(b - a) \tag{7.60}$$

The concentration of N atoms is needed in order to solve this equation. By similar approach, we can get the rate of N atom formation

$$r_N = r_1(1 - ab) + r_2(a - b) + r_3(a - b) \tag{7.61}$$

By the pseudo-steady-state approximation, we can set the left hand side of Eq. (7.61), $r_N = 0$, and it leads to

$$b = \frac{r_1 + r_2 a + r_3 a}{r_1 a + r_2 + r_3} \tag{7.62}$$

Substitute Eq. (7.62) into Eq. (7.60), and we have the corresponding rate of NO formation as

$$r_{NO} = \frac{2r_1(r_2 + r_3)(1 - a^2)}{ar_1 + r_2 + r_3} \tag{7.63}$$

This equation can also be expressed in terms of dimensionless NO concentration, $a = [NO]/[NO]_e$

$$\frac{da}{dt} = \frac{1}{[NO]_e}\left\{\frac{2r_1(r_2 + r_3)(1 - a^2)}{ar_1 + r_2 + r_3}\right\} \quad (7.64)$$

Integration of this equation with an initial condition of $a = 0$ at $t = 0$ leads to the description of the NO concentration at any time.

$$\left(1 - \frac{r_1}{r_2 + r_3}\right)\ln(1 + a) - \left(1 + \frac{r_1}{r_2 + r_3}\right)\ln(1 - a) = \left(\frac{4r_1}{[NO]_e}\right)t \quad (7.65)$$

For easier expression, we can define the characteristic time for thermal NOx formation as,

$$\tau_{NO} = \frac{[NO]_e}{4r_1} \quad (7.66)$$

and the reaction rate ratio,

$$\rho_r = \frac{r_1}{r_2 + r_3} \quad (7.67)$$

By considering τ_{NO} and ρ_r, Eq. (7.65) becomes

$$(1 - \rho_r)\ln(1 + a) - (1 + \rho_r)\ln(1 - a) = t/\tau_{NO} \quad (7.68)$$

In a typical combustion process, the residence time is shorter than the characteristic time. As a result, NO does not reach its equilibrium concentration. Therefore, it is better for the actual NO concentration in the flame to be determined using this equation.

7.7.1.2 Prompt NO

Fenimore [12] found that some of the NO formed during combustion could not be explained by the aforementioned Zeldovich mechanisms. When equivalence ratio is greater than 1, the nitrogen in the air reacts to form hydrogen cyanide (HCN) through the following chemical reaction,

$$N_2 + CH \leftrightarrow HCN + N \quad (7.69)$$

Since there are oxygen-containing compounds in the combustion system, HCN produced in the above reaction and the nitrogen atom reacts further to produce NO through several chain reactions.

$$O_2 + N \leftrightarrow NO + O \quad (7.70)$$

$$HCN + OH \leftrightarrow CN + H_2O \quad (7.71)$$

$$CN + O_2 \leftrightarrow NO + CO \quad (7.72)$$

Prompt NO is formed only in a combustion zone of the flame where the combustion is incomplete and the hydrocarbon radicals are present [3]. These reactions take place very fast, thus it is known as prompt NO. The formation of prompt NO does not depend on temperature as significantly as the thermal NO. The prompt NO is formed mainly under lower temperature conditions during a short residence time [24].

7.7.1.3 NO Through Intermediate Component N_2O

The third mechanism is through N_2O [20, 32], also known as laughing gas, produced by the reaction between oxygen atoms and N_2:

$$O + N_2 + M \leftrightarrow N_2O + M \quad (7.73)$$

$$N_2O + O \leftrightarrow 2NO \quad (7.74)$$

where M can be any coexisting gas compound. Depending on the condition, N_2O could react again either forward to NO or backward to N_2, and the later often dominates. In general, the formation of NO increases with the air to fuel ratio and temperature.

N_2O is one of the greenhouse gases (GHGs). It can also be formed through other mechanisms, including oxidation of HCN and oxidation of char residue. Part of the volatile cyanide compounds of the fuel nitrogen, such as HCN, is oxidized homogeneously to N_2O through the following reactions:

$$HCN + O \leftrightarrow NCO + H \quad (7.75)$$

Table 7.6 Typical values for the nitrogen content of fuels (dry wt%)

Fuel	N content, wt%, dry
Coal	0.5–3
Oil	<1
Natural gas	0.5–20
Light fuel oil	~0.2
Heavy fuel oil	~0.5
Peat	1–2
Petroleum coke	~3
Wood	0.1–0.5
Straw	0.5–1
Bark	~0.5

$$NCO + NO \leftrightarrow N_2O + CO \tag{7.76}$$

The reaction is also very sensitive to temperature and it slows down significantly when the temperature rises. The reaction stops when the temperature is above 950 °C, and the NCO radicals are converted to NO.

7.7.1.4 Fuel NO

Most fuels contain nitrogen element, the NO originated from this part of nitrogen is referred to as fuel NO [19, 33, 37, 44]. The amount of nitrogen in fuel varies with the fuel type. As summarized in Table 7.6 [51], typical coal and oil contain chemically bound organic nitrogen, which is different from that found in natural gas. Depending on the refinery process, some natural gases contain virtually no nitrogen but others have quite a lot of nitrogen in form of N_2. Unlike fossil fuels, biomass is characterized with high nitrogen.

Although the amount of nitrogen in fuel is relatively small, the fuel nitrogen is much more reactive compared to the nitrogen present in the combustion air. Consequently, the formation of fuel NO from a nitrogen-rich fuel is higher than that from a nitrogen-lean fuel. Sometimes, as much as 80 % of the NO in the flue gas of a coal-firing furnace is produced from fuel nitrogen [33]. The fuel NO is sensitive to stoichiometry rather than the temperature because it forms readily at quite low temperatures [7].

For a simple comparison purpose, Fig. 7.4 shows the relative importance of thermal NO, prompt NO and fuel NO formation at different temperatures. Overall NO emissions increase with temperature, mainly due to the increase of thermal NO. Typical NO concentrations in the flue gases produced by the combustion of coal, oil, and natural gas before flue gas cleaning are in the order of hundreds of ppmv. The temperature in a typical industrial furnace is about 1,400–1,500 °C where the magnitude of the NO equilibrium concentration is about 1,000 ppmv. However, the

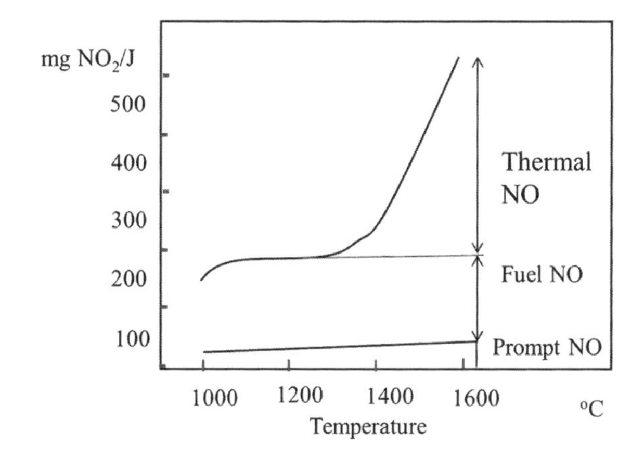

Fig. 7.4 Influence of temperature on *thermal* NO_x, *prompt* NO_x and *fuel* NO_x formation

NO equilibrium concentration drops sharply as the temperature goes down, and becomes negligible below 600 °C.

7.7.1.5 NO*x* Formation in Char Combustion

Char is formed during combustion of solid fuels, especially coal. This char residue forms N_2O under appropriate conditions. The N_2O formed from char nitrogen varies, depending on fuel property and devolatilization conditions. However, the formation mechanism of N_2O from char nitrogen is not yet fully clarified and different mechanisms have been proposed in literature.

Much of the NO formation in coal combustion is by char–nitrogen interaction [48]. Nitrogen atoms can be adsorbed by char, which is porous anyway. The char–nitrogen is oxidized to produce NO, and it can be described with a simplified reaction formula as

$$(\mathrm{N}+\mathrm{C})+\frac{1}{2}\mathrm{O_2}\rightarrow \mathrm{NO}+\mathrm{C_s} \tag{7.77}$$

where (N + C) represents nitrogen atom bounded with carbon by chemisorption on char. NO reduction may take place by the following two reactions

$$2\mathrm{NO}+2\mathrm{CO}\rightarrow \mathrm{N_2}+2\mathrm{CO_2} \tag{7.78}$$

$$\begin{aligned} &2\mathrm{NO}+\mathrm{C_s}\xrightarrow{k_{f6}}(\mathrm{O}+\mathrm{C})+\frac{1}{2}\mathrm{N_2}\\ &k_{f6}=0.21\exp\left(-1.31\times 10^4/T\right)\qquad \mathrm{kmol}/\left(\mathrm{m^2\cdot s\cdot atm}\right)\end{aligned} \tag{7.79}$$

where (O + C) is the oxygen bounded with char carbon by chemisorption, and it reacts with CO to produce CO_2

$$\begin{aligned} &\mathrm{CO}+(\mathrm{O}+\mathrm{C})\xrightarrow{k_{f7}}\mathrm{CO_2}+\mathrm{C_s}\\ &k_{f7}=7.4\times 10^{-4}\exp\left(-9.56\times 10^3/T\right)\qquad \mathrm{kmol}/\left(\mathrm{m^2\cdot s\cdot atm}\right)\end{aligned} \tag{7.80}$$

With the increasing temperature, the oxygen adsorbed onto the surface of carbon may also produce CO directly,

$$\begin{aligned} &(\mathrm{O}+\mathrm{C})\xrightarrow{k_{f8}}\mathrm{CO}\\ &k_{f8}=1.5\times 10^{-2}\exp\left(-2.01\times 10^4/T\right)\qquad \mathrm{kmol}/\left(\mathrm{m^2\cdot s\cdot atm}\right)\end{aligned} \tag{7.81}$$

Then, the overall rate of NO reduction can be determined as follows [48].

$$-r_{\mathrm{NO}}=\frac{k_{f6}P_{\mathrm{NO}}\left(k_{f7}P_{\mathrm{CO}}+k_{f8}\right)}{k_{f6}P_{\mathrm{NO}}+k_{f7}P_{\mathrm{CO}}+k_{f8}} \tag{7.82}$$

The overall rate of NO formation from char-bound nitrogen is also described by Williams et al. [48]

$$r_{NO} = -\frac{[N]_{char}}{[C]_{char}} r_{char} \tag{7.83}$$

where the reaction rate of char is

$$r_{char} = 254 \exp\left(-2.16 \times 10^{-2}/T\right) P_0^n \tag{7.84}$$

The analysis above is also applicable to biomass combustion with the following differences. First of all, most (80 % or more) of the biomass is volatile and there is little char left. The combustion of char fraction is minor compared to volatile combustion.

7.7.2 Nitrogen Dioxide

The oxidation of NO leads to the formation of NO_2 through the overall reaction

$$\begin{aligned} &NO + \frac{1}{2} O_2 \leftrightarrow NO_2 \\ &K_{P,NO_2} = 2.5 \times 10^{-4} \exp\left(\frac{6923}{T}\right) \qquad \text{atm}^{-1/2} \end{aligned} \tag{7.85}$$

The corresponding step reactions are

$$\begin{aligned} &NO + NO + O_2 \leftrightarrow NO_2 + NO_2 \\ &k_f = 1.2 \times 10^{-3} \exp(530/T) \qquad \text{m}^6/\text{mol} \cdot \text{s} \end{aligned} \tag{7.86}$$

$$\begin{aligned} &NO + O + M \leftrightarrow NO_2 + M \\ &\left\{ \begin{aligned} &k_f = 1.5 \times 10^3 \exp(940/T) \qquad \text{m}^6/\text{mol} \cdot \text{s} \\ &k_b = 1.1 \times 10^{10} \exp(-33{,}000/T) \qquad \text{m}^3/\text{mol} \cdot \text{s} \end{aligned} \right\} \end{aligned} \tag{7.87}$$

The conversion of NO to NO_2 can also take place when there are enough hydrogen peroxide radicals (HO_2).

$$\begin{aligned} &NO + HO_2 \leftrightarrow NO_2 + OH \\ &k_f = 2.1 \times 10^6 \exp(240/T) \qquad \text{m}^3/\text{mol} \cdot \text{s} \end{aligned} \tag{7.88}$$

Hydrogen peroxide radical is formed when a hydrogen atom reacts with oxygen in the presence of a third component (M).

$$H + O_2 + M \leftrightarrow HO_2 + M \quad (7.89)$$

This reaction is significant at low temperatures. As a result, considerable HO_2 concentration may be present in the low temperature zones of the flame, and consequently a significant part of the NO present in the cooler zones may react into NO_2 through Eq. (7.88).

In the high temperature zone of the flame, hydrogen and oxygen tends to react directly to form hydroxyl radicals and oxygen atoms:

$$H + O_2 \rightarrow OH + O \quad (7.90)$$

The NO_2 formed at lower temperature decomposes rapidly back to NO when it drifts into the high temperature zone of the flame.

$$NO_2 + H \rightarrow NO + OH \quad (7.91)$$

$$NO_2 + O \leftrightarrow NO + O_2 \quad (7.92)$$

Partially due to this decomposition process, NO_2 remains less than 5 % with 95 % or more NO in a typical flue gas. However, these reactions slow down at low concentrations of O and H. This situation arises when hot and cold streams are mixed rapidly.

7.8 Formation of Particulate Matter

In combustion inorganic minerals in fuel are converted into solid, liquid, and vapors. The solid and liquid contribute directly to the particulate matter formation, while the vapors could condense, solidify, and form secondary particulate matter before and after the emission. The final phase of the ash-forming materials depends on many factors including temperature, pressure, residence time, fuel particle size and size distribution, and the compounds in the combustion system. For solid fuels the ash-forming minerals are mainly converted to oxides of silicon, aluminum, and iron (SiO_2, Al_2O_3 and Fe_2O_3). For liquid and gaseous petroleum fuels, the particulate matter is mainly due to the incomplete combustion and secondary reaction of VOCs and SO_x/NO_x. However, this general knowledge may not apply to alternative and renewable fuels.

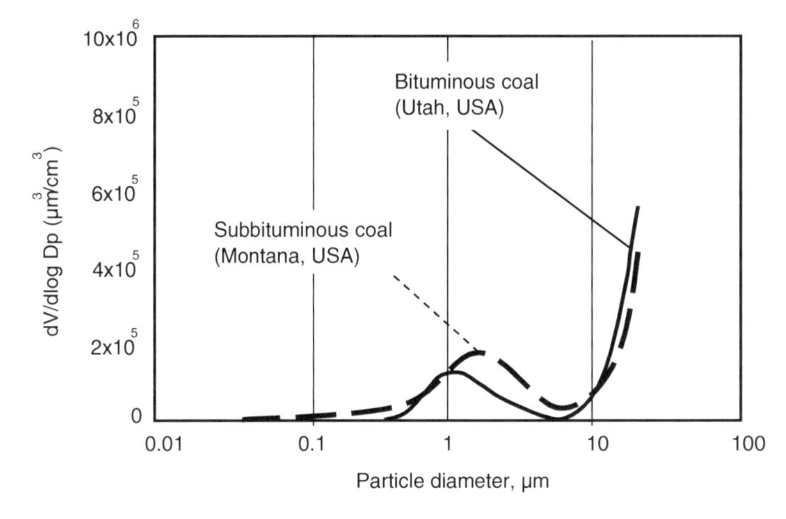

Fig. 7.5 Fly ash particle size distribution (*Data source* Linak et al. [31])

7.8.1 Ash-Forming Elements in Fuels

Fossil fuels contain highly integrated ash-forming matter in form of discrete particles, and bounded combustible compounds. A review of the corresponding particle forming mechanisms can be found in literature [9].

The discrete mineral particles if not removed by fuel cleaning are quickly molten at high temperatures, followed by condensation while traveling in the ducting systems as temperature drops before emitting to the atmosphere. The relative quantities of the bounded minerals increase while hydrocarbon is consumed by oxygen in combustion. Metal oxides can also contribute to the formation of $PM_{2.5}$. The elemental metal vapors are also oxidized and consequently form $PM_{2.5}$ by coagulation. Noncombustible solid compounds in the fuel will form the major part of the fly ash particles, ranging from 10 nm to more than 100 μm.

Most fly ash particles are smaller than 0.1 μm in diameter by number, but larger than 1 μm in diameter by mass. This is similar to the size distribution shown in Fig. 7.5, although the exact size distribution depends on the coal and the design and operation of the boiler. The fly ash size distribution in Fig. 7.5 was measured using bituminous and subbituminous coals in the USA.

The ash formation is also dependent on the combustion device. For example, the fate of ash-forming materials in fluidized bed combustion (FBC) is much different from that of pulverized fuel combustion. In FBC, mechanical abrasion and attrition play a more important role than the mechanism introduced above, and the ash-forming materials remain in the fluidized bed. In circulating fluidized bed combustion, more fly ash is formed due to the higher velocities and smaller fuel particle size.

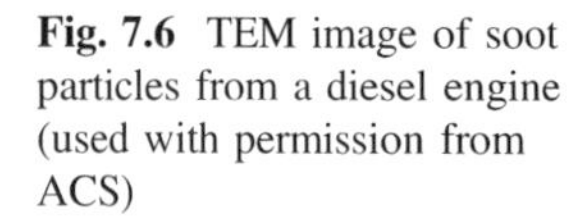

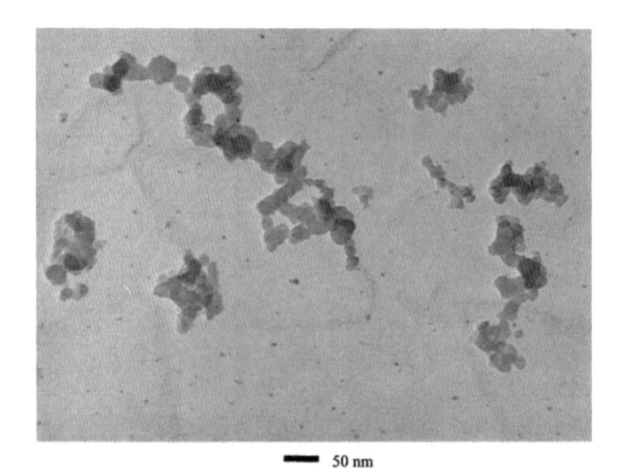

Fig. 7.6 TEM image of soot particles from a diesel engine (used with permission from ACS)

The composition of the ash particles depends strongly on the fuel, although SiO_2, Al_2O_3, Fe_2O_3 and CaO are usually the primary components. Ash from oils contains vanadium (V) and nickel (Ni), plus magnesium (Mg). These metals are added to the fuel as a corrosion inhibitor.

7.8.2 Soot Particles

Carbonaceous particles (10–80 nm) in the combustion system can agglomerate and form larger particles (could be larger than 10 μm) [4, 5, 18]. These clusters are called soot. A soot "particle" could have a family of thousands of carbonaceous particles in it. A transmission electron microscope (TEM) image of diesel engine soot is shown in Fig. 7.6 [41].

Chemically, soot particles are mainly composed of carbon, sulphur, and nitrogen compounds and trace elements. Hydrocarbons can also be adsorbed into soot. These particles could be as small as less than 10 nm in diameter.

There are mainly two mechanisms for the soot formation, depending on the type of fuel [25, 28]. Aliphatics in gaseous or light liquid fuels can be converted to acetylene (C_2H_2) at high temperature, followed by a "polymerization" of C_2H_2 to form soot. This mechanism is also possible for combustion at high temperatures, up to 1,600 °C.

For heavy oil and coal that contain more aromatics than aliphatics, soot is formed by condensation and other processes that have aromatics as the starting point [4, 11, 15, 22, 28, 38, 52]. The overall reaction of soot formation during sub-stoichiometric stages of combustion can be written as

$$C_nH_m + yO_2 \rightarrow 2yCO + {}^1\!/_2 mH_2 + (n - 2y)C(s) \quad \text{with } n > 2y \qquad (7.93)$$

Readers are referred to specialized books for more knowledge.

7.9 Fate of Trace Elements

Chemical elements present in a natural material can be classified as trace elements, minor elements, and the major elements, corresponding to concentrations of <0.1, 0.1–1, and >1 wt%, respectively.

Although these trace elements present in the natural ecosystem, quite amount is relocated after fossil fuel combustion. Within the European Community the 13 elements of highest concern are As, Cd, Co, Cr, Cu, Hg, Mn, Ni, Pb, Sb, Sn, Tl, and V. Eleven of 189 hazardous air pollutants (HAPs) in the USA are metals: As, Be, Cd, Co, Cr, Hg, Mn, Ni, Pb, Sb, Se.

7.9.1 Trace Elements in Fuels

Fossil fuels like coal and heavy oil contain trace elements in the forms of organic salts and inorganic minerals, for example, pyrites, and other sulphides, (alumino-) silicates and carbonates [47]. In typical coal the concentrations of Pb, B, Cr, Ni, and V are high. Ni and V are also found in oils.

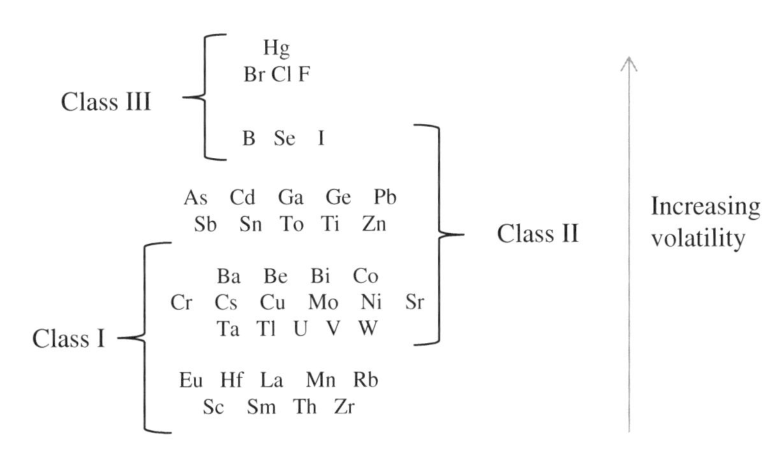

Fig. 7.7 Classification of trace elements

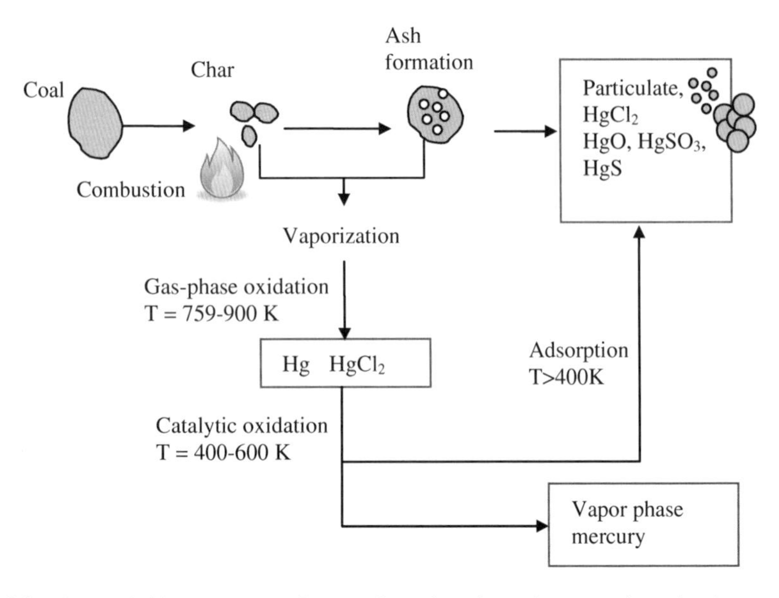

Fig. 7.8 Most probable mercury-species transformations in- and post- coal combustion

7.9.2 Trace Elements in Flue Gases

The fate of trace elements in combustion is influenced by many factors. Temperature is the main factor that determines whether a certain trace element will be volatilized. The other important factor that determines volatility is the air to fuel ratio. Many trace elements are more volatile under fuel rich condition than fuel lean condition. A third important factor is the presence of chlorine. Chlorine often reacts with a significant fraction of the trace elements to form chlorides, which are more volatile than the elemental or oxide form of the trace elements. The fourth factor of importance is total system pressure. Particle size has negligible effect on the vaporization of trace elements [40].

Based on the fate, trace elements may be classified into three categories as shown in Fig. 7.7.

- Class I—these elements do not volatilize during combustion, and they end up being captured with bottom ashes and fly ashes. The injection of sorbent (calcium) for SO_2 capture will expedite this process.
- Class II—these elements are vaporized during combustion and eventually captured by the particulates by condensation and nucleation mechanisms as the temperature drops along the duct of flue gases.
- Class III—these elements are vaporized but cannot be captured by the particulates (fly ash) in the flue gas. They enter the atmosphere, and become hazardous air emissions.

There are some overlaps between these three classes of trace elements. For example, B, Se, and I can be classified into Class II or Class III [39]. Combined with information on toxicity and harmful effects on process equipment, this classification gives first indication on which trace elements will need special attention.

7.9.3 Mercury

Mercury (Hg), a Class III trace element, is very problematic due to its toxicity and volatility (boiling point 357 °C). In combustion more than 90 % of the incoming Hg penetrates through the flue gas cleaning system, and is discharged into the atmosphere as vapor. Due to its high volatility, Hg is released from fossil fuels at about 150 °C, mainly as elemental Hg, $HgCl_2$, Hg_2Cl_2 and HgS. The release of Hg from the fuel will be complete at 500–600 °C [17, 35]. Under oxidizing conditions, in the presence of HCl and Cl_2, elemental Hg is oxidized to $HgCl_2$ at 300–400 °C (Fig. 7.8).

$$Hg + 4HCl + O_2 \leftrightarrow 2\ HgCl_2 + 2H_2O \tag{7.94}$$

$$Hg + Cl_2 \leftrightarrow HgCl_2 \tag{7.95}$$

Elemental mercury may react with NO_2 through the following reactions [6]:

$$NO_2 + Hg \leftrightarrow HgO + NO \tag{7.96}$$

The product of HgO continues to react with HCl as follows.

$$HgO + 2HCl \leftrightarrow HgCl_2 + H_2O \tag{7.97}$$

As a result, three different mercury species must be considered in the flue gas from coal combustion: gaseous elemental mercury (Hg), gaseous oxidized mercury (e.g., $HgCl_2$, HgS, HgO, $HgSO_4$) and particle-bound mercury (Hg-p). It is estimated that Hg emissions from a coal-fired power plant leaving the furnace, entering the flue gas duct are of the order 5–50 $\mu g/m^3$ at STP.

7.10 Greenhouse Gases

By the end of the twentieth century it was widely accepted that CO_2, CH_4, nitrous oxide (N_2O), and water contribute to the global warming. Water vapor is the most abundant greenhouse gas in the atmosphere, and it contributes to 2/3 of the greenhouse effects. The rest of 1/3 effect caused by other GHGs is referred to as the "enhanced greenhouse effect", or the "anthropogenic greenhouse effect". CO_2 is contributing nearly ¾ of the enhanced greenhouse effect.

The major source for CO_2 emission is the combustion of hydrocarbon fossil fuel. CO_2 concentrations in flue gases from natural gas-fired combined cycle power plants are about 4 % by volume, however it increases to 9–14 % from coal fired boilers. These amounts of CO_2 are currently emitted to the atmosphere without control.

Overall, the fossil-fuel combustion produces a variety of air pollutants in a great quantity. A typical 1,000-MW coal-fired power plant can produce about hundreds of thousand tons per year of particulate matter, sulfur dioxide, as well as comparable quantities of nitrogen oxides, carbon monoxide, volatile compounds, and trace metals. These air emissions have to be reduced as much as possible to protect public health and the environment. They are achieved by pre-, in- and post-combustion control approaches. They are covered in the following chapters.

References and Further Readings

1. ASTM International standard (MNL 11271 M, Proximate Analysis). doi:10.1520/MNL11271M
2. ASTM International Standard (MNL11272M, Ultimate Analysis). doi:10.1520/MNL11272M
3. Blevins LG, Renfro MW, Lyle KH, Laurendeau HM, Gore JP (1999) Experimental study of temperature and CH radical location in partially premixed CH_4/air coflow flames. Combust Flame 118:684–696
4. Bockhorn H (1994) Soot formation in combustion. Springer, Berlin
5. Bockhorn H, Schafer T (1994) Growth of soot particles in premixed flames by surface reactions. In: Bockhorn H (ed) Soot formation in combustion. Springer, Berlin/Heidelberg
6. Brown TD, Smith DN, Hargis RA Jr, O'Dowd WJ (1999) Mercury measurement and its control: what we know, have learned and need to further investigate. J Air Waste Manag Assoc 49(12):1469–1473
7. Buhre B, Elliott L, Sheng CD, Gupta RP, Wall TF (2005) Oxy-fuel combustion technology for coal-fired power generation. Prog Energy Combust Sci 31:283–307
8. Cooper CD, Alley FC (2002) Air Pollution control—a design approach, 3rd edn. Waveland Press, Illinois
9. Damle AS, Ensor DS, Ranade MB (1981) Coal combustion aerosol formation mechanisms: a review. Aerosol Sci Technol 1(1):119–133
10. Dryer FL (1972) High temperature oxidation of carbon monoxide and methane in a turbulent flow reactor. AMS report T-1034. March 1972
11. D'Anna A, D'Alessio A, Minutulo P (1994) Spectroscopic and chemical characterization of soot inception processes in premixed laminar flames at atmospheric pressure. In: Bockhorn H (ed) Soot formation in combustion. Springer, Berlin
12. Fenimore CP (1971) Formation of nitric oxide in premixed hydrocarbon flames. In: 13th symposium (international) on combustion. The Combustion Institute, Pittsburgh, p 373
13. Flagan RC, Seinfeld JH (2012) Fundamentals of air pollution engineering. Prentice Hall, Englewood Cliffs
14. Fletcher TH (1985) Sensitivity of combustion calculations to devolatilization rate. Expressions. In: Presentation at the American Flame Research Committee, 1985 Fall Meeting, October 17–18, Sandia National Laboratories, Livermore, California, USA
15. Frenklach M, Ebert LB (1988) Comment on the proposed role of spheroidal carbon clusters in soot formation. J Phys Chem 92:561
16. GEO (2000) Global Environment Outlook. Available at http://www.grid.unep.ch/geo

17. Galbreath KC, Zygarlicke CJ (1996) Mercury speciation in coal combustion and gasification flue gases. Environ Sci Technol 38(8):2421–2426
18. Gelencser A (2004) Carbonaceous aerosol. Springer, Berlin 350
19. Glarborg P, Miller JA, Kee RJ (1986) Kinetic modeling and sensitivity analysis of nitrogen oxide formation in well stirred reactors. Combust Flame 65:177–202
20. Glarborg P (1993) NO*x* chemistry in pulse combustion. In: Workshop on pulsating combustion and its applications. Lund, Sweden, August 1993
21. Habib MA, Elshafei M, Dajani M (2008) Influence of combustion parameters on NO_x production in an industrial boiler. Comput Fluids 37:12–23
22. Harris SJ, Weiner AM, Ashcraft CC (1986) Soot particle inception kinetics in a premixed ethylene flame. Combust Flame 64:65–81
23. Harvey D (2014) UCDavis ChemWiki. http://chemwiki.ucdavis.edu/Analytical_Chemistry/Analytical_Chemistry_2.0/08_Gravimetric_Methods/8C_Volatilization_Gravimetry. accessed 12 June 2014
24. Hayhurst AN, Vince IM (1980) Nitric oxide formation from N_2 in flames: the importance of ‘prompt’ NO, progress in energy combust. Science 6:35–51
25. Haynes BS, Wagner H (1981) Soot formation. Prog Energy Combust Sci 7(4):229–273
26. Heinsohn R, Kabel R (1999) Sources and control of air pollution. Prentice Hall, Upper Saddle River
27. Higman C, van Burgt M (2008) Coal gasification 2nd edn. Gulf Professional Publishers, Houston
28. Homann KH (1984) Formation of large molecules, particulates, and ions in premixed hydrocarbon flames; progress and unresolved questions. Proc Combust Inst 20:857–870
29. Hupa M, Backman F, Bostroem S (1989) Nitrogen oxide emissions of biolers in Finland. J Air Pollut Control Assoc 39:1496–1501
30. Kobayashi H, Howard JB, Sarofim AF (1976) Coal devolatilization at high temperatures. In: 16th symposium (Int’l.) on combustion, The Combustion Institute, Symposium (International) on Combustion (16th), held at The Massachusetts Institute of Technology, Cambridge, Massachusets, 15–20 August 1976
31. Linak WP, Miller CA, Seames WS, Wendt JOL, Ishinomori T Endo Y, Miyamae S (2002) On trimodal particle size distributions in fly ash from pulverized-coal combustion. Proc Combust Inst 29:441–447
32. Malte PC, Pratt DT (1975) Measurement of atomic oxygen and nitrogen oxides in jet-stirred combustion. Proc Combust Inst 15:1061–1070
33. Martin GB, Berkau EE (1971) An investigation of the conversion of various fuel nitrogen compounds to nitrogen oxides in oil combustion. In: 70th National American Institute of Chemical Engineers Meeting, Atlantic City, N. J., August, 1971
34. Mauss F, Schafer T, Bockhorn H (1994) Inception and growth of soot particles in dependence on the surrounding gas phase. Combust Flame 99(3–4):697–705
35. Merdes AC, Keener TC, Khang S-J, Jenkins RG (1998) Investigation into the fate of mercury in bituminous coal during mild pyrolysis. Fuel 77(15):1783–1792
36. Migule AH, Kirchstretter TW, Harley RA, Hering SV (1998) On-road emissions of particulate polycyclic aromatic hydrocarbons and black carbon from gasoline and diesel vehicles. Environ Sci Technol 32:450–455
37. Miller JA, Branch MC, McLean WJ, Chandler DW, Smooke MD, Lee RJ (1984) The conversion of HCN to NO and N_2 in H_2–O_2–HCN–Ar flames at low pressure. In: Twentieth symposium (International) on combustion, The Combustion Institute, pp 673–684
38. Pfefferle LD, Bermudez G, Byle J (1994) Benzene and higher hydrocarbon formation during allene pyrolysis. In: Bockhorn H (ed) Soot formation in combustion. Springer, Berlin
39. Ratafia-Brown J (1994) Overview of trace element partitioning in flames and furnaces of utility coal-fired boilers. Fuel Process Technol 39:139–157
40. Senior CL et al (1998) Toxic substances from coal combustion—a comprehensive assessment. In: Proceedings of the 15th annual international Pittsburgh coal conference, Pittsburgh, PA, USA, Sept 1998

41. Shi JP, Mark D, Harrison RM (2000) Characterization of particles from a current technology heavy-duty diesel engine. Environ Sci Technol 34(5):748–755
42. Sirignano W (1983) Fuel droplet vaporization and spray combustion theory. Prog Energy Combust Sci 9:291–322
43. Stickler D, Becker F, Ubhayakar S (1979) Combustion of pulverized coal at high temperature. AIAA Paper No. 79-0298, The American Institute of Aeronautics and Astronautics (AIAA), Reston, VA, USA
44. Turner DW, Andrews RL, Siegmund CW (1971) Influence of combustion modification and fuel nitrogen content on nitrogen oxide emissions from fuel oil combustion. In: Winter American Institute of chemical engineers meeting, San Francisco, December, 1971. AIChE series no 126, vol 68
45. USEPA (1983) Control techniques for nitrogen oxides emissions from stationary sources, 2nd edn. U.S. EPA, Washington D.C. (revised)
46. Ubhayakar SK, Stickler DB, von Rosenberg CW Jr, Gannon RE (1977) Rapid devolatilization of pulverized coal in hot combustion gases. In: Proceedings of the 16th symposium (international) on combustion, The Combustion Institute, Pittsburgh, Pennsylvania, pp 427–436
47. Vassilev SV, Braekman-Danheux C, Laurent P, Thiemann T, Fontana A (1999) Behavior, capture and inertization of some trace elements during combustion of refuse-derived char from municipal solid waste. Fuel 78:1131–1145
48. Williams A, Pourkashanian M, Jones JM (2001) Combustion of pulverised coal and biomass. Prog Energy Combust Sci 27(6):587–610
49. Yuen M, Chen L (1976) On drag of evaporating droplets. Combust Sci Technol 14:147–154
50. Zeldovich YB (1946) The Oxidation of Nitrogen in Combustion Explosions. Acta Physicochimica USSR 21:577
51. Zevenhoven R, Kilpinen P (2002) Flue gas and fuel gas, 2nd edn. Report TKK ENY-4, The nordic energy research programme, Solid Fuel Committees, Norway, pp 3–4
52. Zhang QL, O' Brien SC, Heath JR, Liu Y, Curl RF, Kroto HW, Smalley RE (1986) Reactivity of large carbon clusters: spheroidal carbon shells and their possible relevance to the formation and morphology of soot. J Phys Chem 90(4): 525–528

Chapter 8
Pre-combustion Air Emission Control

Air emission control is enforced when the ambient air quality or the source air emission rate does not meet certain standards. There are three basic approaches to air emission control: pre-, in-, and post- combustion air emission control. The most cost-effective is to control before combustion, which reduces the load of downstream units. The last choice is add-on devices for flue gas cleaning. These *extra* devices contribute significantly to the capital and operating costs of the plant; they also reduce to some degree the thermal efficiency, because the devices may also reduce some of the power output of the plant. Many of these devices also produce another form of waste, usually solid or liquid being discharged to the environment.

This chapter is focused on the following pre-combustion air emission control technologies.

- Fuel cleaning,
- Fuel substitution,
- Fuel conversion, and
- Alternative energy resources.

8.1 Fuel Cleaning

8.1.1 Coal Cleaning

One of the most cost-effective approaches to the control of air emissions is the removal of the unwanted chemicals from the fuels before combustion.

Two most common engineering practice examples are coal washing, and oil and gas refinery. In the United States, about 50 % of all coals supplied to power plants are washed before delivery. In Germany, almost all coals are washed before firing. The oils and gases are refined globally following the specific government regulations. Fuel cleaning is the most effective way to reduce the contents of sulfur, ash-forming

Z. Tan, *Air Pollution and Greenhouse Gases*, Green Energy and Technology,
DOI 10.1007/978-981-287-212-8_8

compounds, and trace elements. Basic principles of coal cleaning and related technologies have been well documented (e.g., [27]), and they are briefly introduced as follows.

Coal cleaning methods may be classified into conventional physical cleaning and advanced cleaning methods. Advanced cleaning methods include advanced physical cleaning, aqueous phase pretreatment, selective agglomeration, and organic phase pretreatment. Of these alternatives, conventional physical cleaning is widely used for the sake of low cost and relatively high efficiency.

Conventional coal cleaning involves the following steps:

(1) Crushing of the coal: Grinding into smaller particles with diameters less than 50 mm. It also liberates ash-forming minerals and inorganically bound sulfur.
(2) Particle screening: The crushed coal particles are screened into three modes; coarse, intermediate, and fine.
(3) Floatation: It involves the separation of ash and sulfur compounds from the coal before it is pulverized and introduced into combustion chamber. The lighter coal particles float on top while the heavier minerals sink to the bottom in a stream of water.
(4) Drying: In this step wet coal particles are dried using a dewatering device, generally a vacuum filter, centrifuge, or a cyclone, to separate water from the solid, followed by further drying in hot air.

Coal washing can remove about 60 % of ash-forming materials. It is often necessary to reduce the excessive amount of ash-forming materials from the coal especially for steel processing applications. For example, the lignite from India and Greece may contain more than 50 % wt ash-forming material. A significant amount of Pyrite (FeS_x), As, Se, and Hg can be removed by coal pre-cleaning.

Coal washing alone can remove up to 50 % of the pyretic sulfur, which is equivalent to 10–25 % removal of the total sulfur content of the coal. Ninety percent of the inorganic fuel sulfur, especially pyritic sulfur, FeS_2, can be easily removed by coal washing. The organically bound sulfur cannot be removed by physical cleaning methods; however, it can be removed with biological or chemical methods.

Biological treatment can remove both inorganic and organic sulfur. For example, a bacteria, Thiobacillus ferrooxidans is capable of converting FeS_2 into water-soluble $FeSO_4$, whereas a mutant of Pseudomonas, called Coal bug 1 (CB_1) can consume organic sulphur in thiophene groups. However, these techniques are time-consuming and may require very small coal particles.

Chemical methods can remove organically bound sulfur and involves treatment with alkaline or caustic solutions, oxidative leaching, and chlorinolysis with chlorine-based chemicals. However, these chemical methods may change the property of the coal and reduce its potential for use as a fuel.

Like any other process, there are pros and cons for physical coal cleaning. The advantages are as follows

- Reduced SO_2 formation in the flue gas:
  The reduction could be 10–40 % lower than burning coals without pre-cleaning.
- Decrease of ash content:
  This allows the cleaned coal to be used for pulverizer and boiler. It also reduces the load of downstream particle separators such as ESPs and bag houses.
- Lower maintenance costs for boilers:
  The reduction of ash content leads to less wear and tear on coal preparation before combustion and on boiler during combustion.
- Smooth operation:
  Less operational problems such as boiler slagging and fouling.

Meanwhile, one has to bear in mind the disadvantages of coal washing. Intensive energy is required for coal grinding and drying. And moisture added to the coal may reduce the efficiency of the boiler and the entire plant. There may be 2–15 % of energy loss to the preheating of coal.

In addition to these technical issues, one should also consider the environmental regulations and the price of cleaned coal. In some countries, the price of coal is the same regardless of its cleanness. In addition, the waste liquid stream from coal cleaning may contain acidic toxic metals, which pose an extra challenge and cost in waste treatment or disposal. More and more stringent regulations have been introduced in the United States and China, for example, to prevent dumping of these toxic acidic streams into the environment without prior treatment.

8.1.2 Oil and Gas Refinery

Crude oil and raw natural gas contain tens of thousands of kinds of hydrocarbon compounds. By refinery, crude oil is decomposed into various fractions and transformed into fuels including oil and gas and other products.

The oil and gas after refinery process contains specified amount of sulfur, nitrogen, and ash contents. For example, sulfur compounds in the crude oil or raw natural gas can poison many of the catalysts used for the treatment of hydrocarbons in the petrochemical industry. More and more stringent environmental regulations also require reduced sulfur compounds in the final petroleum products. The average sulfur content in Canadian diesel was 350 ppm in 2000, and ultra-low sulfur diesel with 15 ppm sulfur became mandatory in North America for highway vehicles in 2006. In 2009, all EU vehicles will run on 10 ppm sulfur diesel including off-roads.

Due to stringent regulation on sulfur content in fuels, a large amount of sulfur compound is produced as a byproduct of oil refinery. The elemental sulfur is often an important byproduct of oil refining and it is also a major raw material for the productions of fertilizer and sulfuric acid.

Sulfur compounds are first separated from the refinery stream by absorption using amine followed by another separation process to recover the amine and to concentration the H_2S gas stream. The concentrated H_2S streams usually contain >50–60 % of H_2S. H_2S smells like rotten eggs; it is corrosive and toxic. H_2S loaded gas streams must be further treated [5].

8.1.2.1 Claus Process

One of the common practices is to convert H_2S to nontoxic and elemental sulfur by the Claus process, which was first developed by London chemist Carl Friedrich Claus in 1883. A Claus process converts H_2S into elemental sulfur by two steps, thermal step and catalytic step [1, 8, 33, 43]. In the thermal step 1/3 of the H_2S is oxidized into SO_2 and water by the oxygen in the air.

$$H_2S + 1.5O_2 \rightarrow SO_2 + H_2O \tag{8.1}$$

The resultant gas steam contains H_2S and SO_2 in a 2:1 mol ratio. This step requires a high-reaction temperature of 1,000–1,400 °C.

In the following catalytic step, SO_2 reacts with the remaining H_2S in presence of a catalyst to form elemental sulfur. With the catalysts, the reactions in the catalytic step proceed at a much lower temperature of 200–350 °C.

$$2H_2S + SO_2 \leftrightarrow \frac{3}{n} S_n(s) + 2H_2O \quad (\Delta H = -108 \text{ kJ/mole}) \tag{8.2}$$

where typical value of n is 8. Following Eq. (3.27), the equilibrium constant of Eq. (8.2) can be described as

$$K_P = \frac{y_{H_2O}^2 y_{S_8}^{3/8}}{y_{H_2S}^2 y_{SO_2}} P^{\frac{5}{8}} \tag{8.3}$$

The reaction described in Eq. (8.2) is exothermic and equilibrium-limited reaction that calls for low temperatures. However, low temperatures will result in low elemental sulfur yield rate. That is why a catalyst is necessary. Typical catalysts for Claus processes are activated alumina, activated bauxite, or cobalt molybdenum hydrogenation catalyst [33, 43]. In addition, a multistage process with interstage timely removal of elemental sulfur by cooling and sulfur condensation further improves the conversion rate.

By two- or three-stage processes the H_2S conversion efficiencies could reach about 95 or 97 %, respectively.

Alternatively, with the super-Claus process and special catalysts, efficiencies of >99 % can be achieved because it prevents the formation of SO_2. However, a separate hydrogenation reactor has to be employed between the second and third stage. Oxygen enrichment of the air to the burner in the final Claus stage also reduces soot formation and poisoning of the catalysts.

8.1.2.2 Adsorptive Claus Process

Recently, increasing restrictive low-sulfur fuel regulations challenge the conventional Claus process. Depletion of low-sulfur feedstocks for the petroleum industry demands more improved or new technologies to maximize the H_2S conversion. Elsner et al. [13] proposed an improved process called adsorptive Claus process based on the Le Chatelier's principle [47], which implicates that the removal of a reaction product results in an equilibrium displacement to higher conversion.

Unlike the conventional Claus process, where elemental sulfur is removed by in situ condensation, the adsorptive Claus process removes water by selective adsorption. According to the reaction equilibrium constant described in Eq. (8.3), removal of water vapor would have a greater impact on the conversion rate than sulfur removal, because there are 2 mol of water per 3/8 mol of sulfur in the product.

As introduced in Chap. 5, there are several water vapor adsorbents, and one of them is zeolite which can be regenerated by inert sweep gas. A challenge to this process is the chemical resistance of the catalyst and the zeolite to the aggressive gas system, where SO_2 and water may react and produce sulfuric acids.

8.1.2.3 Natural Gas Sweetening

Natural gas sweetening is important to both environment and final product quality. Natural gas contains a large amount of methane (CH_4), and all kinds of impurities as shown in Table 8.1. The mole amount of a substance in a raw natural gas depends highly on the gas field.

In industry, sour gas is referred to high contents of H_2S and CO_2. These acidic gases in the raw natural gas are removed at the gas well to reduce technological challenges to downstream gas transportation in the pipeline, and equally important, to reduce SOx and CO_2 emissions.

First of all, it prevents the formation of gas hydrate (commonly called dry ice). CO_2 hydrate can clog the system during the liquefaction of the natural gas. Furthermore, it reduces the corrosion resulted from H_2S and CO_2 in the presence of

Table 8.1 Typical composition of raw natural gas out of the well

Gases	Mole ratio
CH_4	70–95 %
H_2S	0–15 %
C2+	0–15 %
CO_2	0.1–8 %
N_2	0–0.2 %
Temperature	30–40 °C
Pressure	5–120 atm

Source Ramdin et al. [40]

water and the toxicity of H_2S. Removal of CO_2 also improves the heating value of the natural gas because CO_2 is not combustible and it simply lowers the combustion efficiency.

The great partial pressure of H_2S and CO_2 allows them to be removed by physical adsorption and membrane separation from the natural gas outside of the well. Physical solvents like Rectisol, Purisol, and Selexol are preferred over chemical solvents for high pressure gas purification processes. However, the most popular technology is amine-based absorption. The separated H_2S can be converted into elemental sulfur as explained above. The treatment of separated CO_2 will be elaborated in Sect. 12.6.

8.2 Fuel Substitution

As introduced previously the air emissions are dependent on the fuels. Relatively speaking, natural gas is cleaner than gasoline, and coal is the dirtiest fossil fuel. Solid fossil fuels, biomass, and waste-derived fuels generate a wide spectrum of air pollutants in addition to solid waste. From this point of view gaseous or liquid fuels are considered cleaner because they hardly contain ash-forming elements, which make them suitable for application in internal combustion engines and gas turbines. As a result, the emissions per unit heat or power generated by liquid/gaseous fuels are less than solid fuels.

It is technically feasible to use fuel substitution such as co-combustion of oil and coal, in order to reduce the air emissions. In reality, other factors, likely economical consideration, also contribute to the final decision of which fuel is used for certain process. So far, coal is still the cheapest and most polluted fuel in environment. In the United States, coal is mainly used as an electric utility fuel. Although oil is more expensive than coal to recover, it can be easily transported by pipelines. Like oil, natural gas is more expensive to recover from wells, but contrary to oil, it cannot be easily stored or shipped. Nonetheless natural gas is widely used because of its ease, efficiency, and cleanliness of combustion.

It is obvious that replacing a high-sulfur fuel with a low-sulfur fuel will reduce the amount of SOx in the flue gas. Replacing solid fuel with oil or gas can also reduce the ash formation. In some regions, local authorities may enforce fuel substitutions to reduce the local air pollution. For example, low-sulfur coals must be burned in the plants within the capital of China.

The success of fuel substitution depends on the fuel-flexibility of the burner and the economics of operation with another fuel. It is relatively easy to retrofit a coal-fired boiler with natural gas; however, it may not be the same using oil. Furthermore, we have to consider the operation of downstream air cleaning units. For example, sulfur content in the fuel affects the dust separation performance of an electrostatic precipitator (ESP). In-depth analysis will be introduced Chap. 10 for post-combustion air emission control approaches.

Replacing gasoline with natural gas or propane can reduce the emissions of VOCs. This replacement is not limited to stationary combustion sources, and some engines can be powered by compressed natural gas or propane. Extensive research and development being conducted in the petroleum industry to improve the combustion properties, handling, and use of natural gas as a substitute of gasoline to dominate the auto fuel market, although it is still not the case yet.

Biodiesel, which will be introduced shortly as one type of alternative fuels, is being used as a substitute for petroleum diesel fuels for diesel engines. Intensive experimental and computational studies on biodiesel have been carried out in the past decades. B20, a mixture of 20 % of biodiesel and 80 % of petroleum diesel, is now widely adapted by the US government for transportation industry. It is characterized by near-zero emissions of sulfur and net-nitrogen. But there is controversy data in the literature about the emissions of particulate matter.

8.3 Thermochemical Conversion of Fuels

Solid fuels such as coal or biomass may be converted to liquid or gaseous fuels by a thermochemical conversion (TCC) process. A TCC process is a chemical reforming process in which the depolymerization and reforming reactions of organic matter take place. There are many TCC processes that convert a wide variety of feedstock into different fuels through different types of reactions. They can be small or large in scale. Overall, TCC processes can be divided into three categories.

- Pyrolysis
- Gasification, and
- Liquefaction

8.3.1 Pyrolysis

Pyrolysis is a process where organic matter is degraded by thermal reactions in the absence of added oxidizing agents. Pyrolysis, being as old as civilization, was widely employed in the seventeenth and eighteenth centuries for conversion of charcoal to fuels for the smelting industry.

With wood as a feedstock, pyrolysis is also referred to as carbonization or destructive distillation. Pyrolysis decomposes a fuel with larger molecules into smaller fractions such as C_2H_6, CO, CO_2, H_2, H_2O, oily liquids, and a solid carbonaceous char residue. The actual products depend on the TCC process. The mixture of gases, liquids, and solids require further separation in order to produce the final fuel products. The principles introduced in Chaps. 5 and 6 sets the foundation of the engineering designs of the related equipment and processes.

For conventional pyrolysis, where the heating rate is below 10 °C/s and the residence time is long, the primary products are tar and char. In order to maximize the production of oil, flash pyrolysis with rapid heating of about 100–10,000 °C/sec is employed. External heating is also applied for the process to allow pyrolysis in the absence of combustion. The char and tar can be burned with air for heating.

8.3.2 Gasification and Syngas Cleaning

8.3.2.1 Gasification Chemistry

Gasification is a thermal chemical conversion process in which carbonaceous materials are converted to gases by incomplete oxidation. Gasification is actually fuel rich combustion operating at 25–40 % of the oxygen that would be needed to convert the hydrocarbon fraction of the fuel to CO_2 and H_2O. In this process part of the fuel is combusted to provide the heat needed to gasify the rest.

Major reactions involved in the gasification process are as follows [36].

$$C + \frac{1}{2}O_2 \rightarrow CO \quad \text{(Combustion reaction)} \tag{8.4}$$

$$C + O_2 \rightarrow CO_2 \quad \text{(Combustion reaction)} \tag{8.5}$$

$$C + CO_2 \rightarrow 2CO \quad \text{(Boudouard reaction)} \tag{8.6}$$

$$C + H_2O \rightarrow CO + H_2 \quad \text{(Gasification with steam)} \tag{8.7}$$

Steam is produced by oxidation of hydrogen atoms in the fuel. There are also other minor chemical reactions including gasification with hydrogen, water gas shift reaction, and methanation.

$$C + 2H_2 \rightarrow CH_4 \quad \text{(Gasification with } H_2\text{)} \tag{8.8}$$

$$CO + H_2O \rightarrow H_2 + CO_2 \quad \text{(Water gas shift reaction)} \tag{8.9}$$

$$CO + 3H_2 \rightarrow CH_4 + H_2O \quad \text{(Methanation)} \tag{8.10}$$

A more complicated way to present the gasification of solid or liquid feedstock involves devolatilization to produce volatile hydrocarbons and charts. Then both the chars and volatile hydrocarbon are further gasified to produce syngas and other compounds.

$$C_nH_m + \frac{1}{2}nO_2 \rightarrow \frac{1}{2}mH_2 + nCO \quad \text{(Volatile gasification)} \tag{8.11}$$

$$CH_xO_y + (1 - y)H_2O \rightarrow \left(\frac{x}{2} - y + 1\right)H_2 + CO \quad \text{(Char gasification)} \tag{8.12}$$

If we consider the formula of char containing C, H, O, N, S, and mineral matters, the chemical reactions are much more complicated. With the presence of steam, the sulfur content usually is converted into H_2S. Part of the H_2S can react with CO_2 to produce COS.

$$H_2S + CO_2 \rightarrow COS + H_2O \tag{8.13}$$

The primary gasification products are synthesis gas also known as syngas, which is composed of CO, H_2, CH_4, and many others. This syngas has to be cleaned before H_2, CH_4, and/or CO can be separated as clean fuels.

8.3.2.2 Gasifiers

A gasifier is the main chamber where all the main gasification chemical reactions take place. Various types of gasifiers have been developed over the last centuries. Conventional gasifiers (Fig. 8.1) include

- fixed bed gasifiers,
- fluidized bed gasifiers, and
- entrained flow gasifiers.

These gasifiers are widely used in integrated gasification and combined cycle (IGCC) plants. Most of the coal is gasified in fixed bed gasifiers [36, 42]. Updraft and downdraft moving bed gasifiers are suitable for smaller scale gasification, where fluidized bed gasifiers are often used for solid feedstocks other than coal. The entrained bed gasifiers are most suitable for coal. All these gasifiers have been well commercialized by different companies. Each has its features aiming at different applications. They are only briefly introduced as follows in order to focus on the main scope of this book.

In entrained bed gasifiers, feedstock particles concurrently react with oxidants in suspended (i.e., entrained) fluid flow mode. Entrained bed gasifiers require pulverized feedstocks. The sizes of solid fuel particles have to be less than 1 mm for effective suspension. The gasification temperature may exceed 1,500 °C and the residence time is in the order of 1 s only. The units are usually operated at high pressure (2.94–3.43 MPa). With this high gasification temperature, the syngas stream is almost free of tars, oils, and phenols. On the other hand, the corresponding raw syngas exiting the gasifier usually requires significant cooling before it can be handled by the downstream gas cleaning units. Such a high temperature also requires expensive burners and sophisticated high-temperature heat exchangers to cool the syngas.

In a fixed bed gasifier, preheated feedstock is fed from the top of the gasifier. The feedstock falls through different zones before it reaches the grates. From top to

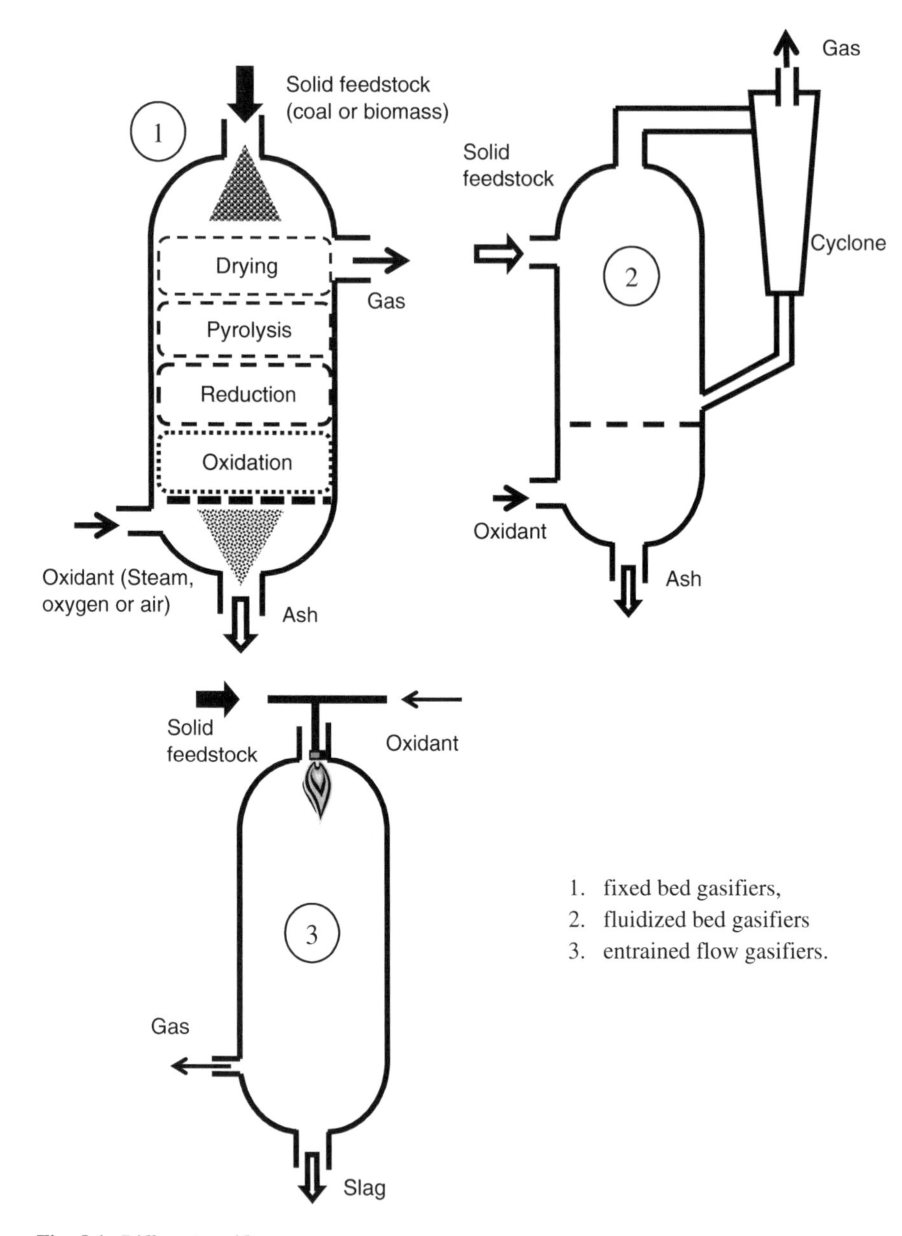

Fig. 8.1 Different gasifiers

bottom, the feedstock is preheated, dried, devolatilized/pyrolysed, gasified, and combusted. Maximum temperature in the combustion zone can be as high as 1,800 °C, mostly greater than 1,300 °C, depending on the reactor design. Post-combustion ash leaves at the bottom and the syngas exits at the top to the syngas cleaning units. The temperature of the syngas leaving the top is about 400–500 °C. The feedstock particle size is normally in the order of tens millimeters and the residence time is about 15–30 min.

In a fluidized bed gasifier, both the feedstock and the oxidants can enter from the bottom of the gasifier. And they travel upward together. As a result of effective mixing, the temperature within the air-based gasifier is relatively uniform and it is in the range of 900–1,050 °C to minimize ash melting. In order to maintain the fluidity of the particles, the feedstock particles are smaller than those in a fixed bed gasifier. It is in the range of 0.5–5 mm. And the residence time of feedstock in the gasifier is typically 10–100 s, which is much shorter than a fixed bed gasifier [19]. Right at the exit of a fluidized bed gasifier, an optional cyclone is usually used to recycle the large unspent feedstock. Large particles will be recycled for further gasification. Smaller ones penetrating through the cyclone become the particulate matter in the syngas stream. As we learned in Sect. 6.4, a cyclone is effective in separating particles of a few micrometers. Therefore, particulate matter in the syngas stream is primarily in the micron range or smaller. Most of these particles can be separated using electrostatic precipitators or filters.

8.3.2.3 Syngas Cleaning and Separation

Raw syngas compounds depend on the feedstock and the gasifier. Among various carbonaceous feedstocks, coal, petroleum coke, and petroleum residues have been used for gasification. Recently, biomass has been tested too as a feedstock.

A sample raw syngas compounds produced from coal, pet coke, and petroleum residues are listed in Table 8.2. Most raw syngases contain considerable amount of CO, H_2, CO_2, H_2S and COS, and particulate matter. CO_2, H_2S and COS are also collectively called acidic gases because they can be converted easily into acids with moisture. CO_2 is also considered as a greenhouse gas.

Table 8.2 Pre-combustion syngas after water gas shift reaction

Gas	Mole fraction (%)
H_2	55.5
CO_2	37.7
N_2	3.9
CO	1.7
H_2O	0.14
H_2S	0.4
Others	0.66

Source Ramdin et al. [40]

The exact syngas properties depend on the type feedstock, gasifier, operating pressure, temperature, and residence time. In general, higher temperature leads to greater carbon conversion. However, overheated feedstock may result in ash fusion and/or ash agglomeration [52]. Most of the commercial gasifiers operate at elevated pressures ($\sim$2.94 MPa) [18] for the equilibrium consideration. However, the pressure does not alter the syngas composition very much.

Particulate removal

Particulate cleaning of the hot syngas is necessary not only to reduce air emissions, but also to prevent corrosion and erosion of downstream gas separation components. For all gasifiers, char materials along with ash can be removed by water spray (quenching) followed by carbon scrubber. Around 95 % of the char carbon can be removed by direct water spray. The residual carbon is handled in the following wet scrubbers. Because of the cooling of the syngas, the thermal efficiency is greatly reduced in the entire process if the syngas will be used immediately, like in an IGCC process.

An alternative approach is hot gas filtration. Particulate filtration at temperatures above 260 °C is called hot gas filtration; sometimes it can reach 900 °C [23]. This high temperature demands special filtration materials that can endure the high temperature as well as the acidic gases in the raw syngas. The filter housing is also expected to be stable against temperature, pressure, and chemical composition of gas and dust.

Common materials in hot gas filtration are ceramic and metallic. These materials allow rigid self-supporting filter elements that can be employed at high temperatures due to their high mechanical strength. The filters can be shaped like candles, as long as a few meters, or honeycomb structure. High costs and system failure due to filter clogging are the main challenges to hot gas filtration.

8.3.2.4 Acidic Gas Removal and Sulfur Recovery

Gas separation follows particulate removal. Engineering designs are based on the principles introduced in Chap. 5, most commonly by adsorption or absorption. There are many options for CO_2 separation from the syngas too. CO_2 capture and storage will be introduced in detail in Chap. 12.

H_2S separation can be achieved by both absorption and adsorption. Physical or chemical adsorption followed by conventional Claus sulfur recovery units has been proven successful in petroleum industry. ZnO/CuO, Cr_2O_3, and Al_2O_3 can adsorb H_2S components. The simplified adsorption and adsorbent regeneration reactions are as follows, using ZnO as an example [48].

$$ZnO + H_2S \rightarrow ZnS + H_2O \quad (\text{Adsorption, } 315\text{–}530\,^{\circ}\text{C}) \tag{8.14}$$

$$ZnS + \frac{3}{2}O_2 \rightarrow ZnO + SO_2 \quad (\text{Regeneration, } 590\text{–}680\,^{\circ}\text{C}) \tag{8.15}$$

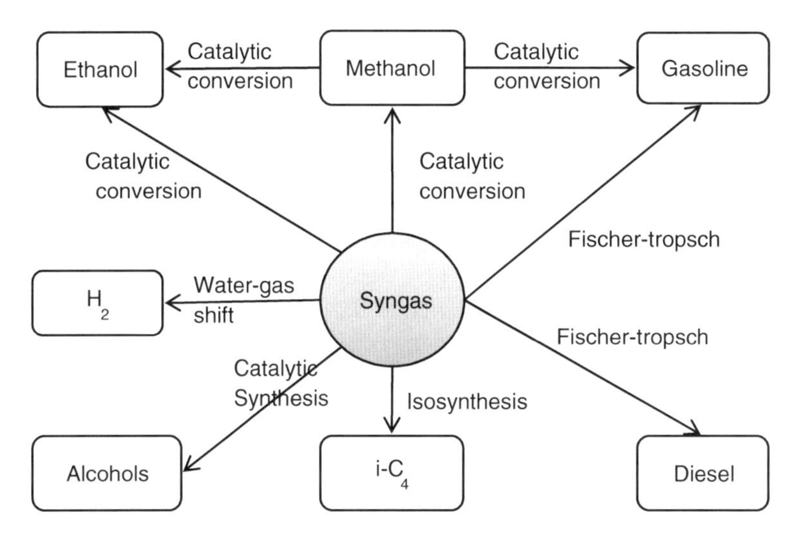

Fig. 8.2 Syngas to fuels

Purified syngas ($H_2 + CO$) is a great building block for energy industry. It can be directly burned for power generation through gas turbines. This is so-called combine cycle technology, which will be introduced in next section. Pure hydrogen produced from syngas can be used in hydrogen-based fuel cells. In addition, syngas can be converted into many other products. Some of these conversions pathways are shown in Fig. 8.2. The most well known is syngas to diesel/gasoline through Fisher-Tropsch process.

$$(2n + 1)\ H_2 + nCO \rightarrow C_nH_{(2n+2)} + nH_2O \quad \text{(Fisher-Tropsch synthesis)} \tag{8.16}$$

8.3.3 Combined Cycle Technologies

Since the 1970s, combined cycle technologies have been developed for gaseous fuels based on gas turbine technology. The fuel gas, which is either natural gas or a syngas from gasification, is burned in a gas turbine and the exhaust is used for a steam cycle. The thermal efficiency of a natural gas fired combined cycles (NGCC) can be in the order of 60 %. In coal-fired integrated gasification combined cycle (IGCC), the thermal efficiency could reach 46 %. The main advantage of combined cycle is that the exhaust is "cleaner" since gaseous fuels are used and less oxygen is used than conventional combustion. Liquid and gaseous fuels do not or hardly contain ash-forming elements. As a result, air emissions per unit heat or power generated are smaller for liquid/gaseous fuels than for solid fuels, however, all this comes with a high cost, especially for IGCC.

8.4 Biofuels

With the depletion of fossil fuels and the growing concern of climate change, renewable energy emerges on to the global stage. On the scale of centuries, the supplies of fossil fuels will be severely depleted. The only sources that could supply energy indefinitely beyond that time horizon are likely to be nuclear fusion and renewable energy. Heat and energy can be produced by solar, fuel cell, wind, geothermal, etc. There is minimal fuel combustion in these energy production processes and consequently are believed to be able to dramatically reduce the air emissions from fuel combustion.

On the other hand, cost is the major barrier preventing these technologies to be commercialized at large scales. Thus far, these alternative energy resources are still at the research and development stage with limited applications. In addition, it also takes energy to produce the devices for energy harvesting and conversion. The net benefit of these technologies should be evaluated based on a life-cycle analysis.

Among all the relevant alternative energy technologies, biofuels are closely related to the scope of this book. Thereby the following is focused on biofuels. Biofuels are produced from biomass, which is biological material derived from living organisms. They grow on the planet earth by converting carbon dioxide (CO_2) with solar energy into HC organic compounds. Biomass can be simply divided into three categories (Table 8.3), lignocellulosic, starch based, and triglyceride-producing biomass.

The type of biofuels produced from biomass depends on the feedstock and the process. Table 8.3 summarizes the platforms for biofuels from biomass. Like fossil fuels, biofuels can also be solid, liquid, and gaseous.

8.4.1 *Solid Biofuels*

As seen from Table 8.3, solid, liquid and gaseous biofuels can be produced from different biomass following different engineering processes. Without deviating too much from the main scopes of this book, typical processes are briefly introduced as follows.

8.4.1.1 Pulverized Biomass

After drying, biomass can be pulverized and mixed with coal for power generation, and it is expected to reduce the net carbon emission because carbon in the biomass is from CO_2 in the atmosphere. Biomass drying consumes energy. Due to the high moisture content in biomass, loss in thermal efficiency is also a concern. Other technical questions about fuel feed, boiler combustion chemistry, and ash deposition and disposal have been raised too [37].

Table 8.3 Biomass and biofuels platforms

Biomass		Biofuel production			
Type	Examples	Technology	Immediate products	Post processing	Final biofuel products
Lignocellulosic biomass	Wood, grass	Pulverization	Powder		Co-firing utility fuel
		Pelletization			Pellets
		Anaerobic digestion	Biogas	Gas separation	H_2, CH_4, CO
		Gasification	Syngas	Fischer-Tropsch Fermentation	FT-diesel, jet fuel, ethanol
		Pyrolysis	Bio-oil	Refining	Bioler oil, diesel
		Liquefaction	Bio-oil	Refining	Bioler oil, diesel
Starch-based plants	Corn, sugarcane	Hydrolysis	Sugar	Fermentation	Bioethanol, biobutanol
Triglyceride-producing plants	Canola, soybean, safflower	Extraction	Edible oil	Transesterfication	Biodiesel
	Waste cooking oil	Purification		Transesterfication	Biodiesel

8.4.1.2 Wood Pellets

Biomass can be processed into wood pellets for effective transportation and efficient combustion too. The feedstock can be any lignocellulosic biomass, and wood is the most widely used one. Wood pellets can be burned in stove for indoor air heater or in specialized furnace for energy production. They burn like high quality coal, with much lower air emissions.

There are several steps that we can follow to make wood pellets. While the exact procedure may be different from plant to plant and the product quality may vary accordingly [30], the common steps are summarized in Fig. 8.3 for guidance only.

- The first step is biomass drying. The moisture content in the raw biomass must be reduced to a level of about 10 % before the pelletizing process begins. Otherwise, it is difficult to pelletize and consequently affects the quality of the final product. This step is especially important when recycled biomass is used as a feedstock.
- The next step is to remove the impurities from the dried wood. Metal, for example, can be removed by magnets and a screen. This step is especially important for recycled biomass too.
- After the removal of the impurities, the raw material is ground in a hammer mill and it becomes wood powder. The powderous particles can be separated using a cyclone or surface filter. This size reduction is necessary for pelletization. Typical powder sizes are below 5 mm in diameter.
- Right before pelletization, the powder is heated to 70 °C or so. Heating and softening ensure that the lignin in the wood is released and the particles can be bonded effectively together in the final product.

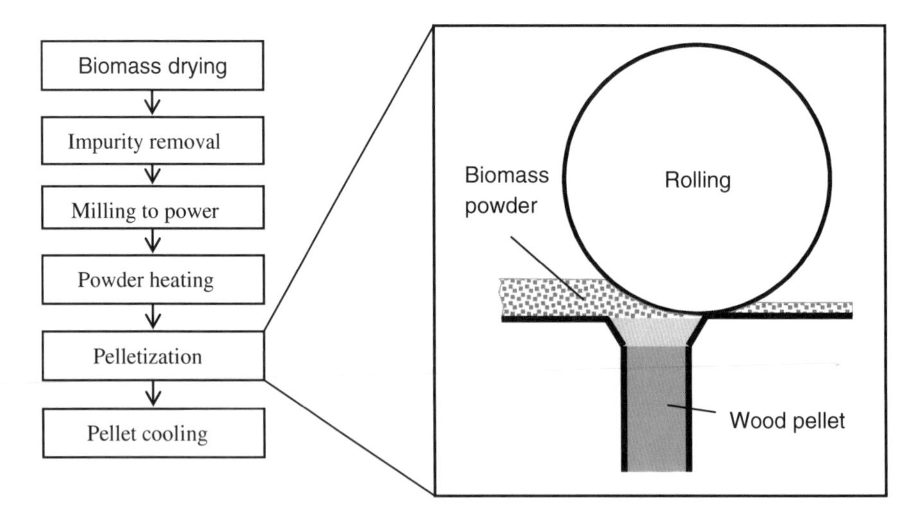

Fig. 8.3 Pellet processing

- Pelletization is a process that reforms the wood powder into certain shapes. As depicted in Fig. 8.3, the powder is pressed down into the die block and is compressed into cylindrical the pellets.
- The last step is to cool the pellets to increase their durability.

The environmental impact assessment for biomass pellet production should be conducted from feedstock harvesting to the delivery of pellets to the end user. Magelli et al. [32] conducted an assessment for wood pellet production in Vancouver, Canada with end users in Sweden by streamline life-cycle analysis. They compared the total emission factors (in g/kg fuel) between wood pellet and natural gas as fuels. The results in Table 8.4 show that wood pellets as a fuel has a great potential in reducing air emissions. The reduction could be greater if the pellets were produced and consumed locally without international transportation. On the other hand, we have to realize that wood pellets cannot replace natural gas because each of them has its uniqueness due to the phase difference.

8.4.2 Biodiesel

Biodiesel has some substantially different properties than petroleum diesel because of its different structure. Biodiesel is described as a fuel comprised of mono-alkyl esters of long chain fatty acids derived from vegetable oils or animal fats (in ASTM Standard D6751-03). Typical biodiesel has the cetane number that is close to the high end No. 2 petroleum diesel. For example, the biodiesel viscosity–temperature relationship is similar to that of No. 2 diesel fuel, which followed the Vogel equation [54].

Table 8.4 Emission factors for pellet and natural gas production

Air emissions		Pellet (g/1,000 kg fuel)	Natural gas (NG) (g/1,000 kg fuel)	Emission factor ratio (Pellet:NG)
Pollutant	CO	1,196	1,213	0.986
	NOx	6,420	6,452	0.995
	VOC	169	384	0.440
	PM	496	505	0.982
	SOx	2,958	3,040	0.973
	NH_3	6	9.8	0.612
GHGs	CO_2	281,550	446,750	0.630
	N_2O	7.8	10.7	0.729
	CH_4	53	972	0.055

Based on the data in Magelli et al. [32]

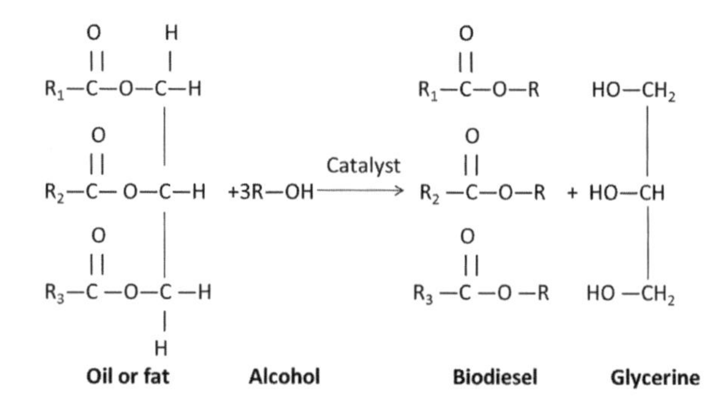

Fig. 8.4 A simplified representation of biodiesel production

$$\ln \nu = A + \frac{B}{T + C} \tag{8.17}$$

where ν (mm^2/s) is the kinematic viscosity of the biodiesel at temperature T (K); A, B, and C are correlation parameters, which can be determined from viscosity measurements at three or more temperatures.

Nowadays most of the biodiesel fuels are produced through the base-catalyzed transesterification of oils or fats. In this process, the vegetable oil or animal fat is combined with alcohol, producing biodiesel and a valuable byproduct glycerin. Figure 8.4 shows a simplified representation of this reaction, where R is the short hydrocarbon chain of an alcohol. R can be either methanol or ethanol or isopropanol. R_1, R_2, and R_3 are fatty acid chains. For naturally occurring oils or fats, these fatty acids are largely palmitic, stearic, oleic, linoleic, and linolenic acids.

The exact chemical compositions of biodiesel depend on feedstock, the source of fatty acids and the alcohol used in production. Since alcohol is relatively constant, fatty acids are the main factor that determines the chemical composition of biodiesel. Most of the fats and oils contain 16 and 18 carbon chains. A large number of vegetable oils contain predominantly unsaturated acids, whilst animal fats like cattle tallow contain 60 % saturated acids and mono unsaturated acids as the remainder.

8.4.2.1 Air Emissions from Biodiesel Combustion

Biodiesel is generally considered to be oxygenated, sulfur-free, biodegradable, and nontoxic.

From an environmental point of view, biodiesel is believed to be biodegradable, renewable, and reduces CO_2 emission. Since most of the carbon in the fuel was originally removed from the air by plants, there is very little net increase in carbon dioxide levels.

The post-combustion exhaust gas from biodiesel combustion is characterized with low carbon monoxide, low unburned hydrocarbons, and low particulate emissions from diesel engines; hence it reduces further already low carbon monoxide and unburned hydrocarbons. Although particulate emissions, especially the black soot portion, are greatly reduced for biodiesel engines, NOx emissions are usually high.

The NOx emissions from biodiesel may be 10 % more than that produced by petroleum diesel. Biodiesel combustion usually has more widespread high-temperature distribution areas than petroleum diesel fuel [53]. As we have learned from combustion chemistry, higher combustion temperature leads to more thermal NOx formation. This partially explains the reason behind the high NOx emissions typically measured from the engine exhaust produced by biodiesel fuels. However, ethanol could act as an effective NOx reducing additive [20]. Strategies for reducing NOx emissions from biodiesel combustion include increasing spray cone angle, retarding start of injection, exhaust gas recirculation (EGR), and charge air cooling [55].

8.4.2.2 Challenges to Biodiesel

The specific gravity of biodiesel ranges from 0.86–0.90, which is higher than that of No. 2 diesel. This may result in a higher fuel mass of biodiesel injection through an unmodified diesel injection system than does a No. 2 petroleum diesel. The high heating value of biodiesel is slightly above 17,000 Btu/lb, and it is lower than that of No. 2 petroleum diesel (19,300 Btu/lb). As a result, the total thermal energy delivered from biodiesel is less than that of No. 2 diesel.

The main barrier for commercialization of biodiesel is its higher cost than petroleum diesel. The price of biodiesel is almost double that of petroleum diesel if subsides are not taken into account.

Fuel stability is another major concern to biodiesel. Biodiesel is less saturated and normally has poorer thermal stability, oxidative stability, and storage stability than petroleum. A study performed at the University of Idaho showed that biodiesel placed in water degraded 95 % by microorganisms in 28 days, while petroleum diesel fuel degraded only 40 % in same time [56]. At low temperatures, biodiesel will gel or crystallize into a solid mass that cannot be pumped; as a result the engine cannot run.

Another emerging challenge is the food crisis as it is unethical to convert edible food to biodiesel to feed the SUVs instead of the starving people. There was a great rise in the price of food (especially corn) and fertilizer from 2006–2008, partially because quite a lot of corn in North American was used for biodiesel production. Research and development should be focused on waste to biodiesel instead.

8.4.3 Bioethanol

Bioethanol (C_2H_5OH) can be produced from starch in corn, sugar in sugar cane, or cellulose from the cellulosic biomass. Starch and sugar based bioethanol is considered as the first generation bioethanol, while cellulosic bioethanol the second.

Corn is the well-developed feedstock in United States, while sugar cane is a tropical and subtropical crop that is the primary feedstock in Brazil, India, and Colombia. Cellulosic bioethanol is a more recent development that is not well commercialized yet. Regardless of the feedstock, the sales price of bioethanol must be competitive with that of petroleum gasoline. However, profit margins in bioethanol production processes are still low. Nonetheless, our focus here is a briefly introduction rather than an in-depth discussion.

Depending on the biomass feedstock, there are several major steps that may apply to bioethanol production [26].

- Feedstock pre-treatment
- Hydrolysis
- Fermentation
- Separation
- Storage.

8.4.3.1 Feedstock Pretreatment

Feedstock pretreatment is necessary to convert most of the carbohydrates in the feedstock into sugars. The first step is size reduction and breakdown of the cell wall structures surrounding the target compounds, mainly cellulose. Some lignin is dissolved in the solution. Sugar cane is crushed followed by juice extraction; corn goes through dry grinding or wet milling according to the downstream conversion technology. The feedstock is then pretreated using chemicals, such as dilute sulfuric acid or ammonia.

After pretreatment, a large amount of water is separated from the hydrolysate slurry. Meanwhile, other unwanted chemicals such as acetic acid and furfural are also removed from the slurry. Then hydrolysate slurry is cooled by dilution water and further chemically conditioned for next step.

8.4.3.2 Hydrolysis and Fermentation

In this step, the starch and cellulose are converted into sugars, primarily being glucose, by cellulose enzymes. It is also called enzymatic hydrolysis. A cellulase enzyme breaks down cellulose fibers and ultimately converts them into glucose monomers.

The resulting glucose and other sugars from feedstock pretreatment are to be converted into bioethanol by fermentation. This process may take several days to complete. The simplified chemical reactions for hydrolysis and fermentation are

$$C_{12}H_{22}O_{11} + H_2O \overset{\text{Enzyme}}{\rightarrow} 2C_6H_{12}O_6 \tag{8.18}$$

$$C_6H_{12}O_6 \overset{\text{Zymase}}{\rightarrow} 2C_2H_5OH + 2CO_2 \tag{8.19}$$

where $C_{12}H_{22}O_{11}$ stands for sucrose, $C_6H_{12}O_6$ glucose, and C_2H_5OH ethanol.

Example 8.1: Conversion rate of bioethanol
According to Eqs. (8.18) and (8.19), what is the stoichiometric mass ratio of conversion by fermentation.

Solution
According to Eqs. (8.18) and (8.19), 1 mol of sucrose $C_{12}H_{22}O_{11}$ can be converted into 4 mol of ethanol, C_2H_5OH, and 4 mol of CO_2. The overall reaction can be described as

$$C_{12}H_{22}O_{11} + H_2O \rightarrow 4C_2H_5OH + 4CO_2 \tag{8.20}$$

The molar weights of $C_{12}H_{22}O_{11}$, C_2H_5OH and CO_2 are 342, 46, and 44 g/mol, respectively. Therefore, the maximum theoretical mass conversion ratio of ethanol is

$$\frac{4 \times M_{C_2H_5OH}}{M_{C_{12}H_{22}O_{11}}} = \frac{4 \times 46}{342} = 53.8\,\%$$

The rest goes to CO_2 and its production ratio is 46.2 %.

This example shows only the theoretical maximum conversion rate. In reality, no more than 47 % of the fermented carbon hydrates is converted to (bio)ethanol.

8.4.3.3 Bioethanol Separation

The product of fermentation is called both or beer. It contains water, ethanol, combustible solids, and much more. Ethanol is separated from this mixture by a few steps. Distillation and molecular sieve adsorption are the core technologies for this purpose.

Distillation can be accomplished in two or three columns depending on the required purity of the ethanol. In the first column, called beer column, water, and dissolved CO_2 are first removed from the mixture. The second column is called rectification column, where ethanol is concentrated to a level of 92.5 % of ethanol. Then it can be further dehydrated to 99.5 % by vapor-phase molecular sieve adsorption.

Both adsorption and desorption principles are employed in the aforementioned separation process. There are two columns in the molecular sieve adsorption process, one works as an adsorption tower while the other is being regenerated. All the other gaseous emissions are treated by absorption using pure water.

Solid-liquid separation devices are employed to recover the solids in the stillage from the beer column. Although the principles were not deliberately introduced in this book, hydraulic cyclones and filters can be used for this purpose. After drying, the solids can be used as alternative fuel for combustion too. The rest of the water goes to a water treatment facility.

8.4.3.4 Bioethanol Combustion

Bioethanol is mainly mixed with petroleum gasoline for combustion in light engines. The overall stoichiometric combustion formula can be described as follows, although the actual chemical reactions are much more complicated.

$$C_2H_5OH + 3O_2 \rightarrow 2CO_2 + 3H_2O \quad (8.21)$$

On a volumetric base, bioethanol (or ethanol) has only 2/3 of the heating value of petroleum gasoline [6]. There is a great power reduction if pure ethanol (E100) is used to power an automobile. Therefore, bioethanol is mixed with gasoline for sale.

Water contamination remains a challenge to the bioethanol industry. The challenge to bioethanol-gasoline mixed fuels is the phase separation induced by water contamination. Ethanol absorbs water readily in a storage tank or transport pipeline. This resultant water in the fuel negative affects the engine operation.

8.4.4 Hydrothermal Conversion of Biomass to Biofuels

8.4.4.1 Properties of Hot Compressed Water

Before we introduce hydrothermal conversion of biomass to biofuel, we have to start with the phase behavior of hot compressed water. Water exhibits unique properties at a critical point of 374 °C, 22 MPa. Above this critical point, water is in a homogeneous phase known as a supercritical fluid. Supercritical water exists simultaneously as both a liquid and a gas. Subcritical water is marked by a higher density and the presence of two phases, one being liquid and the other being vapor [29].

The first special property of hot compressed water is its ionic constant. The ionic products are H^+ and OH^- ions:

$$H_2O \rightarrow H^+ + OH^- \quad (8.22)$$

The ionic constant of subcritical water is considerably higher than those of supercritical water and regular water. As such, subcritical water is a special reaction medium due to its dual acid-base catalysis nature, which enables rapid conversion of biomass compounds to biofuels.

Another special property of hot compressed water is its density. Under low density conditions, diffusion of radicals through the solution is enhanced; under high-density conditions, ionic reaction mechanisms are promoted.

The third parameter of concern is the dielectric constant of hot compressed water. The dielectric constant of water decreases with the increase of temperature. As a result, hot compressed water is much less polar than normal water. It performs more like an organic solvent. The decreasing dielectric constant of hot compressed water is likely to affect organic reactions. One benefit of the changing dielectric constant of water is that it works like an organic solvent during reaction conditions, and it returns to its normal polarity after cooling. This unique property is believed to be beneficial to the production and voluntary separation of alkanes from biomass [11].

The changing reaction environment around critical point approach causes a significant change in reaction mechanisms of hydrothermal conversion of biomass [7, 28]. Both ionic and free radical reactions may take place in hydrothermal conversion of biomass, the latter being preferred above the critical point.

Depending the status of hot compressed water it can convert biomass into both gaseous and liquid fuels. Without catalyst, the higher the water temperature, the more gas, and the less liquid fuels. As a result, hydrothermal conversion processes are divided into hydrothermal liquefaction and hydrothermal gasification.

The overall process of decomposition of biomass is illustrated in Fig. 8.5. Hydrolysis plays an important role in forming glucose/oligomer, which can quickly decompose into, oil, char, and gases. Without catalyst, oil can be converted into char and gases; however, the addition of alkali catalyst will result in more oil production because it inhibits the char production from the oil intermediates. The presence of water in the feedstock is a key factor for the conversion reaction. At temperatures between 250 and 350 °C, organic molecules in liquid water undergo chemical reactions.

Temperature is an important factor that affects the HTC conversion products, liquid, or gas. As shown in Fig. 8.6, more liquid products are produced at lower temperature. As the temperature increases, more gaseous products are produced at the expense of the liquids and because of additional carbon conversion.

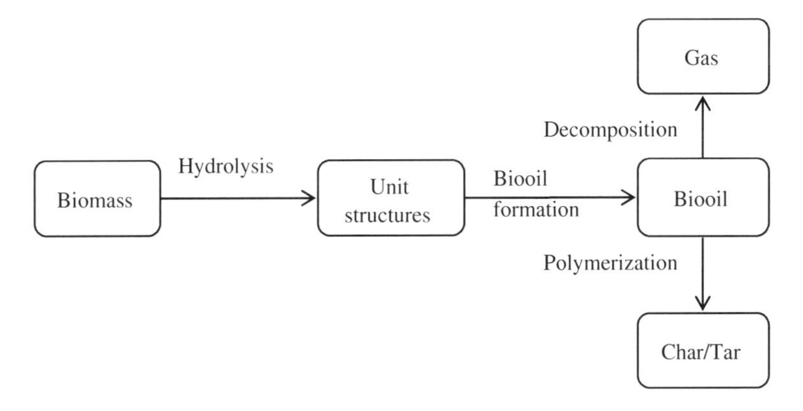

Fig. 8.5 Conversion of cellulose to bio-oil

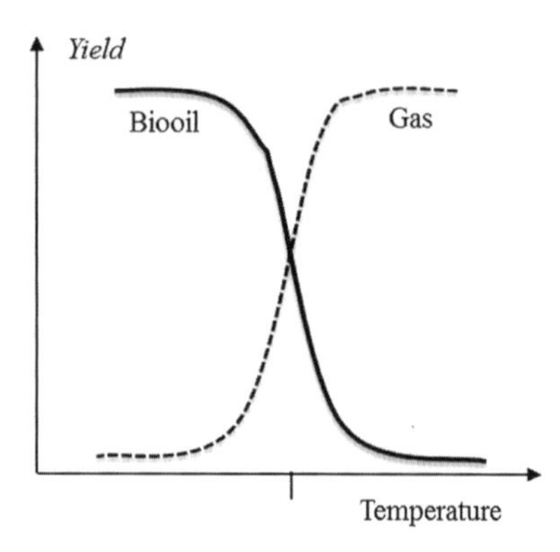

Fig. 8.6 Effect of temperature on the HTC product

Water goes through dramatic changes in both physical and chemical properties when heated. As the temperature rises from 25 to 300 °C under standard pressure, the density and dielectric constant of water decreases from 997 to 713 kg/m^3 and 78.85 to 19.66, respectively, while the ionic product increases from $10^{13.99}$ to $10^{11.30}$. These changes in physical properties make the solvent properties of water at 300 °C roughly equivalent to those of acetone at 25 °C. Ionic reactions of organics are favored by the increase of solubility in water. The increase in the dissociation constant increases the rate of both acid- and base-catalyzed reactions in water far beyond the natural acceleration due to increased temperature. Water itself can also act as an acidic or basic catalyst, and its reactivity can often be reinforced by autocatalysis from water-soluble reaction products.

Under high-temperature high pressure conditions, water shows even more unique properties as a reaction medium, especially with the presence of suitable catalysts. At supercritical condition (385–400 °C, 35 MPa) the reaction for HTC conversion can be completed within a minute.

Residence time is another factor for a successful conversion. The HTC bio-oil yields under acidic, neutral, and alkaline conditions all decrease with increasing residence time. Despite the similar trends of HTC bio-oil yields against conversion temperatures and the residence time, the conversion mechanisms behind them are likely quite different. Under acidic HTC conditions, the decrease in bio-oil yields at high temperatures and long residence time is mainly attributed to the formation of residual solids. Under acidic conditions, 5-HMF, the main component of acidic HTC bio-oil, tends to form hydrothermal char/solid by polymerization. Under neutral conditions, the decrease in HTC bio-oil yield at high temperatures and long reaction residence time is mainly caused by the formation of residual solids and the gas products. Under alkaline HTC conditions, the gas formation from the decomposition of alkaline HTC bio-oil mainly resulted in the bio-oil yield decrease at the high temperatures (>300 °C).

In continuous operation, slurry feeding is a challenge because of the water evaporation at the end of the pipe. Poor conversions were reported because of the loss of water by condensation. Serious erosion and/or corrosion occurred on several spirals of the auger where the reaction was actually took place before the feedstock enters the reactor. Char formation was found to occur downstream because of lack of water brought about by back condensation. Some differences have been found in the chemical composition of the oil from the batch process and the semi-continuous process. A higher percentage of aromatic and/or unsaturated carbon was found in the oil produced using the semi-continuous process.

8.4.4.2 Hydrothermal Liquefaction

Hydrothermal liquefaction employs reactive hydrogen or carbon monoxide carrier gases to produce liquid fuel from organic matter at moderate temperatures. Temperatures between 300 and 400 °C correspond to the maximum yield of oil products [50]. Direct liquefaction involves rapid pyrolysis to produce liquids and/or organic vapors, whereas indirect liquefaction employs catalysts to convert non-condensable, gaseous products of pyrolysis, or gasification into liquid products.

Technically, any matter with high organic content can be converted into another form by thermochemical conversion, but the conversion rate depends on the feedstock and the process itself. A variety of feedstocks are proven technically successful including high-carbon content materials such as coal, peat, and ligno-cellulosic material. Low-quality feedstock includes municipal and industrial wastes and agricultural residues. Studies on thermochemical conversion of biomass were predominantly focused on materials that are highly cellulosic, until recently an emerging interest in liquid waste such as livestock manure (e.g., [50]) and algae [34].

Many factors affect the conversion rate of a HTC process. They include, but not limited to, temperature, pressure, retention time, reactant gases, feedstock, catalysts, if any, and the reactor design. These parameters were found to be most important as they dictate the yield and quality of the products. Regardless of the process, the new fuel is inevitably more expensive than its parent fuel due to the extra energy and equipment required for the conversion.

Hydrothermal conversion is a process that is similar to cooking of food. The major differences between this process and other processes are use of wet feedstock as raw material because it saves energy from drying the feedstock. Also, the presence of water is beneficial for the conversion process.

The primary reactions in the conversion of biomass to oil likely involve the formation of low-molar weight, water-soluble compounds such as glucose. Water also acts as solvent and alkaline catalysts. More importantly, water plays a critical role in the water-gas shift reaction

$$CO + H_2O \rightarrow H_2 + CO_2 \tag{8.23}$$

The resultant hydrogen with other oxygen-containing functional groups, eliminating the oxygen element and yielding hydrocarbon-like compounds.

8.4.5 Biogas

In addition to the thermochemical conversion approach, biomass can also be converted into gaseous fuels by a biological approach called anaerobic digestion. And the resultant gas is sometimes called biogas. It is well known that typical biogas from a well-controlled anaerobic process contains 60 % CH_4 and 40 % CO_2 and other trace compounds.

Anaerobic digestion is a complex biochemical reaction that involves the following four steps.

- Hydrolysis:
  It is the process where complex organic matter is decomposed into simple organic molecules. It is done with the existence of water that splits the chemical bonds of the organic matter.
- Fermentation:
  It is also called acidogenesis, the process where carbohydrates are decomposed by bacteria, enzymes, molds, or yeasts in the absence of oxygen.
- Acetogenesis:
  It is the process where the products of the fermentation process are converted into H_2, CO_2, and acetate by acetogenic bacteria.
- Methanogenesis:
  It involves the formation of CH_4 from acetate, H_2, and CO_2 by methanogenic bacteria.

In the anaerobic digestion process, the methanogenic bacteria are very sensitive and are easily upset by sudden changes in temperature or pH, and toxic substances such as arsenic, copper, and antibiotics.

There are mainly two types of anaerobic digesters:

- Mixed digester
- Plug-flow digester

Mixed digesters are usually employed for liquid feedstock and the latter for semi-solid feedstock (about 13 % solids). In a mixed digester solids are kept in contact with the bacteria for reaction and the mixing is usually maintained either mechanically or by bubbling. Plug-flow digesters eliminate the need for mixing by slowly moving the waste through a tube-shaped vessel. A lagoon is another commonly used method for waste treatment, where aerobic bacteria use oxygen to convert organic matter into CO_2, water, and more bacteria. Therefore, there is no biogas produced in aerobic digestion.

The main drawback of natural anaerobic digestion is its slow reaction. A properly designed anaerobic digester provides better control over the environment for the anaerobic digestion process. A digester is like a reactor, which can be sealed, heated, and agitated to shorten the time needed to stabilize the waste. It also offers a good control of odor emission and easy capture of the gaseous products like methane. Due to the shorter process time requirement, a typical anaerobic digester can be 100 times smaller than a natural anaerobic lagoon.

Regardless of the type of fuels produced from biomass or petrochemical process. Their applications were primarily for energy production by combustion. The combustion processes of these many types of fuels do not always follow the same simplicity, rather they depend on the physical designs of the process.

References and Further Readings

1. Alvarez E, Mendioroz S, Munoz V, Palacios JM (1996) Sulphur recovery from sour gas by using a modified low temperature Claus process on sepiolite. Appl Catal B: Environ 9:179–199
2. Appell HR, Fu YC, Friedman S, Yavorsky PM, Wender I (1980) Converting organic wastes to oil: a replenishale energy source, Report of Investigations 7560. Bureau of Mines, Washington DC
3. ASTM (2000) ASTM Standard D6751-03 Standard Specification for Biodiesel Fuel (B100) Blend Stock for Distillate Fuels. West Conshohocken, Pennsylvania
4. Basüaran Y, Denizli A, Sakintuna B, Taralp A, Yurum Y (2003) Bio-liquefaction/ solubilisation of low-rank Turkish lignite and characterization of the products. Energy & Fuel 17:1068–1074
5. Busca G, Pistarino C (2003) Technologies for the abatement of sulphide compounds from gaseous streams: a comparative overview. J Loss Prev Process Ind 16(2003):363–371
6. Brown RC (2003) Biorenewable Resources, Engineering New Products from Agriculture. Blackwell Publishing, Iowa State Press, Ames
7. Buhler W, Dinjus E, Ederer HJ, Kruse A, Mas C (2002) Ionic reactions and pyrolysis of glycerol as competing reaction pathways in near- and supercritical water. J Supercrit Fluids 22 (1):37–53
8. Chun SW, Jang JY, Park DW, Woo HC, Chung JS (1998) Selective oxidation of hydrogen sulphide to elemental sulphur over TiO_2/SiO_2 catalysts. Appl Catal B 16:235–243
9. Cohen MS, Gabriele PD (1982) Degradation of coal by the fungi polyporus versicolor and poria monticola. Appl Environ Microbiol 44:23–27
10. Cooper CD, Alley FC (2002) Air pollution control—a design approach, 3rd edn. Waveland Press, Inc., Long Grove
11. Dolan R, Yin S, Tan Z (2010) Effect of headspace fraction and aqueous alkalinity on subcritical hydrothermal gasification of cellulose. Int J Hydrogen Energy 35:6600–6610
12. Duun BS, Mackenzie JD, Tseng E (1976) Conversion of cattle manure into useful products. EPA-600/2-76-238. USEPA, Washington DC
13. Elsner MP, Menge M, Müller C, Agar DW (2003) The Claus process: teaching an old dog new tricks. Catal Today 79–80:487–494
14. Garner W, Smith I (1973) The disposal of cattle feedlot wastes by pyrolysis. EPA-R2-73-096. USEPA, Washington DC

15. Graboski MS, McCormick RL (1998) Combustion of fat and vegetable oil derived fuels in diesel engines. Prog Energy Combust Sci 24:125–164
16. Fang Z, Minowa T, Smith Jr RL, Ogi T, Kozinski JA (2004) Liquefaction and gasification of cellulose with Na_2CO_3 and Ni in subcritical water at 350 C. Ind Eng Chem Res 43 (10):2454–2463
17. Fakoussa RM, Hofrichter M (1999) Biotechnology and microbiology of coal degradation. Appl Microbiol Biotechnol 52:25–40
18. Furimsky E (1999) Gasification in petroleum refinery of 21st century. Oil Gas Sci Technol 54:597–618
19. Furimsky E (2006) Gasification of sandcoke: review. Fuel Process Technol 56:262–290
20. Hansen AC, Gratton MR, Yuan W (2006) Diesel engine performance and nox emissions from oxygenated biofuels and blends with diesel fuel. Trans ASABE 49(3):589–595
21. Hayatsu R, Winans RE, McBeth RL, Scott RG, Moore LP, Studier MH (1979) Lignin-like polymers in coals. Nature 278:41–43
22. Hatcher PG (1990) Chemical structural models for coalified wood (vitrinite) in low rank coal. Org Geochem 16:595–968
23. Heidenreich S (2013) Hot gas filtration—a review. Fuel 104:83–94
24. Henry F, Ariman T (1983) A staggered array model of a fibrous filter with electrical enhancement. Part Sci Technol: Int J 1(2):139–154
25. Hofrichter M, Bublitz F, Fritsche W (1997) Fungal attack on coal II: Solubilisation of low-rank coal by filamentous fungi. Fuel Process Technol 52:55–64
26. Humbird D, Davis R, Tao L, Kinchin C, Hsu D, Aden A et al (2011) Process design and economics for biochemical conversion of lignocellulosic biomass to ethanol. Technical Report NREL/TP-5100-47764 May 2011
27. IEA (1991) Advanced Coal Cleaning Technology. IEA Coal Research, London
28. Kruse A, Gawlik A (2003) Biomass conversion in water at 330–410 °C and 30–50 MPa. Identification of key compounds for indicating different chemical reaction pathways. Ind Eng Chem Res 42(2):267–279
29. Kruse A, Dinjus E (2007) Hot compressed water as reaction medium and reactant. Properties and synthesis reactions. J Supercrit Fluids 39(3):362–380
30. Lam PY, Lam PS, Sokhansanj S, Bi XT, Lim CJ, Melin S (2014) Effects of pelletization conditions on breaking strength and dimensional stability of Douglas fir pellet. Fuel 117:1085–1092
31. Machnikowska H, Pawelec K, Podgórska A (2002) Microbial degradation of low rank coals. Fuel Process Technol 77(78):17–23
32. Magelli F, Boucher K, Bi HT, Melin S, Bonoli A (2009) An environmental impact assessment of exported wood pellets from Canada to Europe. Biomass Bioenergy 33:434–441
33. Mendioroz S, Munoz V, Alvarez E, Palacois JM (1995) Kinetic study of the Claus reaction at low temperature using γ-alumina as catalyst. Appl Catal A 132:111–126
34. Menetrez MY (2012) An overview of algae biofuel production and potential environmental impact. Environ Sci Technol 46(13):7073–7085
35. Minowa T, Kondo T, Sudirjo S (1998) Thermochemical liquefaction of Indonesian biomass residues. Biomass Bioenergy 14(5):517–524
36. Mondal P, Dang GS, Garg MO (2011) Syngas production through gasification and cleanup for downstream applications—Recent developments. Fuel Process Technol 92:1395–1410
37. NREL (2014) Biomass cofiring (http://www.nrel.gov/docs/fy00osti/28009.pdf, accessed January 2014), National Renewable Energy Laboratory, USA
38. Ocfemia K (2005) Hydrothermal process of swine manure to oil using a continuous reactor system, PhD dissertation, University of Illinois at Urbana-Champaign, Illinois, USA
39. Ralph JP, Catcheside DEA (1997) Transformations of low rank coal by Phanerochaete chrysosporium and other wood-rot fungi. Fuel Process Technol 52:79–93
40. Ramdin M, de Loos TW, Vlugt TJH (2012) State-of-the-art of CO_2 capture with ionic liquids. Ind Eng Chem Res 51:8149–8177

41. Reiss J (1992) Studies on the solubilisation of German coal by fungi. Appl Microbiol Biotechnol 37:830–832
42. Rodulovic PT, Ghani MU, Smoot LD (1995) An improved model for fixed bed coal combustion and gasification. Fuel 74:582–594
43. Salman OA, Bishara A, Marafi A (1987) An alternative to the claus process for treating hydrogen sulphide. Energy 12:1227–1232
44. Sharp CA, Howell SA, Jobe J (2000) The effect of biodiesel fuels on transient emissions from modern diesel engines, part I regulated emissions and performance. SAE Paper No. 2000-01-1976. Warrendale, PA
45. Shinn JH (1984) From coal to single-stage and two-stage products: A reactive model of coal structure. Fuel 63:1187–1196
46. Spath PL, Dayton DC (2003) Preliminary screening—technical and economic assessment of synthesis gas to fuels and chemicals with emphasis on the potential for biomass-derived syngas. NREL Report. NERL/TP-510-34929
47. Thomsen VB (2000) LeChâtelier's principle in the sciences. J Chem Educ 77(2):173–176
48. Tomás-Alonso F (2005) A new perspective about recovering SO_2 off gas in coal power plants: energy saving. Part II. Regenerable dry methods. Energy Sources 27:1043–1049
49. White RK, Taiganides EP (1971) Pyrolysis of livestock wastes. In: Proceedings of the 2nd International Symposium on Livestock Wastes, ASAE, St. Joseph, MI, USA, pp 190–194
50. Yin S, Dloan R, Harrison M, Tan Z (2010) Subcritical hydrothermal liquefaction of cattle manure to bio-oil: Effects of conversion parameters on bio-oil yield and characterization of bio-oil. Bioresour Technol 101:3657–3664
51. Yin S, Tao X, Shi K, Tan Z (2009) Biosolubilisation of Chinese Lignite. Energy 34:775–781
52. Yong HK (2007) Method of gasification in IGCC system. Hydrogen Energy 32:5088–5093
53. Yuan W, Hansen A, Tat M, van Gerpen J, Tan Z (2005) Spray, ignition, and combustion modeling of biodiesel fuels for investigating NOx emissions. Trans ASAE 48(3):933–939
54. Yuan W, Hansen A, Zhang Q, Tan Z (2005) Temperature-dependent kinematic viscosity of selected biodiesel fuels and blends with diesel fuel. J Am Oil Chem Soc 82(3):195–199
55. Yuan Q, Hansen AC, Zhang Q (2007) Computational modelling of NOx emissions from biodiesel combustion. Int J Vehicle Design 45(1/2):12–32
56. Zhang X, Peterson C, Reece D, Reece D, Möller G, Haws R (1998) Biodegradability of biodiesel in the aquatic environment. Trans ASAE 41:1423–1430

Chapter 9
In-combustion Air Emission Control

In-combustion air emission control is accomplished by proper design and operation of a combustion device, either a burner or an engine. The existing process of in-combustion control is primarily limiting the formation of NO_x by modifying the combustion temperature and other conditions. This chapter starts with an introduction of typical combustion processes followed by specific in-combustion air emission control technologies, including low-NO_x burner, sorbent injection for in furnace SO_2 capture, and approaches to reduce soot formation, and so on.

9.1 Stationary Combustion Devices

9.1.1 Pulverized Coal/Biomass Combustion

Combustion of solid fuels takes place in a variety of stationary systems, such as home heating stove and industrial furnace. In a stove, solid fuels are burned directly without extensive size reduction. In industrial furnaces such as boilers for power generation, their sizes have to be reduced to a certain level for continuous feeding operation. This size reduction is also referred to as pulverization. A comprehensive description of pulverized coal combustion can be found in the book by Smoot and Pratt [14].

In a typical pulverized coal combustion system, coal is ground to fine particles and separated using a mesh screen before being fed to the burner. Pulverized coal particles have a mean diameter of 50 μm with the majority smaller than 200 μm by mass. Then, these coal particles are fed into the furnace by mixing with oxidants (mostly air). And combustion takes place in the flame in the open space of the furnace.

As introduced in previous chapters, ash and other pollutants are formed during solid fuel combustion. The coarse ash particles fall down to the bottom of the furnace as a solid waste, and this is referred to as bottom ash or slag, while the rest of the fine ash particles are carried along the combustion process with flue gas. In general, the

Z. Tan, *Air Pollution and Greenhouse Gases*, Green Energy and Technology,
DOI 10.1007/978-981-287-212-8_9

ash particles formed in a pulverized coal combustion process is very fine and approximately 65–85 % of the ash is fly ash.

Pulverized coal combustion technology is used for the majority of fossil-fuel fired electricity generation. As to be introduced in the coming chapter, flue gas cleaning typically involves the emission controls of particulate (i.e., fly ash), oxides of sulfur and nitrogen, and others. It is characterized with high-combustion efficiency and high-combustion temperature.

9.1.2 Fluidized Bed Combustion

Advanced coal technologies have been developed to improve cost-effectiveness and environmental protection. These emerging technologies differ significantly from the existing conventional combustion technologies such as pulverized coal combustion, cyclone firing, and stoker firing. They are briefly introduced as follows.

Fluidized bed combustion (FBC) is primarily for solid fuel combustion because there is no need to fluidize liquid or gaseous fuels. In FBC, the combustion takes place in a bed where materials are fluidized by air blown from beneath of the layer of particles. The bed materials are fuel particles, sand, ash, char residue, and other solids.

There are two typical FBC systems, bubbling fluidized bed and circulating fluidized bed, which are shown in (Fig. 9.1). In bubbling fluidized bed combustion (BFBC) the bed particles behave like a boiling fluid but remain in the bed, when the gas flows upward at the velocity of 1–3 m/s. In circulating fluidized bed combustion

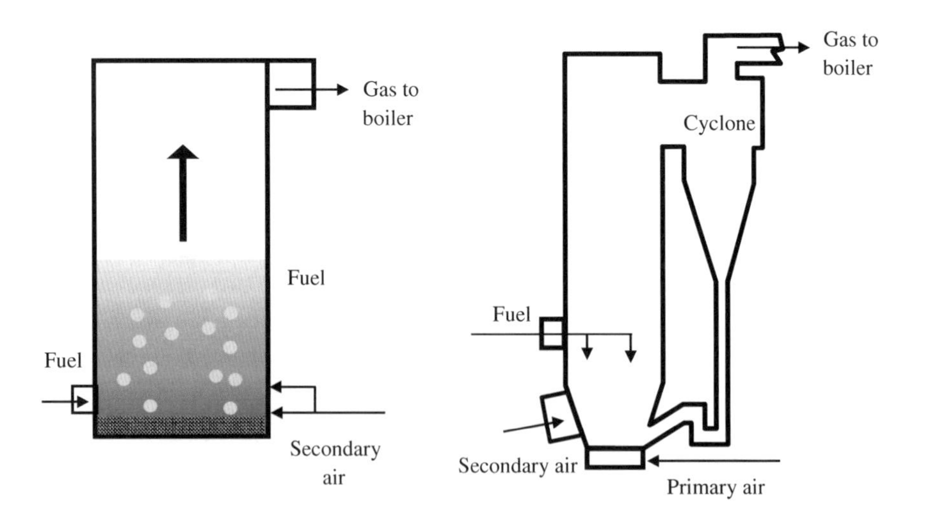

Fig. 9.1 Fluidized bed combustion chambers (*Left* Bubbling fluidized bed, *Right* Circulating fluidized bed)

(CFBC), air velocity is relatively higher than BFBC, being about 5–10 m/s at which the bed material suspension fills the entire combustion chamber.

For either case, the suspended particle size can be easily estimated from the gas speed and aerosol dynamics (Sect. 4.2). It can be determined by the balance between gravity and drag force on the particle

$$F_D = C_D\left[\frac{1}{2}\rho_g\left(V_p - V_g\right)^2\right]\left(\frac{1}{4}\pi d_p^2\right) \tag{9.1}$$

When the particle is suspended in the combustion chamber, the gravitational force equals the drag on the particle. Assuming the particle is spherical and the corresponding diameter is d_{p0}, we have

$$\frac{1}{6}\pi d_{p0}^3 g\rho_p = C_D\left[\frac{1}{2}\rho_g\left(V_p - V_g\right)^2\right]\left(\frac{1}{4}\pi d_{p0}^2\right). \tag{9.2}$$

When the fuel particle is suspended in the gas, $V_p = 0$, and the equation becomes

$$d_{p0} = \frac{3}{4}\frac{\rho_g V_g^2}{\rho_p g}C_D. \tag{9.3}$$

Fuel particles larger than this size will remain in the chamber; smaller ones will be carried away by the gas to the downstream unit. Obviously, the greater the gas speed, the larger particles can penetrate through the combustion reactor and enter downstream unit. The drag coefficient is described in Eq. (9.4).

$$C_D = \begin{cases} \frac{24}{Re_p} & Re_p \le 1 \\ \frac{Re_p}{24}\left(1 + 0.15Re_p^{0.687}\right) & 1 < Re_p \le 1000 \\ 0.44 & Re_p > 1000 \end{cases} \tag{9.4}$$

A simple form can be obtained for $Re_p \le 1$.

$$d_{p0} = \sqrt{\frac{18\mu V_g}{\rho_p g}} \tag{9.5}$$

Example 9.1: Suspended particle size in fluidized bed combustion

Consider a fluidized bed furnace in a coal fired power plant that operates at atmospheric pressure. Assume the coal particle density of 1,000 kg/m^3, and the gas temperature is about 1,100 K. The gas moves upward at a speed of 3 m/s. Use the gas properties using those of the air at the same temperature, and estiamte the suspended particle diameter.

Solution

The properties of the air at 1,100 K and 1 atm are

$\rho_p = 0.3166$ kg/m^3, $\mu = 4.49 \times 10^{-5}$ N.s/m^2.

If we assume $Re_p < 1$

$$d_{p0} = \sqrt{\frac{18\mu V_g}{\rho_p g}} = \sqrt{\frac{18 \times 4.49 \times 10^{-5} \times 3}{1000 \times 9.81}} = 4.97 \times 10^{-4}\,\mathrm{m} = 0.497\,\mathrm{mm}$$

With this size, we can calculate

$$Re_p = \frac{\rho_g V_g d_p}{\mu} = \frac{0.3166 \times 3 \times 4.97 \times 10^{-4}}{4.49 \times 10^{-5}} \gg 1$$

Therefore, our assumption of $Re_p < 1$ was invalid.

Now assume $1 < Re_p \leq 1000$ and

$$\begin{cases} d_{p0} = \frac{3}{4}\frac{\rho_g V_g^2}{\rho_p g} C_D \\ C_D = \frac{Re_p}{24}\left(1 + 0.15 Re_p^{0.687}\right) \end{cases}$$

The calculation becomes complicated, but we can solve it by iteration. Then the suspended particle size is

$$d_{p0} = 6.055\mathrm{mm}$$

And the corresponding particle Reynolds number is $Re_p \cong 128$

Under steady operation condition, a large proportion of the bed materials leave the chamber via an exit on top of the chamber and are collected by a particle separator, most likely a cyclone for material recirculation to the bed. Cyclones are being used at temperatures of 1,000 °C in PFBC systems for solid recycling. Penetrated fine particles are called fly ash and join the flue gas. Due to the high particle concentrations in this application, particle agglomeration may occur, which favor particle separation.

The fuel particle size is larger while the furnace temperature is lower in FBC (800–950 °C) compared to pulverized coal combustion (>1,000 °C). Due to the relatively low-combustion temperature, FBC can handle low-grade fuels such as wet sludge or waste solid fuels with a relatively low-NO_x emission. The SO_x emissions can be reduced by addition of sorbent like a limestone or lime to the bed. On the other hand, the emissions of N_2O may be high.

Another drawback of CFBC is the increased fly ash, which is mainly the result of the higher velocity, smaller fuel particle size, and more intense attrition and abrasion. The fate of ash-forming material in fluidized bed is much different from that in a pulverized coal combustion chamber. Again, temperatures in CFBC are much lower and particles are larger, but mechanical stresses are stronger due to strong turbulence and the impact between particles. Although ash-forming material remains in the bed, fine particles are produced due to attrition and abrasion.

9.2 Internal Combustion Engines

An internal combustion engine is a device where atomized liquid fuels are burned to produce thermal energy (heat), which is converted to mechanical energy to drive the transportation vehicles. Combustion in engines takes place at high pressure with a variable volume, most commonly a piston for liquid fuels. The most common internal combustion engines are spark ignition engines used in automobile industry and diesel engines primarily for trucking industry. Large diesel engines are also used in off road power generation and trains. Another type of internal combustion engine is gas turbines for air craft or power generation. It is excluded from this book.

9.2.1 Spark Ignition Engines

A schematic diagram of a basic piston engine (or reciprocating engine) is shown in (Fig. 9.2). It is a metal block containing a series of chambers. The core of the engine is a cylinder that houses a piston. The inner diameter of the cylinder is also referred to as bore. The upper part of the block consists of outer walls that form hollow jackets around the cylinder walls. The hollow jackets contain the coolants to prevent the engine from being overheated. The lower part is called the crankcase, which provides rigid mounting points for the bearings to fix the crank.

The crank drives the piston up and down through the connecting rod. The angle between the crank and the centerline of the cylinder is defined as crank angle, θ. When the piston is at the top of the cylinder, it reaches its top dead center (TDC) and $\theta = 0$; when the piston is at the lowest position, bottom dead center (BDC), $\theta = \pi$. The distance between TDC and BDC is referred to as stroke.

A typical operating cycle of this type of engine involves four stokes (Fig. 9.3):

1. intake
2. compression
3. combustion/expansion
4. exhaust

When the piston moves downward, the intake camshaft opens the intake duct and fuel and air mixture is drawn into the cylinder. When the piston starts moving upward, both the intake valve and the exhaust valve are closed forming an enclosure. The rising piston compresses the air-fuel mixture. When the piston

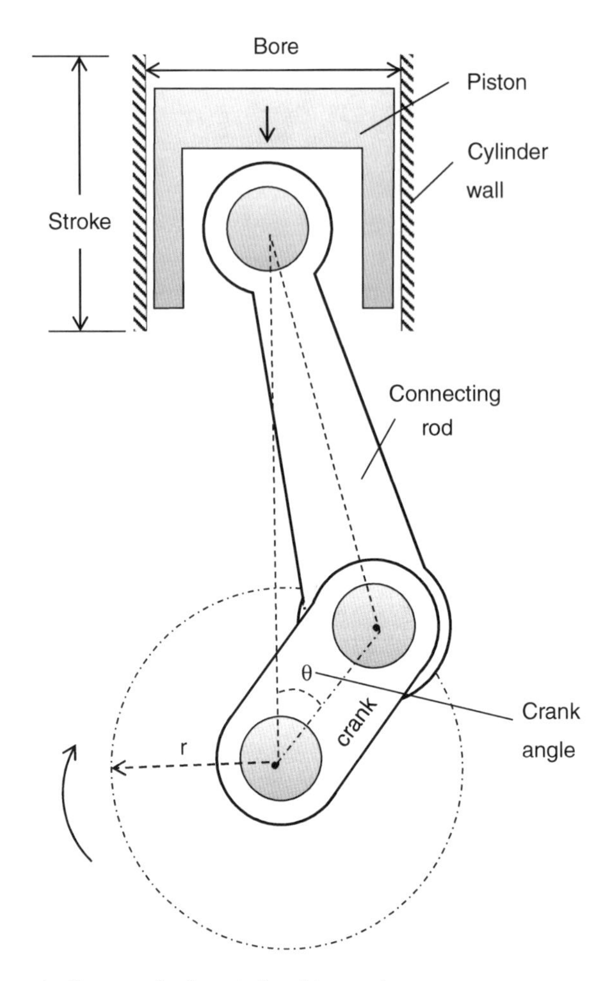

Fig. 9.2 Schematic diagram of a four-stroke piston engine

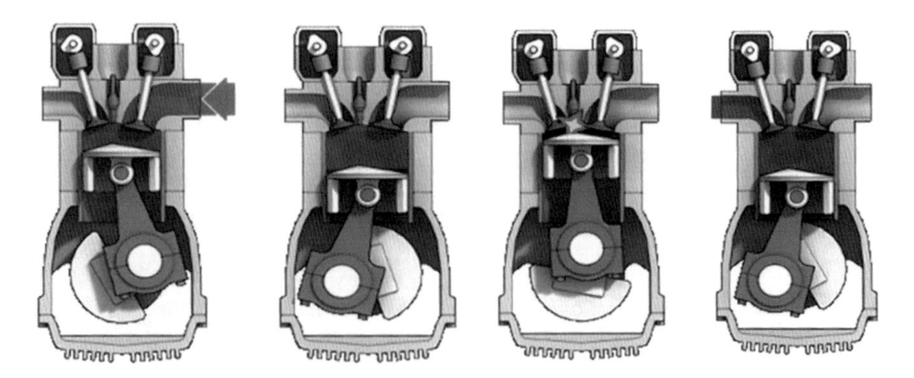

Fig. 9.3 Four-stroke piston engine cycle (Wihipedia.org)

reaches the top of the stroke, the spark plug is fired to ignite the air-fuel mixture and starts the combustion at the right instant. As the combustion proceeds, the burning mixture expands at increasing temperature and drives the piton downward to produce mechanical energy. Combustion is expected to complete when the piton reaches BDC, when the exhaust valves opens. The exhaust is pushed out of the cylinder. A new cycle starts when the rising piston reaches top of the stroke.

9.2.1.1 Carburetor

The air and fuel are premixed in the carburetor, a device that controls the flow of air and fuel. The mixture is usually at an equivalence ratio of $0.7 < \Phi < 1.4$ to match the residence time for air emission control. A cold engine is started with a fuel rich mixture to secure successful ignition. This results in highly incomplete combustion and high CO and HC emissions. Evaporation of the liquid fuel results in evaporative air emissions from the carburetor after the engine is turned off. Vapor recovery system and the adsorption-desorption system can be employed to reduce the evaporative air emissions. Activated carbon is a common adsorbent for the vapor recovery. Recycled gasoline is purged by air and the mixture is delivered to the engine for combustion without reentering the tank.

9.2.1.2 Flame in Engine

The local equivalence ratio in flame depends on the spot in the cylinder. After the ignition of the air-fuel mixture, the gas mixture is not burned instantaneously. The turbulent flame moves downward as the cylinder volume expands. The speed of the flame depends on the engine design and the operating conditions such as the equivalence ratio and the speed of the piston. As combustion continues in the cylinder, the temperature and pressure of the gas rise. This is different from the combustion in stationary systems, where pressure seems to be stable.

9.2.2 *Diesel Engines*

Diesel engines do not have carburetors for air-fuel premixing. Instead, air is drawn into the cylinder through the intake valve and diesel is injected into the cylinder of the engine. Diesel injection starts when it approaches the end of the compression stroke. At the moment, the high-temperature high-pressure compressed air ignites the vapor of the diesel droplets. Combustion in a diesel engine is unsteady and it varies with the fuel injection mode and mixing with the air.

In a direct injection (DI) engine, both air and fuel are turbulent and they are not homogeneously mixed. A prechamber can be used to improve the mixing effect and as a result, this type of engine is called prechamber diesel engine, or indirect injection engine (IDI). Combustion takes place in the prechamber and the burning gas enters the diesel engine cylinder through a passageway. Air emissions, especially fine particulate (soot particles) are reduced from an IDI diesel engine at the cost of lower engine efficiency.

Diesel is injected into the cylinder by multiple small nozzles. The droplets move at great initial Reynolds number because of the great relative velocity between droplets and the surrounding gas. As a result, the droplets may be further broken down to smaller ones.

Diesel engines are usually run at fuel-lean condition and the corresponding gaseous air pollutants such as CO and HC are reduced. However, the particulate matter (soot) emission is much higher than the gasoline engine because of the slower air-fuel mixing.

9.3 SO_2 Capture by Furnace Sorbent Injection

SO_2 can be captured by injection of proper sorbent such as limestone or lime into the furnace of a stationary system. The sorbent can be injected into the furnace or the hot part of the flue gas channel. This works well for older boilers with a relatively short remaining lifetime. As solid sorbent is injected into the furnace, more particles will join the fly ash and increase the load of downstream particulate control devices. Additional soot-blowing device is needed to remove solids accumulated on the inner surfaces of the furnaces, which is not a big technical problem.

9.3.1 SO_2 Capture by FSI in Pulverized Coal Combustion

The principle of furnace sorbent injection (FSI) is shown in Fig. 9.4. The concentration of SO_2 in typical flue gases from coal firing can be up to 5,000 ppmv. The efficiency of SO_2 removal by sorbent injection depends on the temperature where the sorbent is injected. When the sorbent is injected at a low-temperature area, the SO_2 removal efficiencies can be in the range of 60–75 % with Ca and S molar ratios from 2 to 4. This efficiency can be further increased at a relatively higher cost, by spraying water downstream into the flue gas duct before the particulate control devices, to reactivate the spent sorbent to capture more SO_2.

The relation between temperature and the SO_2 removal is shown in Fig. 9.5 for a sorbent such as limestone ($CaCO_3$) or hydrated lime ($Ca(OH)_2$). With other conditions the same, $Ca(OH)_2$ gives higher efficiency than $CaCO_3$.

The main solid product of the desulfurization reaction is $CaSO_4$ when temperature is below 1,200 °C. The corresponding mechanism is illustrated in Fig. 9.6. When exposed to high-temperature gases, the sorbent first decomposes to CaO, which then reacts with SO_2 and O_2 to form $CaSO_4$. A white shell of $CaSO_4$ formed by desulfurization reactions surrounds the inner unreacted CaO, which slows down

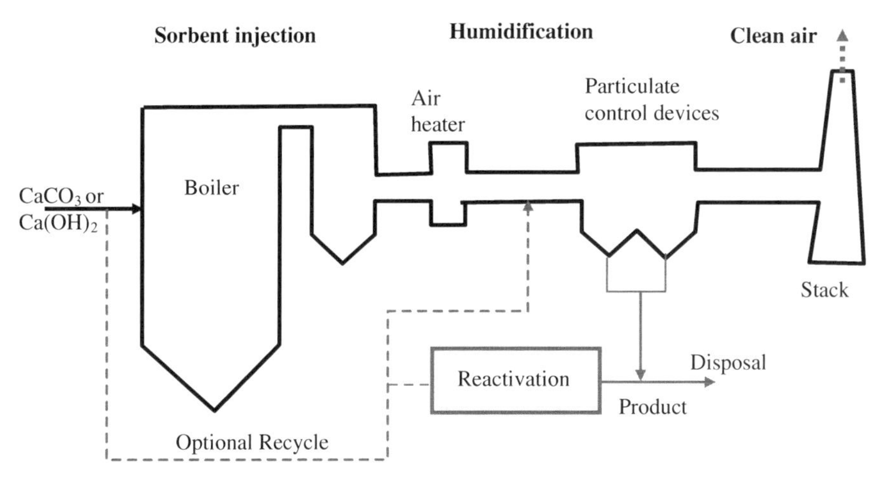

Fig. 9.4 A typical furnace sorbent injection process

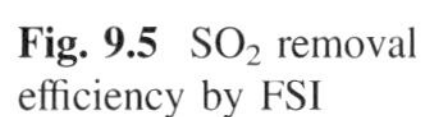
Fig. 9.5 SO_2 removal efficiency by FSI

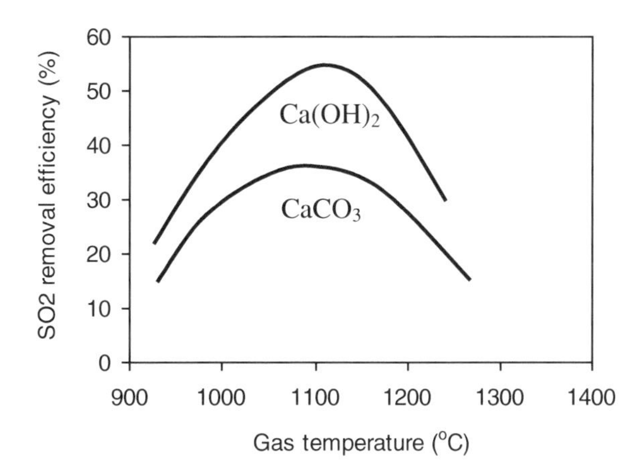

the further reaction by plugging and blocking as it is difficult for the gases to penetrate through this shell. Molar volumes of $CaCO_3$, CaO, and $CaSO_4$ are 36.9, 16.9, and 46.0 cm^3/mole, respectively; consequently, a large portion of CaO is trapped and not consumed. Trapped CaO can be released for further reaction by reactivation process.

One classic example is the Limestone Injection into the Furnace and Activation of unreacted Calcium (LIFAC) process developed by Kvaerner Pulping Power Division in Finland in the 1970s and 1980s. The desulfurization efficiencies were reported to be 65–85 % at Ca/S ratios of 2–2.5. In the LIFAC process, limestone is injected into the upper part of furnace near the superheater where it calcines into CaO:

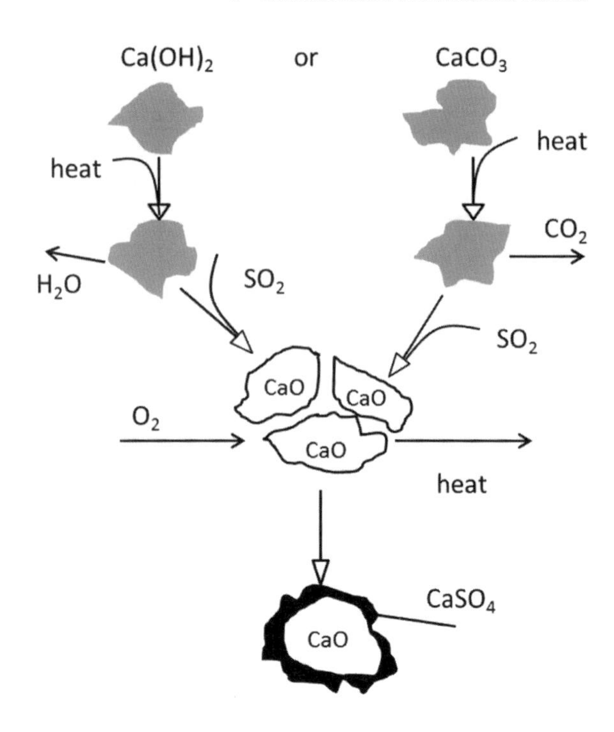

Fig. 9.6 Chemical processes during in-duct sorbent injection for SO_2 removal

$$CaCO_3 \rightarrow CaO + CO_2 \tag{9.6}$$

which then react with SO_2 and O_2 to form $CaSO_4$

$$CaO + {}^1\!/_2 O_2 + SO_2 \rightarrow CaSO_4 \tag{9.7}$$

The spent sorbent is reactivated using water spray downstream in an activation reactor, where the temperature is about 400 °C.

$$CaO + H_2O \rightarrow Ca(OH)_2 \tag{9.8}$$

The newly formed calcium hydroxide ($Ca(OH)_2$), which is more reactive than CaO, reacts with more SO_2 from the flue gas stream by

$$Ca(OH)_2 + SO_2 \rightarrow CaSO_3 + H_2O \tag{9.9}$$

At low temperature the solid product is $CaSO_3$ in the flue gas duct, which is collected together with fly ash in the downstream particulate control device, an electrostatic precipitator or a filter baghouse.

9.3.2 SO_2 Capture in Fluidized Bed Combustion

As mentioned above in FBC SO_2 sorbent can be mixed with other bed materials for desulfurization, since the temperature in chamber is low, $CaSO_4$ formed remains a stable compound. In addition to relatively low-NO_x emissions due to low combustion temperature, FBC can also capture H_2S, if any, forming CaS, which can be oxidized to $CaSO_4$, before it is oxidized into SO_2.

In addition to Reactions (9.6) and (9.7), direct sulfation may also take place. Similar to fixed bed combustion, the formation of $CaSO_4$ in FBC also results in the partially used sorbent, and the reactivation of spent sorbent will give higher efficiency at lower cost of sorbent.

$$CaCO_3 + {}^1/_2O_2 + SO_2 \rightarrow CaSO_4 + CO_2 \quad (9.10)$$

This reaction is relatively slow because calcined limestone, CaO, is more reactive than uncalcined limestone, $CaCO_3$. In reality, the calcareous materials mined have different purity and chemical compositions, leading to large differences in SO_2 removal efficiencies, even though they are tested at identical conditions.

Desulfurization by injection of sorbent into an atmospheric FBC performs the best at 800–850 °C, which is more sensitive for BFBC than CFBC. This maximum temperature was mainly related to the stability of the $CaSO_4$. It becomes unstable at a temperature above 850 °C, when part of the desulfurization product, $CaSO_4$, react with CO and/or H_2 to produce CaS, CaO or $CaCO_3$, depending on temperature, and partial pressures of the reactants, via a complex chemistry that involves complicated step reactions [18, 21], and they are not listed herein.

The comparison among the desulfurization efficiencies in CFBC, BFBC, and pulverized coal combustion firing with FSI is shown in Fig. 9.7. It can be seen that

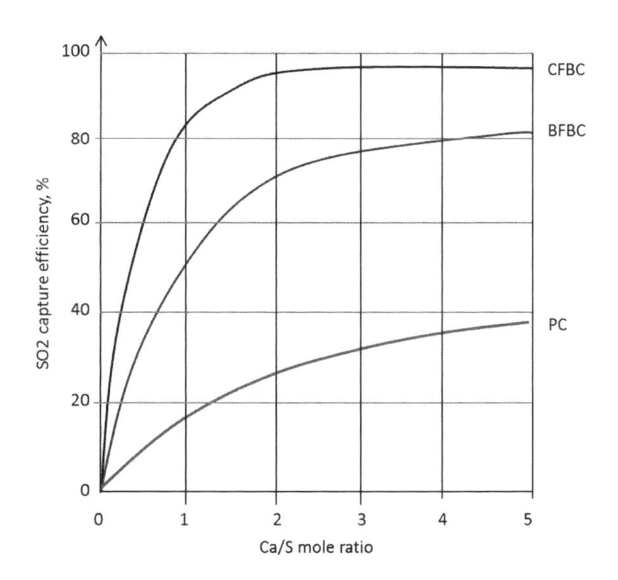

Fig. 9.7 SO_2 capture in CFBC, BFBC and PC furnaces

among the three, CFBC performs the best in terms of desulfurization efficiency while pulverized coal combustion is the worst. On the other hand, the solid residues from BFBC find limited use due to the high lime content whereas PFBC residues have better properties for concrete or cement, which is the main use of desulfurization solid residues.

9.4 In-combustion NO_x Control

Removal of nitrogen oxides during combustion process is much more favorable than post-combustion by flue gas cleaning. With the recent advances in NO_x control during the combustion stage, conditions in the boiler furnaces are no longer the same as those in the 1980s. The NO_x emissions are dramatically reduced with the development of advanced combustion techniques, such as FBC, where the combustion temperatures are low.

As introduced in combustion basics, the conversion of nitrogen in the combustion air to nitric oxide is temperature sensitive. As such, the formation of thermal NO may be reduced by lowering the combustion temperature and by minimizing the flue gas residence time. The formation rate of thermal NO appears to be practically low if the combustion temperature is below 1,400 °C, where at the temperatures above 1,600 °C, the formation of NO is strongly accelerated [22].

A variety of technologies have been developed to lower the combustion temperature. A few examples include, air staging, fuel staging, exhaust gas recirculation (EGR), reducing temperature of preheated combustion air, reducing the flame temperature by a long flame, and reducing the excess air. However, the efficiencies of the methods are case-specific, as NO_x is not the only concern. NO_x reduction is often a matter of optimization against the falling overall thermal efficiency due to the lower flame temperature and the increase of combustibles in ash and flue gas. In general, the NO reduction efficiencies of above-mentioned methods remain lower than 70 %.

9.4.1 Air Staging

NO_x formation may be substantially reduced by rearranging the combustion air supply, which is referred to as air staging [2, 10, 16]. As illustrated in Fig. 9.8, part of the air (primary air) is supplied as oxidizer at the root of the flame, where fuel rich combustion takes place and most of the HCN and NH_3 are oxidized to molecular nitrogen. The remaining air needed for combustion is supplied to the flame from the flame periphery, where little HCN or NH_3 is left to produce nitric oxide. As a result of this air staging, the peak flame temperatures in burners remain lower than the conventional burners, and the formation of thermal NO is reduced too.

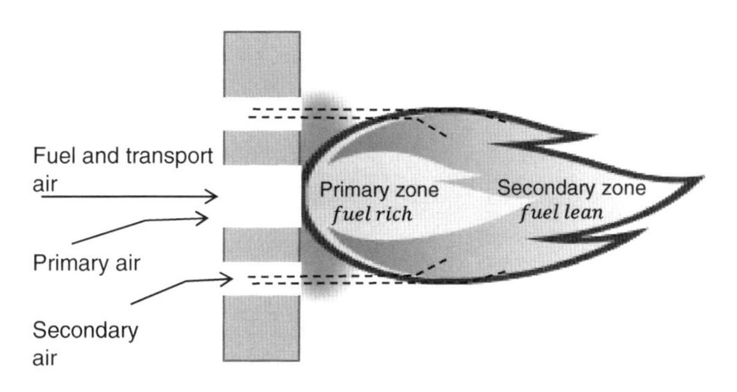

Fig. 9.8 Air staging in a burner

Air staging may also be applied to the entire furnace as shown in Fig. 9.9. The primary air is supplied at lower rows of burners, where combustion is under fuel rich condition. The secondary air or the remaining air is supplied at the upper level of the furnace, in the middle or top of the furnace. Similar principle may also be applied to grate furnaces and fluidized combustors.

While successful air staging reduces the NO_x emission, one has to carefully monitor the possible unburned air emission compounds, for example, CO, C_xH_y, and unburned carbon in ash. They are resulted from the fuel rich combustion in the primary zone.

The NO reduction by air staging varies from 10 to 50 % [22], depending on the relative quantity of volatile compounds in the fuels. For fuels with low volatiles, the fixed chars retain a considerable part of the fuel nitrogen. The nitrogen in char can form NO that cannot be effectively controlled by air staging. The share of char nitrogen in the NO emission of air-staged combustion is actually greater than that from conventional pulverized coal combustion without air staging. For pulverized coal combustion, the conversion rate of char nitrogen to NO varied from 20 to 80 %, depending primarily on the property of the coal.

Fig. 9.9 Air staging in a furnace

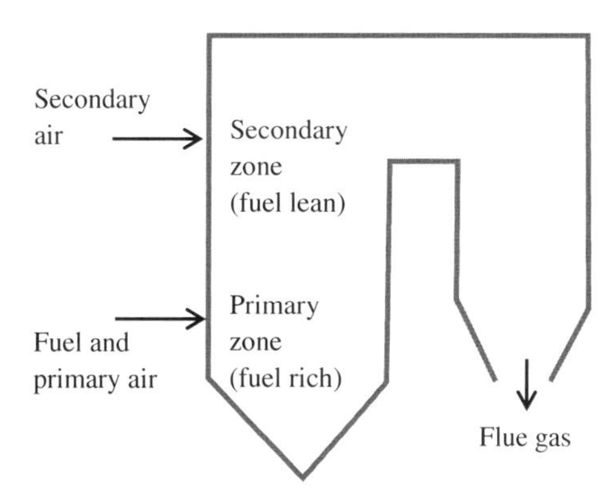

9.4.2 Fuel Staging

Fuel staging is also known as reburning or three-stage combustion; it utilizes fuel to reduce the formation of nitric oxide. Like the air staging technology, fuel staging may also be applied either to a low-NO_x burner or to the entire furnace [11–13]. The principles are illustrated in Fig. 9.10. In the low-NO_x burner, the same fuel is used as primary and secondary fuel. The method includes three stages, being primary, secondary stage, and final combustion.

In the primary combustion zone, the primary fuel such as coal or oil is oxidized by excess air. Both fuel NO and thermal NO are formed in this zone. Secondary fuel is usually natural gas and it is injected to the furnace at the secondary stage. It is typically corresponding to 10–20 % of the energy content of the primary fuel. At this stage, the NO formed in the primary stage is reduced to molecular nitrogen by complicated chain reactions initiated by hydrocarbon radicals (CH_i), which originate from the secondary fuel. High temperature (>1,000 °C) is favorable for these reactions, which may be written as follows.

$$NO + CH_i \rightarrow HCN + O + OH \rightarrow H_iNCO + H \rightarrow NH_i + NO \rightarrow N_2 \quad (9.11)$$

Air is added to the final combustion stage for the completion of fuel combustion. The nitrogen compounds (e.g., NO, HCN and NH_3) are oxidized back to NO and/or N_2. Different from the secondary stage, a low temperature (<1,000 °C) is preferred during the final combustion stage for the formation of molecular nitrogen. However, the temperature cannot be too low in order to minimize the formation of laughing gas (N_2O) and to maximize the oxidation of carbon monoxide. In demonstration plants the NO reduction by fuel staging could reach 30–70 % [22], with some practical problems such as increased unburned components in flue gas and increased corrosion of the furnace.

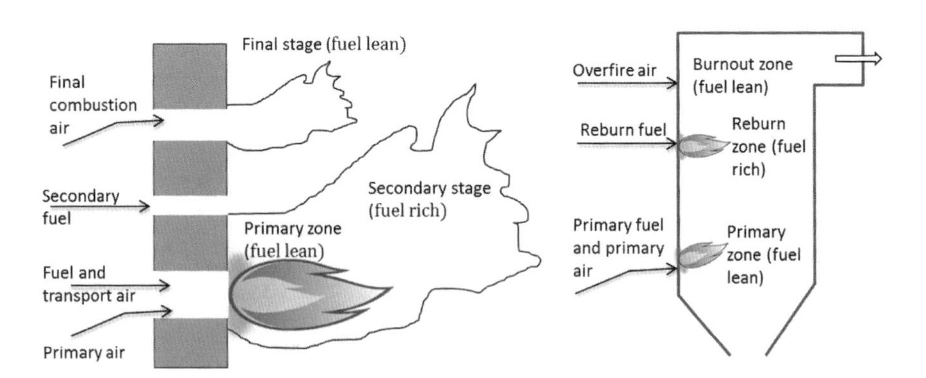

Fig. 9.10 Principle of fuel staging (*Left* low-NO_x burner, *Right* furnace)

The reduction of NO_x formation also depends on the unit size and the integration of the burners with the furnace. There are three important types of low-NO_x burners that are integrated in pulverized coal combustion processes. They are,

- wall firing,
- tangentially firing
- opposed wall firing

Each of them may be dry bottom firing and wet bottom firing, where the bottom ash is taken out in solid or liquid form, respectively. The former is preferred for high ash-content fuels. The well-known cyclone furnace is a special case of wet bottom firing furnaces [15].

9.4.3 Flue Gas Recirculation

Flue gas recirculation (FGR) is used primarily for combustion of oil or gas fuels [9]. As illustrated in Fig. 9.11, part of the flue gas is redirected to the primary combustion zone. Since the specific heat of the flue gas components, especially, H_2O, is higher than their counterparts in the furnace, FGR results in a lower temperature in the primary combustion zone and consequent low-NO_x formation.

Similar technology is being widely used for engine NO_x emission control and it is called EGR. An EGR system is effective in reduction of NO_x formation by lower combustion temperature. The main challenges are effective control system and other air pollutants due to lower temperature, such as HC and CO. Details are introduced in Sect. 9.6.

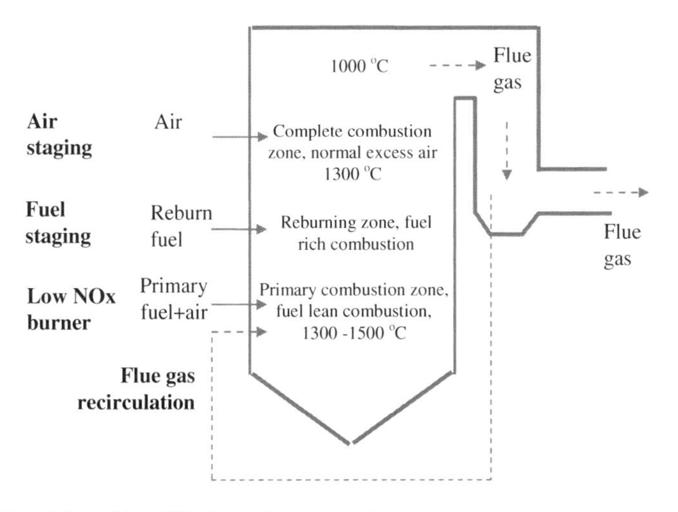

Fig. 9.11 Principles of low-NO_x formation approaches

Table 9.1 Comparison of in-furnace Low-NO_x technologies

Technology	Advantage	Disadvantage
Low excess air		Fuel burnout decreases
Low-NO_x burner (i.e., fuel staging in flame)	Always for NO_x reduction	Minor decrease in fuel burn-out
Air staging	Always for NO_x reduction	Increased risk for corrosion, slagging
Fuel staging in furnace	Always, especially when the reburn fuel is the same as the fuel in first stage	High capital cost
Flue gas recirculation	Effective for high-temperature oil or gas fired furnaces	Low efficiency

9.4.4 Combined Low-NO_x Technologies

Table 9.1 summarizes the comparison of the in-furnace low-NO_x technologies. Each of the methods has its pros and cons. In order to achieve the best NO_x reduction performance, a combination of the low-NO_x methods described above can be employed. A typical combination is illustrated in Fig. 9.11 above.

The actual selection and combination of low-NO_x combustion technologies depends on the fuel and the boiler itself. When combining several in-furnace measures for coal fired boilers, only 70 % or less of the total NO_x reduction can be achieved. The major reason for the limited NO_x reduction is that these in-combustion technologies mainly control the NO_x related to the nitrogen compounds released from the fuel at the early stage of the combustion, or in short, volatile-N.

Not all the fuel nitrogen is released during this stage, and 20–80 % may remain in the fuel forming char-*N*, which is converted to NO_x in the flue gas. The char-N content is largely unknown and varies from fuel to fuel and from furnace to furnace. Current low-NO_x methods are capable of controlling almost 100 % of the NO_x from volatile-N. Therefore, additional post-combustion NO_x control is necessary to further remove the NO_x from the flue gas before being discharged into the atmosphere.

9.5 In-combustion Soot Control

Soot formation can be reduced by controlled turbulent mixing, which promotes more complete combustion [3, 7, 8]. Additives to the fuels such as iron, nickel, manganese, or cobalt act as catalyst, can further improve the oxidation of soot, however, sometime these additives may also increase the production of soot. Feitelberg et al. [6] reported that additives of iron and manganese increased the production of soot up to a factor of 3 under all combustion conditions. They found that these additives did not affect soot particle collision rate or inception.

9.6 Engine Exhaust Gas Recirculation

The major air emissions from an engine are CO, total hydrogen carbon (HC), NO_x, and soot, which take up only 1 % of the exhaust gas. The rest are CO_2, H_2O, N_2, and O_2. The motivation of engine EGR is to further reduce these trace level emissions with minimal compensation in power and thermal efficiency.

Similar to FGR approach for stationary combustion processes, engine exhaust can also be recirculated for the reduction of NO_x emission from engines [1, 23]. With more and more stringent limits for NO_x emissions, further reductions in NO_x emissions from mobile sources without notable sacrifice to engine power have become more and more challenging. These challenges have led to recent growing R&D projects in engine EGR systems. EGR has become one of the essential techniques for engine emission reduction.

When EGR system is installed the engine intake consists of fresh air and recirculated exhaust. The percentage EGR can be defined as

$$r_{\mathrm{EGR}} = \frac{\mathrm{EGR}}{\mathrm{Air} + \mathrm{Fuel} + \mathrm{EGR}} \tag{9.12}$$

where EGR, Air and Fuel are the amount, mostly in terms of mass, of EGR, air and fuel, respectively. They can be measured in mass or volume. By mass up to 30 % of the engine exhaust can be recirculated whereas the ratio can reach 50 % by volume.

A more practical EGR ratio is based on the CO_2 concentration.

$$r_{\mathrm{EGR}} = \frac{c_{\mathrm{CO_2,in}}}{c_{\mathrm{CO_2,ex}}} \tag{9.13}$$

where $c_{\mathrm{CO_2,in}}$ is the intake CO_2 concentration and $c_{\mathrm{CO_2,ex}}$ is the exhaust CO_2 concentration.

Fresh air contains negligible amount of CO_2 compared to that in the engine exhaust. Since CO_2 is primarily a combustion product. The recycled engine exhaust carries a substantial amount of CO_2 that increases with EGR flow rate and engine loads.

CO, HC, NO_x and soot account for less than 1 % of the engine exhaust and their relative abundance depends on the engine load. The corresponding oxygen level varies from 5 % at full load to 20 % during idling. Therefore, the effectiveness of NO_x reduction by EGR also depends on engine load.

Diesel engines can operate with a high EGR ratio because the exhaust contains a high concentration of O_2 and low concentrations of CO_2 and H_2O. In addition, test results showed that cooled EGR reduces NO_x more effectively than hot EGR [20].

Unlike the exhaust from other engines with premixture aiming at certain air-fuel ratios, the relative amount of gases in a diesel exhaust depends very much on the diesel engine load. A diesel engine adjusts its fuel injection rate according to the engine load. As a result, the oxygen concentration in the diesel engine exhaust varies with the engine load. The oxygen concentration may vary from 5 to 20 %

when a diesel engine load changes from full load to idling state. And, the CO_2 concentration changes with an opposite direction. As a result, the effect in NO_x reduction by EGR depends on the diesel engine load.

9.7 Practice Problems

1. A coal having 4 % of ash content and 1.0 % of sulfur content by weight is burned by power plant at a rate of 5,000 tons/day. Calculate the amount of calcium carbonate ($CaCO_3$) needed to capture 90 % of the sulfur dioxide. Assume molar ratio Ca:S = 2.
2. A power plant burns coal at a rate of 6,000 tons/ day. The coal has 5 % of ash content and 2 % of sulfur content by weight. Calculate the amount of calcium oxide (CaO) needed to capture 100 % of the sulphur dioxide. Assume molar ratio Ca:S = 3.
3. The emission rate of NO_x from a truck is 2.5 g/km. Determine the conversion efficiency of a catalytic converter in order to meet an emission standard of 0.25 g/km.
4. In-combustion NO_x control can be achieved by air staging burner, where combustion is fuel rich in

   a. the primary zone
   b. the secondary zone
   c. both primary and secondary zones
   d. neither primary or secondary zone.

5. There are three steps in fuel staging combustion, and the combustion follows the order of

   a. fuel rich, fuel lean, fuel rich
   b. fuel lean, fuel rich, fuel lean
   c. fuel rich, fuel rich, fuel lean
   d. fuel rich, stoichiometric, fuel lean.

6. In an in-combustion SO_2 capture by lime stone injected into the furnace, the Ca: S molar ratio is 2, because

   a. 2 mol of Ca is needed in order to react with 1 mol of SO_2
   b. 2 mol of Ca is needed in order to react with 1 mol of SO_3 and 1 mol SO_2
   c. the sorbent cannot be used effectively
   d. more sorbent is needed to continue the reaction along the flue when the gas exist the furnace.

7. A practical gasoline engine emission control approach is
   a. spray finer gasoline droplets into the cylinder to reduce complete combustion products
   b. engine exhaust recirculation into air intake to reduce HC emissions
   c. optimized Stroke-Bore ratio to reduce particulate emissions
   d. fuel lean operation when the engine starts.

References and Further Readings

1. Abd-Alla GH (2002) Using exhaust gas recirculation in internal combustion engines: a review. Energy Convers Manag 43:1027–1042
2. Beer JM (2000) Combustion technology developments in power generation in response to environmental challenges. Prog Energy Combust Sci 26:301–327
3. Bockhorn H (1994) Soot formation in combustion. Springer, Berlin
4. Bockhorn H, Schafer T (1994) Growth of soot particles in premixed flames by surface reactions. In: Bockhorn H (ed) Soot formation in combustion. Springer, Heidelberg
5. Cooper CD, Alley FC (2002) Air pollution control: a design approach, 3rd edn. Waveland Press, Long Grove
6. Feitelberg AS, Longwell JP, Sarofim AF (1993) Metal enhanced soot and PAH formation. Combust Flame 92(3):241–253
7. Gelencser A (2004) Carbonaceous aerosol. Springer, Berlin, pp 350
8. Haynes BS, HGg Wagner (1981) Soot formation. Prog Energy Combust Sci 7(4):229–273
9. Kim HK, Kim Y, Lee SM, Ahn KY (2007) NO reduction in 0.03–0.2 MW oxy-fuel combustor using flue gas recirculation technology. Proc Combust Inst 31:3377–3384
10. Lyngfelt A, Leckner B (1993) SO_2 capture and N_2O reduction in a circulating fluidized-bed boiler: influence of temperature and air staging. Fuel 72:1553–1561
11. Maly PM, Zamansky VM, Ho L, Payne R (1999) Alternative fuel reburning. Fuel 78:327–334
12. Pratapas J, Bluestein J (1994) Natural gas reburn: cost effective NO_x control. Power Eng 98:47–50
13. Smoot LD, Hill SC, Xu H (1998) NO_x control through reburning. Prog Energy Combust Sci 24:385–408
14. Smoot LD, Pratt DT (1979) Pulverized-coal combustion and gasification. Plenum Press, New York
15. Soud H, Fukasawa K (1996) Developments in NO_x abatement sand control. Report IEACR/ 89, IEA coal research, London, UK
16. Staiger B, Unterberger S, Berger R, Hein KRG (2005) Development of an air staging technology to reduce NO_x emissions in grate fired boilers. Energy 30:1429–1438
17. US DOE (1999) Technologies for the combined control of sulphur dioxide and nitrogen oxides emissions from coal-fired boilers. Clean coal technology, topical report 13, May 1999
18. Yrjas P, Hupa M (1997) Influence of periodically changing oxidising and reducing environment on sulfur capture under PFBC conditions. In: Proceedings of the 14th international conference on fluidised bed combustion, ASME, May 1997, Vancouver, pp 229–236
19. Yrjas KP, Iisa K, Hupa M (1993) Sulphur absorption capacity of different limestones and dolomited under pressurised fluidised bed combustion conditions. In: Proceedings of the 12th international conference on fluidised bed combustion, May 1993, San Diego. ASME, New York, pp 265–271
20. Zelenka P, Aufinger H, Reczek W, Catellieri W (1998) Cooled EGR: a key technology for future efficient HD Diesels. SAE paper 980190

21. Zevenhoven R, Yrjas P, Hupa M (1999) Sulphur capture under periodically changing oxidising and reducing conditions in PFBC. In: Savannah GA (ed) Proceedings of the 15th international conference on fluidised bed combustion, ASME, May 1999, pp 102
22. Zevenhoven R, Kilpinen P (2002) Flue gas and fuel gas, 2nd edn. Report TKK – ENY-4, The nordic energy research programme, soild fuel committees, Norway, pp 3–4
23. Zheng M, Reader GT, Hawley JG (2004) Diesel engine exhaust gas recirculation: a review on advanced and novel concepts. Energy Convers Manag 45:883–900

Chapter 10
Post-combustion Air Emission Control

10.1 Introduction

While some of the air emissions are greatly reduced by pre- and in- combustion technologies, there are always various pollutants remain in the flue gas. Their concentrations have to be further reduced to certain levels to meet local air emission standards. Air emission control at this stage is referred to as post combustion control in this book.

Post-combustion technologies include separation from the gas stream based on the principles introduced in Part I. Sometimes it also involves phase change by condensation of vapor to liquid followed by liquid-gas separation. Air emissions in the flue gas or exhaust gas can also be transported from gas phase to liquid or solid phase by sorption, or converted to less hazardous or benign species by incineration, or catalytic conversion. Dilution by atmospheric air, which is also called air dispersion, is the last step, through which the air pollutants enter the atmosphere. Air dispersion will be introduced shortly in the coming chapter.

10.2 Control of Particulate Matter Emissions

Cyclone, electrostatic precipitation, and filtration technologies are widely used for post-combustion particulate matter emission control. Cyclone is primarily employed as a precleaner or a fuel recycling device in a fluidized bed combustion system. Electrostatic precipitation and filtration are two most widely commercialized technologies for post-combustion PM emission control. Sometimes, water spray tower can be used for sticky particulate removal too. Their engineering designs are introduced as follows.

Z. Tan, *Air Pollution and Greenhouse Gases*, Green Energy and Technology,
DOI 10.1007/978-981-287-212-8_10

Typical ESP has a cut size of 0.5 μm in diameter and airflow velocity ranges 1–1.5 m/s [22]. There has been a vast market of ESPs; however, fabric filters work best for certain particle due to its independence of particle resistivity. Also the dust cakes on the filters remove more pollutant gases than ESPs. ESPs are cheaper than filter and wet scrubber due to low operational cost, but much more expensive than cyclones. ESPs are not applicable for gaseous pollutants and require well-trained specialized personnel to operate due to the high voltage (tens of kilovolts) required for corona discharging.

10.2.1 Electrostatic precipitator Designs

Electrostatic precipitators (ESPs) are widely used for particulate emission control from combustion of fossil fuels, mainly coal. As introduced in Part I, the basic principle of ESPs involves the following three steps in particle separation:

- Charging of particles
- Particle movement relative to the gas flow
- Particle deposition on a collection surface.

In an engineering practice, we shall also consider the removal of the deposited particles, or dust cake, from the collection surface for continuous operation. Dust cake on the surface of the collection plates can be removed by mechanical vibration or a knock off hammer. This usually results in re-entrainment of particles.

Alternatively wet removal by water spray can be considered. Wet ESPs have very high efficiencies because they are operated with a stream of water that continuously removes the dust from the collector surfaces as slurry to reduce the particle re-entrainment. Many of them are found near Japanese cities where the particulate emission standard is the most stringent (10 mg/m^3 or lower at STP). Their footprints are also smaller than regular ESPs for higher gas velocities and the absence of the rapping devices that are required to remove the particles from cold-side ESP collector surfaces. Disadvantages of Wet ESPs include reduced gas temperature, corrosion concerns due to high dust and high sulfur flue gases, and secondary water pollution to be treated.

ESPs can be installed at cold side or hot side of the flue gas. Most ESPs are "cold-side" ones located between the air pre-heater and the flue gas desulfurization (FGD) system, if any, where the temperature is about 120–200 °C. Alternatively, "hot-side" ESPs are installed at upstream of the air pre-heater, where flue gas temperature is between 300 and 450 °C. In this area, the particle resistivity is less sensitive to the composition of the flue gas. The problem of hot-side ESPs is the significant heat losses. In addition, they are more sensitive to temperature changes when the furnace or boiler is at partial load.

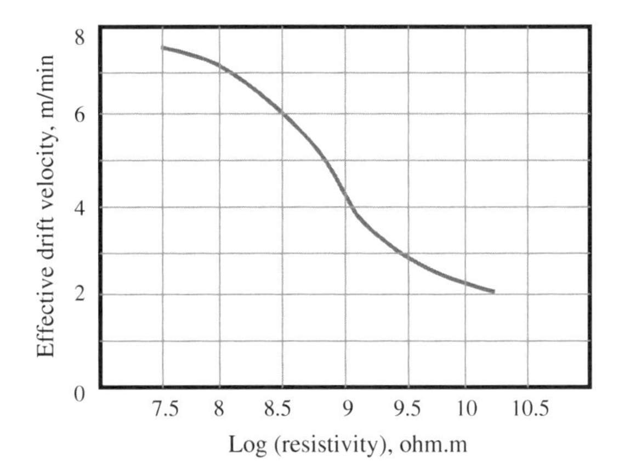

Fig. 10.1 Typical effect of resistivity of fly ash particles on the electric drift velocity

When a selective catalytic reduction (SCR, to be introduced shortly) unit is employed for NO_x control at 350–400 °C, it is better to put the ESP upstream of the SCR, because after the ESPs, fly ash particles are removed and it can improve SCR catalyst lifetime, and thereby reduce SCR operation and maintenance problems.

Besides the size of particles, other properties of the particles also affect the particle charging behavior and the consequent performance of ESPs. The actual process is further complicated by interactions between the particle and the air. The resistivity of the particles is the main factor that affects the performance of the ESPs [25]. Figure 10.1 shows the relationship between the resistivity and the precipitation speed of the particles. For the same device, the greater particle precipitation speed, the greater particle removal efficiency. Overall, Fig. 10.1 shows that the particle removal efficiency drops with the increase of particle resistivity [9].

For the effective separation of particles using an ESP, the resistivity of the particles is preferably in the range 10^7–10^{12} Ωm. Particles with too high resistivity cannot be charged effectively. In addition, a spark from the collection plate to the discharge electrode may take place due to high intensity of the electric field building up in the collected dust cake. On the other hand, particles with very low resistivity may lose their charges rapidly to water in the gas or to other particles. More importantly, these particles may be easily re-entrained from the collection electrodes or the collected particle layer because they may rapidly lose their charges or switch polarity of the charges.

The resistivity of particles depends on

- temperature
- moisture content
- chemical composition.

The resistivity of the particles first increases with the temperature and then drops. Depending on the chemical composition of the fly ash particles considered, the maximum resistivity is sustained between 140 and 170 °C.

Particle resistivity is also affected by the moisture content in the gas, especially at temperatures below 200 °C. The moisture content may lower the particle resistivity by several orders of magnitude. This can be explained that water molecules are very active in removing electric charge from the particles in addition to the formation of sulfuric acid in the gas.

Sulfur content in the coal also affects the resistivity of the coal-based fly ash. For low sulfur coal, the resistivity of the corresponding fly ash particles is likely to be too high (>10^{12} ohm-m). As a result lower fuel sulfur is likely lead to charging and more serious spark problems. Therefore, switching from high sulfur coal to low sulfur coal could result in unexpected lose in ESP performance. For high sulfur coal, part of the fuel sulfur is oxidized to SO_2 and then to SO_3. With the presence of water vapor in the flue gas, sulfuric acid is produced and water condensates on the surface of the fly ash particles. This reduces the resistivity and increases the coagulation of the fly ash particles. Typically, switching from a 1 %-wt sulfur coal to a 0.6 %-wt sulfur coal will require a 20 % larger ESP collection area in order to compromise the loss of performance [9].

In addition to sulfur content, many other chemical compounds in the coal used as fuel also affect the resistivity of the fly ash particles. Iron (Fe_2O_3), sodium (Na_2O), and water can decrease the resistivity and hence are favorable to ESPs, whereas calcium (CaO), magnesium (MgO), silicon (SiO_2) and aluminum (Al_2O_3) increase the resistivity of the particles, thus worsening the performance of ESPs.

Industrial field evaluations have shown that industrial ESPs are effectively for only large particles, and they have limited capability in capturing fine particles. It is mainly because of the low precipitation speed of these small particles. These fine particles have to be captured by filtration in order to meet more and more stringent fine particulate matter emission standards.

10.2.2 Filtration System Designs

10.2.2.1 Filter Media

In the engineering designs of industrial air filtration systems, one first needs to consider the materials of filter media according to the gas properties. Typical filter media include, but are not limited to, bag filters made of fabric fiber materials, textile, plastics, and ceramics. Rigid barrier filters are made of metal or sintered ceramic, powder or fibers; Granular filters based on layers of granular solids are widely employed in liquid-solid separation for water treatment [28], and they can also be used for air purification.

The selection of the media is determined by many factors such as the operation temperature, property of the particles, and availability of the media. Fabric filters are widely used for environment where temperatures are relatively low. Cotton may be used for the temperatures below 80 °C while Teflon and glass fiber work for up to 260 °C. For applications up to 450 °C stainless steel can be used under unfriendly corrosive environment. At higher temperatures, ceramic materials are the best choice.

10.2.2.2 Bag-house Filters

Typical filtration systems offer collection efficiencies greater than 99 % over a large size range. The advantage of the filtration over ESP is its independence on the electric resistivity of the particles. This characteristic makes filter very competitive for particles with high-resistivity. A disadvantage of a filter compared with an ESP is the larger pressure drop and the allowable gas speed. Typically the face speed through a filter is in the range of 0.5–5 cm/s. Otherwise, filter flooding will occur and result in low filtration efficiency.

In order to handle a high airflow rate while maintaining the low face speed, a high filter surface area is necessary. In industry, hundreds of cylindrical or tubular filter bags of fabric materials are confined in a "bag-house" to create a high surface area to allow certain amount of air flow to pass through the filters at an acceptable low face speed.

Figure 10.2 shows two typical operation modes of bag-house filters. For an inside-out filtration system, the gas passes through the filters from the inside and exit from the outside of the filter bags. This "blows up" the bag filters to their maximum volume and produces the dust cake on the inside of the bags. Outside–in

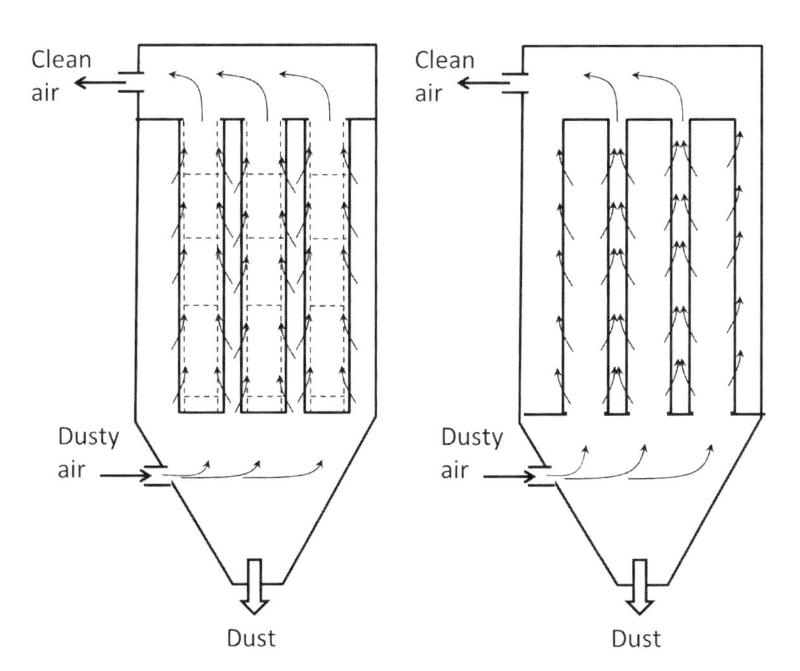

Fig. 10.2 Bag-house filter systems based on outside–in (*left*) and inside–out (*right*) operation

operation allows the gas to enter the filters from the outside surface where the dust cake builds up. In this case a wire or perforated support structure is needed inside the filter bags to prevent the filters from being distorted.

Selection of the mode of operation of a filter bag-house depends mainly on the mechanical properties of the filter medium and the method that is used to remove the dust cakes on the filters before the critical pressure drop is reached.

10.2.2.3 Dust Cake

For either operation mode, captured particles are retained on the filter surface, gradually forming the so-called "dust cake" as illustrated in Fig. 10.3. Generally, this filter cake is equally important to the actual filtration process as the media filter. This dust cake will increase the filter efficiency, but even more for the pressure drop. Both the pressure drop and the filtration efficiency are at the lowest for a clean filter. A pre-coat and pre-heat procedure is often used to prevent filter medium from acid condensation and from becoming "blinded" by the finest particles present in the process gas.

Dust cake and the filter are both porous media, and their corresponding pressure drops can be estimated using the Darcy's law, named after Henry Darcy, a French engineer. According to the Darcy's law, the face speed is related to the permeability of the porous media by

$$U_0 = \frac{Q}{A} = \frac{\Delta P}{\mu} \frac{k}{\Delta x} \tag{10.1}$$

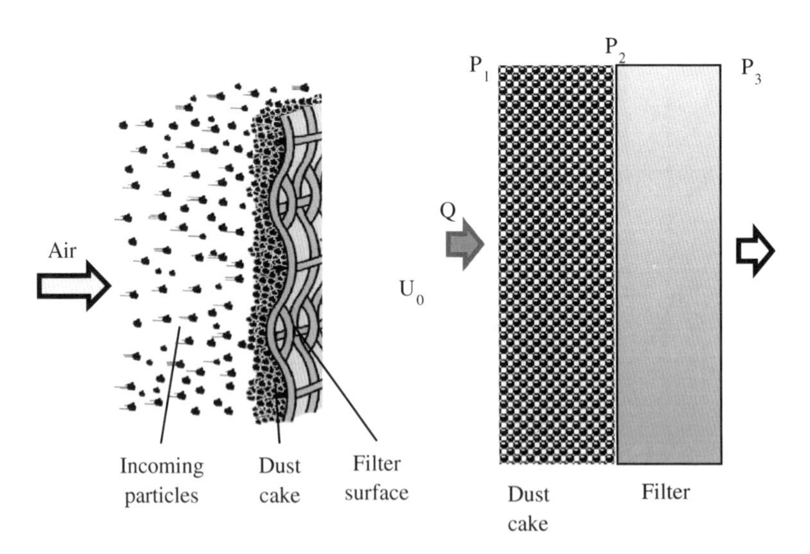

Fig. 10.3 Dust cake buildup model

where k is the permeability (m^2), and it is often determined experimentally, ΔP is the pressure drop across a porous media (a positive number), μ is the air viscosity, and Δx is the thickness of the porous media.

Consider a filter with a dust layer on the surface as shown in Fig. 10.3. The pressure at the surface of the dust cake, between the cake and the filter, and after the cake is P_1, P_2 and P_3, respectively. Applying Eq. (10.1) to the filter media leads to

$$U_0 = \frac{P_2 - P_3}{\mu}\left(\frac{k}{\Delta x}\right)_f. \tag{10.2}$$

Similarly, applying Eq. (10.1) to the dust cake leads to

$$U_0 = \frac{P_1 - P_2}{\mu}\left(\frac{k}{\Delta x}\right)_{ck}. \tag{10.3}$$

For the entire filter with cake, Eqs. (10.2) and (10.3) together give the total pressure drop

$$\Delta P = P_1 - P_3 = \mu U_0\left[\left(\frac{\Delta x}{k}\right)_{ck} + \left(\frac{\Delta x}{k}\right)_f\right]. \tag{10.4}$$

Example 10.1: Dust Cake

A set of bag-house surface filters have an active area of 60 m^2. The pressure drop through a freshly cleaned bag-house is 125 Pa and the bag-house is expected to be cleaned to remove dust cake when the pressure drop reaches 500 Pa in 1 h. The gas being cleaned has a flow rate of 300 m^3/min with a particle loading of 10 g/m^3. If the average mass filtration efficiency is 99 %, and the dust cake has a solidity of 0.5. Assume the real dust material is 2,000 kg/m^3. Estimate

a. the thickness of the cake when the filter is ready for cleaning
b. the permeability of the dust cake when it is due to be removed.

Solution

The mass of dust collected on the surface of the working filters in 1 h is

$$m = cQt\eta = 0.01\frac{\text{kg}}{\text{m}^3} \times \frac{300\ \text{m}^3}{\text{min}} \times 60\ \text{min} \times 0.99 = 178.2\ \text{kg}$$

(a) The thickness of the cake collected in 1 h is

$$\Delta x_{ck} = \frac{m}{\rho_{ck} A\alpha} = \frac{178.2\ \text{kg}}{2000\frac{\text{kg}}{\text{m}^3} \times 60\ \text{m}^2 \times 0.5} = 0.003\ \text{m or 3 mm}$$

When the dust cake is to be removed, the pressure drop cross the dust cake is

$$(\Delta P)_{ck} = P_1 - P_2 = \Delta P - (\Delta P)_f = 500 - 125 = 375\,Pa$$

The face speed of the gas flow is

$$U_0 = \frac{Q}{A} = \frac{(300\ \mathrm{m^3/min}) \times (1\ \mathrm{min}/60\ \mathrm{s})}{60\ \mathrm{m^2}} = 0.083\ \mathrm{m/s}$$

(b) The dust cake permeability can be derived from Eq. (10.3)

$$k_{ck} = \frac{\mu U_0 (\Delta x)_{ck}}{P_1 - P_2} = \frac{1.81 \times 10^{-5}\ \mathrm{Pa.s} \times 0.0083\ \mathrm{m/s} \times 0.003\ \mathrm{m}}{375\ \mathrm{Pa}} = 1.19 \times 10^{-12}\ \mathrm{m^2}$$

This value is close to that of permeable sandstone.

Dust Cake Removal

There are mainly three types of fabric filter bag-house cleaning methods:

(1) reverse air cleaning,
(2) pulse-jet cleaning, and
(3) shake/deflate systems.

Their respective principles are shown in Fig. 10.4.

During a bag-house cleaning using reverse-gas and shake/deflate methods, the air emission stream must be temporarily interrupted or bypassed, for example, using an offline operation. The pulse-jet method operates on-line on a few bags while the rest of the bags continue working without being interrupted. Reverse gas systems use cleaned gas from another filter unit to remove the dust cake from the filter. Shake/deflate systems employ both a mechanical shaking force and reverse air to remove dust cakes.

The filter face speed or air to cloth ratio depends on the cleaning methods to be employed. It is about 1, 1.5–2, and 3–4 cm/s for reverse air systems, pulse-jet systems, and shake/deflate systems, respectively, at a comparable pressure drop. The corresponding dust cake loads also vary from 1 to 2.5 kg/m^2 for shake/deflate systems, 2.5–7.5 kg/m^2 for reverse air systems, and 5–10 kg/m^2 for pulse-jet filters. A typical filter bag has a length of 5–10 m, and a diameter of 0.2–0.3 m, where corresponding surface area is 3–10 m^2 per bag. Pulse-jet units operate with somewhat smaller bags.

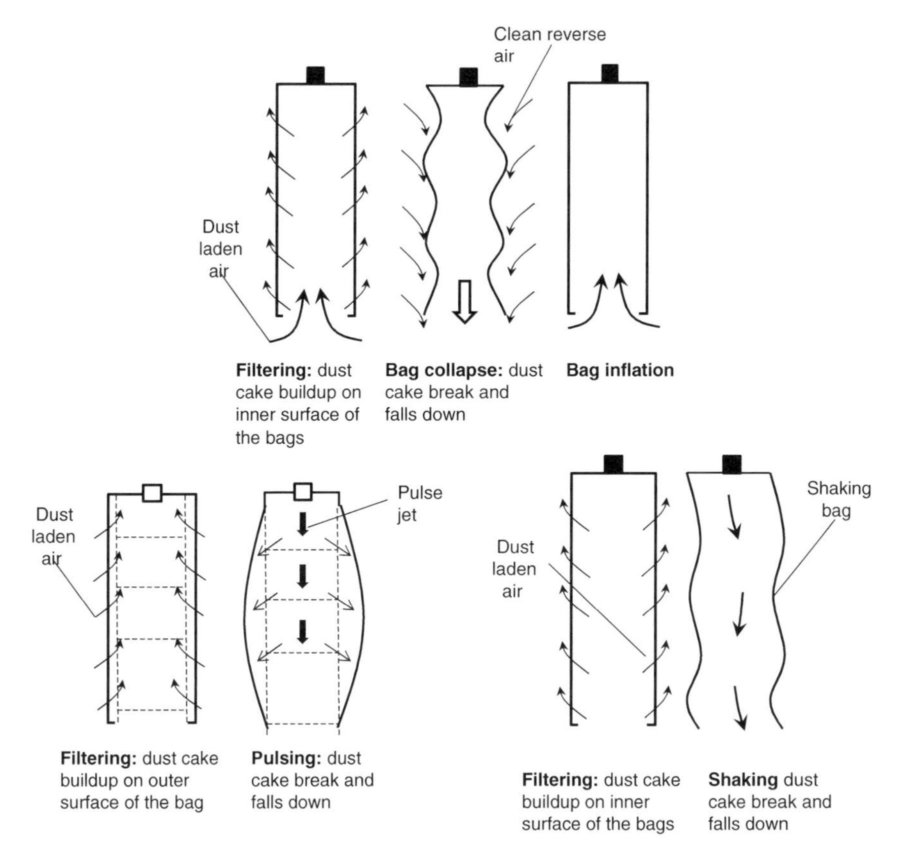

Fig. 10.4 Three bag-house cleaning methods: reverse-air (*top*), pulse-jet (*bottom left*) and shake/deflate (*bottom right*)

10.2.3 Wet Scrubbing

Wet scrubbers can be applied to control the emissions of both gaseous pollutants and particulate matter. They remove air pollutants by inertial of diffusional impaction, reaction with a sorbent or reagent slurry, or absorption into a liquid solvent. These types of scrubbers can be used for the control of hazardous air pollutants (HAP), inorganic fumes, vapors, and gases (e.g., chromic acid, hydrogen sulfide, ammonia, chlorides, fluorides, and SO_2). They may also occasionally be used to control volatile organic compounds (VOC).

Wet scrubbers are comparable to but not as effective as filters or ESPs for the removal of sub-micron particles. However, wet scrubbers were noticeable when certain gaseous components are to be removed simultaneously, or the particles are too sticky or corrosive to be accepted by other alternatives.

In Sect. 5.2, we only analyzed counter flow packed bed scrubber. There are actually many other types of wet scrubbers in engineering practice. Despite of their differences in configuration, the ultimate objective is to maximize the gas-liquid contact surface area. The larger the contact area, the greater mass transfer speed, and the faster air cleaning process is.

In choosing a wet scrubber, we have to consider the fabrication costs, operating costs, and environmental impact. The capital cost of a wet scrubber is lower than that of a bag-house filter or an ESP but higher than a cyclone. Operation and maintenance costs, however, are much higher due to consumption of liquid, accessories required to handle high pressure drops and slurries, and problems related to corrosion, abrasion, solids buildup, failure of rotating parts and restart problems after a shutdown. Most importantly, wet scrubber transforms an air pollution problem to a water treatment problem.

Schematic diagrams of typical wet scrubbers are shown in Fig. 10.5. Typical wet scrubber designs include

- Bubble column
- Packed bed scrubber
- Spray tower
- Venturi scrubber.

A bubble column is an apparatus used for gas–liquid absorption. Liquid is stored in a vertical column. Gas is introduced from the bottom of the column and distributed through a bubble generator, called a sparger. Large bubbles can be generated by passing through the holes in the sparger; millimeter range or micron-sized bubbles can be generated by porous media. Gas-liquid mixing can be done by buoyancy, or it can be mechanically mixed by stirring. The liquid can be static or parallel flow or counter-current with respect to the gas phase.

A bubble column is useful in reactions where the gas-liquid reaction is slow in relation to the absorption rate. It is useful for small quantity of gas. However, when the gas flow rate is high, it would require extreme power to pass this large volume of gas through a liquid tower. And, the dust accumulation may plug the column too. Therefore, other types of wet scrubbers should be considered for large volume of air cleaning.

In a spray tower, liquid (water) droplets are sufficiently large to overcome the upward gas speed. The droplet sizes are controlled to optimize particle contact and to provide easy droplet separation. The liquid to gas ratio is much smaller than that of a bubble column.

One special type of spray tower is called cyclonic spray tower. The body of a cyclonic spray tower is just like a cyclone. When air or gas enters the cyclone, the centrifugal forces are created by either a tangential inlet or vane-induced inlet. The liquid sprays are aligned in the center of the tower. Centrifugal forces increase

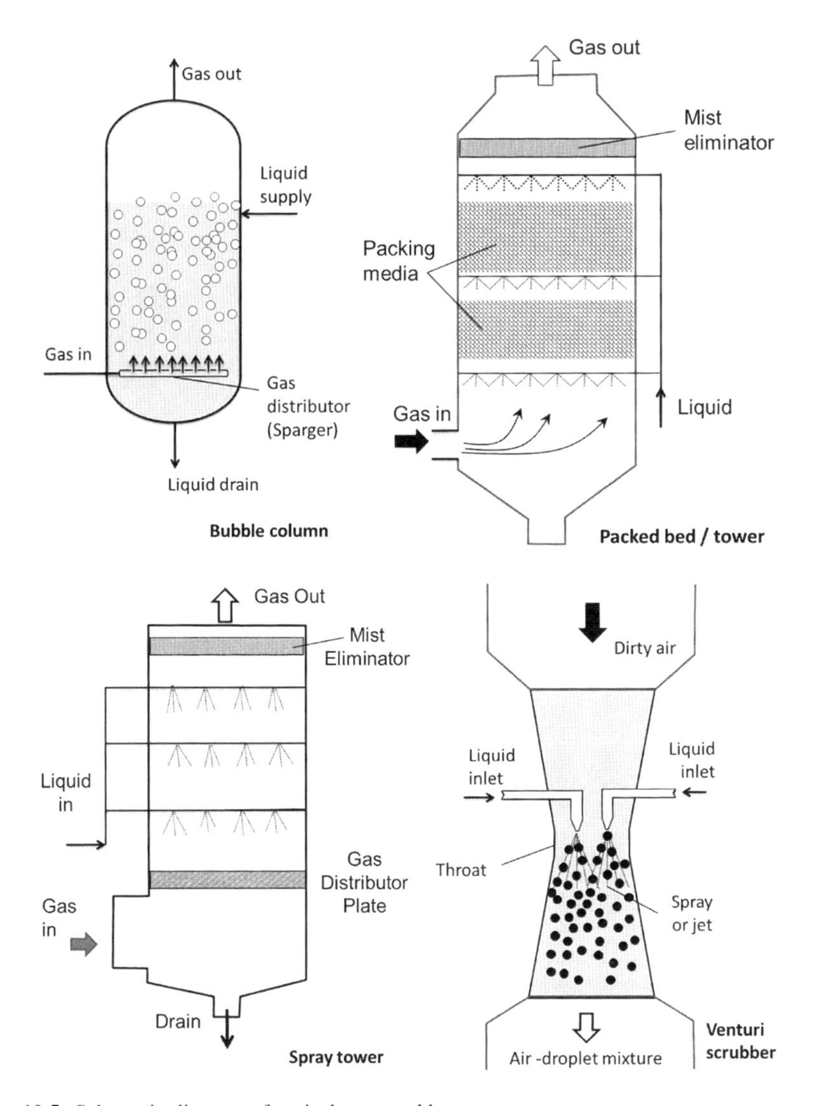

Fig. 10.5 Schematic diagram of typical wet scrubbers

the forces of collision and relative velocity of the droplets with respect to gas stream. A well designed cyclonic spray tower greatly increases the collection of particles smaller than 10 μm when compared to simple countercurrent spray towers.

In a venturi scrubber gas meets the liquid at the throat. Atomized water droplets are added at throat as sprays or jets. Gas first accelerates and then deaccelerates before and after the throat. As a result, the gas pressure drop is expected to be high due to the converging and diverging flow pattern. The mixture of gas and liquid droplets has to be separated again using a downstream unit. It can be a cyclonic collector or a filter to separate water droplets from air stream.

Table 10.1 Particle size collection efficiencies of various wet scrubbers

Type of scrubber	Pressure drop in Pa	Minimum collectable particle diameter in μm
Gravity spray tower	125–375	10
Cyclonic spray tower	500–25,00	2–6
Packed bed scrubbers	500–4,000	1–5
Venturi scrubber	2,500–18,000	0.5–1

A minimum particle size collected with approximately 85 % efficiency

Source http://www.epa.gov/ttn/caaa/t1/reports/Sect.5-4.pdf

Table 10.1 shows the comparison of size collection efficiencies of various wet scrubbers for particulate separation. Obviously a gravity spray tower has the lowest pressure drop as well as the largest minimum collectable particle diameter of 10 μm. Among them, a venturi scrubber seems to be the most effective and it also has the highest pressure drop.

10.3 Flue Gas Desulfurization

FGD, one major post-combustion SO_x control technology, has been developed for three temperature ranges, low (<200 °C), medium (200–400 °C) and high (>600 °C). The former two are usually in duct and the last one right after the furnace. FGD can also be classified into wet, dry, and semi-dry processes, depending on the presence of water in the sorbent and the flue gas.

Data on worldwide FGD applications show that wet limestone FGD have been predominantly selected over other FGD technologies. Wet FGD is a well-established technology with high sorbent conversion rate and high desulfurization efficiency. The main challenges facing the wet FGD system are the high capital costs and extensive post processing of liquid FGD byproducts. As a promising alternative to wet FGD, dry or semi-dry FGD at medium temperatures have been developed for potential cost and energy savings and relative ease of operation compared to wet FGD. At a medium temperature, the reaction rate of $Ca(OH)_2$ with SO_2 should be very fast, allowing higher conversions because the sinterization of the sorbents can be avoided and a better contact between the solid and the flue gas can be reached.

10.3.1 Wet FGD

In a wet FGD process, a solution of lime ($Ca(OH)_2$) with a high content of calcium and magnesium are sprayed into an absorption tower to capture the SO_x in the flue gas. CaO is produced by calcining the $CaCO_3$ at temperatures above 800 °C with the release of CO_2.

$$CaCO_3 \rightarrow CO_2 + CaO \tag{10.5}$$

The simplified reactions for desulfurization are

$$Ca(OH)_2 + SO_2 \rightarrow CaSO_3 + H_2O \tag{10.6}$$

The liquid after reaction is collected at the bottom of the scrubber and is partially recycled. The rest of the used solution stream is fed to the thickener to convert calcium sulfite to calcium sulfate by oxidation.

$$CaSO_3 + 2H_2O + {}^1\!/_2O_2 \rightarrow CaSO_4 + 2H_2O \rightarrow CaSO_4 \cdot 2H_2O \tag{10.7}$$

Limestone/lime FGD is sensitive to acidic or alkaline components in the flue gas. Acidic components reduce the pH value and change the equilibrium for SO_2 absorption (Eq. (10.6)). Therefore, a pre-scrubber is useful to remove the acidic components, fly ash, and mercury from the flue gas. It also improves the quality of the gypsum. Alternatively, organic buffers can be applied to maintain a constant pH of the solution. This also lowers the capital cost by lowering the absorber size and pumping accessories.

10.3.1.1 Other Chemical Sorbents

There are other sorbents that can also react with SO_2 in the flue gas. For example, $Mg(OH)_2$ (magnesium hydroxide) can react with SO_2 and produce $MgSO_3$ (magnesium sulfite):

$$Mg(OH)_2 + SO_2 \rightarrow MgSO_3 + H_2O \tag{10.8}$$

Similarly, caustic (NaOH) can react with SO_2 producing sodium sulfite

$$2NaOH + SO_2 \rightarrow Na_2SO_3 + H_2O \tag{10.9}$$

Seawater is a natural alkaline to absorb SO_2. When SO_2 is absorbed into the seawater, it reacts with oxygen and water to form sulfate ions SO_4- and free H^+. And surplus of H^+ is offset by the carbonates in seawater pushing the carbonate equilibrium to release CO_2 gas:

$$SO_2 + H_2O + {}^1\!/_2O_2 \rightarrow SO_4^{2-} + 2H^+ \tag{10.10}$$

$$HCO_3^- + H^+ \rightarrow H_2O + CO_2 \tag{10.11}$$

Table 10.2 The most important FGD systems with non-regenerable sorbents

Process	Sorbent	By-product
Wet scrubbers	Lime/Limestone	Gypsum, calcium sulfate/sulfite
	Lime/Fly ash	Calcium sulfate/sulfite/fly ash
Spray-dry scrubbers	Lime	Calcium sulfate/sulfite
Dual-alkali	Primary: sodium hydroxide	Calcium sulfate/sulfite
	Secondary: lime	
Seawater	Primary: seawater	Waste seawater
	Secondary: lime	
Walther	Ammonia	Ammonia sulfate

Source http://www.iea-coal.org.uk/

Aqueous ammonia can also be an effective SO_2 absorbent. The corresponding overall reactions can be described using Eqs. (10.12) and (10.13) that follow.

$$SO_2 + 2NH_3 + H_2O \rightarrow (NH_3)_2SO_3 \quad (10.12)$$

And the product of $(NH_3)_2SO_3$ can further react with SO_2 to produce ammonium hydrogen sulfite

$$SO_2 + (NH_3)_2 + H_2O \rightarrow 2NH_4HSO_3 \quad (10.13)$$

The resultant products are usually used as fertilizer feedstock.

The most important FGD processes as listed by IEA Coal Research are given in Table 10.2.

The Walther process is based on scrubbing with ammonia water where the product is mainly $(NH_4)_2SO_4$. However, the sulfur content in the fuel has to be below 2 %-wt in order to minimize the risk of formation of ammonium sulfate aerosols.

10.3.1.2 Renewable Sorbents

Over 80 % of the power utilities with sulfur emission control use a non-regenerable sorbent based on calcium to remove the sulfur from the flue gas. However, in most FGD processes, the sorbent is not completely used due to the nature of shell formation surrounding the core lime [35]. It is a waste of resources to throw away the "used" sorbents. In addition, this throwaway sorbent creates also extra cost for landfill. Therefore, regeneration of "used" sorbent has been developed accordingly. Important commercial FGD processes with regenerable sorbents are listed by IEA Coal Research and summarized in Table 10.3.

Table 10.3 The most important FGD systems with regenerable sorbents

Process	Sorbent/principle	End/by—product
Wellman—Lord	Sodium sulfite (Na_2SO_3)	Concentrated SO_2
Bergbau Forschung/Uhde	Activated carbon	Concentrated SO_2
Linde SOLINOX	Physical absorption (amine)	Concentrated SO_2
Spray-dry scrubbing	Sodium carbonate (Na_2CO_3)	Elemental sulfur
MgO process	Magnesium oxide (MgO)	Concentrated SO_2

Wellman-Lord process [23] is the most widely used wet renewable sorbent FGD process. Sodium carbonate, Na_2CO_3 is the actual sorbent and $NaHSO_3$ is the product of reaction.

$$Na_2CO_3 + SO_2 + H_2O \rightarrow 2NaHSO_3 \tag{10.14}$$

The sulfur is released again as SO_2, which is extracted as a mixture of approximately 85 % SO_2 and 15 % water. The mixture can be further processed to sulfuric acid.

$$NaHSO_3 \rightarrow Na_2SO_3 + SO_2 + H_2O \tag{10.15}$$

$$SO_2 + {}^1/_2O_2 + H_2O \rightarrow H_2SO_4 \tag{10.16}$$

Some sodium sulfite/sulfate is formed as a by-product; make-up soda or trona is needed to balance this loss. Make-up Na_2CO_3 (soda) or $Na_2CO_3 \cdot NaHCO_3$ (trona) is necessary to compensate the loss of sodium from the system with the $CaSO_3$/$CaSO_4$ product. Therefore, an actual process includes a more concentrated sodium-based liquid with limestone/lime scrubbers, combined with precipitation and regeneration in separate devices.

10.3.2 Steam Reactivation of Calcium Based Sorbents

Since the molar volume of $CaSO_3$ or $CaSO_4$ is higher than molar volume of CaO or $Ca(OH)_2$, the volume of sulfation products shall be greater than the volume of sorbent consumed. This result in a progressive pore blockage that gradually eliminates the access of SO_2 to the active CaO surface. Due to this pore eliminating phenomena, only a fraction of the CaO is utilized in sulfation. The low conversion rate of sorbent results in not only a waste of resource and low efficiency in FGD, but also increased particulate contaminants in the gas stream and consequently the increased burden for the particulate separation units and sludge treatment.

To increase the conversion rate of the spent sorbent, steam reactivation technology was first proposed by Shearer et al. [26]. Steam reactivation appeared more economical and promising as compared with the other technologies that use additive or catalyst.

There are two main mechanisms of steam reactivation of spent Ca-based sorbent. One mechanism is the water penetration theory, according to which during a steam reactivation process, water penetrates the sulfation product layer and reacts with the unconverted CaO trapped inside the sorbent and produced $Ca(OH)_2$. The mole volume of $Ca(OH)_2$ is greater than CaO, the fresh $Ca(OH)_2$ produced inside expands and increases the specific surface and specific volume of the spent sorbent. When reactivated sorbent is used for the second sulfation at higher temperatures, $Ca(OH)_2$ decomposes and results in fresh pores. These fresh pores make more SO_2 accessible to the internal CaO. This theory indicates that the existence of unconverted CaO inside the spent sorbent is required for a successful steam reactivation.

However, Couturier et al. [10] studied steam reactivation for spent sorbent with particle sizes smaller than 1 mm in diameter. Without trapped CaO, the conversion rate of the spent sorbent was still increased by reactivation from 45 to 80 % or higher. They proposed that there were unconverted CaO between the produced sulfate crystals, and the CaO was converted into $Ca(OH)_2$ through the reactivation process, resulted in volume expansion. This volume expansion produced not only fresh pores but also the fresh surfaces due to the breakdown of the sorbent particles.

Wang et al. [35] reported the mechanism of steam reactivation of spent Ca-based sorbent at 200–800 °C using multiple techniques, including mercury porosimeter, X-ray diffraction (XRD), scanning electron microscopy (SEM), and weight change analysis. Compared to the conversion rates of 10 and 12 % after first sulfation and direct second sulfation reaction, the conversion rate with sorbent regeneration reached a high range of 20–45 %. Regeneration temperature played a more important role than the retention time for regeneration. For example, the conversion rate reached the highest value of 45 % at 200 °C, and then decreased to about 30 % at 300 °C. There was no big difference between 400–800 °C with a conversion rate between 20 and 25 %.

Being reactivated at 200 °C, the sorbent particles broke down obviously and the reactivated sorbent was almost entirely $Ca(OH)_2$. While at 300–500 °C, reactivated particles did not change much in size, and both CaO and $Ca(OH)_2$ exist in the reactivated sorbent. At 600–800 °C, negligible $Ca(OH)_2$ was observed in the reactivated sorbent, but mainly in the form of CaO.

The overall mechanisms of the steam reactivation are summarized in Fig. 10.6. Because the molar volume of $Ca(OH)_2$ is greater than that of CaO, the produced $Ca(OH)_2$ also resulted in the volume expansion of the sorbent particles. Depending on the quantity of produced $Ca(OH)_2$, the volume expansion results in:

(1) break-down of the particles or pore size increase
(2) penetration of part of the $Ca(OH)_2$ through the surface of the particles, or
(3) the lumps on the surfaces of the particles.

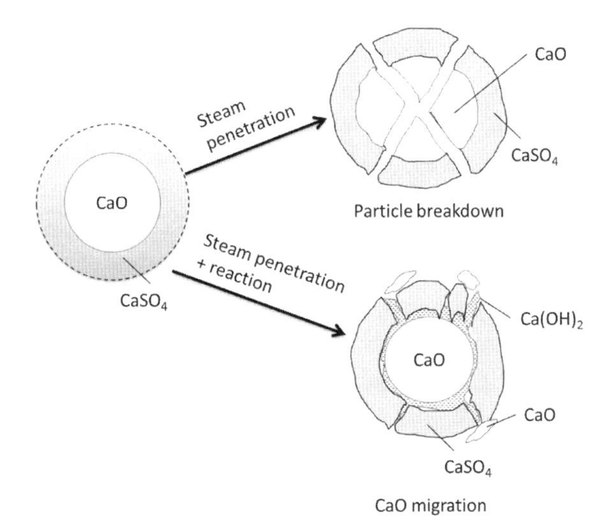

Fig. 10.6 Mechanisms of steam reactivation of spent sorbent

Breakdown of the particles and the production of $Ca(OH)_2$ lead to increased rate of conversion. The migration of CaO from inside outward is the main mechanism of steam reactivation at the temperatures over 300 °C.

At reactivation temperature of 200 °C, the water vapor molecules diffused into the sorbent through the product layer, and then contacted with trapped CaO and reacted with CaO to produce fresh $Ca(OH)_2$. Since the molar volume of $Ca(OH)_2$ is 95.9 % larger than that of CaO, this expansion of the volume resulted in the breakdown of the sorbent particles, producing smaller loose porous particles with more reactive surfaces. As the particle size becomes smaller, the specific volume and the specific surface increase. All these changes are favorable for the diffusion of SO_2 into the sorbent particles, and consequently increased the chance of the chemical reactions between SO_2 and CaO. The change in chemical composition was the conversion of CaO into $Ca(OH)_2$. The reactivity between $Ca(OH)_2$ and SO_2 is greater than CaO and SO_2.

Sulfation process decomposed $Ca(OH)_2$ to CaO at the temperature of 600 °C, which is higher than the decomposition temperature of $Ca(OH)_2$. This change resulted in unstable interval products such as CaO^*. CaO^* could react with SO_2 and produce $CaSO_4$ or $CaSO_3$, and some of the CaO^* could be converted into stable CaO.

At 300 °C or higher, water vapor molecules also penetrated the surface and reached inside the sorbent particles. The CaO reacted with water and produced $Ca(OH)_2$. Different reactivation temperature resulted in different conversion rate of CaO to $Ca(OH)_2$, and the amount of produced $Ca(OH)_2$ decreased with the steam-reactivation temperature, however, no particle breakdown observed during the steam-reactivations at 300 °C or higher. The porosimeter results showed that both the specific volume and the specific surface of the sorbent decreased, while the reactivity of the sorbent increased. This indicates that the performance of a steam-reactivation process cannot be evaluated solely based on the pore structure. The XRD and SEM analyses showed that the migration of CaO from inside outward played a critical role in reactivation of the spent sorbent.

10.3.3 Dry FGD

Dry sorbent, typically lime powder, can be injected into the flue gas duct where the gas temperatures are about 150–400 °C. In a power plant it is located before the emission stack, and likely to be located between the air pre-heater and the particulate control devices.

Dry FGD can also be an add-one device that is referred to as a dry scrubber; it does not use water or any liquid solvent. The scrubber is usually filled with caustic granulates or pellets to create chemical reactions that remove the SO_x emission compounds. A dry scrubber can be placed before an exhaust gas economizer (EGE) or used in conjunction with SCR units, which typically require exhaust gas temperatures above 350 °C to reduce both SO_x and NO_x emissions.

Using caustic lime as an example, in a dry scrubber, the caustic lime ($Ca(OH)_2$) reacts with sulfur dioxide (SO_2) to form calcium sulfite, which is then oxidized to form calcium sulfate dehydrate or gypsum:

$$SO_2 + Ca(OH)_2 \rightarrow CaSO_3 + H_2O \tag{10.17}$$

$$CaSO_3 + {}^1\!/\!_2O_2 \rightarrow CaSO_4 \tag{10.18}$$

The oxygen is usually supplied from the air to reduce the operating cost. The final product is likely to be gypsum ($CaSO_4 \cdot 2H_2O$) and there is no CO_2.

The desulfurization product includes not only $CaSO_4$ particles but also unused lime powder. They are then removed from the flue gas by the particulate air cleaning devices downstream. This approach is relatively cheap and can easily be added on to the older facilities without proper sulfur control.

10.3.4 Semi-Dry FGD

Since solid-gas reaction is slow and lime is not reactive enough, additional humidification is needed in order to improve the desulfurization efficiency and the effectiveness of sorbent usage. This approach is referred to as semi-dry FGD, where a calcium-based sorbent is sprayed into the flue gas duct as a water slurry. The lime in the spray will be converted to calcium hydroxide, and it is more reactive to SO_2 than CaO

$$CaO + H_2O \rightarrow Ca(OH)_2 \quad (10.19)$$

Consequently, the desulfurization reaction becomes

$$Ca(OH)_2 + SO_2 \rightarrow CaSO_3 + H_2O \quad (10.20)$$

The reaction would be even faster before complete water evaporation due to the ions in liquid phase.

A potential problem associated with the semi-dry FGD is the corrosion of duct under humidified environment. Typically these systems are operated with a Ca/S ratio near 2, with typical SO_2 concentrations of a few 1000 ppmv from coal combustion.

Another important parameter is the approach temperature, which is defined as how far the gas temperature is above the saturation temperature for the water in the gas. The lower the approach temperature, the longer it takes for water to be vaporized completely, and consequently, a higher sulfur uptake by the sorbent.

10.4 NO_x Reduction Using SCR and SNCR

NO_x control from the flue gas can be completed by dry and wet approaches too. Dry approaches include (SCR), selective noncatalytic reduction (SNCR), and adsorption, whereas wet approaches are, similar to wet FGD, by wet scrubbers. This section focuses on dry approach of SCR/SNCR.

10.4.1 Selective Catalytic Reduction

SCR is one of the most successful technologies for NO_x removal from flue gases [7, 24, 34]. Typical catalysts include V_2O_5 or WO_3 on a TiO_2 support. SCR operates at temperatures of 350–400 °C. For a power plant the location is right before the air pre-heater, where ammonia is injected into the furnace and reacts with NO to form water and nitrogen as products:

$$NH_3 + NO \rightarrow N_2 + H_2O \quad (10.21)$$

Typical NO_x reduction efficiency is in the range of 90–95 % when SCR is operated at an ammonia injection molar ratio of NH_3/NO of about 0.8. The NH_3 concentration after removal is below 5 ppm. It was reported that commercial SCR units reduce NO_x emissions by 40–70 %, depending on upstream NO_x concentrations and the local allowable NO_x emissions [32].

Meanwhile, the SCR catalyst also catalyze the oxidation of SO_2 to SO_3, which may cause corrosion problems downstream at lower temperatures where water is condensed and sulfuric acid is formed.

Like many catalytic reactions, catalyst degradation is a big concern here as well. Ammonium sulfate can also be formed in the SCR unit which causes corrosive deposits. Therefore, SCR is not recommended for coal-fired plants when the sulfur content in the fuel is above 0.75 %-wt.

The SCR degradation is also caused by catalyst poisoning by As and other trace elements, loss of active catalyst by evaporation, in addition to the corrosion and erosion and the buildup of solid deposits in the catalyst structure. In order to minimize the particle deposition in SCR, the flue gas is preferably passed downward through a series of 2–4 layers of catalyst beds in the SCR unit. This allows the fly ash particles to settle down by gravitational settling.

There is typical a loss of 10–20 % of the initial efficiency within 1–1.5 years of operation. Therefore, a minimum catalyst life of 2 years is currently considered acceptable. This frequently replacement leads to a major economic challenge to SCR because typically SCR catalysts take 30–40 % of the cost of the entire SRC system.

10.4.2 SNCR

Like many air pollution control technologies, there is always a balance between the economic costs and the effectiveness. Similarly selective non-catalytic NO reduction (SNCR) process offers less efficiency but often also less cost for NO_x reduction. In a SNCR process, also referred as the Thermal De-NO_x process, ammonia is added to the flue gas where the temperature is about 900 °C. At this temperature, nitric oxide is converted to molecular nitrogen where water is a by-product. The step reactions can be described as follows [2, 3].

$$NH_3 + OH \rightarrow NH_2 + H_2O \tag{10.22}$$

$$NH_2 + NO \rightarrow NNH + OH \rightarrow N_2 + H_2O \tag{10.23}$$

$$NH_2 + NO \rightarrow N_2 + H + OH \tag{10.24}$$

The presence of O and OH radicals in the flue gas is necessary for ammonia to decompose to amino radicals (NH_i), which reacts with nitric oxide to produce N_2 and water.

SNCR is only practical within a narrow temperature range from 850 to 1000 °C thus making it very temperature sensitive. The optimum temperature for the SNCR process is 950 °C. At a higher temperature, NH_3 starts to react to nitric oxide, while at lower temperature, the NH_3 decomposes slowly and results in significant waste of NH_3. This problem can be solved by additives, such as, hydrogen peroxide (H_2O_2) and hydrocarbons (C_xH_y), which can shift the optimum temperature by about 200 °C.

The optimal temperature also drops with the decrease of NO concentration and the increase of residence time or the CO concentration. At firing of pulverized coal, the NO reduction efficiency usually falls to 40–80 %. Another potential problem of the SNCR is the increase of N_2O and CO emissions. Instead of ammonia, other compounds may be used as reduction chemicals, such as, urea or cyanuric acid.

10.4.3 Reagents

Both ammonia and urea have been successfully employed as reagents for SCR and SNCR. Ammonia is generally less expensive than urea. However, the choice of reagent is based on not only cost but physical properties and operational considerations. A popular replacement of ammonia is urea, which can produce ammonia by heating.

$$NH_2CONH_2 + (1 + x)H_2O \rightarrow NH_4COONH_2 + xH_2O \rightarrow 2NH_3 + CO_2 + xH_2O \quad (10.25)$$

The reaction occurs in two steps and overall it is endothermic. The first reaction involves the production of ammonium carbamate from the combination of urea and water. Ammonium carbamate then breaks down in the hot flue gas to produce ammonia and carbon dioxide.

Urea is a nontoxic, less volatile liquid that can be stored and handled more safely than ammonia. Urea solution droplets can penetrate farther into the flue gas when injected into the boiler. This enhances mixing with the flue gas, which is challenging for large boilers. Because of these advantages, urea is more commonly used than ammonia in large boiler applications of SNCR systems. Use of the urea, however, usually results in higher emissions of N_2O than when using ammonia. Furthermore, SNCR does not remove NO_2, while SCR does.

10.5 Simultaneous Removal of SO_x and NO_x

Several technologies have been developed to remove NO_x and SO_2 simultaneously for the sake of low cost and small footprint. They include electron beam flue gas treatment [3] and activated carbon-based adsorption [36]. Dry adsorption on activated carbon can be completed at a temperature as high as 220 °C. Principles of absorption/adsorption methods have been introduced in Part 1. SO_x/NO_x removal by electron beam radiation is briefly introduced by [13]. SO_2/NO_X can also be captured by copper oxide spray mixed with ammonia. SO_2 is first captured CuO at 400 °C.

$$CuO + SO_2 + \frac{1}{2}O_2 \rightarrow CuSO_4 \tag{10.26}$$

$$CuO + SO_2 \rightarrow CuSO_3 \tag{10.27}$$

$$CuO + SO_3 \rightarrow CuSO_4 \tag{10.28}$$

Meanwhile, NO is reduced by ammonia to nitrogen, for which copper oxide is a very good catalyst.

$$4NO + 4NH_3 + O_2 \rightarrow 4N_2 + 6H_2O \tag{10.29}$$

$$2NO + 2NH_3 + O_2 + H_2 \rightarrow 2N_2 + 4H_2O \tag{10.30}$$

The copper sulfate ($CuSO_4$) product is transported to the two-stage heater/regenerator section where it is reduced to Cu at 500° C. SO_2 is released and it also can be used for sulfuric acid production.

More recently, researchers are trying to use wet scrubbers to absorb both SO_2 and NOx simultaneously, in this approach, the caustic solution contains strong oxidizers [39] and the oxidants include Fe(II)-EDTA and H_2O_2, penta- and hex-amminecobalt (II) chloride [37, 38]. Penta- and hexa-amminecobalt(II) chloride seemed to be superior over other two oxidants, but they are not stable and cannot be stored for long time. More importantly, waste disposal is a major challenge when cobalt is mixed with fly ash in the sludge and dissolved in the liquid solvent.

10.6 Control of Volatile Organic Compounds

The properties of the VOCs determine the suitable methods to capture or to oxidize the VOCs in a gas stream. The most important parameter is the flammability limits, also referred to as flammable limits of the VOCs/Air mixture. The lower flammable limit (LFL) describes the mixture with the smallest fraction of combustible gas. The upper flammable limit (UFL) defines the richest flammable mixture. These limits define whether the air-VOCs mixture may ignite or not [5]. When the mixture exceeds 25 % of the LFL, there is a high chance of ignition and it has to be handled with caution. The values of LFL and UFL for several VOCs and other gaseous organic compounds are available in literature [18]. For gas mixtures with VOC contents below LFL, the most important options are VOC combustion, condensation, and carbon absorption. When the concentration of VOC is above the UFL the gas can be oxidized in flares or boilers with air or steam.

10.6.1 Volatile Organic Compounds Adsorption

As introduced in Part I, low-concentrations VOCs can be physically adsorbed by activated carbon and other adsorbents. Activated carbon can be produced from a char or coke material by oxidation with air at 500–600 °C or by oxidation with boiling nitric acid.

Activated carbon adsorption of VOCs works well in the range of 0.05–30 m^3/s for gases with VOC concentrations of 20–5000 ppmv. The removal efficiencies are in the order of 90–98 %. Adsorption of VOCs is also not very sensitive to LFL or UFL values for the gas.

The adsorption bed will have to be replaced or regenerated when it reaches its break-through point after a certain time of operation, when the bed becomes saturated. In industry, most operators set the breakthrough value to be the same as that of the local emission standard. For an activated carbon bed used for adsorption of VOCs, the regeneration can be done with steam, hot air or hot nitrogen at 400–1000 °C. The VOCs released from the regeneration process are more concentrated and can easily be oxidized or recycled for other purposes.

Repeated loading and regeneration reduce the quality of the activated carbon and it results in lowered adsorption efficiency. In addition, the presence of other species such as halogenated organics and high moisture contents also reduce the performance of the activated carbon bed. For example, carbon bed adsorption is less suitable for ketones such as acetone, since exothermic polymerization reactions deactivate the bed and may even result in fires. Eventually, a fresh bed will replace a deactivated carbon bed and the activated carbon is subjected to disposal.

10.6.2 Oxidation of VOCs

Some air pollutants like VOCs, HC, and CO can be oxidized and converted into less toxic chemicals. There are three types of VOC oxidation processes, according to their applicable VOC concentrations.

- Flaring: Flares are usually used for gas streams that have an organic vapor concentration greater than 2–3 times the LFL.
- Thermal oxidation: Thermal oxidizers are used for contaminated gas streams that have an organic vapor concentration 25–50 % of the LFL.
- Catalytic oxidation: Catalytic oxidizers are used for gas streams that have VOC concentrations <25 % of the LFL.

10.6.3 Flaring

Flaring is more often employed for gas flaring in oil and gas refinery plants than the flue gas or exhaust gas from a combustion source. For gases with VOC concentrations above the UFL level, flaring at the stack outlet can burn the VOCs into other less problematic pollutants. Addition of steam is necessary for many VOCs flaring, otherwise only a few VOCs can be oxidized without major problems.

There are two types of flares, being elevated flare and ground flare. The most commonly used type in refineries and chemical plants are elevated flares. They are characterized with large capacities. The waste gas stream is fed through a stack with a height of from tens to a hundred meters. Combustion takes place at the top of the stack.

The elevated flare, can be steam assisted, air assisted or non-assisted. Steam/air injection promote smokeless combustion; adequately elevated flare has the best dispersion characteristics for malodorous and toxic combustion products, which are mainly caused by incomplete combustion due to the cross wind. On the other hand, steam injection/air injection cause noise pollution.

Ground level flare can achieve smokeless operation without noise, but it is poor in dispersion of combustion products because its short stack is close to ground. This may result in severe air pollution or hazard if the combustion products are toxic or in the event of flame-out. Ground level flares are often used for the destruction of landfill gases, which contain a large amount of methane and other odorous compounds.

10.6.4 Thermal Oxidizers

A thermal oxidizer burns VOC-containing gas streams in an enclosed refractory-lined chamber that contains one or more burners. The design of thermal oxidation systems for VOCs has been well documented in the literature (e.g., [19]). It is summarized as follows.

An example of a thermal VOC incinerator is shown in Fig. 10.7. The waste gas is oxidized before entering the stack, and meanwhile the heat is recovered. The thermal oxidizer consists of a refractory-brick lined chamber that has one or more gas- or oil-fired burners. The contaminated gas stream does not usually pass through the burner itself, unless a portion of the gas stream is used to provide the oxygen needed to support combustion of the fuel. Instead, the burners are used to heat the gas stream to the temperature necessary to oxidize the organic contaminants. That temperature is based primarily on the auto ignition temperature of the most difficult to destroy compound in the gas stream. Auto ignition temperatures for most organic compounds range from 400 to 540 °C. Operation at temperatures near the auto ignition value will result in destruction of the contaminants; however, the

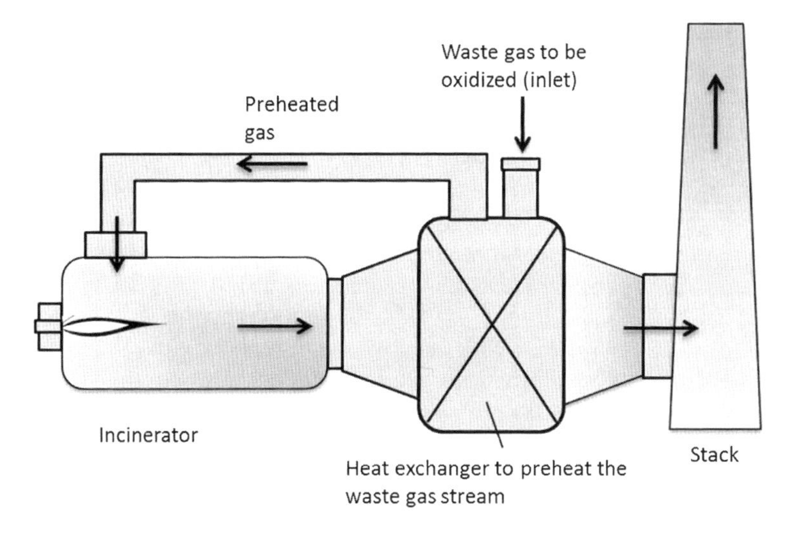

Fig. 10.7 A typical thermal VOC incinerator with recuperative heat recovery

carbon in the compounds will be left primarily as CO, even if sufficient oxygen is available.

To fully oxidize the carbon to CO_2, incineration should be conducted at a minimum temperature of 700 °C. Most thermal oxidizers operate at temperatures of 700–1,000 °C. When catalyst is used (catalytic oxidation) the temperatures can be lowered to a range of 200–500 °C.

The combustion chamber is sized to provide sufficient residence time to complete the oxidation reactions. The time needed for destruction depends on the temperature and mixing within the chamber, with shorter times required at higher temperatures. Typical residence times are 0.3–0.5 s, but may exceed 1 s. Residence time is typically calculated by dividing the volume of the combustion chamber by the actual volumetric flow rate through the combustion chamber. However, it should be recognized that not all of the combustion chamber may be at the effective oxidation temperature. Thus, residence times calculated in this manner should be somewhat higher than the minimum desired value.

Typical destruction and removal efficiency of a thermal oxidizer is above 99 % at typical capacities of 0.5–250 m^3/s. There are two options for heat recovery, a regenerative and a recuperative recovery system with the thermal efficiencies of 95 and 70 %, respectively [7]. Selection of heat recovery system depends on the property of the VOC containing gas as well as flexibility of operation.

A recuperative thermal oxidizer uses a shell-and-tube type heat exchanger to recover heat from the exhaust gas and to preheat the incoming VOC rich gas, thereby reducing supplemental fuel consumption. Recuperative heat exchangers with a thermal energy recovery efficiency of up to 80 %, mostly 50–60 %, are in common commercial use.

10.6.4.1 Regenerative Thermal Oxidizer

As depicted in Fig. 10.8, a regenerative thermal oxidizer uses ceramic beds to absorb heat from the exhaust gas and uses the captured heat to preheat the incoming process gas stream. The destruction of VOCs is accomplished in the combustion chamber, which is always fired and kept hot by a separate burner. This system provides high heat recovery of up to 98 %, and can operate with VOC lean gas streams because supplemental heat requirements are kept to a minimum with the high heat recovery. The gas steam may contain less than 0.5 % VOC, and have a low heat value.

The valves in a two-chamber regenerative thermal oxidizer switch the direction of flow so that the incoming gas passes through the freshly warmed bed. A typical cycle is about 30–120 s and the change of temperature is usually less than 70 °C. As such it requires large, rapid-cycling valves and extensive ductwork. The valves must be designed for very low leakage since any leakage contaminates the treated exhaust gases with untreated process gas. Critical high-efficiency systems use zero-leakage valves with an air purge between double-seal surfaces. If the VOC emissions from a two-chamber bed are measured, the concentration profile would show

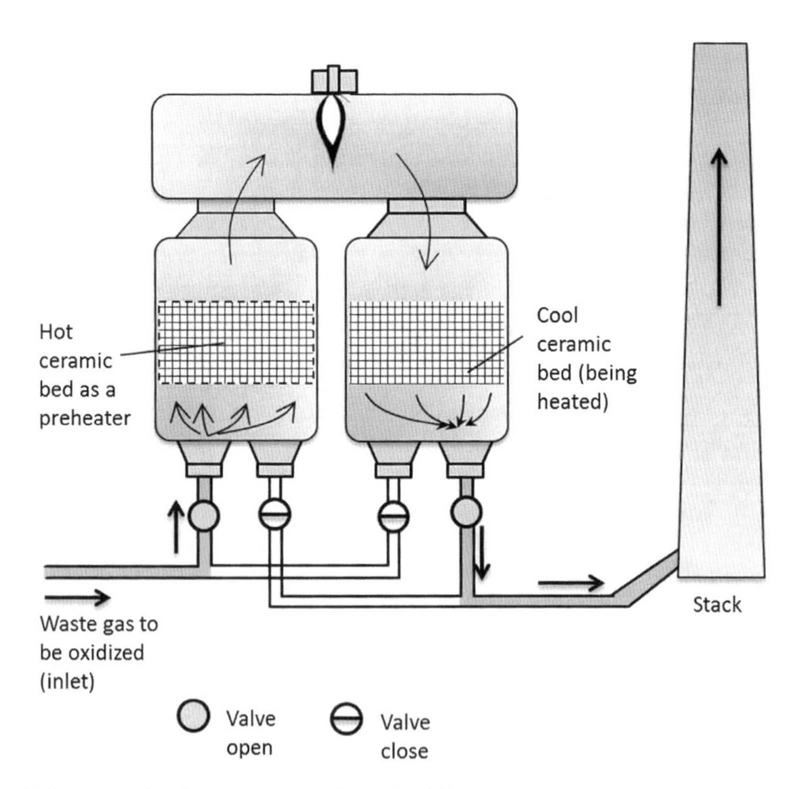

Fig. 10.8 A two-chamber regenerative thermal oxidizer

intermittent spikes each time the valves switch the direction of flow, because the untreated VOC in the gas present in the inlet bed when it is suddenly switched to the outlet.

Again, the advantage of thermal oxidation in an incinerator is the high destruction efficiency that can be obtained by proper control of the combustion chamber design and operation. If temperatures are maintained above 980 °C, greater than 99 % hydrocarbon destruction is routinely achievable. This efficiency depends on residence time, temperature, and turbulence (the three Ts) in the combustion chamber.

Thermal oxidizers can be costly to install because of required support equipment, including high pressure fuel supplies (for example, natural gas), and substantial process-control and monitoring equipment. In addition, public perception of a new "incinerator" can make it difficult to locate and permit a new unit.

10.6.5 Catalytic Oxidation

In a catalytic oxidation process, catalysts such as Pt or Pd on a Al_2O_3 support may give destruction and removal efficiencies of up to 95 % with small size. The catalytic oxidation units are, however, much more expensive in operation. The high cost is caused by frequent replacement of the catalyst after being poisoned by other pollutants such as soot, particles, chlorine, sulfur, silicon, vanadium, lead, and/or hydrocarbons with great molecular weight. Temperature excursions also reduce the lifetime of catalysts. Furthermore, secondary air pollution may be produced by catalytic oxidation. Gases other than VOCs may be converted to hazardous compounds that require further treatment downstream. From this point of view, catalytic VOC oxidization is most suitable for the cleaning gases with stable VOC properties.

As seen in Fig. 10.9, catalysts are loaded on a catalyst bed in the incinerator. The support structure of catalysts is arranged in a matrix that provides high geometric surface area, low pressure drop, and uniform flow. Structures providing these characteristics include honeycombs, grids, and mesh pads. Either a monolith or a beads/pellets configuration can be employed, depending on the kinetics of VOC oxidation and the presence of other pollutants. Monolith is a cluster of parallel tubes and it is used for fast kinetics. Bead or pellet bed is preferred for slow kinetics and it is less sensitive to fouling or poisoning. The performances of thermal and catalytic oxidizers are often affected by the presence of CO. Oxidation of CO will be addressed separately.

Thermal oxidation is not effective for engine exhaust gases mainly due to low gas temperatures and low concentration of the VOC emission from liquid fuel combustion. VOC emissions from engines are successfully oxidized in catalytic converters, where the catalysts are Pt, Pd, or Rh. Pd is more sensitive to poisoning by lead and sulfur. Most of the time, it is integrated in a three-way catalyst with CO oxidation and NO_x reduction.

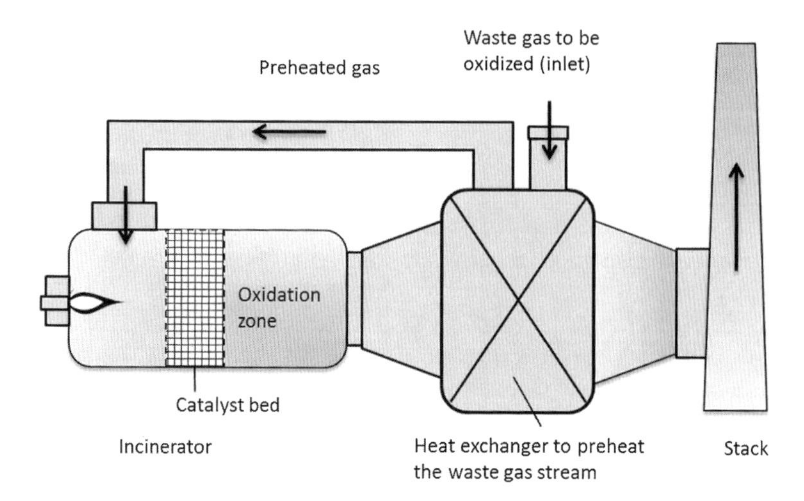

Fig. 10.9 VOC destruction by catalytic oxidation with recuperative heat exchanger

Catalytic oxidizers are coupled with both recuperative and regenerative heat exchangers. The catalyst enables the oxidation reaction to occur at much lower temperatures than thermal oxidation. Typical standard catalysts for VOC oxidation operate at 350–500 °C. For lower temperatures, 200–300 °C, special catalysts may be needed for successful oxidation.

A typical minimum inlet temperature is in the range of 260–315 °C. For each 1 % LFL concentration that is oxidized in the catalyst bed, the gas stream temperature will need to increase about 15 °C. Catalytic oxidizers are usually used only on gas streams with an organic vapor concentration <25 % of the LFL. Extremely high temperature will damage the catalysts.

High cost and performance problems are related to physical and chemical deterioration of the catalyst material. Certain metals can react or alloy with the catalyst, resulting in permanent catalyst deactivation. Fast acting poisons include mercury, phosphorus, arsenic, antimony, and bismuth. Slow acting poisons include lead, zinc and tin. At temperatures above about 535 °C, even copper and iron are capable of alloying with platinum catalyst, reducing its activity. Some materials, principally sulfur and halogen compounds, have a high adsorptive affinity for some catalytic surfaces, reducing the active sites available to the organic compounds.

10.6.6 Other Approaches to Volatile Organic Compounds Control

When the concentrations of VOCs are relatively high, about 5000 ppmv, they can be separated by VOC condensation, i.e., by cooling or pressurizing. It is cost-effective for high concentration; however, for VOCs with low boiling point, process costs may become excessive. Gases with VOC concentrations above 25 % of the LFL can be processed with a wide range capacity of 0.05–10 m^3/s. Readers are referred to literature (e.g., [11]) for the principles and designs of VOC condensers.

Control of VOCs can also be done with a liquid solvent by absorption in a wet scrubber. It is based on the principle of absorption introduced above. The choice of wet scrubber depends on the presence of particulates in the gas stream. Absorption using liquid solvent is suitable for gas streams with high moisture contents ($RH > 50$ %) and the concentrations of VOCs in the range of 500-5000 ppmv. It can function at a higher flow capacity of 1–50 m^3/s.

10.7 Control of Soot Particles

Ideally soot should be oxidized completely before leaving a combustion device (e.g., engine or furnace). However, soot is a big problem for combustion of solid fuel and diesel due to lack of knowledge about the formation and oxidative destruction of soot. Recent studies indicate that a lot of soot is also generated from bio-diesel fueled diesel engines.

A particulate control system can separate and collect some of the soot as part of the total particulate matter. Other options for the control of soot after combustion are thermal oxidation, catalytic oxidation and carbon filtration. In most of the cases, they are integrated into one unit. An example of a popular soot control technology is the continuously regenerating trap (CRT) system. Briefly, NO in the exhaust gas is converted to NO_2 by catalytic oxidation right before the gas entering the filter. Then the carbonaceous particles collected on a filter are oxidized at 250 °C by NO_2. Meanwhile, 80–90 % of hydrocarbon and CO are also oxidized. The removal efficiency of soot particles can be 90 % or higher.

10.8 Control of Trace Metals

As introduce above, trace elements that are found in fossil fuels can be grouped into three classes, Class I, Class II (Pb, Cd, Sb, Ni), and Class III (Hg, Se, As). As mentioned in the pre-combustion control technologies, some of these chemicals can be removed from the fuels by fuel cleaning. Crushing and milling of coal particles can also remove As, Se, FeS_x, and significant amount of Hg from the fuel. The elements related to sulfide minerals (pyrite, sphalerite) can be largely removed by coal pre-cleaning: As, Cd, Co, Cu, Hg, Mo, Ni, Pb, Se, and Zn [20].

Class I and Class II elements are removed together with the particulates by the dust control system such as ESP and bag-house filter at efficiencies of >99 and 90–99 %, respectively. Trace elements that penetrate through an ESP or bag-house filter may be captured downstream by a wet or dry scrubber, likely for SO_2 control. Wet scrubber, if any, also can capture some of the submicron particles downstream of the dry particulate control devices. These captured particles also contain a large amount of metals, including copper, zinc, silver, and gold. Therefore, it is not necessary to design a specific device for the control of Class I and Class II elements.

The majority of Class III elements in the flue gas will, however, penetrate through the particulate control devices and wet scrubbers. Systems have to be designed specifically for the removal of Class III trace elements, which include the notorious Pb, As, Se, Cd, and Hg. Sorbents can be injected into the flue gas upstream of the particulate control system, unless a separate reactor is employed. Usually a large amount of sorbent is needed to compensate low removal efficiencies. Frequently, cost is the concern for large-scale applications.

Various sorbents such as activated carbon, silica, alumina, kaolinite, limestone, emathlite, bauxite, and titania have been tested for trace elements control at temperatures of 400–1,000 °C, by direct injection into the flue gas or passing the flue gas through packed or fluidized beds. Alumina may be effective for As, Be and fairly active for Cd and Ni, silica is effective for Cd, Pb, and Hg, while titania can be a good sorbent for Cd and Pb. Limestone can be used to capture Pb, Cd, Sb, Hg, Se, and As species [4].

At 350–600 °C, As and Se can be removed from flue gases by calcium-based sorbents, forming $CaSeO_3$ and $Ca_3(AsO_4)_2$. Since the temperature is favorable for desulfurization as well, the concentration of SO_2 in the flue gas stream has to be relatively low to prevent the competitive sulfation of the sorbent. Selenium can also be removed from gases by physisorption using activated carbon at 125–250 °C [1].

Control of mercury emissions is a challenging task because of the complicated transformations of mercury in combustion. All three types of Hg species must be considered for effective mercury emission control:

(1) insoluble gaseous mercury
(2) soluble gaseous oxidized mercury, mainly $HgCl_2$, and
(3) particle-bound mercury, Hg-p.

Existing particulate control and FGD devices or separated sorption units using activated carbon, metal oxides, and fly ashes can be used to for the capture of mercury from flue gas.

10.8.1 Mercury in Particulate Control and FGD Devices

Some particulate control devices also help to reduce Hg emissions. Typical Hg removal efficiencies for particulate emission control equipment at US coal-fired power plants with ESPs and fabric filters are about 32 and 44 %, respectively. The majority of coal combustion plants employ ESPs for particulate control with more and more bag-houses being installed, which will help in mercury reduction. For example, the activated carbon injected upstream of an ESP yield lower Hg removal efficiency than when it is injected upstream of a bag-house filter.

Further mercury removal may be accomplished by the wet scrubber desulfurization systems. A typical wet scrubber FGD shows Hg removal of 34 %, while dry scrubber shows an efficiency of about 30 % with injected sorbent collection on a fabric filter [6]. In a wet scrubber, water may capture some $HgCl_2$ due to its solubility in water. In addition, the low temperature in a wet FGD system is favorable for volatile Hg to condense and for the removal from the flue gas stream. However, increased concentrations of Hg vapor were found too when the flue gas passed through a wet FGD [17].

Total Hg removal depends on the arrangement of particulate control and FGD devices. For a coal-fired power plant, a bag-house/FGD combination can give Hg removal efficiencies of 88–92 %, but the efficiencies were much lower (23–54 %) for an ESP/FGD combination [31]. The reason behind this dramatic difference in the efficiency is that oxidized mercury is reduced to elemental Hg in ESPS. This does not happen in filters.

A filter bag-house is preferred over an ESP for the removal of mercury from the flue gas due to more interactions between the gaseous mercury in the flue gas, and the dust cake on the filters help capture Hg too. This explains partially why bag-houses are progressively being installed for new plants.

10.8.2 Mercury Adsorption by Activated Carbon

For coal-fired power plants, especially those use lignite and peat, the control of mercury may present a serious and expensive problem in the near future. Developed countries, including Canada, will follow the implementation of Hg regulations for coal-fired power plants as being done in the USA.

There are several unique characters that relate Hg emission to coal combustion. One is the low concentration of Hg in high volume of flue gas streams typically in the order of 0.01 ppmm. Another factor is the low Cl/Hg ratio and the small fraction of oxidized Hg, indicating the relative importance of elemental Hg in the flue gas. For coal, over 50 % of the Hg is in the form of elemental Hg when the chlorine content of the fuel is less than 0.1 % by weight, while less than 20 % of the Hg elemental in the flue gas when the chlorine content of the fuel is above 0.2 % by weight.

Hg can be removed from the coal flue gases either by injection of activated carbons in flue gas ducts or by passing the flue gas through beds packed with activated carbons. For a packed bed system, the C/Hg ratio by mass is in the range of 5,000–1,00,000 due to short residence time of approximately 1 s of the flue gas in air pollution control devices. The actual C/Hg ratio depends mainly on the temperature of the flue gas [6]. Sorbent particles larger than 20 μm in diameter may be too large for the typical contact times of 1 s.

It was found that the capacity or reactivity of the activated carbon is limited at a temperature of 180 °C or higher [8]. The decrease in temperature favors the removal of Hg, indicates that physical sorption is the main mechanism for Hg capture by activated carbon.

Meanwhile, the activated carbon is not adsorbing Hg exclusively, as there are many other species present in the flue gas, such as H_2O, NO_2, SO_2, and other trace elements. For example, the presence of NO_2 and SO_2 has a negative effect on mercury adsorption. In addition, the presence of HCl, if any, supports the transformation of physisorbed elemental Hg to stronger bound chemisorbed Hg. A 90 % of Hg reduction can be achieved by activated carbon sorption.

Much higher Hg removal can be achieved when activated carbons are impregnated with sulfur or iodide. In sulfur-impregnated activated carbons, Hg is strongly bound as HgS, sorption capacities of which may be 100–1000 times higher than those of activated carbon alone. It is especially effective at high temperatures [33]. Iodide-impregnated activated carbons show the improved performance by forming stable Hg–I complexes [6].

10.8.3 Mercury Captured by Metal Oxides, Silicates, and Fly Ashes

Alternative Hg sorbents using metal oxides, silicates, fly ashes, etc., have been developed to replace activated carbons. Metal oxides such as MnO_2, Cr_2O_3 and MoS_2 showed moderate capacities, making them possible alternatives for activated carbons [15]. As an example, in a Finish cement plant, the Hg emission was reduced by approximately 90 % when 10 % of the coal/petcoke fuel was replaced by car tire scrap. This benefit is due to the metals such as Mn and Cr in the tires that were introduced to the combustion system [21].

Fly ash may also absorb or adsorb Hg [30]. The performance seems to depend on many factors such as the carbon content of fly ash, temperature, gas phase composition, as well as the property of the fossil fuel.

10.9 Proper Layout for Post-combustion Air Pollution Control Devices

After fossil fuel combustion and in-combustion controls, multiple air pollutants present in the flue gas must be resolved. The air pollutants of concern depend on the fuel and local regulations on emission control. NO_x is the major concern for natural gas combustion, while SO_2 may also need attention when oil is used for combustion. For solid fuels, particulate, SO_2 (except for biomass) and NO_x emissions must all be reduced to certain levels before the flue gases being discharged to the atmosphere for dispersion. Similarly, solid fuel-fired IGCC processes involve hot gas cleanup to upgrade the fuel gas to meet the gas turbine specifications.

Since more than one piece of equipment is installed to control different air pollutants, they have to be installed in a proper sequence in order for them to function properly. As a general rule, particulate matter has to be removed as much as possible before dealing with gaseous pollutants. However, when limestone is injected in the duct to remove SO_x or catalyst is used to remove NO_x, particulate control devices should be arranged even further upstream in order to capture the excessive particles.

A typical air pollution control system for stationary sources like power plants is composed of a dry sorbent injection followed by filtration devices. Selection of the exact equipment depends on the fuel quality, government regulations of concerns and process type. The gas temperature plays an important role in the positioning of the equipment; so do the cross-effects between air pollution control processes. An attempt to list the most important of interactions is given in Table 10.4. Considering the interactions between different air emissions, a good planning is needed for optimized air emission control.

10.10 Practice Problems

1. A power plant burns coal at a rate of 5000 tons/day. The coal has 5 % of ash content. Assume 25 % of ash falls to the bottom of the furnace and 75 % becomes fly ash. In order to meet a particulate matter emissions limit of 2 tons/day find out the efficiency required of a final control device.
2. Calculate the number of bags required in a bag-house to filter 17,000 m^3/min of air at a temperature of 200 °C and pressure of 1 atm. Each bag is 3.5 m long and 0.3 m in diameter and air to cloth ratio is 0.9 m/min.
3. A power plant is using a large bag-house to control air pollution. The air is flowing through at a rate of 60 m^3/s with a dust loading of 230 mg/m^3. Calculate the collection efficiency for the bag-house if emission regulation for the plant is 100 μg/m^3.
4. A power plant produces a flue gas with a mass flow rate of 1.75 million lb per hour. Assume that the mass flow rate of sulfur dioxide in flue gas is 2000 lb/hr. This flue gas is passed through a wet scrubber to meet emission standards. The efficiency of scrubber is 90 %. Determine the rate of sulfur dioxide emitted to the atmosphere.
5. The pressure drop through a freshly cleaned bag-house is 125 Pa and the bag is expected to be cleaned to remove dust cake when the pressure drop reaches 500 Pa in 1 h. The gas being cleaned has a flow rate of 300 m^3/min, and the a particle loading is 10 g/m^3. If the average mass filtration efficiency is 99 %, and the dust cake has a thickness of 5 mm, a solidity of 0.5 and permeability of $1 \times 10^{-12} m^2$. Assume the real dust material is 2,000 kg/m^3. Estimate the total filtration area.

Table 10.4 Interaction of air emission control approaches

	SO_x	NO_x	Particulate matter	Trace elements
SO_x		The SCR catalyst converts $SO_2 \rightarrow SO_3$	Ash particles possibly catalyze $SO_2 \rightarrow SO_3$	Removal of vanadium reduces $SO_2 \rightarrow SO_3$
			Fly ash in a wet FGD scrubbers deposits, low quality gypsum	Na_2S_4 for Hg trapping in HCl scrubber may increases H_2S
NO_x	In situ calcium based sorbents in FBC reduce NO but may increase N_2O emissions	SCR for de-NOx may increase NH_3 emissions	Hot-side ESPs before SCR for de-NO_x protects the catalyst	Vanadium oxide may act as an SCR catalyst
	High SO_2 concentrations may result in NH_4HSO_4 and $(NH_4)_2SO_4$ deposits in SCR for de-NO_x	SNCR may increase NH_3 and N_2O emissions		Some metals kill SCR catalyst
Trace elements	Wet FGD and dry scrubbing (duct injection or furnace injection of sorbent) also removes trace elements	SCR converts $Hg \rightarrow Hg^{2+}$	Filters and ESPs remove most Classes I, II and some Class III trace elements	
Particulate matter	In-furnace lime injection adds load to filter/ESPs Low sulfur fly ashes decrease the efficiency of ESP	Low NO_x combustion may increase carbon ash		Particulate matter adsorb some trace elements

6. A selective catalytic reduction system is designed to remove 80 % of NO_x in the flue gas. Calculate the stoichiometric amount of ammonia needed if flue gas has 500 ppmv of NO_x and is flowing at a rate of 10,000 m^3 per minute at 300 °C and 1 atm.
7. The flue gas from an industrial power plant contains 1100 ppmv of NO and is emitted at a rate of 1200 m^3/s at temperature of 573 K and pressure of 1 atm. A selective catalytic reduction system is designed to remove 75 % of NO. Calculate the quantity of ammonia needed in kg/hr.
8. Particles with low conductivity passing through an ESP are likely to

   a. form a dust cake that can be easily removed from the collecting plate
   b. restrain the gas stream
   c. lose ions quickly through the collection plate
   d. have low migration speed because of the resistance from the dust cake on the collection plate.

9. In a power plant, SCR is used for NO emission control by conversion of NO into N_2. Urea is preferred over ammonia because

   a. urea has a great inertia and cover a wider range of flue gas
   b. urea costs less and less evaporative
   c. urea based SCR is not sensitive to dust
   d. urea based SCR is not sensitive to temperature.

10 Which one of the following statements is most accurate?

   a. High dust SCR is less frequently maintained than a low dust SCR
   b. The gas temperature passing through a high dust SCR is lower than that of a low dust SCR
   c. The catalyst in a high dust SCR costs less than that in a low dust SCR
   d. None of the above.

References and Further Readings

1. Agnihotri R, Chauk S, Jadhav R, Gupta H, Mahuli S, Fan L-S (1998) Multifunctional sorbents for trace metal capture: fundamental sorption characteristics. In: Proceedings of the 15th annual international Pittsburgh coal conference, Pittsburgh, PA, USA, Sept 1998
2. Bae SW, Roh SA, Kim SD (2006) NO removal by reducing agents and additives in the selective non-catalytic reduction (SNCR) process. Chemosphere 65:170–175
3. Basfar AA, Fageeha OI, Kunnummal N, Al-Ghamdi S, Chmielewski AG, Licki J, Pawelec A, Tyminski B, Zimek Z (2008) Electron beam flue gas treatment (EBFGT) technology for simultaneous removal of SO_2 and NO_x from combustion of liquid fuels. Fuel 87:1446–1452
4. Biswas P, Wu CY (1998) Control of toxic metal emissions from combustors using sorbents: a review. J Air Waste Manag Assoc 48:113–127
5. Borman GL, Ragland KW (1998) Combustion engineering. McGraw-Hill, New York

6. Brown TD, Smith DN, Hargis RA Jr, O'Dowd WJ (1999) Mercury measurement and its control: what we know, have learned and need to further investigate. J Air Waste Manag Assoc 49:1469–1473
7. Broer S, Hammer T (2000) Selective catalytic reduction of nitrogen oxides by combining a non-thermal plasma and a V_2O_5-WO_3/TiO_2 catalyst. Appl Catal B 28:101–111
8. Chen S, Rostam-Adabi M, Chang R (1996) Mercury removal from combustion flue gas by activated carbon injection: mass transfer effects, vol 41, no 1. The 211th ACS National Meeting, New Orleans, LA, pp 442–446
9. Cooper CD, Alley FC (2002) Air pollution control—a design approach, 3rd edn. Waveland Press Inc., Long Grove
10. Couturier MF, Marquis DL, Steward FR (1994) Reactivation of partially-sulphated limestone particles from a CFB combustor by hydration. Can J Chem Eng 72:91–97
11. Dunn RF, El-Halwagi MM (1994) Selection of optimal VOC-condensation systems. Waste Manag 14(2):103–113
12. Farthing GH (1998) Mercury emissions control strategies for coal-fired power plants. In: Presentation to the 23rd international technical conference on coal utilization and fuel systems, Clearwater, FL, March 1998
13. Flagan RC, Seinfeld JH (1988) Fundamentals of air pollution engineering. Prentice Hall, Englewood Cliffs
14. Flagan RC, Seinfeld JH (2012) Fundamentals of air pollution engineering, 2nd edn. Dover Publishing, Englewood Cliffs
15. Granite EJ, Pennline HW, Hargis RA (2000) Novel sorbents for mercury removal from flue gas. Ind Eng Chem Res 39:1020–1029
16. Javed MT, Irfan N, Gibbs BM (2007) Control of combustion-generated nitrogen oxides by selective non-catalytic reduction. J Environ Manage 83:251–289
17. Krissmann J, Siddiqi MA, Peters-Gerth P, Ripke M, Lucas K (1998) A study on the thermodynamic behaviour of mercury in a wet flue gas cleaning process. Ind Eng Chem Res 37(8):3288–3294
18. LaGrega MD, Buckingham PL, Evans J (1994) Hazardous waste management, chapter 12 thermal treatment. McGraw-Hill, New York
19. Lewandowski DA (1999) Design of thermal oxidation systems for volatile organic compounds. CRC Press LLC, Boca Raton. ISBN 1-56670-410-3
20. Miller SF, Wincek RT, Miller BG, Scaroni AW (1998) Trace elements emissions when firing pulverised coal in a pilot-scale combustion facility. In: Proceedings of the 23rd international technical conference on coal utilization and fuel systems, Clearwater, FL, USA, March 1998, pp 953–964
21. Mukherjee AB, Kääntee U, Zevenhoven R (2001) The effects of switching from coal to alternative fuels on heavy metals emissions from cement manufacturing. In: The 6th international conference on the biogeochemistry of trace elements, Guelph, ON, Canada, 29 Jul–2 Aug 2001
22. Navarrete B, Canadas L, Cortes V, Salvador L, Galindo J (1997) Influence of plate spacing and ash resistivity on the efficiency of electrostatic precipitators. J Electrostat 39:65–81
23. Neumann U (1991) The Wellman Lord process. Sulphur dioxide and nitrogen oxides in industrial waste gases: emission, legislation and abatement, eurocourses: Chem Environ sci 3:111–137
24. Qi G, Yang RT, Rinaldi FC (2006) Selective catalytic reduction of nitric oxide with hydrogen over Pd-based catalysts. J Catal 237:381–392
25. Ray TK (2004) Air pollution control in industries—theory, selection and design of air pollution control equipment vol. 1. TechBooks International, New Delhi
26. Shearer JA, Smith GW, Myles KM, Johnson I (1980) Hydration enhanced sulfation of limestone and dolomite in the fluidized-bed combustion of coal. J Air Pollut Control Assoc 30:684–688
27. Sun H, Zhang Y, Quan X, Chen S, Qu Z, Zhou Y (2008) Wire-mesh honeycomb catalyst for selective catalytic reduction of NO_x under lean-burn conditions. Catal Today 139:130–134

28. Tien C, Ramarao BV (2007) Granular filtration of aerosols and hydrosols, 2nd edn. Elsevier, Oxford
29. Ruddy EN, Carroll LA (1993) Select the best VOC control strategy. Chem Eng Process 89 (7):28–35
30. Sakulpitakphon T, Hower JC, Trimble AS, Schram WH, Thomas GA (2000) Mercury capture by fly ash: study of the combustion of a high-mercury coal at a utility boiler. Energy Fuels 14:727–733
31. US EPA (2005) Control of mercury emissions from coal fired electric utility boilers: an update. http://www.epa.gov/ttn/atw/utility/ord_whtpaper_hgcontroltech_oar-2002-0056-6141.pdf. U. S. Environmental Protection Agency, Research Triangle Park, NC, USA
32. US DOE (1997) Control of nitrogen oxide emissions: selective catalytic reduction (SCR). Clean coal technology, topical report 9. U.S. Department of Energy, July 1997
33. Vidic RD, Liu W, Brown TD (1998) Application of sulphur impregnated activated carbons for the control of mercury emissions. In: Proceedings of the 15th annual international Pittsburgh coal conference, Pittsburgh, PA, Sept 1998
34. Wan Y, Ma J, Wang Z, Zhou W, Kaliaguine S (2004) Selective catalytic reduction of NO over Cu-Al-MCM-41. J Catal 227:242–252
35. Wang A, Tan Z, Qi H, Xu X (2008) Steam reactivation of used sorbent for FGD. J Coal Oil Gas Technol 1(3):330–344
36. Wang Y, Liu Z, Zhan L, Huang Z, Liu Q, Ma J (2004) Performance of an activated carbon honeycomb supported V_2O_5 catalyst in simultaneous SO_2 and NO removal. Chem Eng Sci 59:5283–5290
37. Yu H, Tan Z (2013) Determination of equilibrium constants for reactions between nitric oxide and ammoniacal cobalt(ii) solutions at temperatures from 298.15 to 309.15 K and pH Values between 9.06 and 9.37 under atmospheric pressure in a bubble column. Ind Eng Chem Res 52 (10):3663–3673
38. Yu H, Tan Z (2014) On the kinetics of the absorption of nitric oxide into ammoniacal cobalt(ii) solutions. Environ Sci Technol 48:2453–2463
39. Yu H, Zhu Q, Tan Z (2012) Absorption of nitric oxide from simulated flue gas using different absorbents at room temperature and atmospheric pressure. Appl Energy 93:53–58

Chapter 11
Air Dispersion

For the air pollutants penetrating though the post-combustion air pollution control devices, dispersion is the last step to minimize the environmental impact. An air dispersion models can be employed for predicting concentrations downwind of the source for environmental impact assessments, risk analysis, emergency planning, and development and implementation of air emission standards.

A great amount of models have been developed and they can be categorized into four generic classes:

- Gaussian models
- Numerical models
- Statistical models
- Physical models

Many air dispersion models are available at the US EPA websites for free use. In general, the more complex models are more reliable. In this chapter, we only introduce three kinds of models with different levels of complexity. Advanced models for special applications can be found in the book by de Visscher [21].

This chapter starts with an introduction to some basic concepts that are critical to the understanding of different air dispersion models. Then basic models will be introduced. All of the models introduced herein are based on conservation of mass (or material balances) in a control volume. Most models can be used for predicting the concentrations of several different pollutants, but they must be applied separately to one pollutant at a time. At the end, comments on advanced models will be given, although they are not discussed in a great depth.

11.1 Box Model

A box model is the simplest one for air dispersion in a ground-level community. It is useful in such well-defined environments as tunnels, streets with high rise buildings, community in a valley, and indoor environments. It is assumed that air in

Z. Tan, *Air Pollution and Greenhouse Gases*, Green Energy and Technology,
DOI 10.1007/978-981-287-212-8_11

these boxes is well mixed in the calculation domain, and moving along one dimension that is perpendicular to the inlet surfaces.

A schematic diagram of this model is shown in Fig. 11.1, in which C_i = the concentration of the pollutant in entering the box (kg/m^3); $\dot{m}$ is the source of in-box generation (kg/s); u = the speed of air (with pollutant); C = concentration of the pollutant within the box and the exit.

Conservation of mass of air leads to the volume flow rates

$$Q_i = Q_o = Q = uZY. \tag{11.1}$$

Conservation of mass of the pollutant gives

$$\begin{aligned} ZYX\frac{dC}{dt} &= (\dot{m} + uZYC_i) - uZYC \\ \frac{dC}{(\dot{m} + uZYC_i) - uZYC} &= \frac{1}{ZYX}dt. \end{aligned} \tag{11.2}$$

The concentration in the box can be determined by integration if the volume of the box ZYX is fixed.

$$\int_{C_\infty}^{C} \frac{dC}{(\dot{m} + uz_xYC_i) - uhwC} = \frac{1}{z_xYX}t \tag{11.3}$$

When u, $\dot{m}$, C_i are constants, the steady-state concentration C_{ss} in the box can be determined from Eq. (11.2) with $dC/dt = 0$.

$$C_{ss} = C_i + \frac{\dot{m}}{uZY} \tag{11.4}$$

Integrating Eq. (11.3) from time zero to any time gives,

$$\frac{C(t) - C_i}{C_{ss} - C_i} = 1 - \exp\left(-\frac{ut}{X}\right). \tag{11.5}$$

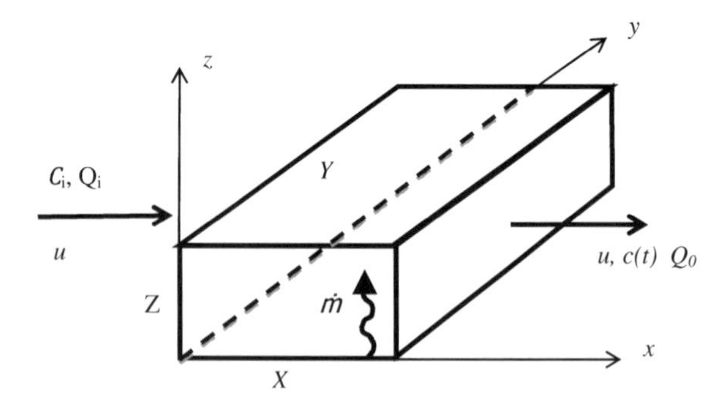

Fig. 11.1 Schematic diagram of box model

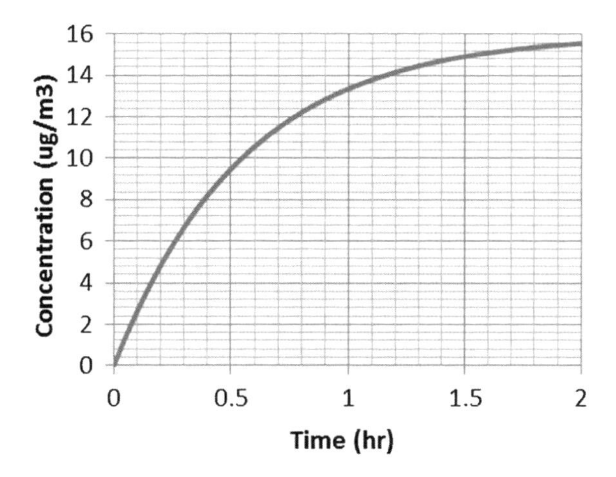

Fig. 11.2 Box model example

Example 11.1: Box model example

A city street is $Y = 25$ m wide with high rise buildings of 100 m high, which traps pollutants below 100 m. In a rush hour, cars line up on the $X = 1$ km long street and emit air pollutants continuously. The particulate air pollutant emission rate is $20\,\mu\text{g/s}$ per meter of street and constant wind blowing at a steady speed of $u = 0.5\,\text{m/s}$. Plot the concentration of the air pollutant in the street against time.

Solution

The source of generation is

$$\dot{m} = 20\,\mu\text{g/s.m} \times 1{,}000\,\text{m} = 20{,}000\,\mu\text{g/s}.$$

Ignoring the air pollutant in the incoming air, the steady state concentration is

$$C_{ss} = \frac{\dot{m}}{uYZ} = \frac{20{,}000}{0.5 \times 25 \times 100} = 16\,\mu\text{g/m}^3.$$

Then the concentration over time is

$$C(t) = C_{ss}\left[1 - \exp\left(-\frac{ut}{X}\right)\right] = 16\left[1 - \exp\left(-\frac{0.5 \times t}{1{,}000}\right)\right].$$

The plot is shown in Fig. 11.2. It shows that over extended period of time, the air pollutant concentration would approach the steady state concentration of 16 μm/m^3.

11.2 General Gaussian Dispersion Model

While the box model is useful for certain applications, it cannot be used for predicting the concentration in the atmosphere where air is subjected to mixing and concentration changes with time and location. For this kind of problem, we have to use other complex air dispersion models by considering the effect of meteorology. Before we start the mathematical expression of the models, we first put forward a few concepts.

11.2.1 Atmosphere

The temperature and pressure of the Earth's atmosphere change with elevation. Based on the variation of the average temperature profile with altitude, atmosphere can be divided into five distinguished layers (Fig. 11.3). From near Earth surface upward to the space, they are called [21]

- *Troposphere,*
- *Stratosphere,*
- *Mesosphere,*
- *Thermosphere,*
- *Exosphere.*

Troposphere is the lowest layer of the atmosphere, which is below 8–18 km altitude, depending on latitude and time of year, from the Earth's surface; In this region, temperature decreases with height, and there is a rapid vertical mixing of air.

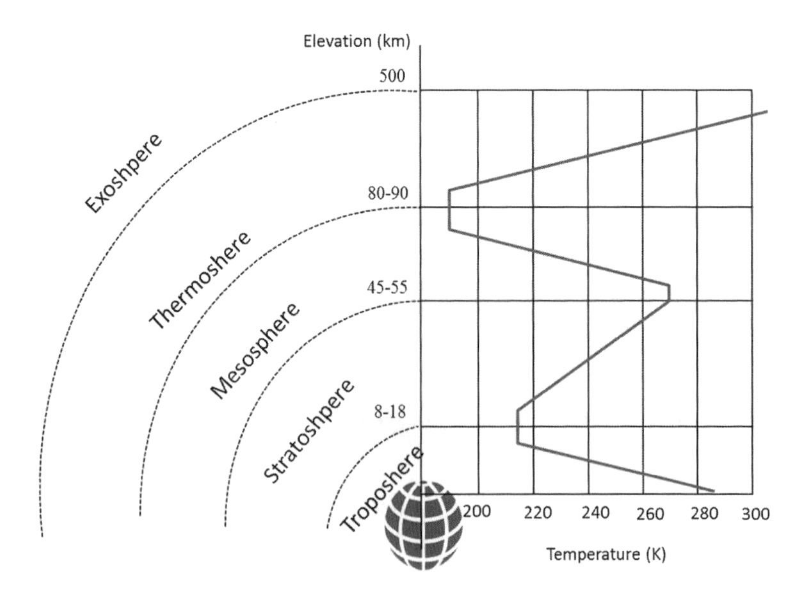

Fig. 11.3 Schematic diagram of atmosphere layers and temperature profile (not in scale)

The next region is *stratosphere*, and it is below 45–55 km altitude above the ground; temperature increases with altitude and vertical mixing of air is slow.

Mesosphere extends to 80–90 km altitude where temperature decreases again with altitude to the coldest point in the atmosphere; Vertical mixing becomes rapid again.

Thermosphere is the region above the mesosphere characterized by high temperatures and rapid vertical mixing. The next layer, *ionosphere,* is characterized by ions that are produced by photoionization. The outermost region of the atmosphere, *Exosphere is* >500 km altitude.

The boundary between troposphere and stratosphere is called tropopause. Its height changes with location. For example, the average heights of the tropopause over the equator and over the poles are about 8 and 18 km, respectively.

The atmosphere is a sink. It is an imperfect sink with limited ability to carry away (transport), dilute (dispersion) and remove (deposition) pollutants. Air motions carry pollutants from one region of the atmosphere to another. On the way to its destination, air and pollutants are dispersed by—mixing of pollutants with air.

Air dispersion takes place primarily in the lower layers of the atmosphere which interacts with the surface of the Earth. Sometimes referred to as ground boundary layer, the planetary boundary layer (PBL) is the lowest layer of the troposphere where wind is influenced by friction. The thickness (depth) of the PBL is not constant and it is dependent on many factors. At night and in the cool season the PBL tends to be lower in thickness while during the day and in the warm season it tends to have a greater thickness. This is because the wind speed and air density change with temperature. Stronger wind speeds enable more convective mixing, which cause the PBL to expand. At night, the PBL contracts due to a reduction of rising air from the surface. Cold air is denser than warm air; therefore, the PBL tends to be thin in the cool season.

Other conditions include, solar heating and cooling, temperature, pressure of the air, and the wind speed and direction. They affect the result of most commonly used air dispersion models, because these parameters contribute to the vertical motion of air pollutant in the atmosphere. They affect the atmosphere stability, which will be introduced soon.

Most of the motions of the atmosphere are actually horizontal as a result of uneven heating of the Earth's surface (most to the equator and least to the poles), the Earth's rotation (Carioles force) and the influence of the ground and the sea. The surface of the land and the oceans is a well-defined lower boundary for dispersion modeling in atmosphere. Major mountain ranges like the Himalayas, Rockies, Alps, and Andes are major barriers to horizontal winds. Even smaller mountains and valleys can strongly influence wind direction but on a smaller scale. The surface of the ground and seas also changes the temperature at these boundaries depending on the surface properties.

11.2.2 Atmospheric Motion and Properties

The density of any part of the atmosphere can be determined by the ideal gas law described in Eq. (11.6).

$$\rho = \frac{MP}{RT} \tag{11.6}$$

where the ideal gas constant is $R = 8.314\,\mathrm{J/(mol.K)}, M$ is the molar weight of the air in this case. Using this equation, the density of air is determined by M and T at one particular altitude, if P is fixed. Notice that M changes with the water vapor content; increased amount of water vapor content decrease the value of M.

When P is a variable, the air density at any point in the atmosphere can be calculated by

$$\rho = -\frac{1}{g}\frac{dP}{dz}. \tag{11.7}$$

Combining these two equations leads to

$$\frac{dP}{dz} = -\frac{gMP}{RT}. \tag{11.8}$$

When temperature and molar weight of air are both constant, integration of this equation from P_0 at z_0 to P at z leads to

$$P(z) = P_0 \exp\left(-\frac{gM}{RT}z\right). \tag{11.9}$$

11.2.3 Air Parcel

We need to put forward an important term that is used in air dispersion analysis, which is air parcel. An air parcel is an imaginary body of air to which may be assigned any or all of the basic dynamic and thermodynamic properties of atmospheric air. The air parcel is large enough to contain a great number of molecules, but its volume is small enough to be assigned with a uniform property. In air dispersion modelling, it is most likely part of the "air" from the emission source.

There are many factors affecting the motion of an air parcel and its dispersion after being emitted from a certain source. One of them is atmosphere environment, which contributes to the initial and boundary conditions and many other input parameters.

Air dispersion is greatly affected by the interaction between atmosphere and the pollutant containing air parcel. Most commonly considered factors are wind, temperatures, and humidity. Less obvious, but equally important, are vertical motions that influence air parcel motion. Atmospheric stability affects the vertical motion of air parcels. The temperature difference between the air parcel and atmosphere causes vertical motion, at least near the emission source, but the convective circulation thus established is affected directly by the stability of the atmosphere. Winds tend to be turbulent and gusty when the atmosphere is unstable, and this type of weather causes air pollutants to disperse erratically. Subsidence occurs in larger scale vertical circulation as air from high-pressure areas replaces that carried aloft in adjacent low-pressure systems. This often brings very dry air from a high altitude to a low level.

11.2.4 Adiabatic Lapse Rate of Temperature

In the analysis above, we assumed constant air temperature and molar weight, however, both temperature T and molar weight M change with elevation, and the change in M is not as important as that of temperature. For an adiabatic air parcel in the atmosphere, it may produce work to the surroundings. The rate of temperature change over elevation can be derived from the first law of thermodynamics [21] as

$$\frac{dT}{dz} = -\frac{g}{c_{p,a} + h_{fg}(dw/dT)} \tag{11.10}$$

where $c_{p,a}$ is the air heating capacity, w is the water vapor mass fraction in the air (kg vapor/air). Since most air contains less than a few percent of water vapor, except for those in cloud or fogs, the effect of water vapor can be ignored and we can use a simpler equation that is meant to be for dry air

$$\frac{dT}{dz} = -\frac{g}{c_{p,a}} \tag{11.11}$$

This is also called the adiabatic lapse rate of dry air temperature. Temperature decreases with elevation and it is a straight line in a plot of elevation *versus* temperature (Fig. 11.4). With a typical air heating capacity of $c_{p,a} = 1{,}006$ J/kg.K and $g = 9.81$ m/s^2, we can estimate the adiabatic lapse rate of dry air temperature $\frac{dT}{dz} = 0.975$ or 9.75 K/km. This is an estimation for guidance only. The actual value changes with location and time. As dry air moves vertically, its temperature changes at about 1 °C per 100 m [21]. For air with moisture, the temperature change could be 5–10 °C per km, depending on its water content, and air up in the sky reaches saturation easily to form clouds.

The change of temperature over elevation is the main reason behind atmospheric stability, which is an important factor in air dispersion.

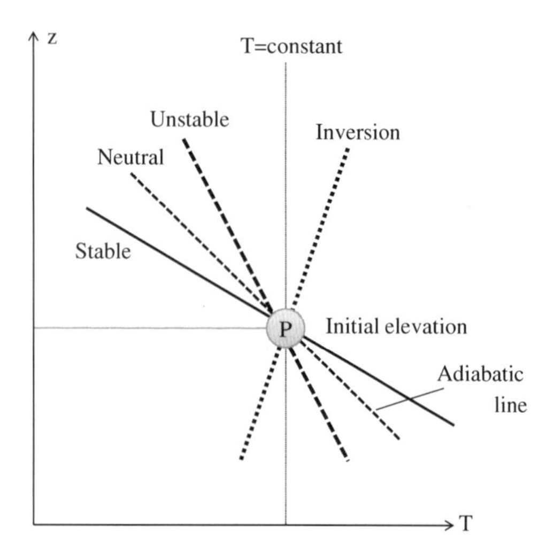

Fig. 11.4 Atmosphere stability

11.2.5 Atmospheric Stability

Consider an air parcel moving slowly in the atmosphere. It is subjected to gravity, friction, and buoyancy. We can ignore friction at low velocity, the total force exerted on the air parcel and the motion of the parcel is described using Newton's second law, with positive direction upward, as

$$F = (\rho_a - \rho_p)Vg = \rho_p V \frac{dv}{dt} \tag{11.12}$$

where ρ_p and ρ_a are the densities of the air parcel and the surrounding air, respectively; V is the volume of the air parcel and (dv/dt) is the acceleration of the air parcel. Simplification of Eq. (11.12) leads to

$$\frac{dv}{dt} = \left(\frac{\rho_a}{\rho_p} - 1\right)g. \tag{11.13}$$

Both the air parcel and the surrounding air can be assumed ideal gases, and the densities can be described using Eq. (11.6); with same atmospheric pressure P, same molar weight M and same ideal gas constant R, Eq. (11.13) becomes

$$\frac{dv}{dt} = \left(\frac{T_p}{T_a} - 1\right)g \tag{11.14}$$

This equation shows that, when $T_p > T_a$, the acceleration of the air parcel is positive, which means that it moves upward, and vice versa. When $T_p = T_a$, the air parcel acceleration will be zero. Depending on the change rate of the air parcel temperature with respect to that of the surrounding air, the air parcel may sink, move upward or remain still with respect to the surrounding air. The atmosphere is called stable, unstable and neutral atmosphere in terms of stability. They are depicted in Fig. 11.4.

- *Stable atmosphere*: Consider an air parcel in an atmosphere with the same temperature at its initial position. When the air parcel temperature elapse along elevation is greater than that of the surrounding air, the air parcel is colder than the surrounding air when it moves up or hotter while it goes down. As a result, the surrounding air exerts a total force to move the air parcel back to its original position. This total force is a result of the combination of buoyancy, friction, and gravity.
- *Unstable atmosphere*: When the air parcel temperature elapse along elevation is weaker than that of the surrounding air, the air parcel is colder than the surrounding air when it moves down and hotter when it moves up. As a result, the surrounding air exerts a total force to drive the air parcel away from its original position and convection is produced.
- *Neutral atmosphere*: When the air parcel temperature elapse along elevation is the same as that of the surrounding air, the air parcel will remain still with respect to the surrounding air. There will be no relative motion between the air parcel and the surrounding air in the atmosphere.

Atmosphere stability is affected significantly by so called temperature inversion, when atmosphere temperature increases with elevation. Temperature inversion leads to extremely stable atmosphere and sinking air emission parcel. As a result, poor air dispersion causes accumulation of pollutants at the ground level.

The stratification of air temperature in any control volume leads to the air parcel movement vertically. In the same place, atmosphere could be stable, neutral, or unstable depending on the time of the day and weather condition. For example, the ground surface and the air above it are cooled overnight. At dawn, the temperature increases with height below 300 m or so, and the atmosphere is stable; any vertical disturbances are strongly damped out.

Ground level stability is also affected by the heat transfer between the air and the Earth surface. The direction of net heat transfer depends on the temperature difference between the air and the surroundings usually from high temperature to low temperature. At the ground level, when the Earth surface temperature is higher than the nearby air, the heat transfer from the Earth surface to the air leads to unstable conditions and promotes air convection. Vice versa, air is cooled by a cooler Earth surface results in a stable atmosphere. When air and Earth surface temperatures are the same, there is no heat transfer between them and it is likely a neutral condition.

This ground-level stability changes over the hours in a day. It is a cyclic behavior, and this cycle is qualitatively depicted in Fig. 11.5 for guidance only. The rising sun in the morning heats the Earth surface and the air above it. The warmed air near the ground rises to a certain elevation until it reaches the cold air at higher elevations. Over time, this rising air gradually changes the temperature profile in the near-ground atmosphere. This heat transfer from ground level to atmosphere continues during the day until it reaches mid afternoon. At this moment, the heated air

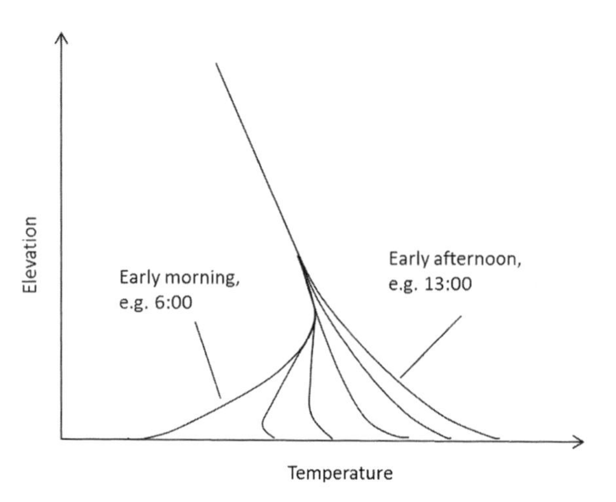

Fig. 11.5 Diurnal cycle of air temperature above ground

may extend to as high as thousands of meters, where the atmosphere above is very stable. Now the ground level starts to cool down before sunset. And another cycle begins.

When the meteorological conditions are unknown, Pasquill classes A–F can be determined from the weather conditions and the wind speed measured at 10 m above the ground, u_{10} [18]. They are shown in Table 11.1.

11.2.6 Wind Speed

Wind speed and direction affect the dispersion of air pollutants. Air pollutants are better dispersed in strong winds owing to the strong mixing effect, both horizontally and vertically. Typical wind speed at ground level is no less than 1 m/s. Air movement below this speed is referred to as calm air. Usually wind speed is lower at the ground level than that at a higher elevation. In an unstable planetary boundary layer vertical motion of air is significant, and it increases the ground level wind in the early afternoon as a result of self-limiting instability.

Let's start our analysis with a simple case, where wind is developed in an open area smooth surface in an adiabatic atmosphere. Similar to the boundary layer concept (see Chap. 2), there is also a boundary layer above the ground, which can be as high as 500 m. The friction is negligible at higher elevation. In this case, the region where ground friction plays a significant role is within the planetary boundary layer. The ground level (surface) shear stress τ_0 is defined as

$$\tau_0 = \rho_0 u_*^2 \tag{11.15}$$

where u_* is called friction speed and ρ_0 is the ground-level air density. Although the term "*friction velocity*" is widely used in air dispersion modeling, rigorously

Table 11.1 Pasquill stability classes and corresponding weather conditions

P class	A	B	C	D	E	F
Stability	Extremely unstable	Moderately unstable	Slightly unstable	Neutral	Slightly stable	Stable

u_{10} (m/s)	Day time			Night time	
	Incoming solar radiation			Thinly overcast or >4/8 cloud	Clear or cloud
	Strong	Moderate	Slight		
0–2	A	A–B	B	–	–
2–3	A–B	B	C	E	F
3–5	B	B–C	C	D	E
5–6	C	C–D	D	D	D
>6	C	D	D	D	D

speaking, it is speed, the magnitude of the velocity. In air dispersion modeling, we deal with rough surfaces more than the smooth surfaces. However, the term of friction velocity may appear in the analysis that follows.

11.2.6.1 Wind Speed Profile in Neutral Atmosphere

The wind speed profile above a rough ground surface in neutral atmosphere can be calculated using Eq. (11.16) [21],

$$\frac{u}{u_*} = \frac{1}{k}\ln\left(\frac{z}{z_0}\right) \tag{11.16}$$

where z_0 is the surface roughness height (m) and k is the Karman constant. Although there are many values for the Karman constant, the most widely used one is $k = 0.4$. [1, 15].

Surface roughness height is not a physical height but an indicator of stability: the wind speed is zero at this height. Typical values of surface roughness height are available in Table 11.2. It could be as high as tens of meters for Rocky Mountains and as low as few millimeters for ice and ocean surfaces. Over-ocean surface roughness can be estimated using the equation proposed by Hosker [14]:

$$z_0 = 2 \times 10^{-6} u_{10}^{2.5} \tag{11.17}$$

where u_{10} is the wind speed measured at 10 m heigh (m/s).

Table 11.2 Typical surface roughness heights in urban and rural areas

Terrain	Description	z_0 (m)	Source
Urban	Roughly open with occasional obstacles	0.1	[5]
	Rough area with scatter obstacles	0.25	
	Very rough areas with low buildings or industrial tanks as obstacles	0.5	
	Skimming areas with buildings of similar height	1	
	Chaotic city center with buildings of different heights	2	
Rural	Agricultural land	0.25	[17]
	Range land	0.05	
	Forrest land, wet land, forest wet land	1	
	Water body	0.001	
	Perennial snow or ice	0.20	
Mountains	Rocky mountains	50–70	[9]
	Mountains	5–70	
Ocean		Eq. (11–17)	[14]

In order to use Eq. (11.16), the wind speed at one elevation has to be known. This pair of data, given notations of (u', z'), allows us to determine the friction speed using Eq. (11.18).

$$u_* = \frac{ku'}{\ln(z'/z_0)} \tag{11.18}$$

Example 11.2: Wind speed profile
In a rural area, the friction height is $z_0 = 0.25$ m, and the wind speed measured at 10 m height is 4 m/s under neutral condition. Plot the vertical wind speed profile.

Solution
Equation (11.18) gives

$$\frac{u_*}{k} = \frac{u_{10}}{\ln(z_{10}/z_0)} = \frac{4}{\ln(10/0.25)} = 1.084.$$

Then we have the velocity as a function of elevation:

$$u = \frac{u_*}{k} \ln\left(\frac{z}{z_0}\right) = 1.084 \ln(4z)$$

The plot is shown in Fig. 11.6. This profile is similar to what we saw in boundary layer analysis in Chap. 2. With the decrease in speed change rate along increasing elevations, the friction effect becomes negligible at high elevation.

11.2.6.2 Wind Speed Profile in Stable Atmosphere

For non-neutral conditions, the wind speed depends strongly on the stability of the atmosphere, which in turn depends on the heat transfer q between the atmosphere and the ground. We have to put forward a new but important parameter, Obukhov Length, after the Russian scientist A.M. Obukhov. He set the foundation of modern micrometeorology by introducing a universal length scale for exchange processes in the surface layer in 1946 [9].

$$L = -\left(\frac{\rho_0 c_p T_0}{g}\right) \frac{u_*^3}{qk} \tag{11.19}$$

Like surface roughness height, Obukhov length is not a physical length either. It is related to the stability indicator at different elevations. Researchers in the area of air dispersion modeling have developed a variety of equations for the calculation of Obukhov Length, however, the simple yet practical equation given by Seinfeld and Pandis [17] is widely used in air dispersion models.

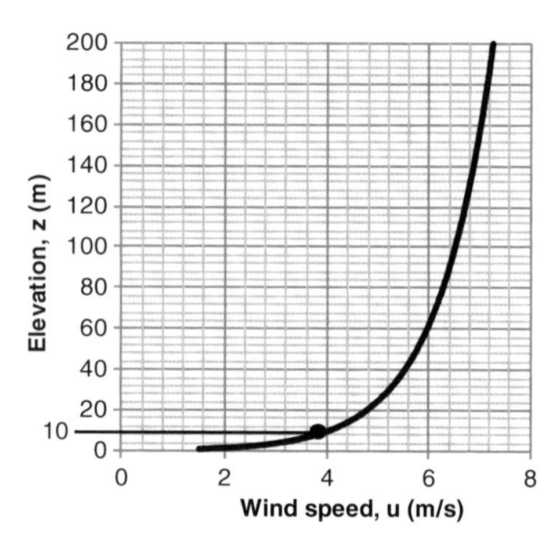

Fig. 11.6 Calculated wind speed profile under neutral condition

Table 11.3 Coefficients a and b for different Pasquill stability classes

P class	A	B	C	D	E	F
Stability	Extremely unstable	Moderately unstable	Slightly unstable	Neutral	Slightly stable	Stable
a	−0.096	−0.037	−0.002	0	0.004	0.035
b	0.029		0.018	0	−0.018	−0.036

$$\frac{1}{L} = a + b \times \log_{10}(z_0) \tag{11.20}$$

where the coefficients a and b are listed in Table 11.3.

With the availability of Obukhov Length, we can calculate the stability indicator

$$s_z = \frac{z}{L} = z[a + b \times \log_{10}(z_0)] \tag{11.21}$$

A positive value of s_z corresponds to stable atmosphere and a negative s_z means unstable condition. With the Obukhov length determined using Eq. (11.20), the wind speed under stable and unstable conditions can be determined now as follows. For stable conditions ($s_z > 0$),

$$\frac{u}{u_*} = \frac{1}{k}\ln\left(\frac{z}{z_0}\right) + \frac{5}{k}\ln\left(\frac{z - z_0}{L}\right) \tag{11.22}$$

Readers can find a more complex yet accurate approach in literature (e.g. [8]).

11.2.6.3 Wind Speed Profile in Unstable Atmosphere

The procedure for the calculation of wind speed profile under unstable atmosphere is more complex; one widely used approximation is the one given by Benoit [2]

$$\frac{u}{u_*} = \frac{1}{k}\ln\left(\frac{z}{z_0}\right) + \underbrace{\frac{1}{k}\left\{\ln\left[\frac{(\beta_0^2+1)(\beta_0+1)^2}{(\beta^2+1)(\beta+1)^2}\right] + 2[\arctan(\beta) - \arctan(\beta_0)]\right\}}. \quad (11.23)$$

With $\beta_0 = (1 - 15z_0/L)^{0.25}$ and $\beta = (1 - 15z/L)^{0.25}$. The symbol arctan in the last term is the arctangent function with a unit of radian.

11.2.6.4 Windrose

In reality, both speed and direction of the wind change over time in a region. Wind roses provided by meteorological services can be used to take this variation into consideration. It summarizes the incoming direction, speed and frequency of wind at certain location. Note that the direction marked on the wind rose is the direction from which wind blows.

11.3 Gaussian-Plume Dispersion Models

Gaussian air dispersion models are the most widely used for estimating the impact of nonreactive air pollutants. A Gaussian-plume model can be used to predict the downwind concentration resulting from the point source under a specific atmospheric condition. It is a material balance model for a point source such as a power plant stack. Admittedly it is a source emission from a small area, but this area is small enough to be considered as a point comparing to the atmospheric environment of concern.

Gaussian model is a statistical model that shows the Gaussian distribution of pollutant concentration over a period of time, 15 min or longer. It does not predict the concentration in a plume at any instant, but rather the statistical distribution of the pollutant concentration about the plume center line, which is a Gaussian distribution. As illustrated in Fig. 11.7, the instant plume appears like the shaded area, but the time-averaged concentration may be different from what it appears to be at that instant.

Now let's derive the equation to show that the distribution is indeed a Gaussian distribution. Referring to Fig. 11.7, consider a stack with an effective height of H and the plume rise is ($\Delta h = H - h$). The plume is subjected to a cross wind with a speed of u. Set the origin of the coordinate system at the base of the stack, with the x axis aligned in the downwind direction. The plume rises from the stack and then travel in the x direction. Meanwhile it disperses along y and z directions.y direction is not shown but pointed into the paper. To simplify the problem, the velocity u is for now assumed to be independent of time, location, or elevation. Assume steady state condition in the plume and the source emission rate is a constant $\dot{m}$ (kg/s).

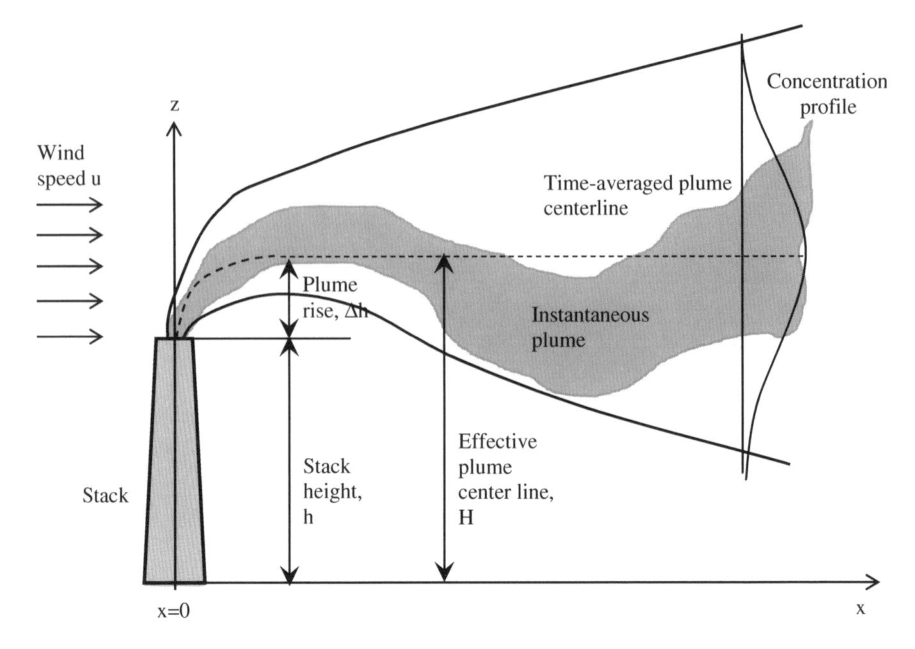

Fig. 11.7 Schematic representation of Gaussian plume dispersion

11.3.1 General Gaussian Dispersion Model

Consider a cubic control volume $\Delta x \Delta y \Delta z$ along the center of the plume ($z = H$). Due to the complex nature of air dispersion by turbulent mixing, we may approximate the flux of air pollutant being mixed across any surface by

$$j = -D\frac{\partial C}{\partial n} \tag{11.24}$$

where $j =$ flux of mass flow per unit area (kg/s·m^2), C = concentration (kg/m^3), n = distance in the direction considered (normally x, y, or z) (m), D = turbulent dispersion coefficient (m^2/s).

The dimension of D is the same as that for molecular diffusivity or thermal diffusivity. However, this does not mean that air dispersion is a result of thermal diffusion or molecular diffusion. Therefore, the turbulent dispersion coefficient, D, is also referred to as the eddy diffusivity. And, they are not necessarily the same for x, y and z directions.

Then the net mass flow rate along x direction is described as the difference through surfaces x and $x + \Delta x$

$$\Delta J_x = (C_x - C_{x+\Delta x})u\Delta y\Delta z + \left[\left(-D_x\frac{\partial C}{\partial x}\right)_x - \left(-D_x\frac{\partial C}{\partial x}\right)_{x+\Delta x}\right]\Delta y\Delta z \quad (11.25x)$$

ΔJ_x has a unit of kg/s.

The first term on RHS stands for the effect of wind and the second term for eddy diffusion effect. Similar equations apply to y and z directions without wind effect, and we have

$$\Delta J_y = \left[\left(-D_y\frac{\partial C}{\partial y}\right)_y - \left(-D_y\frac{\partial C}{\partial y}\right)_{y+\Delta y}\right]\Delta x\Delta z. \quad (11.25y)$$

$$\Delta J_z = \left[\left(-D_z\frac{\partial C}{\partial z}\right)_z - \left(-D_z\frac{\partial C}{\partial z}\right)_{z+\Delta z}\right]\Delta y\Delta x. \quad (11.25z)$$

The mass balance leads to increment of pollutant in the cubic volume over a small period of time as

$$\Delta C(\Delta x\Delta y\Delta z) = \left(\Delta J_x + \Delta J_y + \Delta J_x\right)\Delta t \quad (11.26)$$

Substituting Eqs. (11.25x), (11.25y), (11.25z) into (11.26) leads to

$$\frac{\Delta C}{\Delta t} = \frac{(C_x - C_{x+\Delta x})u + \left[\left(-D_x\frac{\partial C}{\partial x}\right)_x - \left(-D_z\frac{\partial C}{\partial z}\right)_{x+\Delta x}\right]}{\Delta x} + \frac{\left[\left(-D_y\frac{\partial C}{\partial y}\right)_y - \left(-D_y\frac{\partial C}{\partial y}\right)_{y+\Delta y}\right]}{\Delta y} + \frac{\left[\left(-D_z\frac{\partial C}{\partial z}\right)_z - \left(-D_z\frac{\partial C}{\partial z}\right)_{z+\Delta z}\right]}{\Delta z} \quad (11.27)$$

Taking the limit of an infinitesimally small cube and time interval, Eq. (11.27) becomes

$$\frac{\partial C}{\partial t} = -u\frac{\partial C}{\partial x} + D_x\frac{\partial^2 C}{\partial x^2} + D_y\frac{\partial^2 C}{\partial y^2} + D_z\frac{\partial^2 C}{\partial z^2} \quad (11.28)$$

Since D_i is not necessarily the same for all the three directions, this equation contains three D's as D_x, D_y and D_z for x, y, and z directions, respectively. Consequently, the Gaussian plume equation above can be applied to one, two- or three-dimensional analyses, whereas 1-D analysis is of less meaningful application, and only 2-D and 3-D models will be introduced as follows.

If we assume, which is the cases we deal with most of the time,

- steady state ($\frac{\partial C}{\partial t} = 0$)
- x-direction transport by wind is much greater than that by eddy diffusion ($u\frac{\partial C}{\partial x} \gg D_x\frac{\partial^2 C}{\partial x^2}$)

Then Eq. (11.28) is further simplified as

$$u\frac{\partial C}{\partial x} = D_y\frac{\partial^2 C}{\partial y^2} + D_z\frac{\partial^2 C}{\partial z^2} \tag{11.29}$$

Integrating Eq. (11.29) with the following boundary conditions

$$\begin{aligned}
&C = 0 \text{ as } x, y, z \to \infty \\
&C \to \infty \text{ at } x, y, z \to 0 \\
&D_z\frac{\partial C}{\partial z} = 0 \text{ at } z \to 0 \text{ and } x, y > 0 \quad \text{(wall boundary)} \\
&\int_{-\infty}^{\infty}\int_{0}^{\infty} uC(y, z)dzdy = \dot{m} \text{ at } x > 0 \quad \text{(conservation of mass)}
\end{aligned}$$

we can get the air pollutant concentration at any point (x, y, z) as

$$C(x, y, z) = \frac{\dot{m}}{2\pi x\sqrt{D_y D_z}}\exp\left[\left(-\frac{u}{4x}\right)\left(\frac{y^2}{D_y} + \frac{z^2}{D_z}\right)\right] \tag{11.30}$$

If we define

$$\sigma_y^2 = \frac{2xD_y}{u} \text{ and } \sigma_z^2 = \frac{2xD_z}{u} \tag{11.31}$$

where σ_y and σ_z are the dispersion coefficients in the transverse (y) and vertical (z) direction, respectively.

Equation (11.30) can be rearranged as

$$C(x, y, z) = \frac{\dot{m}}{2\pi\sigma_y\sigma_z}\frac{1}{u}\exp\left[-\frac{y^2}{2\sigma_y^2} - \frac{z^2}{2\sigma_z^2}\right] \tag{11.32}$$

In the preceding analysis, we have assumed the source of emission is at the origin of the coordinate (z = 0). In reality, the actual emission source is at z = H, therefore Eq. (11.32) shall be corrected as

$$C(x, y, z) = \frac{\dot{m}}{2\pi\sigma_y\sigma_z}\frac{1}{u}\exp\left[-\frac{y^2}{2\sigma_y^2} - \frac{(z-H)^2}{2\sigma_z^2}\right] \tag{11.33}$$

Table 11.4 Briggs parameterization for the dispersion coefficients

Stability class	Open/Rural sites		Urban/Industrial sites	
	$\sigma_y(m)$	$\sigma_z(m)$	$\sigma_y(m)$	$\sigma_z(m)$
A	$\frac{0.22x}{\sqrt{1+0.0001x}}$	$0.20x$	$\frac{0.32x}{\sqrt{1+0.0004x}}$	$0.24x\sqrt{1+0.001x}$
B	$\frac{0.16x}{\sqrt{1+0.0001x}}$	$0.12x$		
C	$\frac{0.11x}{\sqrt{1+0.0001x}}$	$\frac{0.08x}{\sqrt{1+0.0002x}}$	$\frac{0.22x}{\sqrt{1+0.0004x}}$	$0.20x$
D	$\frac{0.08x}{\sqrt{1+0.0001x}}$	$\frac{0.06x}{\sqrt{1+0.0015x}}$	$\frac{0.16x}{\sqrt{1+0.0004x}}$	$\frac{0.14x}{\sqrt{1+0.0003x}}$
E	$\frac{0.06x}{\sqrt{1+0.0001x}}$	$\frac{0.03x}{1+0.0003x}$	$\frac{0.11x}{\sqrt{1+0.0004x}}$	$\frac{0.08x}{\sqrt{1+0.0015x}}$
F	$\frac{0.04x}{\sqrt{1+0.0001x}}$	$\frac{0.016x}{1+0.0003x}$		

Note that the dispersion coefficients are different at different distances from the source. For the ease of programming, Briggs' parameterization [5] for different Pasquill stability classes is widely used in air dispersion modeling. They are summarized in Table 11.4, where units of both x and σ are meter.

These equations are best applicable to $x < 10{,}000$ m and become unrealiable for longer distances. They are not supposed to be used for distances greater than 30,000 m. The corresponding roughness lengths (z_0) are 3 cm and 1 m for rural and urban sites, respectively [11]. These equations also show that Gaussian dispersion coefficients along horizontal and vertical directions are not constants, and that they vary at the distances downwind of a stack as a function of atmospheric stability.

Continue from Eq. (11.33), the air pollutant concentration at the plume centerline can be determined by substituting $y = 0$ and $z = H$, into Eq. (11.33)

$$C(x, y = 0, z = H) = \frac{\dot{m}}{2\pi\sigma_y\sigma_z u} \tag{11.34}$$

Example 11.3: Gaussian plume model

In a bright sunny day, the wind speed is assumed to be 6 m/s and horizontal. A power plant in a rural area with a stack of 100 m high continuously discharges SO_2 into the atmosphere at a stable rate of 0.1 kg/s. The plume rise is 20 m. Ignoring the chemical reactions in the atmosphere,

(a) estimate the SO_2 concentration at the center of the plume 5 km downwind from the stack.
(b) estimate the ground level SO_2 concentration 5 km downwind
(c) plot the ground level concentration right under the plume along wind direction from $x = 2{,}000$ m to $x = 6{,}000$ m

Solution

First, determine the atmosphere stability using the Pasquill stability class described in Table 11.1. It can be determined as class C stability. Note that this is still an approximation because we are considering the wind speed to be uniform, which may not be not the case in reality.

Anyways, with the class C stability the corresponding air dispersion coefficients for rural area can be calculated using the equations in Table 11.4 as follows.

$$\sigma_y = \frac{0.11x}{\sqrt{1+0.0001x}} = \frac{0.11 \times 5000}{\sqrt{1+0.0001 \times 5000}} = 449.1\,\text{m}$$
$$\sigma_z = \frac{0.08x}{\sqrt{1+0.0002x}} = \frac{0.08 \times 5000}{\sqrt{1+0.0002 \times 5000}} = 282.8\,\text{m}$$

(a) The concentration at the plume center 5 km downwind can be determined using Eq. (11.33) with z-H = 0 and y = 0

$$C(x, 0, H) = \frac{\dot{m}}{2\pi\sigma_y\sigma_z u}$$
$$= \frac{0.1}{2\pi \times 449.1 \times 282.8 \times 6} = 2.09 \times 10^{-8}(\text{kg/m}^3) = 20.9(\mu\text{g/m}^3)$$

(b) The concentration on the ground right below the plume center at x = 5000 m downwind is calculated using Eq. (11.33) with $z = 0$, $y = 0$ and $H = 120$:

$$C(x, y, z) = \frac{\dot{m}}{2\pi\sigma_y\sigma_z u}\exp\left[-\frac{H^2}{2\sigma_z^2}\right]$$
$$= 20.9 \times \exp\left[-\frac{1}{2}\left(\frac{120}{282.8}\right)^2\right] = 20.9 \times 0.914 = 19.1(\mu\text{g/m}^3)$$

(c) The concentration at the ground right below the plume center at 2,000–6,000 m downwind is calculated using Eq. (11.33) with $z = 0$, $y = 0$ and $H = 120$:

$$C(x, y, z) = \frac{\dot{m}}{2\pi\sigma_y\sigma_z u}\exp\left[-\frac{H^2}{2\sigma_z^2}\right]$$

where

$$\sigma_y = \frac{0.11x}{\sqrt{1+0.0001x}}$$
$$\sigma_z = \frac{0.08x}{\sqrt{1+0.0002x}}$$

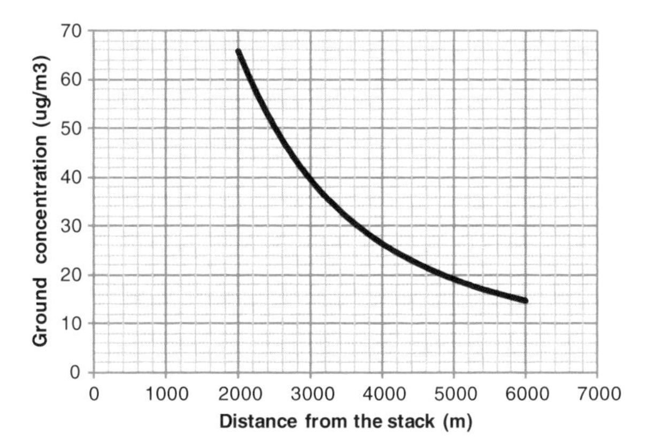

Fig. 11.8 Calculated ground level SO_2 concentration based on Gaussian dispersion model

The plot is shown in Fig. 11.8.

The aforementioned analyses are applicable to simple cases where the following factors are not considered.

- Varialbe plume rise
- Variable wind
- Mixing height
- Unstable release from source, i.e., puff effect

Improved models that take one or more of these factors into consideration are introduced as follows.

11.3.2 Plume Rise

In addition to the effect of the meteorology on the plume dispersion itself, the plume rise of a plume also depends on the meteorological parameters. As seen in Fig. 11.7, the plume rises gradually and the centerline reaches its highest value eventually. Several equations have been developed for plume rise, and the most widely used ones are, again by Briggs [4] as follows.

Plume rise is a result of buoyancy and momentum. They are charaterized with the following two parameters called buoyancy flux (F_B) and momentum flux(F_M), respectively.

$$F_B = \left(1 - \frac{\rho_s}{\rho_a}\right)\frac{d_s^2}{4} g v_s \tag{11.35}$$

$$F_M = \left(\frac{\rho_s}{\rho_a}\right)\frac{d_s^2}{4} v_s^2 \tag{11.36}$$

where the subscript s stands for stack, ρ_s and ρ_a are the densities of the stack emission gas and the surrounding air, respectively. $g = 9.81$ m/s^2 is the gravitational acceleration; v_s is the vertical discharge speed of the emission gas from the stack (m/s), which is assumed along $+z$ direction. d_s is the inner diameter of the stack (m). The units of F_B and F_M are m^4/s^3 and m^4/s^2, respectively.

A practical parameter is the flue gas temperature instead of the air density. Both the stack emission gas and the surrounding air can be considered as ideal gases, and both are under atmospheric pressure. From the relationship between density and temperature described in Eq. (11.6), we have

$$\frac{\rho_s}{\rho_a} = \frac{M_s}{M_a}\frac{T_a}{T_s} \approx \frac{T_a}{T_s} \tag{11.37}$$

Despite the difference in molar weights of stack emission gas and the surrounding air in the atmosphere, the difference of molar weights is much less than that of temperature. Therefore, we can simplify the density ratio by ignoring the molar weight ratio. In such a case, Eqs. (11.35) and (11.36) become

$$F_B = \left(1 - \frac{T_a}{T_s}\right)\frac{d_s^2}{4} g v_s \tag{11.38}$$

$$F_M = \left(\frac{T_a}{T_s}\right)\frac{d_s^2}{4} v_s^2 \tag{11.39}$$

When both buoyancy and momentum determine the plume rise, the transitional plume rise is described as

$$\Delta h = \left(\frac{25}{3}\frac{F_M}{u^2}x + \frac{25}{6}\frac{F_B}{u^3}x^2\right)^{1/3} \tag{11.40}$$

where u is the average wind speed at the stack height (m/s), x is the downwind distance away from the stack (m).

When one is dominating over another, the equation can be further simplified. When the plume temperature is much greater than that of the surrounding atmosphere temperature, the plume is mostly buoyancy-dominant, especially those from a power plant because the emission stream is hotter than the ambient air ($T_s > T_a$). For a buoyancy dominating plume, the transitional plume rise is

$$\Delta h = \left(\frac{25}{6}\frac{F_B x^2}{u^3}\right)^{1/3} \tag{11.41}$$

In reality, the plume rise stops at certain height, and the maximum plume rise is achieved at a critical distance of x_c. The critical distance can be estimated using

$$x_c = \begin{cases} 49F_B^{5/8} & \text{for} \quad F_B < 55\,\text{m}^4/\text{s}^3 \\ 119F_B^{2/5} & \text{for} \quad F_B > 55\,\text{m}^4/\text{s}^3 \end{cases} \tag{11.42}$$

And the corresponding maximum plume rise is

$$\Delta h_m \cong \begin{cases} 21.4\frac{F_B^{3/4}}{u} & \text{for} \quad F_B < 55\,\text{m}^4/\text{s}^3 \\ 38.7\frac{F_B^{3/5}}{u} & \text{for} \quad F_B > 55\,\text{m}^4/\text{s}^3 \end{cases} \tag{11.43}$$

In **Example 11.3**, we used the maximum plume rise for calculation. The actual ground-level concentration can now be predicted with improved accuracy if we consider the local plume rise.

Example 11.4: Plume rise

Consider a power plant stack with a diameter of $d_s = 1.2$ m and the stack emission gas is discharged at the speed of $v_s = 5\,\text{m/s}$. Assume horizontal wind speed $u = 1.1$ m/s, and surrounding air temperature is $T_a = 300\,\text{K}$. Plot the plume rise downwind the emission source for discharge temperature of $T_s = 500\,\text{K}$.

Solution

Since the from a power plant is buoyancy-dominant plume, we only consider the buoyancy flux

$$F_B = \left(1 - \frac{T_a}{T_s}\right)\frac{d_s^2}{4} g v_s = \left(1 - \frac{300}{500}\right)\left(\frac{1.2}{2}\right)^2 9.81 \times 1.1 = 7.063\,\text{m}^4/\text{s}^3$$

Since $F_B < 55\,\text{m}^4/\text{s}^3$ the corresponding maximum plume rise is calculated using

$$\Delta h_m = \frac{21.4}{u} F_B^{3/4} = 84.3\,\text{m}$$

For $T_e = 500\,\text{K}$, the transitional plume rise can then be determined using

$$\begin{aligned} \Delta h &= \left(\frac{25}{6}\frac{F_B}{u^3}x^2\right)^{\frac{1}{3}} = \left(\frac{25}{6} \times \frac{7.063}{1.1^3}x^2\right)^{\frac{1}{3}} \\ &= \left(22.11x^2\right)^{1/3} \end{aligned}$$

Figure 11.9 is produced using $\Delta h = (22.11x^2)^{1/3}$ with a cap of Δh_m.

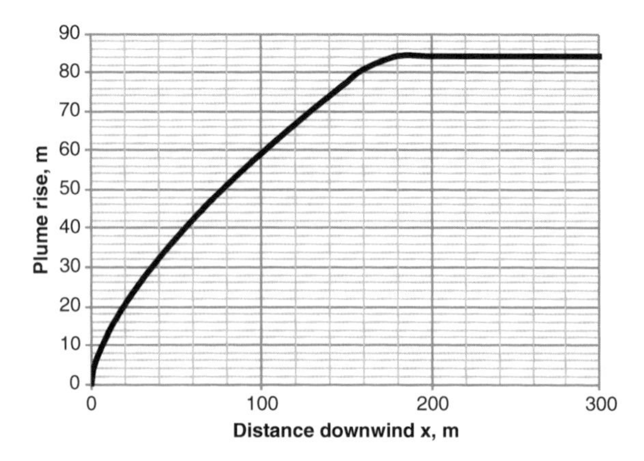

Fig. 11.9 Plume rise versus distance

11.3.3 Plume Downwash

Opposite to plume rise, a plume may drop due to the interaction between the plume and the atmosphere near the stack. This is called plume downwash. Plume downwash may result in an increase of ground-level air pollutant concentrations, because of the lower final plume height and decreased buoyancy in case of buoyant emissions. Canepa [7] gave a comprehensive overview about the studies of downwash effects in air dispersion.

As shown in Fig. 11.10, a stack downwash is a result of the wake downwind of an emitting stack due to the stack itself. Stack downwash is not a big problem for tall and large utility and industrial stacks, but it is important for short stacks because of the low wind speed at lower elevation.

There are many models developed over the past 70 years for plume downwash, and the basics are introduced as follows. Generally speaking, stack downwash

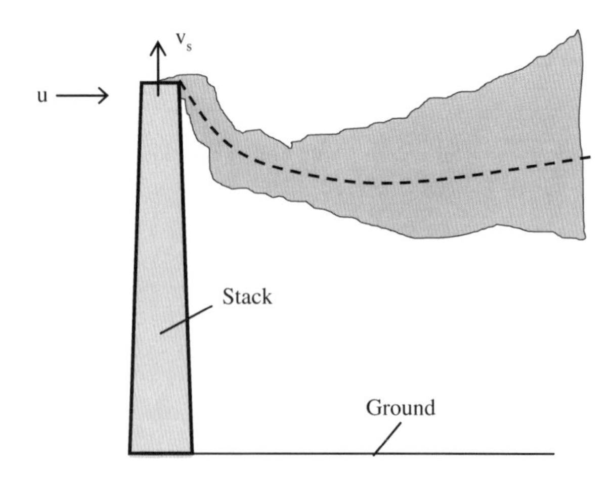

Fig. 11.10 Schematic representation of stack downwash

effect becomes important when the exit gas speed is less than 1.5 times the wind speed, $v_s < 1.5\,u$. However, low effluent speed does not necessarily cause stack downwash. Bjorklund and Bowers [6] proposed the following procedure to calculate the final plume rise of a buoyant plume with stack downwash.

$$\Delta h'_m = f \Delta h_m \tag{11.44}$$

where Δh_m is the final plume rise without stack downwash effect determined using Eq. (11.40). f is the correction factor to the plume rise due to stack downwash. The correction factor depends on the Froude number (Fr) of the stack emission gas and the square of F_r is

$$Fr^2 = \frac{v_s^2}{gd_s} \frac{T_a}{(T_s - T_a)} \tag{11.45}$$

where T_a is the temperature of ambient air surrounding the top of the stack. The correction factor can be determined using Eq. (11.46).

$$f \cong \begin{cases} 1 & \text{for} \quad v_s > 1.5\,u \text{ OR } Fr^2 < 3 \\ 3\left(1 - \frac{u}{v_s}\right) & \text{for} \quad u < v_s \leq 1.5\,u \text{ AND } Fr^2 \geq 3 \\ 0 & \text{for} \quad v_s \leq u \text{ AND } Fr^2 \geq 3 \end{cases} \tag{11.46}$$

11.3.3.1 Building Downwash

A building downwash occurs when the plume is near a building and is brought downward by the flow of air over and around the building. To understand the building downwash and related plume drop, we have to understand some basic fluid dynamics. As illustrated in Fig. 11.11, consider a building block attacked by a horizontal air flow, there are aerodynamic cavity zones produced around the building: one is the separation zone on the roof, and another cavity zone behind the building. Sometimes they may merge into one large cavity covering both roof and downwind the building, depending on the wind speed and the surrounding

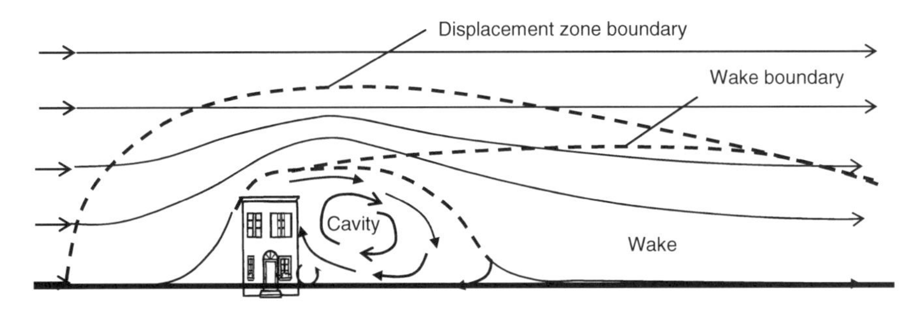

Fig. 11.11 Schematic representation of building downwash

environment. A vortex at the front side of the building does not affect the plume as much as the roof-top and rear-side separation zones.

Mathematically, empirical equations are given by Hanna et al. [12] for predicting the wake effect on the plume entrance. As a widely accepted safe engineering practice, it is simple to follow the rule of thumb that stack height (h) is 2.5 times the height of the heightest building nearby.

$$h > 2.5H_B \tag{11.47}$$

where H_B is the height of the building.

11.3.4 Ground Surface Reflection

We did not consider the boundaries in the analysis above. Most of the time, we are interested in concentrations at ground level for the protection of human beings and their properties. Although mathematically we could continue the calculation for $z<0$, it is physically wrong because air pollutants cannot enter underground by eddy dispersion. In this case, the ground acts as a wall in the computational domain and we have to consider the wall effect.

If we ignore the deposition at the ground surface, it is commonly assumed this wall to be reflective like a mirror. As illustrated in Fig. 11.12, the reflects at the surface as if there was an underground mirrored source. Mathematically, the contribution of the mirror source is calculated using Eq. (11.33) by replacing $(z-H)^2$ with $(z+H)^2$ (Fig. 11.12).

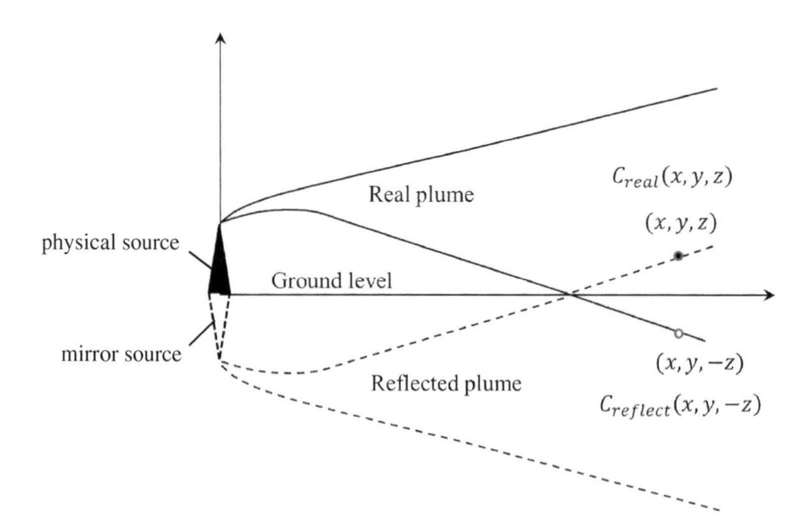

Fig. 11.12 Plume reflection on the ground surface

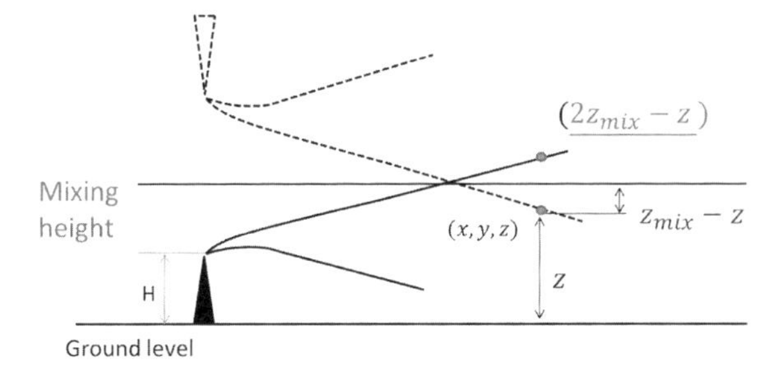

Fig. 11.13 Plume reflection on the mixing height

Combining both the real and the mirrored plumes, Eq. (11.33) becomes

$$C(x,y,z) = \frac{\dot{m}}{2\pi\sigma_y\sigma_z u}\exp\left[-\frac{y^2}{2\sigma_y^2}\right]\left\{\exp\left[-\frac{(z-H)^2}{2\sigma_z^2}\right] + \exp\left[-\frac{(z+H)^2}{2\sigma_z^2}\right]\right\}. \tag{11.48}$$

Again, if we are interested in ground-level concentrations and we can substitute $z = 0$ into this equation, then we have

$$C(x,y,0) = \frac{\dot{m}}{\pi\sigma_y\sigma_z u}\exp\left[-\frac{y^2}{2\sigma_y^2}\right]\exp\left[-\frac{H^2}{2\sigma_z^2}\right]. \tag{11.49}$$

Similarly we can estimate the ground-level concentration under the centerline of the plume with $y = z = 0$. Then the equation is further simplified as

$$C(x,0,0) = \frac{\dot{m}}{\pi\sigma_y\sigma_z}\frac{1}{u}\exp\left[-\frac{H^2}{2\sigma_z^2}\right]. \tag{11.50}$$

11.3.5 Mixing Height Reflection

Mixing height is another important factor that affects air dispersion; it sets the upper boundary limit to the dispersion of air pollutants. Air pollutants released at ground level will be mixed up to the mixing height, but not above it because of the extremely stable atmosphere above the mixing height. There is no upward air motion above the mixing height. The troposphere-stratosphere boundary in atmosphere is a typical natural mixing-height as a result of temperature inversion. It varies with location and time of the year.

Typical values of mixing heights are in the order of 100–1,000 m. For example, Table 11.5 shows typical values of the mixing height for contiguous United States [13].

Table 11.5 Typical mixing heights for the contiguous United States

Time	Mixing height (m)		
	Minimum	Maximum	Average
Summer morning	200	1,100	450
Summer afternoon	600	4,000	2,100
Winter morning	200	900	470
Winter afternoon	600	1,400	970

Local mixing heights can be measured using special devices, although they are not done as frequently as needed. Therefore, empirical equations are proposed for air dispersion modeling purpose as follows.

For neutral atmosphere, the mixing height can be estimated using Eq. (11.51)

$$z_{\text{mix}} = C_0 \frac{u_*}{2\Omega \sin\Phi} \quad \text{(Neutral atmosphere)} \tag{11.51}$$

where C_0 is a coefficient that varies from 0.2 to 0.4 [10]; u_* is the friction speed. The term $(2\Omega \cdot \sin\Phi)$ in the denominator stands for the Coriolis force because of the rotation of the Earth. $\Omega = 7.27 \times 10^{-5}\text{rad/s}$ [18] is the angular speed of the Earth and Φ is the latitude where the air is of concern.

There are a few options for non-neutral atmosphere over the time, one simple yet practical empirical equation was proposed by Venkatram [20] for stable conditions

$$\mathrm{z_{mix}} = \mathrm{C_s u_*^{1.5}} \quad \text{(Stable atmosphere)} \tag{11.52}$$

where $\mathrm{C_s} = 2{,}400\ \mathrm{m^{0.5}s^{1.5}}$ with $\mathrm{u_*}$ in m/s and $\mathrm{z_{mix}}$ in m. The calculation of the mixing height for unstable conditions can be calculated using the following equation [22].

$$z_{\text{mix}} = C_u \frac{u_*^{1.5}}{\sqrt{L(2\Omega \sin\Phi)^3}} \quad \text{(Unstable atmosphere)} \tag{11.53}$$

This equation shows that $z_{\text{mix}} \propto u_*^{1.5}$ under unstable conditions. The trend agrees with that for stable condition described in Eq. (11.52). However, the coefficient C_u requires the knowledge of heat transfer q from the ground to the air. The analysis is very complex and readers are referred to state-of-the-art literature for in-depth analysis.

As the plume moves downwind, it eventually spreads wide enough to reach the mixing height z_{mix}, which is the upper limit of the computation domain. Then the air pollutant will no longer spread vertically, but transport horizontally only. However, at a location that is close to the mixing height, we can consider it as a refection wall (Fig. 11.13). And the actual air pollutant concentrations along the mixing height should be higher than one would get by using Eq. (11.33).

To take both mixing height and the ground surface effects into consideration, the term for vertical dispersion is further refined in the Gaussian dispersion model and it leads to

$$C(x,y,z) = \frac{\dot{m}}{2\pi\sigma_y\sigma_z u}\exp\left[-\frac{y^2}{2\sigma_y^2}\right] \underline{\sum_{j=-\infty}^{j=\infty}\left\{\exp\left[-\frac{(z-H+2jz_{\text{mix}})^2}{2\sigma_z^2}\right]+\exp\left[-\frac{(z+H+jz_{\text{mix}})^2}{2\sigma_z^2}\right]\right\}} \tag{11.54}$$

where z_{mix} = mixing height; in practice, Eq. (11.54) can be limited to $j = -1, 0, +1$ for values from σ_z to z_{mix}. When $\sigma_z > z_{\text{mix}}$, the plume reflects between the mixing height and the ground surface. After a few reflections, the air can be considered as completely mixed along vertical direction. Then we can estimate the perfectly mixed concentration by

$$C(x,y) = \frac{\dot{m}}{\sqrt{2\pi}\sigma_y z_{\text{mix}} u}\exp\left[-\frac{y^2}{2\sigma_y^2}\right] \quad \text{for } \sigma_z > z_{\text{mix}}. \tag{11.55}$$

11.4 Gaussian Puff Models

We have assumed in the Gaussian models that the wind speed is constant (or used average wind speed) and the emission source is continuous and steady. When the wind is variable or the emissions source is not steady, we have to employ another type of model for air dispersion analysis. It is called Gaussian Puff Model. Unlike the plume models, puff models work for low wind speed conditions. They also can handle the change of wind directions over time by coupling with wind rose.

The Gaussian puff models can be derived in a Lagrangian frame reference, a frame that is attached to the center for the puff. Consider a cubic control volume along the center of the plume ($z = H$). Equation (11.25x) becomes

$$\Delta J_x = \left[\left(-D_x\frac{\partial C}{\partial x}\right)_x - \left(-D_x\frac{\partial C}{\partial x}\right)_{x+\Delta x}\right]\Delta y\Delta z \tag{11.56}$$

Equations (11.25y) and (11.25z) remain the same. With Eqs. (11.25y), (11.25z), and (11.56), Eq. (11.28) becomes

$$\frac{\partial C}{\partial t} = D_x\frac{\partial^2 C}{\partial x^2} + D_y\frac{\partial^2 C}{\partial y^2} + D_z\frac{\partial^2 C}{\partial z^2} \tag{11.57}$$

Consider an instantaneous short-term release of air pollutant from a stack, where the mass of air pollutant released is m(kg). Integrating Eq. (11.57) with the boundary conditions

$$C \to 0 \text{ as t} \to \infty; x, y, z \geq 0$$
$$C \to 0 \text{ as t} \to 0; x, y, z > 0$$
$$\int_0^{\infty}\int_{-\infty}^{\infty}\int_{-\infty}^{\infty} Cdxdydz = m \quad \text{(conservation of mass)}$$

leads to

$$C(x,y,z,t) = \frac{m}{(4\pi t)^{\frac{3}{2}}(D_x D_y D_z)^{\frac{1}{2}}} \exp\left[-\frac{1}{4t}\left(\frac{x^2}{D_x}+\frac{y^2}{D_y}+\frac{z^2}{D_z}\right)\right] \tag{11.58}$$

Similar to Eq. (11.31), we define

$$\sigma_i^2 = 2D_i t \qquad i = x, y, z \tag{11.59}$$

and Eq. (11.58) can be rewritten in another form as

$$C = \frac{m}{(2\pi)^{3/2}\sigma_x\sigma_y\sigma_z}\exp\left[-\frac{x^2}{2\sigma_x^2}\right]\exp\left[-\frac{y^2}{2\sigma_y^2}\right]\exp\left[-\frac{(z-H)^2}{2\sigma_z^2}\right]. \tag{11.60}$$

When not available, the x-direction dispersion coefficient can be approximate using $\sigma_x \approx \sigma_y$ because they both are for horizontal directions.

The Gaussian puff model is useful in safety analysis of accidental release of air pollutants and other chemicals rather than a continuous release of air pollutants. Readers are referred to the literature for in-depth understanding of these topics.

Corresponding computer programs have been developed for different models and they are widely available at government agencies and consulting firms, case by case. However, users of any air dispersion models must be advised that they are for estimates with differences from actual observations as a result of inversion aloft, short-term fluctuations, inversion breakup fumigation, etc. Advanced dispersion models aiming at these additional topics are available in literature and readers are suggested to explore them as needed.

11.5 Practice Problems

1. An air parcel temperature is 300 K and the surrounding atmosphere temperature is 280 K, what is the acceleration of this air parcel at this location? Assume air pressure p = 1 atm.

2. On a clear day at night, the wind speed measured at 10 m above the ground is 4 m/s, what is the stability class of the atmosphere? And calculate the wind speed at 100 m high.
3. In a city center with different buildings, the wind speed measured at 10 m height is 4 m/s under neutral condition. What is the wind speed at 50 m high?
4. The wind speed at 10 m high under neutral condition is 5.5 m/s, estimate

   (a) the wind speed at the stack height of 200 m?
   (b) the mixing height with $C_0 = 0.3$

5. Consider a power plant stack with a diameter of $d_s = 2$ m and the stack emission gas is discharged at a speed of $v_s = 5$ m/s. Assume wind speed u = 2 m/s, and surrounding air $T_a = 290$ K. Plot the plume rise downwind the emission source for discharge temperature of $T_s = 450$ K.
6. Same as that described in Problem 4 above, the power plant in a rural area has a stack of 200 m high with an inner diameter of 2 m. continuously discharge SO_2 into the atmosphere at a concentration of 100 mg/m^3. The discharge air flow rate is 2 million m^3/hr. On a slightly sunny day, the wind speed at 10 m high is about 5.5 m/s. Ignore the chemical reactions in the atmosphere, and estimate

   (a) SO_2 concentration at the center of the plume 4 km downwind from the stack.
   (b) ground level SO_2 concentration 4 km downwind

References and Further Readings

1. Andreas EL (2009) A new value of the von Karman constant: implications and implementation. J Appl Meteorol Climatol 48:923–944
2. Benoit R (1977) On the integral of the surface layer profile-gradient functions. J Appl Meteorol 16:859–860
3. Bjorklund JR, Bowers JF (1982) User's instruction for the SHORTZ and LONGZ computer programs, vol I–II. EPA Document EPA-903/9-82-004A and B. US EPA, Middle Atlantic Region III, Philadelphia, Pennsylvania, USA
4. Briggs GA (1965) A plume rise model compared with observations. J Air Pollut Control Assoc 15(9):433–438
5. Briggs GA (1973) Diffusion estimation for small emissions. Annual Report of Air Resources Atmospheric Turbulence and Diffusion Laboratory, NOAA Oak Ridge, Tennessee, USA
6. Britter RE, Hanna SR (2003) Flow and dispersion in urban areas. Annu Rev Fluid Mech 2003 (35):469–496
7. Canepa E (2004) An overview about the study of downwash effects on dispersion of airborne pollutants. Environ Model Softw 19(2004):1077–1087
8. Cheng Y, Brutsaert W (2005) Flux-profile relationship for wind speed and temperature in the stable atmospheric boundary layer. Bound-Layer Meteorol 114:519–538
9. Foken T (2006) 50 years of the Monin-Obukhov similarity theory. Bound-Layer Meteorol 119:431–447
10. Garrant JR (1990) The internal boundary layer—a review. Bound-Layer Meteorol 50:171–203

11. Griffiths RF (1994) Errors in the use of the briggs parameterization for atmospheric dispersion coefficients. Atmos Environ 28(17):2861–2865
12. Hanna SR, Briggs GA, Hosker RP (1982) Handbook on atmospheric diffusion, NTIS DE 81009809 (DOC/TIC-22800), Springfield, VA, USA
13. Holzworth GC (1972) Mixing heights, wind speeds, and potential for urban air pollution throughout the contiguous United States. US EPA Report #: EPA450R72102; AP-101
14. Hosker RP (1974) A comparison of estimation procedures for over-water plume dispersion. Proceedings of symposium on atmospheric diffusion and air pollution. Santa Barbara, California, 9–13 September 1974
15. McKoen BJ, Sreenivasan KR (2007) Introduction: scaling and structure in high: reynolds number wall-bounded flows. Philos Trans Roy Soc A 365:635–646
16. Scire JS, Robe FR, Fernau ME, Yamartino RJ (2000) A user's guide for the CALMET meteorological model. Earth Tech, USA (myjsp.src.com)
17. Seinfeld J, Pandis S (2006) Atmospheric chemistry and physics. In: (Chap .8) Properties of the atmospheric aerosol, 2nd edn. Wiley-Interscience, Hoboken
18. Serway RA, Beichner RJ, Jewett JW (2000) Physics for scientists and engineers with modern physics, 5th edn. Saunders College Publishing, Orlando, FL, p 317 Q10
19. Turner DB (1970) Workbook of atmospheric dispersion estimates (US EPA AP-20)
20. Venkatram A (1980) Estimating the Monin-Obukhov length in the stable boundary layer for dispersion calculations. Bound Layer Meteorol 19:481–485
21. de Visscher A (2013) Air dispersion modeling, foundations and applications. Wiley, Hoboken, NJ, USA
22. Zilitinkevich SS (1972) On the determination of the height of the Ekman boundary layer. Bound-Layer Meteorol 3(2):141–145

Part III
Special Topics

Chapter 12
Carbon Capture and Storage

12.1 Background Information

It has been accepted globally that carbon dioxide and several other gases known as greenhouse gases (GHGs) are causing global climate change by changing the physical and chemical processes in the Earth's upper troposphere and stratosphere. An independent record of the global average surface temperature shows that global warming is a fact of the past 130 years [5]. Although existing data published by different researchers differ from each other as a result of their data selection, processing, and bias corrections, they are leading to the same conclusion that global surface temperature has increased by 0.6–0.7 °C over the past century.

GHGs allow solar energy to enter the atmosphere freely. When sunlight strikes the surface of the Earth, some of the solar energy returns to space by reflection of infrared radiation. GHGs also absorb this infrared radiation and trap the heat in the atmosphere. If the amount of heat (or solar energy) from the Sun to the Earth's surface is the same as that leaving the Earth's surface to space, then the temperature of the Earth's surface remains stable. This perfect balance allows life to sustain on planet Earth.

However, there is an uncertainty in how the climate system varies naturally and reacts to extra GHGs. Making progress in reducing uncertainties in projections of future climate will require an understanding of the buildup of GHGs in the atmosphere and the behavior of the climate system.

The most important GHGs include carbon dioxide (CO_2), methane (CH_4), and nitrous oxide (N_2O). However, the most abundant greenhouse gas in the atmosphere is actually water vapor, which doubles the greenhouse effects caused by all the other GHGs. Some of the GHGs exist in nature and they include water vapor, carbon dioxide, methane, and nitrous oxide; others are exclusively human-made such as fluorinated gases are created solely by human activities. This is referred to as the "enhanced greenhouse effect" or the "anthropogenic greenhouse effect" as it is primarily due to the human activities.

Z. Tan, *Air Pollution and Greenhouse Gases*, Green Energy and Technology,
DOI 10.1007/978-981-287-212-8_12

Table 12.1 gives six leading GHGs that are under discussion in international climate change negotiations: CO_2, N_2O, CH_4, hydrofluorocarbons (HFC), perfluorocarbons (PFC), and sulfur hexafluoride (SF_6). All the GHGs do not contribute equally to the global climate change, therefore, global warming potential (GWP) is used to quantify the contribution to global warming for a unit mass of greenhouse gas, taking carbon dioxide as a reference with GWP = 1. More detailed information such as the calculation of GWP and lifetime of GHGs can be found in the IPCC [32] Report. The GWPs of other GHGs are higher than CO_2, due to difference in lifetime and radiation absorption behavior in the atmosphere. For example, the GWP of SF_6 is 23,900 times that of CO_2 due to its great stability and persistence. Its lifetime in atmosphere is estimated to be 3,200 years, making it the strongest GHG known today.

However, over all CO_2 is responsible for most of the greenhouse effect, only second to water vapor, due to its large total quantity. SF_6 contributes to approximately 0.1 % of the enhanced greenhouse effect.

CH_4 emissions mainly result from landfills, agriculture, coal mining, oil, and gas handling and processing. One of the solutions to this problem is to capture the CH_4 emission in a controlled environment, e.g., an anaerobic digester, and burn it into CO_2 to reduce its greenhouse effect. The chemical reaction is described as

$$CH_4 + 2O_2 \rightarrow CO_2 + 2H_2O \tag{12.1}$$

This reaction shows that one mole of CH_4 produces the same mole amount of CO_2. It, however, reduces the greenhouse effect by 20 times. And this is also the motivation behind flaring in the oil and gas industry.

N_2O emissions result from agriculture, chemical plants such as nitric acid or nylon processing units, and combustion processes. Fluidized bed combustion (FBC) is the most problematic with respect to N_2O emissions. Coal-fired FBC emits 50–150 times higher N_2O than pulverized coal firing plants. Vehicles emit more N_2O than stationary sources; a typical passenger car emits 20 mg/km N_2O in addition to 50 mg/km CH_4.

HFCs and PFCs are synthetic chemicals, produced as alternatives for the ozone-depleting chlorofluorocarbons (CFCs) in response to the "phase out" of CFCs under the Montreal protocol of 1987. HFC-134a is the major substitute for CFCs in refrigerators.

Table 12.1 Six leading GHGs and their GWPs (excluding water vapor)

GHG name	Formula	GWP
Carbon dioxide	CO_2	1
Methane	CH_4	21
Nitrous oxide	N_2O	310
Hydrofluorocarbons	HFC	140–11,700
Perfluorocarbons	PFC	7,400
Sulfur hexafluoride	SF_6	23,900

Source EIA [21]

Levels of GHGs have increased dramatically since the industrial revolution. For example, about 75 % man-made carbon dioxide emissions were from burning fossil fuels during the past 20 years. About 3.2 billion metric tons is added to the atmosphere annually. The U.S. produces about 25 % of global CO_2 emissions where 85 % of the US energy is produced through fossil fuel combustion [32]. Global carbon dioxide emissions continue increasing annually between 2001 and 2025, and the emerging economies (China and India, for example) contribute to much of the enhanced GHG effect. These developing countries' GHG emissions are expected to grow at 2.7 % annually by 2025 [22].

Combustion is the major source for the increase of CO_2 in the atmosphere. It can be concluded that reduction of CO_2 emissions from fossil fuel combustion will have the largest impact on GHG emissions. In this chapter, we focus primarily on approaches to CO_2 emission control. Readers are referred to the literature for the approaches to the GHG emission control.

12.2 CO_2 Generation in Combustion

According to IPCC [30–32] data, CO_2 emissions from large fossil fueled power plants account for half of the total carbon emissions. Other sources include industrial processes such as cement production, integrated steel mills, and oil-gas refinery. The mechanisms of CO_2 generation in combustion processes have been introduced in Parts I and II, and it is briefly summarized as follows for readers who are interested in this chapter only.

Stoichiometric combustion process of a hydrocarbon fuel $C_\alpha H_\beta$ perfectly mixed with oxygen can be described as

$$C_\alpha H_\beta + \left(\alpha + \frac{\beta}{4}\right) O_2 \rightarrow \alpha CO_2 + \frac{\beta}{2} H_2O \tag{12.2}$$

Thus the stoichiometry of a general hydrocarbon $C_\alpha H_\beta$ mixed with dry air perfectly can be described by the following formulas:

$$C_\alpha H_\beta + \frac{1}{\phi}\left(\alpha + \frac{\beta}{4}\right)(O_2 + 3.76N_2) \rightarrow \alpha CO_2 + \frac{\beta}{2} H_2O + \frac{3.76}{\phi}\left(\alpha + \frac{\beta}{4}\right) N_2 \tag{12.3}$$

where ϕ is the equivalence ratio.

The fuel-lean reaction formula for the combustion of $C_\alpha H_\beta$ perfectly mixed with excess air is

$$\begin{aligned} &C_\alpha H_\beta + \frac{1}{\phi}\left(\alpha + \frac{\beta}{4}\right)(O_2 + 3.76N_2) \\ &\rightarrow \alpha CO_2 + \frac{\beta}{2} H_2O + \frac{3.76}{\phi}\left(\alpha + \frac{\beta}{4}\right) N_2 + \left(\frac{1}{\phi} - 1\right)\left(\alpha + \frac{\beta}{4}\right) O_2 \end{aligned} \tag{12.4}$$

As introduced in combustion chemistry above, it is very challenging in engineering practices to achieve perfect mixing in the entire combustion device. Stoichiometric, fuel lean, and fuel rich combustion take place at different spots in the combustion device. As a result, the actual combustion formula is very complicated. One example formula is

$$\begin{aligned} & C_\alpha H_\beta + \frac{1}{\phi}\left(\alpha + \frac{\beta}{4}\right)(O_2 + 3.76\ N_2) \\ & \rightarrow [xCO_2 + (1-x)CO] + [yH_2O + (1-y)H_2] + \frac{3.76}{\phi}\left(\alpha + \frac{\beta}{4}\right)N_2 \\ & + \left[\frac{1}{\phi}\left(\alpha + \frac{\beta}{4}\right) - \frac{1+x+y}{2}\right]O_2 \end{aligned} \tag{12.5}$$

where x and y in Eq. (12.5) can be determined by considering the chemical equilibrium reactions. The general chemical equilibrium formula for the reaction $aA + bB \leftrightarrow cC + dD$ is described in Eq. (12.6)

$$K_p = \frac{n_C^c n_D^d}{n_A^a n_B^b}\left(\frac{P}{n}\right)^{\Delta n} \tag{12.6}$$

where $\Delta n = c - d - a - b$. The chemical equilibrium constants based on partial pressure for chemical reactions with CO_2 and CO, $\ln(K_P)$, can be found in Table 12.2.

With the products described in Eq. (12.5), we can consider chemical equilibrium reactions to solve the unknowns of x and y.

Example 12.1: CO_2 emission rate calculation

Natural gas is fed into a burner at a rate of 1,000 m^3/h at 1 atm and 25 °C. Assuming that the air is premixed with $\phi = 1$, and the final products at equilibrium under 1,000 K and 1 atm contain O_2, N_2, CO_2, CO, H_2O, and H_2, determine the emission rate of CO_2 generation.

Solution

Assuming the natural gas is pure methane, with $\phi = 1$, $\alpha = 1$ and $\beta = 4$ Eq. (12.5) becomes

$$\begin{aligned} & CH_4 + 2(O_2 + 3.76N_2) \\ & \rightarrow [xCO_2 + (1-x)CO] + [yH_2O + (1-y)H_2] + 7.52N_2 + \left(\frac{3+x+y}{2}\right)O_2 \end{aligned}$$

Table 12.2 Equilibrium constants based on partial pressure for chemical reactions with CO_2 and CO

T (K)	$\ln(K_p)$	
	$CO_2 + H_2 \leftrightarrow CO + H_2O$	$CO_2 \leftrightarrow CO + \frac{1}{2}O_2$
298	−11.554	−103.762
500	−4.9252	−57.616
1,000	−0.366	−23.529
1,200	0.3108	−17.871
1,400	0.767	−13.842
1,600	1.091	−10.83
1,800	1.328	−8.497
2,000	1.51	−6.635
2,200	1.648	−5.12
2,400	1.759	−3.86
2,600	1.847	−2.801
2,800	1.918	−1.894
3,000	1.976	−1.111
3,200	2.022	−0.429
3,400	2.061	0.169
3,600		0.701
3,800		1.176
4,000		1.599

For each mole of CH_4, the total mole amount of products is

$$n = 11.02 + \frac{x + y}{2}$$

Consider two equilibrium reactions as follows:

$$CO_2 + H_2 \leftrightarrow CO + H_2O \qquad (1)$$

$$CO_2 \leftrightarrow CO + \frac{1}{2}O_2 \qquad (2)$$

The chemical equilibrium constants at 1,000 K can be found in Table 12.2.

$$K_{P1} = e^{-0.366} = 0.6935 = \frac{(1 - x)y}{x(1 - y)}$$

$$K_{P2} = e^{-23.529} = 6.046 \times 10^{-11} = \frac{1 - x}{x} \frac{\left(\frac{3 + x + y}{2}\right)^{1/2}}{\left(11.02 + \frac{x + y}{2}\right)^{1/2}}$$

Reorganizing the equations leads to

$$\begin{cases} x = \dfrac{1}{1 + 0.6935\left(\frac{1-y}{y}\right)} \\ \dfrac{1-x}{x}\sqrt{\dfrac{3 + x + y}{22.04 + x + y}} = 6.046 \times 10^{-11} \end{cases}$$

Solving these equations with the assistance of software we can get

$$x \approx y \approx 1; 1 - x = 1.32 \times 10^{-10}; 1 - y = 1.91 \times 10^{-10}$$

The extremely low CO and H_2 concentration is a result of stoichiometric combustion at high temperature. Anyway, for each mole of CH_4, there is about 1 mol of CO_2 produced. At 1 atm and 25 °C, one mole of gas corresponds to 0.248 m^3/mol. Thus, the CH_4 feeding rate is 4032.26 mol/h. As a result, the CO_2 production rate is the same as 4032.26 mol/h, or

$$\dot{m}_{CO_2} = 44\frac{\text{kg}}{\text{kmol}} \times 4.032\frac{\text{kmole}}{\text{h}} = 177.42\,\text{kg/h}$$

12.3 General Approaches to Reducing GHG Emissions

Enhanced global warming effects can be reduced by reducing the emission of CO_2 and other GHGs. Emission reductions of CO_2 can be accomplished by a combination of several of the following approaches:

- Improved energy conversion efficiency in stationary and mobile combustion processes.
- Supply and end-use efficiency improvement and conservation.
- Shift to alternative energy sources, which have been introduced in Chap. 8. They are effective in air pollution control as well as carbon emission reduction.
- Carbon capture and storage (CCS).

Carbon emissions of CO_2 are inevitable as long as hydrocarbon fuels are used for energy conversion. However, more efficient combustion processes may produce less CO_2 per unit power generated. The CO_2 emissions per kWh power generated from a low efficiency coal-fired boiler may be 4 times that from a natural gas-fired gas turbine combined cycle plant (NGCC).

End-use efficiency improvements and energy conservation are the simplest and most cost-effective approaches to reduce carbon emissions and other air pollutants. For example, in the residential and commercial sectors, it can be achieved by reducing heating and air conditioning consumptions, better insulations, lowering hot water consumption, replacement of incandescent with fluorescent lighting, and so on. It is estimated that in the residential-commercial sector in USA, by the year 2010, carbon emissions could be reduced by 10.5 % below 1990 levels with cost-effective conservation measures.

The follow sections focus on CCS.

12.4 Carbon Capture Processes

CCS refers to a number of technologies that capture CO_2 at some stage from processes such as combustion (most for power generation), gasification, cement manufacture, iron and steel making, and natural gas treatment. We introduce CCS in a generic way using combustion as main examples. Similar approaches can be taken for other industrial processes as well.

Like approaches to other air emission control, carbon capture can be achieved by pre-, in-, and post-combustion gas separation. The following topics are introduced in the section that follows.

- Pre-combustion carbon capture
  - Gasification and IGCC
- In-combustion carbon capture
  - Oxyfuel combustion
  - Chemical-looping combustion
- Post-combustion carbon capture
  - CO_2 capture from flue gas
  - CO_2 capture from atmosphere

12.4.1 Pre-combustion Carbon Capture

Pre-combustion carbon capture process is associated with the integrated gasification combined cycle (IGCC) [49]. Figure 12.1 shows the simplified process of IGCC with pre-combustion CO_2 capture. It is achieved by converting primary fossil fuels into hydrogen fuel. Hydrogen can be produced by partial oxidation of primary fuel for syngas followed by water shift and syngas purification, where CO_2 and other impurities are removed.

12.4.1.1 Syngas Production

Partial oxidation of a primary fuel is a highly fuel rich combustion process as described in Eq. (12.5). Since the process aims at the conversion of solid fuels into gaseous

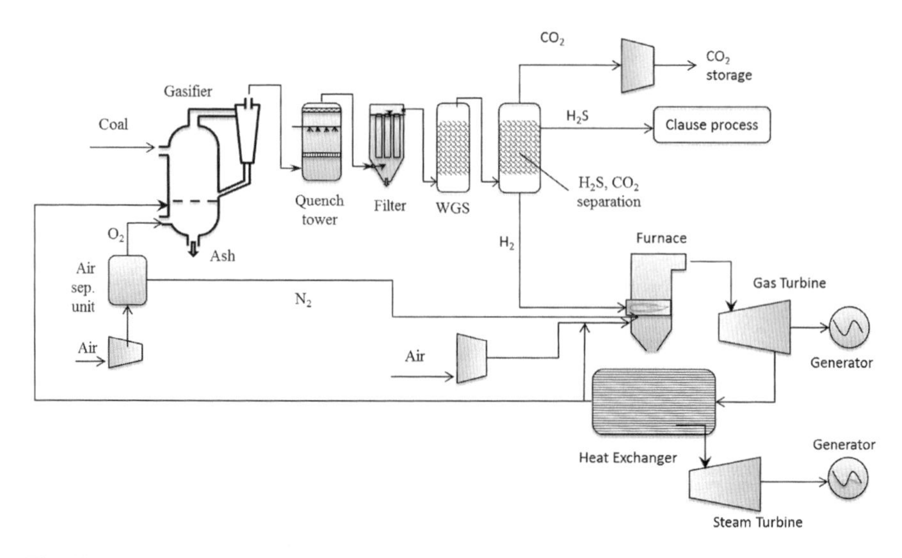

Fig. 12.1 Simplified IGCC with pre-combustion CO_2 capture process

products, it is commonly called gasification. Unlike a conventional combustion process where we expected as little CO and H_2 as possible, the gasification process aims to maximize the production of CO and H_2, which are the main compounds of syngas. Since the partial oxidation temperature is much lower than that required for thermal NO_x formation, the produced syngas is mainly a mixture of CO and H_2 with other trace amounts of impurities such as NH_4, H_2S, SO_2, and particulates.

Advanced gasification process employs oxygen as an oxidant in order to reduce nitrogen-introduced impurities, and the overall reaction can be described as

$$C_\alpha H_\beta + \frac{1}{\phi}\left(\alpha + \frac{\beta}{4}\right)O_2 \rightarrow [xCO_2 + (1-x)CO] + [yH_2O + (1-y)H_2] + zO_2 \tag{12.7}$$

Oxygen is usually produced by separation of nitrogen from the air in an air separation unit (see Fig. 12.1). In return, the formation of thermal NO_x is minimized.

One way or another, an ideal partial oxidation process should aim at $x, y \rightarrow 0$.

In practice, solid fuels (e.g., coal and biomass) and methane are used for partial oxidation. When solid fuels are used, the corresponding simplified chemical reactions are described as follows:

$$C(s) + \frac{1}{2}O_2 \rightarrow CO \qquad \text{Carbon partial oxidation} \tag{12.8}$$

$$C(s) + H_2O \rightarrow CO + H_2 \qquad \text{Carbon steam reaction} \tag{12.9}$$

$$C(s) + CO_2 \rightarrow 2CO \qquad \text{Boudouard reaction} \tag{12.10}$$

Partial oxidation of methane can be described as

$$CH_4 + \frac{1}{2}O_2 \rightarrow CO + 2H_2 \qquad \text{Methane partial oxidation} \tag{12.11}$$

$$CH_4 + H_2O \rightarrow CO + 3H_2 \qquad \text{Steam methane reaction} \tag{12.12}$$

The steam-methane reaction is effective in the temperature range of 700–950 °C at 1.4–4 MPa. Under these conditions, 70–80 % of the methane can be converted into hydrogen in one single reactor.

The syngas can be directly burned for energy production where CO is converted into CO_2 by oxidation, but water–gas shift (WGS) is necessary to facilitate pre-combustion carbon capture. The CO produced in partial oxidation process eventually is converted into CO_2 by the WGS reaction

$$CO + H_2O \rightarrow H_2 + CO_2 \tag{12.13}$$

Typical syngas composition after WGS is shown in Table 12.3. The high concentration of CO_2 favors effective CO_2 capture.

By the WGS reaction over 90 % of the carbon exist as CO_2. For the processes aiming at high purity H_2, CO_2 removal efficiency has to be improved. This is achieved commonly by multiple CO_2 removal units.

Table 12.3 Composition of typical gases subjected to pre-combustion and post-combustion CO_2 separation

Gases	Mole fraction	
	Pre-combustion syngas after WGS reaction	Post-combustion flue gas
CO_2	37.7 %	10–15 %
H_2O	0.14 %	5–10 %
H_2	55.5 %	
NO_x		<1,000 ppm
SO_x		<1,000 ppm
O_2		3–4 %
CO	1.7 %	20 ppm
N_2	3.9 %	70–75 %
H_2S	0.4 %	
Temperature	40 °C	40–75 °C
Pressure	30 atm	1 atm

Sources D'Alessandro et al. [17] and Ramdin et al. [49]

12.4.2 In-combustion Carbon Capture

12.4.2.1 Oxyfuel Combustion

In-combustion carbon capture is achieved primarily by oxyfuel combustion. Oxygen rather than air is used as oxidizer in combustion. Then the main combustion products, as described using Eq. (12.2), are CO_2 and water. Depending on the fuel composition, there may be SO_2, NO_x and others, but thermal NO_x is greatly reduced. As a result, the CO_2 can be readily separated from water and other trace compounds for transport and storage.

The oxyfuel combustion process is a promising concept but still under research and development (R&D). The fuels fed into an oxyfuel combustion system can be natural gas, biomass, or coal. A simplified schematic diagram of oxyfuel combustion with CO_2 capture process is shown in Fig. 12.2. Unlike conventional combustion technologies, this process utilizes oxygen instead of air as the oxidant, thereby eliminating nitrogen in the downstream separation. The corresponding simplified combustion stoichiometry is described using

$$C_\alpha H_\beta + \left(\alpha + \frac{\beta}{4}\right)O_2 \rightarrow \alpha CO_2 + \frac{\beta}{2}H_2O \tag{12.14}$$

Example 12.2: Oxyfuel computation flame temperature
In Sect. 3.5.1, we calculated the adiabatic flame temperature of CH_4 burned with air at 298 K, when they are premixed perfectly with an equivalence ratio of 1.0 and the combustion is complete. Now let us redo the calculation using pure oxygen instead of air. Determine the corresponding constant pressure adiabatic flame temperature.

Solution
First of all, set up stoichiometric combustion reaction equation, using the methods introduced in Sect. 3.4,

$$CH_4 + 2O_2 \rightarrow CO_2 + 2H_2O$$

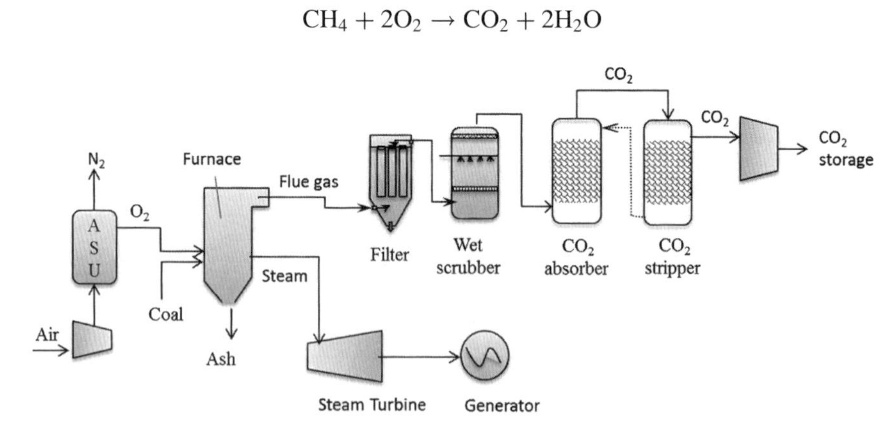

Fig. 12.2 Schematic diagram of the oxyfuel combustion process for CO_2 capture

For an adiabatic constant pressure system, with the reactant temperature of $T_R = T_0 = 298$ K, the left-hand side (LHS) of Eq. (12.15)

$$\sum_R n_i\left[\left(a_i(T_R - 298\,\mathrm{K}) + \frac{b_i}{2}\left(T_R^2 - (298\,\mathrm{K})^2\right)\right) + h^o_{f,i}\right] = \sum_P n_i\left[\left(a_i(T_a - 298\,\mathrm{K}) + \frac{b_i}{2}\left(T_a^2 - (298\,\mathrm{K})^2\right)\right) + h^o_{f,i}\right] \quad (12.15)$$

becomes

$$\mathrm{LHS} = h^o_{f,CH_4} + 2h^o_{f,O_2} \quad (1)$$

The right-hand side (RHS) of the equation is simplified as

$$\mathrm{RHS} = \left[\left(a_{CO_2}(T_a - 298\,\mathrm{K}) + \frac{b_{CO_2}}{2}\left(T_a^2 - (298\,\mathrm{K})^2\right)\right) + h^o_{f,CO_2}\right] + 2\left[\left(a_{H_2O}(T_a - 298\,\mathrm{K}) + \frac{b_{H_2O}}{2}\left(T_a^2 - (298\,\mathrm{K})^2\right)\right) + h^o_{f,H_2O}\right] \quad (2)$$

Looking into Table A.4, we can get the parameters needed (Table 12.4)

Substituting these values into Eq. (1) and Eq. (2) leads to

$$\mathrm{LHS} = -74{,}980\,(\mathrm{J/mole})$$

$$\mathrm{RHS} = \left[\left(44.3191(T_a - 298) + \frac{0.0073}{2}\left(T_a^2 - (298\ \mathrm{K})^2\right)\right) - 394{,}088\right] + 2\left[\left(32.4766(T_a - 298\,\mathrm{K}) + \frac{0.00862}{2}\left(T_a^2 - (298\,\mathrm{K})^2\right)\right) - 242{,}174\right]$$

Simplification of the equation gives

$$0.01227T_a^2 + 109.2723\,T_a - 837109 = 0$$

Solving this equation we can get, $T_a = 4930.8$ K. This calculated value is much higher that air-fuel combustion: $T_a = 2341$ K.

Table 12.4 Parameters used in Example 1.22

Species	$h^o_{f,i}$ (J/mole)	$C_p(T)$ (J/mole · K)	
		a_i	b_i
CO_2	−394,088	44.3191	0.0073
H_2O	−242,174	32.4766	0.00862
O_2	0	30.5041	0.00349
CH_4	−74,980	44.2539	0.02273

Although it is only estimation, the comparison does show that the oxyfuel combustion flame temperature is higher than that of conventional combustion with air. As a result, the boiler of the oxyfuel combustion process requires special materials that can survive extreme temperature.

Another concern of the oxyfuel combustion process with sulfur containing fuels is the high SO_x concentration without the dilution of nitrogen, resulting in high corrosion on the ducts. Extra costs are associated with concentrated oxygen production by costly air separation units.

The benefit is a simple process for carbon capture after combustion. Without nitrogen and NO_x in the flue gas, it contains mainly H_2O and CO_2. After the removal of soot and SO_2, if any, CO_2 can be readily separated from water vapor by condensation in a cooler. This highly concentrated CO_2 is ready for transportation and storage.

12.4.2.2 Chemical Looping Combustion

Similar to oxyfuel combustion, chemical-looping combustion is an emerging combustion process that is attractive for the benefit of carbon capture. This idea was first introduced by Lewis and Gilliland [38] as a way to produce pure CO_2 from fossil fuels. Thirty years later, Ishida et al. [33] proposed the use of chemical-looping combustion for power generation with climate mitigation. Figure 12.3 shows an example schematic diagram of chemical-looping combustion based on circulating fluidized bed principle. Most of the state of the art focuses on gaseous fuels reacting with oxygen carrier. Similar principles can be applied to the oxidation of the vapors of liquid fuels and volatiles released from solid fuels.

The combustion in the fuel reactor takes place following different reaction steps. The oxygen-needed is released from the oxygen-carrier (Me_xO_y) at high temperature.

$$2Me_xO_y \leftrightarrow 2Me_xO_{y-1} + O_2 \tag{12.16}$$

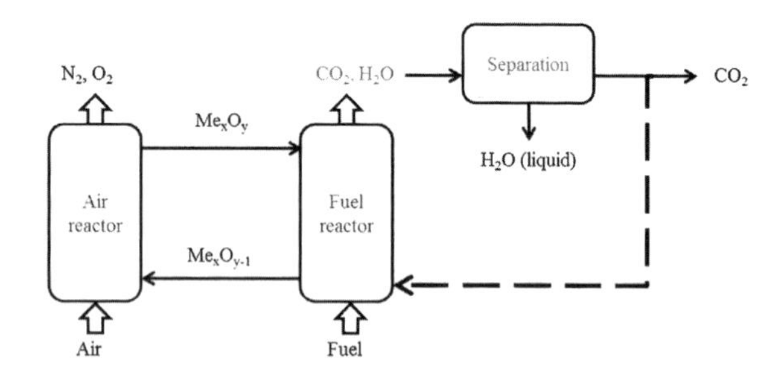

Fig. 12.3 Schematic diagram of the chemical looping combustion

The reduced oxygen-carrier Me_xO_{y-1} is regenerated in the air reactor according to the reverse reaction in Eq. (12.16), where the oxygen is from the air. Effluent gases from the air-reactor containing air with reduced amount of oxygen can be discharged without negative environmental impact, and the regenerated solid oxygen carrier is transported to the fuel reactor, and another cycle starts.

The key of this technology is the special oxygen carrier! About one thousand oxygen carrier materials have been studied in laboratory [2]. The first-generation oxygen carriers focused on mainly oxides of four metals: Ni, Fe, Mn, and Cu. Recent advances focused on low cost materials by combing different metal oxides and oxygen releasing materials aiming at solid fuel combustion. Among many combined materials tested in laboratory, very few have been tested in continuous operation. Most researchers have used ilmenite ($FeTiO_3$) [8, 15] because of its low cost and reasonably high reactivity towards syngas.

This process must enable the release of oxygen from the oxygen-carriers. Thereby, only those metal oxides that have a suitable equilibrium partial pressure of oxygen at temperatures of interest for combustion (800–1,200 °C) can be used as oxygen-carriers. These materials include, but are not limited to CuO/Cu_2O, Mn_2O_3/Mn_3O_4, and Co_3O_4/CoO [34]. Operating conditions in the air-reactor and fuel-reactor would be determined for specific oxygen-carrier [50].

Three example metal oxides are as follows:

$$4CuO \leftrightarrow 2Cu_2O + O_2 \tag{12.17}$$

$$6Mn_2O_3 \leftrightarrow 4Mn_3O_4 + O_2 \tag{12.18}$$

$$2Co_3O_4 \leftrightarrow 6CoO + O_2 \tag{12.19}$$

Then the oxygen reacts with the gas fuel or char and volatiles (mainly H_2, CO, CH_4) from the solid fuels following the combustion principles.

$$\text{Gases} + \text{Volatiles} + \text{Char} + O_2 \rightarrow CO_2 + H_2O \tag{12.20}$$

If we take CH_4, H_2 and CO as examples, the overall chemical reactions are

$$CH_4 + 2O_2 \rightarrow CO_2 + 2H_2O \tag{12.21}$$

$$H_2 + \frac{1}{2}O_2 \rightarrow H_2O \tag{12.22}$$

$$CO + \frac{1}{2}O_2 \rightarrow CO_2 \tag{12.23}$$

As described in Eq. (12.20), the ultimate goal of chemical-looping combustion is to convert fuel into carbon dioxide and water without NO_x. The degree of fuel conversion to CO_2 and H_2O is quantified using the gas yield; which is defined as the fraction of the fuel oxidized to CO_2 or H_2O. Similar to the approaches in oxyfuel combustion, water vapor is removed from highly concentrated CO_2 stream that can be compressed for transport and storage.

12.4.3 Post-combustion Carbon Capture

Post-combustion CO_2 capture is mainly achieved by flue gas cleaning for stationary sources such as a power plant and an industrial facility. Unlike pre-combustion and oxyfuel combustion approaches, post-combustion carbon capture technologies are more complex due to more air powered combustion-related air contaminants. Capture of most of the contaminants has been introduced in the other chapters of this book.

CO_2 separation is enabled after all the post-combustion air cleaning processes introduced in the chapter for post-combustion air emission control. As shown in Table 12.3, the cleaned flue gas usually contains 10–15 % of CO_2 that is diluted by nitrogen gas. Therefore, an effective post-combustion CO_2 separation technology should be able to cope with low CO_2 partial pressure. In general, physical sorption is not effective for separation of gases with low partial pressure. Therefore, we should focus on chemisorption technologies for post-combustion CO_2 capture.

If we consider low CO_2 partial pressure as a disadvantage of post-combustion CO_2 capture, this disadvantage is compensated with its flexibility in retrofitting an existing facility without major modifications. Capture ambient CO_2 has been proposed as well. It can be considered as an alternative CO_2 capture process, but at temperatures lower than those of combustion flue gases before discharge.

Despite the differences in pre- and post-combustion approaches, CO_2 must be separated from the carrier gas. They share the fundamentals introduced in Chap. 5, primarily the separation of gas (CO_2 specifically) from the mixture. However, at the time of writing this book, all the CCS projects are under pilot tests. No single large-scale plant with CCS is known to be operational continuously.

The major barrier to the commercialization of CCS process at large scale is the high costs associated with carbon separation, transport, and injection. Currently, the most widely used technology for CO_2 separation is based on amine-solvents, for example, monoethanolamine (MEA). A large amount of energy is required for solvent regeneration. The energy consumption rate, assuming a 30 wt% MEA (aqueous) solution and 90 % of removal efficiency, was estimated to be in the range of 2.5–3.6 GJ per ton of CO_2. Additional energy consumption for compressing the captured CO_2 to the required pressure of 150 bar for transportation and storage is 0.42 GJ/ton CO_2. These numbers can be transferred to extra costs of $50–150 per ton CO_2 removed using amines [18]. The existing CCS technologies are far from being cost-effective and unattractive for large-scale applications. Much more research is needed for design of low-cost sorbents and optimized process design CO_2 capture. Our focus that follows is on the sorbent and related process design.

12.5 CO_2 Separation by Adsorption

CO_2 can be separated from a gas stream by both physical adsorption and chemical adsorption. The general differences between a physical and chemical adsorption has been introduced in Chap. 5. Physical CO_2 adsorption operates at temperatures lower than 100 °C; whereas the chemical counterpart operates at a high range of 400–600 °C.

12.5.1 Physical Adsorption

A good CO_2 physical adsorbent is expected to be characterized with high affinity with CO_2 compared to other gases in the stream, high adsorption capacity, low heat of adsorption, low adsorption hysteresis, and steep adsorption isotherm. These features ensure the cost-effectiveness of the operation with high efficiency, low energy consumption, and low material cost.

The best adsorbent is expected to be characterized with high CO_2 capacity at low pressure, high selectivity for CO_2, fast adsorption/desorption kinetics, good mechanical properties, high hydrothermal and chemical stability, as well as low costs of synthesis. Unfortunately, these criteria are too ideal for any single adsorbent.

The most widely investigated low temperature CO_2 adsorbents are zeolites and activated carbons. Compare to Zeolite-13X and natural zeolite, activated carbon showed higher carbon adsorption capacity and steepest isotherm, but high hysteresis challenges the desorption process.

Metal–organic frameworks (MOFs) have recently attracted intense research interest in CO_2 adsorption due to their large porous volume and surface areas [42, 43]. In general, the CO_2 adsorption in MOFs varied with CO_2 pressure. At high pressures, CO_2 adsorption capacities depend on surface areas and pore volumes of the MOFs. Otherwise, the capacities depend on the heat of adsorption. In addition, many MOFs have shown high CO_2/N_2 and CO_2/CH_4 selectivity. However, there are two important challenges to MOFs into practical applications of CO_2 capture:

(1) the mass production of MOFs at low cost
(2) the stabilities of MOFs toward moisture, other acid gases, and heat for regeneration.

Interested readers are encouraged to conduct a state-of-the-art literature review in this area of research.

Continuous CO_2 adsorption can be implemented by a pressure swing adsorption process. It is widely used for air drying using synthetic zeolites or activated alumina

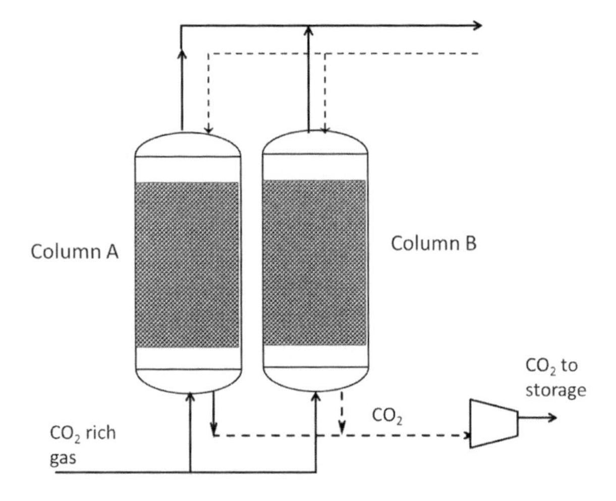

Fig. 12.4 Schematic diagram of pressure swing adsorption process

for moisture removal. Advances in pressured swing adsorption can be found in the state of the art provided by Grande [24]. A pressure swing adsorption (PSA) process has at least two columns, when one is working on CO_2 adsorption from the high pressure gas, the used sorbent in other column is regenerated at low pressure. A simple PSA process for CO_2 capture is shown in Fig. 12.4.

12.5.2 Chemical Adsorbents

Chemical adsorption at high temperature (400–600 °C) has a potential for pre-combustion CO_2 capture and hydrogen production. High temperature sorbent can be employed for carbon separation without temperature quenching and consequently do not comprise much with overall energy efficiency. The working principles are similar to those of calcium based adsorption.

12.5.2.1 Calcium-Based Adsorbent

CaO is one of the oldest sorbent for CO_2 capture by chemical adsorption. Calcium oxide reacts with CO_2 at high temperatures to produce carbonates:

$$CaO_{(s)} + CO_2 \leftrightarrow CaCO_{3(s)} \quad (12.24)$$

where (s) stands for solid state. The forward reaction takes place at about 600–800 °C, and the backward reaction at higher temperature (>870 °C) may result in sorbent regeneration by decarbonation and releasing CO_2.

Calcium-based sorbents are commonly produced from limestone. Sometimes, sorbents that are produced from calcined dolomite ($CaCO_3 \cdot MgCO_3$) or huntite ($CaCO_3 \cdot 3MgCO_3$) contain MgO. MgO does not react with CO_2 at 600–800 °C, but its presence helps with the lifetime and durability of the calcium-based sorbent. Since CaO degrades quickly after a few cycles of regeneration, more reliable sorbents have been developed and tested at high temperatures; they include Calcium aluminate ($CaAl_2O_4$), Sodium Zirconate (Na_2ZrO_3), Lithium zircanate (Li_2ZrO_3), and Lithium orthosilicate (Li_4SiO_4).

12.5.2.2 Temperature Swing Adsorption Process

By taking advantage of this reversible reaction (12.24) between CO_2 and CaO at different temperatures, a temperature swing adsorption-desorption system allows continuous chemical adsorption of CO_2. The schematic diagram for syngas cleaning by separating CO_2 from syngases is shown in Fig. 12.5. Similar process can be employed for the post-combustion carbon capture system. Alternatively, the sorbent can also be lithium silicate, which can be regenerated at a lower temperature of 800 °C.

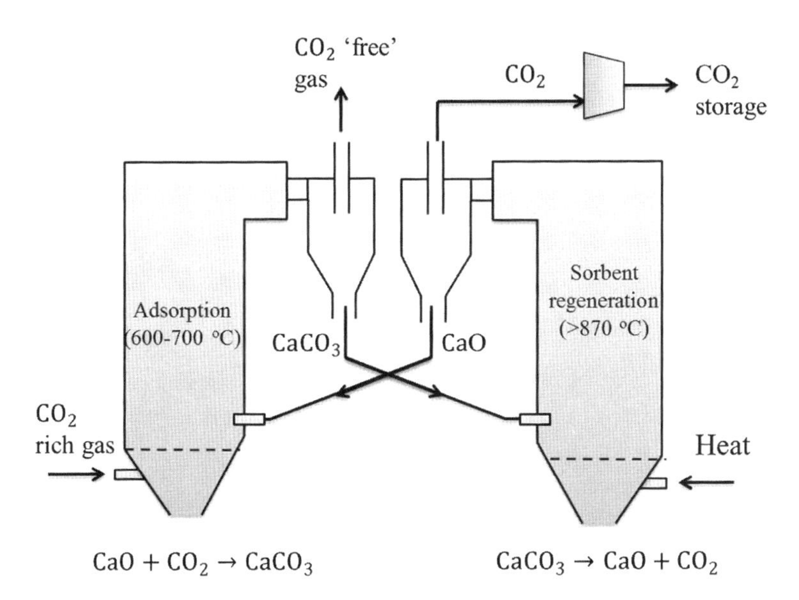

Fig. 12.5 Schematic diagram of temperature swing carbon capture process

12.6 CO_2 Separation by Absorption

CO_2 absorption can also be classified into physical and chemical absorptions. Physical absorption requires much lower energy to regenerate the solvent than chemical absorption. A conceptual comparison between chemical and physical absorption capacity at different absorbate partial pressures is shown in Fig. 12.6. Physical absorption is not economical for absorption of gases with a low partial pressure, say from the combustion flue gas. It aims at compressed gases.

12.6.1 Physical Absorption

Physical absorption processes use organic or inorganic solvents to absorb CO_2 from the carrier gases. The process capacity is governed by Henry's law described in Eq. (2.76); the equilibrium solubility of CO_2 in a solvent is

$$c_{CO_2} = \frac{P_{CO_2}}{\mathrm{H}} \tag{12.25}$$

where c_{CO_2} is the equilibrium CO_2 concentration in the solvent, P_{CO_2} is the CO_2 partial pressure in the gas phase, and H is the corresponding Henry's constant. Practically speaking, solvent with a great CO_2 solubility is preferred at a reasonable cost for physical CO_2 absorption.

Physical absorption is primarily used for high pressure CO_2 separation to increase the solubility in the solvent. Refrigerated methanol (CH_3OH) was considered an effective CO_2 solvent for CO_2 sequestration at low temperature. The solubility of CO_2 in methanol at −10 °C and P_{CO_2} = 1 atm is 10 L of CO_2 per liter of methanol, which is 4 times that of water.

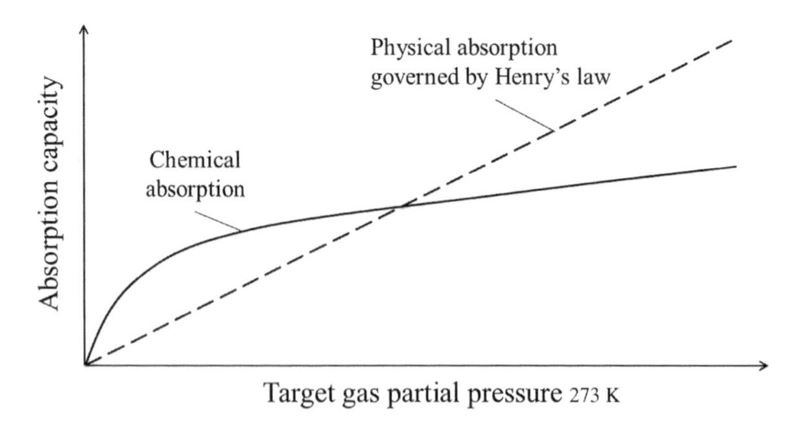

Fig. 12.6 Relative solvent loading versus partial pressure of the absorbate gas

12.6.2 Amine-Based Chemical Absorption

Chemical absorption is suitable for CO_2 capture at low pressure and low temperature. The sorbents used for chemical absorption include carbonate-based, sodium hydroxide-based, aqueous ammonia-based, and amine-based sorbents.

Amine-based CO_2 absorption is a relatively mature technology used in the ammonia process, steam reforming process, and the natural gas sweetening process. Amines are ammonia-derived organic compounds when one or more hydrogen atom(s) of ammonia are replaced with organic substituents. Amines are classified as primary, secondary, and tertiary amines based on the number of replaced hydrogen atoms.

The choice of amine for CO_2 absorption depends on three key factors: rate of reaction, regeneration energy, and loading capacity. The ideal solvent is characterized with a great rate of reaction, low regeneration energy, and great loading capacity. However, none of the alkanolamines meets all these three requirements. In general, the rate of reaction follows the order of *primary > secondary > tertiary*; the order is reversed for regeneration energy and loading capacity [35]. Ethanolamine (C_2H_7NO or $2R\text{-}NH_2$), also called *2-aminoethanol* or *monoethanolamine* (*MEA*) is the most commonly used amine for CO_2 absorption. It is a primary amine, and a alcohol too.

Among all the amines tested, MEA showed its high reactivity with CO_2, and is predominantly used for CO_2 capture in the industry. Aqueous amine solutions are bases and the pKa values of typical alkanolamines are listed in Table 12.5. Thereby, amines can react with not only CO_2 but also other acidic gases like SO_2 and NO_2 at great rates of reactions. Therefore, SO_2 and NO_2 have to be effectively removed first before CO_2 absorption if it is to be used for post-combustion carbon capture. It also reduces the contamination to the regenerated solvent and recovered CO_2.

12.6.2.1 Kinetics of Amine-CO_2 Reactions

As the most popular CO_2 absorbent, the kinetics of CO_2 reaction with alkanolamines deserves an in-depth discussion. The reaction of CO_2 with primary and secondary

Table 12.5 pKa values of base amines at 298 K

Amine		pKa
MEA	Monoethanolamine	9.50
DEA	Diethanolamine	8.88
DIPA	Diisopropanolamine	8.80
TEA	Triethanolamine	7.76
MDEA	Methyldiethanolamine	8.57
AMP	2-Amino-2-methyl propanol	9.70
DEMEA	Diethylmonoethanolamine	9.82

Source: [54]

amines is usually described by the *zwitterion mechanism*, whereas the reaction with tertiary amines is described by the base-catalyzed hydration of CO_2 [54].

According to the zwitterion mechanism [16] the reaction between CO_2 and the amine (AmH) proceeds through the formation of a zwitterion as an intermediate (R1, Eq. (12.27)).

$$CO_2 + AmH \underset{k_{b1}}{\overset{k_{f1}}{\leftrightarrow}} AmH^+COO^- \quad (R1) \tag{12.27}$$

The zwitterion reacts with a base or bases (Bs) to form carbamate as the final product (R2, Eq. (12.28)).

$$AmH^+\ COO^- + Bs \overset{k_B}{\rightarrow} AmCOO^- + BsH^+ \quad (R2) \tag{12.28}$$

When the base is MEA its self, Eqs. (12.27) and (12.28) lead to an overall reaction. It shows that the stoichiometric CO_2 to MEA mole ratio is 0.5 mol CO_2 per mole of MEA.

With water in the solution, the following reactions may also take place:.

$$H_2O \leftrightarrow H^+ + OH^- \tag{12.29}$$

$$CO_2 + OH^- \overset{k_{OH}}{\leftrightarrow} HCO_3^- \quad (R3) \tag{12.30}$$

$$CO_2 + H_2O \overset{k_{H_2O}}{\leftrightarrow} HCO_3^- + H^+ \quad (R4) \tag{12.31}$$

Assuming zwitterion concentration at quasi-steady state, the overall rate of reaction for R1 and R2 is

$$r_{1,2} = k_{f1}[CO_2][AmH] - k_{b1}[AmH^+COO^-] = k_B[AmH^+\ COO^-][Bs] \tag{12.32}$$

Reorganizing Eq. (12.32) leads to

$$\frac{r_{1,2}}{[CO_2][AmH]} = \frac{k_{f1}}{1 + k_{b1}/(k_B[Bs])} \tag{12.33}$$

When R2 is instantaneous, $k_{b1} \ll k_B$, as it is for some amines. Almost every zwitterion is deprotonated before it can revert to CO_2 and amine through the backward reaction in R1. Mathematically, the denominator in Eq. (12.33) can be simplified by $k_{b1}/(k_B[Bs]) \rightarrow 0$ and Eq. (12.33) becomes

$$r_{1,2} = k_{f1}[CO_2][AmH] \tag{12.34}$$

The rate of reaction for R3 and R4 are, respectively,

$$r_3 = k_{OH}[CO_2][OH^-] \tag{12.35}$$

$$r_4 = k_{H_2O}[CO_2][H_2O] \tag{12.36}$$

Combination of Eqs. (12.33), (12.35) and (12.36) leads to the overall CO_2 reaction rate by summation of $r_{1,2}$, r_3 and r_4:

$$r_{CO_2} = \left\{ \frac{k_{f1}[AmH]}{1 + k_{b1}/(k_B[Bs])} + k_{OH}[OH^-] + k_{H_2O}[H_2O] \right\} [CO_2] \tag{12.37}$$

Denote the part in the { } bracket as *observed reaction rate* with respect to CO_2, k_{CO_2}:

$$k_{CO_2} = \frac{k_{f1}[AmH]}{1 + k_{b1}/(k_B[Bs])} + k_{OH}[OH^-] + k_{H_2O}[H_2O] \tag{12.38}$$

where the first term is called *apparent reaction rate constant*

$$k_{app} = \frac{k_{f1}[AmH]}{1 + k_{b1}/(k_B[Bs])} \tag{12.39}$$

Similar to the denotation of $k_{OH}[OH^-] + k_{H_2O}[H_2\ O]$, we can define the reaction rate constant with respect to amine as

$$k_{AmH} = \frac{k_{f1}}{1 + k_{b1}/(k_B[Bs])} \tag{12.40}$$

Then Eq. (12.38) becomes,

$$k_{CO_2} = k_{AmH}[AmH] + k_{OH}[OH^-] + k_{H_2O}[H_2O] \tag{12.41}$$

And Eq. (12.37) becomes

$$r_{CO_2} = k_{CO_2}[CO_2] \tag{12.42}$$

Many researchers have reported k_{AmH} for different amines at different conditions and some are listed in Table 12.6.

12.6.2.2 CO_2 Absorption Rate

The actual rate of CO_2 absorption into amine is an important indicator of energy consumption of the solvent. A fast rate of CO_2 absorption minimizes energy use in

Table 12.6 k_{AmH} for different amines

Amine	Reaction rate constant (m^3/kmole · s)	T (K)	Conc. (kmole/m^3)	References
AEEA	$k_{AEEA} = 6.07 \times 10^7 \exp\left(-\frac{3030}{T}\right)$	305–322	1.19–3.46	[44]
DEMEA	$k_{DEMEA} = 9.95 \times 10^7 \exp\left(-\frac{6238}{T}\right)$	298–313		[40]
MDEA	$k_{MDEA} = 4.61 \times 10^8 \exp\left(-\frac{5400}{T}\right)$	303–313		[36]
MEA	$k_{MEA} = 4.61 \times 10^9 \exp\left(-\frac{4412}{T}\right)$	293–333	3–9	[1]
EMEA	8,000	298	0.028–0.082	[40]
AEPD	378	303	5–25 wt%	[57]
AMP	810.4	298	0.25–3.5	[56]
DEA	2,375	298	0.25–3.5	[56]
DIPA	2,585	298	0.25–3.5	[56]

an optimized system. As explained in Chaps. 2 and 5, it depends on the rate of mass transfer from gas to liquid phases. CO_2 is typically absorbed by the process of diffusion with fast reactions in the liquid films. An optimized absorber design requires 90 % CO_2 removal with a reasonable amount of packing materials in the scrubber.

According to the double film theory introduced in Sect. 2.3.4, the overall gas side mass transfer coefficient (K_G) is related to the gas film mass transfer coefficient k_g and the liquid film mass transfer coefficient k'_g as

$$\frac{1}{K_G} = \frac{1}{k'_g} + \frac{1}{k_g} \tag{12.43}$$

Then the CO_2 flux by absorption can be described as

$$CO_2 \text{ Flux} = K_G\left(P_{CO_{2,g}} - P^*_{CO_2}\right)_{LM} = k'_g\left(P_{CO_{2,\text{interface}}} - P^*_{CO_{2,\text{bulksolution}}}\right) \tag{12.44}$$

Assume that the concentration of free amine in the liquid film is the same as the bulk liquid, the liquid film mass transfer coefficient, k'_g, can be estimated using [9, 41].

$$k'_g \approx \frac{(k_{AmH} D_{CO_2}[AmH])^{\frac{1}{2}}}{H_{CO_2}} \tag{12.45}$$

where k'_g has a unit of mol/($m^2 \cdot$ Pa); it is sometimes also referred to as the normalized absorption flux of CO_2. D_{CO_2} is the diffusivity of CO_2 in the liquid (amine), k_{AmH} is the reaction rate constant of CO_2, [AmH] is the free amine concentration in the bulk solution, and H_{CO_2} is the Henry's law constant of CO_2 over the solvent.

Table 12.7 k'_g for different amines [9]

Amine	$k'_g \times 10^7$ mole/(m^2 · Pa)	Capacity (mole/kg)
Piperazine (PZ)	8.5	0.79
PZ/bis-aminoethylether	7.3	0.67
2-Methyl PZ/PZ	7.1	0.84
2-Methyl PZ	5.9	0.93
2-Amino-2-methyl propanol (AMP)	2.4	0.96
PZ/aminoethyl PZ	8.1	0.67
PZ/AMP	7.5	0.70
Hydroxyethyl PZ	5.3	0.68
PZ/AMP	8.6	0.78
2-Piperidine ethanol	3.5	1.23
Monoethanolamine (MEA)	3.6	0.66
MEA	4.3	0.47
Methydiethanolamine (MDEA)/PZ)	8.3	0.99
MDEA/PZ	6.9	0.80
Kglycinate	3.2	0.35
Ksarconinate	5	0.35
MEA/PZ	7.2	0.62

The measured k'_g for different amines and amine alternatives are summarized in Table 12.7. There is no obvious correlation between the mass transfer coefficients and the capacities of different amines.

12.6.2.3 Amine-Based CO_2 Capture Process

MEA is considered the baseline solvent for CO_2 capture. The design of a MEA based CO_2 absorption tower follows the principles introduced in Sect. 5.2. In a typical CO_2 scrubber, the flue gas at 40–60 °C enters the tower from the bottom while a 20–30 wt% MEA solution flows downward continuously from the top. Because MEA is corrosive, diluted instead of concentrated MEA is used. After selective absorption of CO_2 from the flue gas, the CO_2-rich amine solution is drained off from the bottom of the absorber.

The rich solvent is regenerated in a stripper that operates at 100–140 °C. The energy required for solvent regeneration can be from waste heat recovery from a steam and/or a reboiler. A high-efficiency heat exchanger can also be employed to recycle the heat from the rich solvent from the absorber tower. The recovered gas phase contains steam and CO_2, which are separated from each other by condensation. The final concentrated CO_2 stream is ready for CO_2 transport and storage.

As the most widely used solvent for CO_2 capture, MEA is still not an ideal solvent yet. Throughout the process, solvent degradation may take place in the presence of oxygen. Furthermore, secondary air emission is produced due to the high volatility of the solvent. And its corrosivity and energy intensive regeneration result in high costs in capital and operation.

Improved amines, such as the secondary amines (e.g., DEA) and tertiary amines (e.g., MDEA) have been considered as an alternative for MEA. Primary and secondary amines react with CO_2 quickly to form carbamate through the zwitterion mechanism. For MEA, the corresponding heat of absorption is $Q = -\Delta H_R = 2.0\,\text{MJ/kg} - CO_2$. The reaction of CO_2 with secondary amines has a lower enthalpy of reaction [54], which favors regeneration of the solvent by stripping.

Tertiary amines react with CO_2 following a base-catalyzed hydration mechanism, which is different from the zwitterion mechanism to form bicarbonate instead of carbamate. The overall reaction indicates a theoretical CO_2 loading capacity of 1 mol of CO_2 per mole of tertiary amine. However, the reactivity of tertiary amines with respect to CO_2 is lower than that of primary or secondary amines. The corresponding enthalpy of reaction for the bicarbonate formation is lower than that for the carbamate formation. This means lower energy consumption for solvent regeneration.

Piperazine (PZ) has been used in blended systems as an additive to increase the rate of absorption of CO_2 in systems with low absorption rates but otherwise attractive solvent characteristics [23]. Adding Piperazine into the solvent can improve the rate and capacity of CO_2 absorption by the following reactions.

$$KHCO_3 \rightarrow K^+ + HCO_3^- \tag{12.46}$$

$$PZ + HCO_3^- \leftrightarrow PZCOO^- + H_2O \tag{12.47}$$

$$PZCOO^- + HCO_3^- \leftrightarrow PZ(COO^-)_2 + H_2O \tag{12.48}$$

Reactions (12.47) and (12.48) also require low energy for regeneration. On the other hand, thermal degradation of Piperazine in CO_2 capture is a technical challenge that deserves further investigation.

An example of an optimized process for post-combustion CO_2 capture by amine scrubbing is available in the DOE-NETL report [19]. The CO_2 absorption rate of Piperazine (PZ) doubles that of 30 wt% MEA with 1.8 times the intrinsic working capacity. The incoming flue gas is cleaned by water spray to remove fine particulates and cooled down to 40–60 °C. Filtered lean solvent enters the amine scrubber, usually packed bed from upper level and flows downward by gravity to absorb CO_2 from the counter flow flue gas. Cleaned flue gas is further cold-water washed to recover the penetrating solvent droplets or vapor before discharging to the atmosphere through the stack. The washing water joins the lean solvent to form the rich solvent exiting the scrubber at the bottom of the tower. In addition to the main absorber, the process may also need SO_2 removal before the PZ absorber.

Recently, sterically hindered amines have also attracted considerable attention for their low regeneration costs. A sterically hindered amine is a primary amine where the amino group is attached to a tertiary carbon atom; it can also be a secondary amine where the amino group is attached to a secondary or tertiary carbon atom. For example, 2-Amino-2-methyl-1-propanol (AMP) is a sterically hindered primary amine and 2-piperidineethanol (PE) is a sterically hindered secondary amines.

12.6.3 Non-amine-Based Chemical Absorption

12.6.3.1 Sodium Hydroxide-Based Chemical Absorption

The chemical reactions between CO_2 and sodium hydroxide can be described as follows:

$$CO_2 + H_2O \rightarrow H_2CO_3 \tag{12.49}$$

$$H_2CO_3 + NaOH \rightarrow NaHCO_3 + H_2O \tag{12.50}$$

$$NaHCO_3 + NaOH \rightarrow Na_2CO_3 + H_2O \tag{12.51}$$

Water and CO_2 first react to form carbonic acid, which reacts with sodium hydroxide to produce bicarbonate ($NaHCO_3$). Bicarbonate also reacts with sodium hydroxide to produce a more stable product of carbonate. There are $NaHCO_3$ and Na_2CO_3 in the final product. Their relative quantity depends on the pH of the liquid in the last two reactions.

Regeneration of NaOH is usually achieved by adding lime (CaO) into the final product converting Na_2CO_3 to $CaCO_3$, followed by calcination at 870 °C or higher to release the pure CO_2 for storage of other applications. The corresponding chemical reactions are

$$CaO + H_2O \rightarrow Ca(OH)_2 \tag{12.52}$$

$$Ca(OH)_2 + Na_2CO_3 \rightarrow 2NaOH + CaCO_3 \tag{12.53}$$

$$CaCO_3 \rightarrow CaO + CO_2 \tag{12.54}$$

The calcination reaction Eq. (12.54) is energy intensive and the corresponding enthalpy of reaction is about $\Delta H_R = 250\,\mathrm{kJ/mol\text{-}CO_2}$. It makes this process economically challenging. Alternatively, sodium hydroxide can be regenerated using sodium trititanate ($Na_2O \cdot 3TiO_2$), which requires half the energy for regeneration of NaOH.

12.6.3.2 Carbonate-Based Chemical Absorption

Limestone reacts with CO_2 and water as follows:

$$CO_2 + H_2O \rightarrow H_2CO_3 \tag{12.55}$$

$$CaCO_3 + H_2CO_3 \rightarrow Ca(HCO_3)_2 \tag{12.56}$$

These reactions take place in the processes that are used for power plant or cement plant flue gas cleaning. Usually crushed limestone is packed in a reactor and wetted by a continuous flow of water. The CO_2-laden flue gas is pumped through the reactor to enable the above chemical reactions.

Example 12.3:
How much water and $CaCO_3$ is needed to capture 1,000 kg of CO_2 from the flue gas? How much calcium bicarbonate $\left(Ca(HCO_3)_2\right)$ is produced?

Solution
By considering the following chemical reactions,

$$CO_2 + H_2O \rightarrow H_2CO_3$$

$$CaCO_3 + H_2CO_3 \rightarrow Ca(HCO_3)_2$$

the overall reaction formula is

$$CO_2 + H_2O + CaCO_3 \rightarrow Ca(HCO_3)_2$$

It indicates that the mole ratio of the compound in this reaction is 1:1:1:1. Therefore, in order to capture 1,000 kg of CO_2, which is (1000/44) kmole, the same mole amounts of $CaCO_3$ and water are consumed to produce the same mole amount of $Ca(HCO_3)_2$.

By considering their molar weight, the corresponding mass can be determined as: 2,300 kg of $CaCO_3$ 400 kg of water and 3,700 kg $Ca(HCO_3)_2$.

Unlike other processes, the bicarbonate-rich effluent stream can be disposed in the ocean instead of regeneration. It is environmentally beneficial in that it can counteract ocean acidification. The challenges are the (fresh) water consumption and the costs for transportation related to the feedstock and product disposal.

Alternatively, potassium carbonate (K_2CO_3) can be used as a regenerable chemical absorbent. The corresponding CO_2 absorption reactions are

$$K_2CO_3 + H_2O \rightarrow KOH + KHCO_3 \tag{12.57}$$

$$KOH + CO_2 \rightarrow KHCO_3 \tag{12.58}$$

These reactions are slow at low CO_2 partial pressure, although the regeneration of the absorbent can be achieved at relatively low desorption energy of about 0.9–1.6 MJ/kg-CO_2. Therefore, it is not recommended for post-combustion CO_2 capture.

12.6.3.3 Aqueous Ammonia-Based Chemical Absorption

An aqueous ammonia solution reacts with water and CO_2 to form ammonium carbonate and ammonium bicarbonate. The overall reaction can be described as follows:

$$NH_3 + CO_2 + H_2O \rightarrow NH_4HCO_3 \tag{12.59}$$

Ammonia based chemical absorption/scrubbing also can capture SO_2 and NO_2 from a flue gas. While it allows multiple air emission control, it is challenging to produce pure acidic gases. The best option is to use the ammonium salts as fertilizer feedstock.

A typical ammonia-based CO_2 capture process is as follows. In a counter flow absorption tower flue gas flows upward with aqueous ammonia downward. The absorption process operates at near-freezing conditions (0–10 °C) with the flue gas cooled by the upstream de-SO_2 process. The low temperature allows high absorption capacity and reduces ammonia evaporation, which is also called "*ammonia slip.*"

Downstream the absorption tower, ammonia slip is further reduced by cold-water washing [39]. The effluent gas contains mainly nitrogen, oxygen and low-concentration penetrated CO_2. The solvent is regenerated at a temperatures of >120 °C pressures of >2 MPa. Cold-water washing is also employed to reduce ammonia slip in the generation process.

12.6.4 Ionic Liquids as CO_2 Solvents

Ionic liquids (ILs) have been developed for the physical and chemical absorption of CO_2. ILs are melting organic slats with unique properties. They comprise a large organic cation and a small inorganic anion. The use of ILs as CO_2 solvents is believed to have many advantages over conventional amine-based solvents, such as potentially lower energy consumption in the solvent regeneration step, lower volatility, lower vapor pressure, non-flammability, more thermally stable, and easier recycling. However, all these features are subjected to further R&D evaluation, and, they come with an unusually high manufacturing cost [9].

A great number of ILs have been developed and tested in laboratory. Among these ILs, the imidazolium class is the most widely investigated and reported. ILs without functional groups can be used as physical sorbent for CO_2 capture. ILs prepared with specific function groups enable selective gas absorption. Amine functional groups have been used effective CO_2 chemical absorption.

After CO_2 sorption, the used ILs can be regenerated by heating the spent solvent to release the absorbed gas(es). Since regeneration usually takes place at a higher temperature than the sorption stage, thermal stability is a key factor in selecting the right ILs.

12.6.4.1 CO_2 Solubility in Physical Ionic Liquids

Earlier research on CO_2 capture with ILs focused primarily on physical absorption without chemical reactions. In this section we concentrate on properties that affect the CO_2 physical absorption including CO_2 solubility and selectivity.

The actual mechanisms behind the high CO_2 solubility in IL are still not well understood [49]. Scientists and engineers are actively working on understanding and increasing the solubility of CO_2 in ILs. A review of the different approaches that have been used to model the phase behavior of gas-IL systems is provided by Vega et al. [55]. Molecular simulation and experimental data have shown that anion and cation play an important role in the dissolution of CO_2 [6, 11]. Most researchers agree that the anion dominates the solubility of CO_2 in an IL and that the cation plays a secondary role. Nonetheless, anion-fluorination and cation-fluorination can improve the solubility much [3].

CO_2 solubility in ILs increases with increasing molecular weight, molar volume, and the free volume of ILs. CO_2 solubility in ILs is dominated by entropic effects rather than solute–solvent interactions [12]. As such, it cannot be calculated based on mole fraction (like in Henry's law or Raoult's law in Sect. 2.3) because of the strong molecular weight (molar volume) of ILs. It should be determined as a function of molality, that is, the mole amount of solubility per kilogram of solvent (mole CO_2/kg IL).

Before more sound theories are available, we can use the following empirical correlation proposed by Carvalho and Coutinho [12] to calculate the CO_2 solubility in ILs by physical absorption:

$$P = m_i^0 \times \exp\left(6.8591 - \frac{2004.3}{T}\right) \tag{12.60}$$

where m_i^0 is the molality in (mole of CO_2/kg of IL), P is the pressure in MPa, and T is temperature in K. This equation was obtained using experimental data and is deemed valid for pressures up to 5 MPa, molalities up to 3 mol/kg, and temperatures ranging from room temperature to 363 K. Figure 12.7 shows the calculated solubility versus pressure at three different temperatures. For a fixed temperature, the solubility is in linear relationship with the pressure.

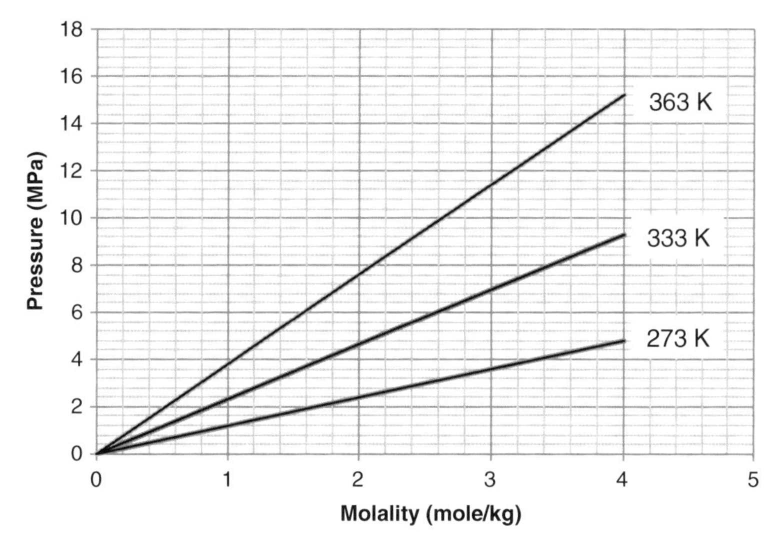

Fig. 12.7 CO_2 solubility in physical ionic liquids

Example 12.4: CO_2 solubility in physical IL
Estimate the amount of CO_2 in kg that can be absorbed into 1,000 kg of IL at 2 MPa and 300 K using the Carvalho and Coutinho [12] model.

Solution
From Eq. (12.60) we can get

$$m_i^0 = \frac{P}{\exp\left(6.8591 - \frac{2004.3}{T}\right)} = \frac{2}{\exp\left(6.8591 - \frac{2004.3}{300}\right)}$$
$$= 1.6742(\text{mole } CO_2/\text{kg IL})$$

The molar weight of CO_2 is 44 g/mole, and that gives the solubility as

$$c_{CO_2} = 1.6742 \times 0.044(\text{kg } CO_2/\text{kg IL}) = 0.07357(\text{kg } CO_2/\text{kg IL})$$

As seen in the result of this example, CO_2 solubility is low in ILs, which means extremely high cost is needed to use ILs to absorb CO_2 from industrial sources.

12.6.4.2 Gas Selectivity

Another important indicator of the right adsorbent is its selectivity from a gas mixture. Since there are always other gases, more or less, present besides CO_2. CO_2 selectivity affects the performance of CO_2 adsorption into ILs. Therefore, CO_2

Table 12.8 Henry's law constants of different gases in [hmpy][Tf_2N] ionic liquid

Gas	Temperature	
	283 K	298 K
SO_2		1.54 ± 0.01
CO_2	25.4 ± 0.1	32.8 ± 0.2
CH_4		300 ± 30
O_2	422 ± 220	463 ± 104
N_2		3390 ± 2310

solubility along is not sufficient to judge the separation performance of a solvent. We must also consider selectivity.

For pre- and post-combustion CO_2 capture, we are mostly concerned with the selectivity of CO_2/H_2, CO_2/CH_4 and CO_2/N_2. Depending on the actual gas compounds, selectivity like CO_2/H_2S, CO_2/SO_x, H_2S/CH_4 and CO_2/CO might also become important.

Ramdin et al. [49] summarized the scarce selectivity data reported in the literature. Anderson et al. [4] reported the solubility of different gases in different ILs in terms of Henry's law constant. Table 12.8 shows the Henry's law constants of [hmpy][Tf_2N]. In general, N_2 and O_2 solubility are lower compared to CO_2 leading to a high CO_2/N_2 or CO_2/O_2 selectivity.

On the other hand, SO_2 is very competitive to CO_2 absorption; SO_2 should be removed from stream first before carbon capture.

12.7 CO_2 Transportation

While CO_2 capture is accomplished at the source of emission, its destination storage sites are usually far away from the source. Transportation infrastructure is necessary to deliver CO_2 from one point to another. Although other options like rail and trucking have been studied at small scales, realistic CO_2 transportation technologies for large-scale CO_2 delivery are

- Pipeline transportation
- Ship transportation

Transporting CO_2 via pipeline is deemed to be cost-effective on land, whereas ship transportation is more economical when there is a large body of water between the source and destination. For either case, transported CO_2 is not a gas but a liquid.

12.7.1 *Pipeline Transportation*

Prior to transportation via pipeline, CO_2 is compressed to a supercritical fluid or liquid state for efficient pipeline transportation [51]. The critical point of CO_2 is 31.1 °C and 73 atm. However, temperature and pressure drops along the pipeline, therefore, prior to delivery CO_2 is compressed to a pressure that is more than 73 atm.

The pressure drop per unit length of pipeline can be described as

$$\frac{\Delta P}{\Delta L} = f_D\left(\frac{\rho U^2}{2d}\right) \tag{12.61}$$

where $\frac{\Delta P}{\Delta L}$ is the pressure drop per unit length of pipeline (Pa/m), f_D is the dimensionless Darcy friction factor, d is the diameter of the pipeline, ρ is the density of the fluids, CO_2 in this case, and U is the average speed of the fluid.

The Darcy friction factor can be calculated using the Colebrook-White equation [14].

$$f_D^{-1/2} = -2\log_{10}\left(\frac{\varepsilon}{3.7d} + \frac{2.51}{\mathrm{Re} f_D^{1/2}}\right) \tag{12.62}$$

where ε is the pipeline inner surface roughness, which is about 46 μm, for typical commercial steel pipes. Re is the Reynolds number in pipeline and it can be calculated using Eq. (12.63)

$$\mathrm{Re} = \frac{\rho U d}{\mu} = \frac{4\dot{m}}{\mu \pi d} \tag{12.63}$$

Example 12.5: Pressure drop in a pipeline
Consider a pipeline made of commercial steel with an inner diameter of 40 cm for transporting CO_2 at a flow rate of 3.5 Mega tons per year. Use CO_2 properties under 11 MPa and 25 °C as follows: $\rho = 877\,\mathrm{kg/m^3}$ and $\mu = 7.73 \times 10^{-5}\,\mathrm{Pa \cdot s}$. Estimate the pressure drop over a distance of 100 km.

Solution
Convert the unit of the CO_2 flow rate as

$$\dot{m} = 3.5\frac{\mathrm{Mt}}{\mathrm{year}} = \frac{3.5 \times 10^6 \times 1000\,\mathrm{kg}}{365 \times 24 \times 3600\,\mathrm{s}} = 111\,\mathrm{kg/s}$$

The average speed in the pipe can be determined using

$$U = \frac{4\dot{m}}{\rho \pi d^2} = \frac{4 \times 111}{877 \times \pi \times 0.4^2} = 1.007\,\text{m/s}$$

The corresponding Reynolds number is determined using Eq. (12.63).

$$\text{Re} = \frac{4\dot{m}}{\mu \pi d} = \frac{4 \times 111}{7.73 \times 10^{-5} \times \pi \times 0.4} = 4.57 \times 10^6$$

With $\varepsilon = 4.6 \times 10^{-5}$ m, $d = 0.4$ m and $\text{Re} = 4.57 \times 10^6$, the Darcy friction factor can be determined using Eq. (12.62) as follows:

$$\begin{aligned} f_D^{-\frac{1}{2}} &= -2\log_{10}\left(\frac{\varepsilon}{3.7d} + \frac{2.51}{\text{Re} f_D^{\frac{1}{2}}}\right) \\ &= -2\log_{10}\left(\frac{4.6 \times 10^{-5}}{3.7 \times 0.4} + \frac{2.51}{4.57 \times 10^6 f_D^{\frac{1}{2}}}\right) \\ &= -2\log_{10}\left(3.108 \times 10^{-5} + \frac{5.492 \times 10^{-5}}{f_D^{\frac{1}{2}}}\right) \end{aligned}$$

By iteration, we can get $f_D = 0.0217$.

Then we can calculate the pressure drop over 100 km distance length using Eq. (12.61).

$$\begin{aligned} \Delta P &= f_D\left(\frac{\rho U^2}{2d}\right)\Delta L \\ &= 0.0217\left(\frac{877 \times 1.007^2}{2 \times 0.4}\right) \times 100 \times 10^3 = 2.41 \times 10^6\,\text{Pa} \end{aligned}$$

So the pressure drop is about 2.41 MPa over a distance of 100 km. Due to this kind of great resistance, intermediate pumping (or booster) stations are required at certain intervals along the pipeline.

We have to be careful that the CO_2 properties are assumed constant in the above example in order to simplify the calculation. In reality, they cannot remain constant. Glilgen et al. [25] experimentally determined the pressure, density, and temperature relationship of CO_2 in the homogeneous region for pressures up to 13 MPa and temperatures in the range of 220–360 K. For $P < 9$ MPa and $T > 298$ K, CO_2 density varies considerably with P and T.

In the engineering practice, it is critical to ensure enough pressure to maintain above vapor–liquid equilibrium conditions to avoid liquid slugs and other operational problems resulted from a two-phase (gas–liquid) flow. A typical operating pressure is in the range of 84.9–207.3 atm where CO_2 is a dense-phase fluid over a wide range of temperatures. Nonetheless, the gas properties also depend on impurities such as SO_x, NO_x, and other air contaminants. A complex model is required to accurately predict the fluid property along the pipeline. Interested readers have to dig deeper into the literature for in-depth knowledge.

12.7.2 Ship Transportation

For CO_2 transport by ship, it is most efficient to transport CO_2 as a cryogenic liquid. The best condition for CO_2 transport by ship is about 6.4 atm and −51.2 °C [7]. As it is stored in containers during transportation, there is no pressure drop concern.

The optimal transport pressure depends on the state of the fluid. In short, large-scale transport of CO_2 by ship could be achieved by semi-pressurized vessels of around 20,000 m^3 at pressures of 6.415 atm and −52 °C in order to use existing ship and infrastructures. This condition also allows low unit cost in transport. The total costs of ship-based transport are estimated to be $20–30 per ton CO_2.

12.8 CO_2 Storage

It would be ideal to convert CO_2 into valuable products, such as methanol. A small portion can be used for production of soda drinks. Unfortunately, applications and consumptions of CO_2 are very limited compared to its production rate in fossil fuel-fired power plants and vehicles.

More practically, the captured CO_2 must be then stored properly without escape from its sinks into the atmosphere. Options listed below are considered relatively feasible both economically and environment-friendly.

(1) Geological storage

- Enhanced oil recovery (EOR),
- Deep saline reservoirs and aquifers,
- Unmineable coal beds.

(2) Deep ocean storage
(3) Ecosystem storage

Geological storage is the only CO_2 storage method that has been well developed globally at a commercial scale. Ocean storage is under extensive research and development owing to its potentially large capacity. Ecosystem storage is another promising approach that is under investigation. These options are introduced as follows.

12.8.1 Enhanced Oil Recovery and Enhanced Gas Recovery

Carbon dioxide (CO_2) can be employed for EOR by CO_2 flooding for reservoirs deeper than 800 m. This depth allows CO_2 to be stored as a dense supercritical fluid. In the EnCana Weyburn field project that started September 2000, for example, over 2.2 Mt supercritical CO_2 per year is injected to a depth of 1,400–1,500 m for OER [22].

A schematic diagram of an EOR process is shown in Fig. 12.8. CO_2 is injected into the deep reservoir from the injection well, and it forms a CO_2 bank. A CO_2 miscible oil bank is formed as a result of the miscible CO_2-oil mixing. The mixture of CO_2 and crude oil is delivered to the ground surface through the production well. They are separated at the ground facilities, producing liquid and gaseous fuels with recovered CO_2.

For immiscible CO_2 operation, CO_2 bank pushes oil forward from the reservoir to the production well. Depending on the reservoir, the incremental oil recovery by EOR ranges from 5 to 15 %. This additional economic benefit makes EOR competitive over other options.

Eventually, a significant amount of CO_2 injected for EOR is trapped underground by residual oil, water, or mineral. It is expected that this part of CO_2 can remain underground for thousands of years. However, the supercritical CO_2 is mixed with water may be transported underground out of the reservoirs. Long term CO_2 distribution in the reservoirs can be predicted by numerical simulations for long-term risk assessment.

Similar to EOR, CO_2 can also be injected into depleting gas reservoir for enhanced gas recovery. However, the economical return is not as high as EOR in that most primary gas recovery rate is pretty high already, and current gas price is low.

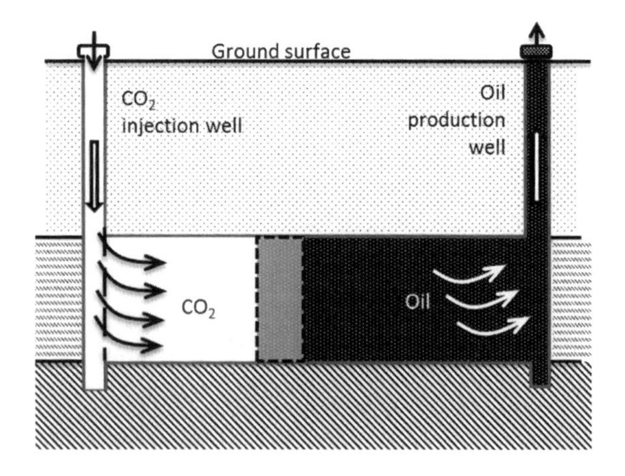

Fig. 12.8 EOR by CO_2 injection

12.8.2 Coal Bed Methane Recovery

Coal bed methane is the methane adsorbed in underground coals. Coal bed methane recovery is then a methane desorption process based on pressure swing desorption. The process is similar to EOR above; liquefied CO_2 is injected into a deep well, which is drilled into the coal bed, to reduce the hydrostatic pressure. As a result, methane is released from coal bed and transported to gas pipeline.

CO_2 injection also results in enhanced methane recovery from coal bed methane by the competitive adsorption between CO_2 and methane. According to physical adsorption mechanisms, CO_2 and methane compete for the adsorption sites in the pores of the coal. As seen Fig. 12.9, coal adsorbs CO_2 more effectively than methane [37], this adsorption preference results in desorption of methane from the coal when CO_2 is injected into the coal bed. Liberated methane from the coal bed is also delivered to the surface facilities.

Meanwhile, the replacement of methane with CO_2 on the pores of the coal also results in swelling of the coal matric and this may lead to reduced permeability, and consequent coal softening and more CO_2 injection in order to recover more methane.

Cost-effective coal bed methane recovery requires high initial concentration of methane and adequate permeability for gas flow. Both depend on the pressure or depth of the coal bed: the deeper, the higher methane concentration but the lower permeability. This contradictory effect requires economical coal bed methane recovery to be executed at a depth in the range of 300–1,000 m or so.

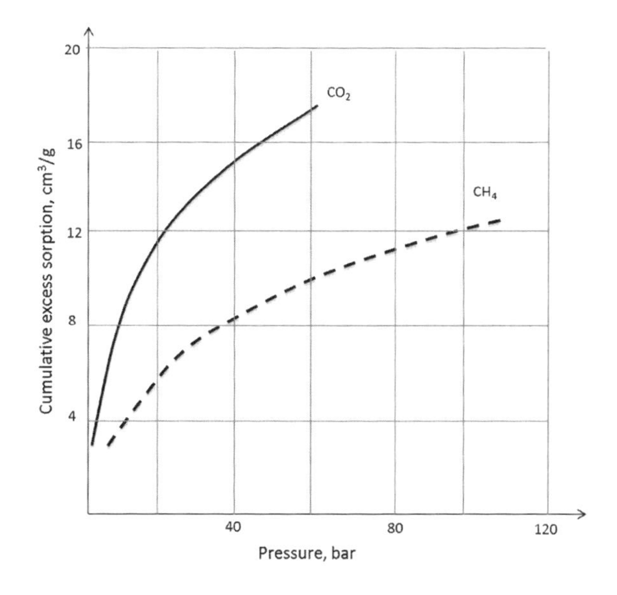

Fig. 12.9 Excess sorption isotherms of CO_2 and CH_4 in a coal (*Data source* Krooss et al. [37])

12.8.3 Saline Aquifer Storage

Saline aquifer storage is a CO_2 storage option for regions without energy resources recovery opportunities [13, 53]. Similar to EOR, CO_2 is injected into the saline aquifer as a supercritical fluid, which forms a deeply underground plume. The storage potential and security risks depend on the geological characteristics and usually they are predicted by model simulations.

Figure 12.10 shows a schematic diagram of deep saline aquifer CO_2 storage. CO_2 is injected down to the underground and sealed with a cap rock. Over time, stored CO_2 migrates slowly out of the CO_2 reservoir as a result of dissolution or chemical reactions with the cap rock.

There are four primary mechanisms for saline aquifer CO_2 storage, in order of increasing security

- Structural and stratigraphic trapping
- Residual trapping
- Solubility trapping, and
- Mineral trapping

Structural and stratigraphic trapping CO_2 is trapped under a cap rock. This is simply an initial and temporary physical trapping of CO_2 for 1–100 years, depending on the structural characteristics.

Residual trapping As the CO_2 plume rises through a water-saturated rock, high pressure buoyant CO_2 drives water out of the pores of the rock. Some CO_2 is trapped in the rock pores as residue. The residual trapping over time declines as CO_2 is dissolved in the formation brine and diffuses into the surrounding unsaturated aquifer.

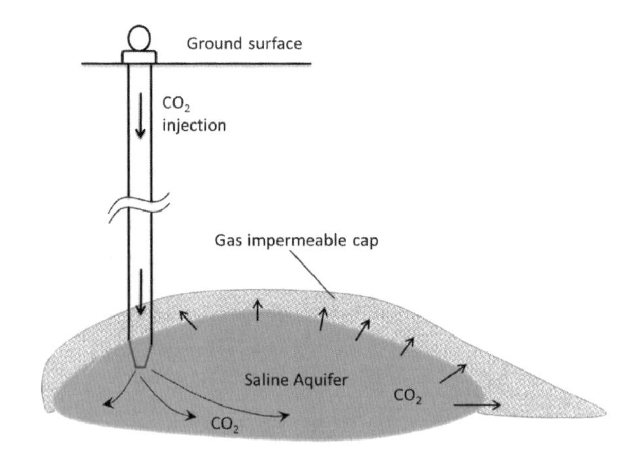

Fig. 12.10 Saline aquifer CO_2 storage

According to Darcy's law, the volume flow rate of a fluid through the porous medium is

$$J = -\frac{kA}{\mu}\frac{dp}{dx} \tag{12.64}$$

where J is the volume flow rate in m^3/s, k is he permeability of the porous medium (m^2), A is the area (m^2), μ is the fluid viscosity (kg/m · s or Pa · s), dp/dx stands for the pressure gradient in Pa/m over flow direction. One can estimate the bottom hole pressure corresponding to a certain CO_2 injection rate or leaking rate.

Solubility trapping Solubility trapping is one major trapping mechanism in saline aquifer storage of CO_2. Dissolution of CO_2 in formation water can last as long as 1,000 years, depending on the brine composition and pH as well as the mineralogy of the reservoir.

Mineral trapping Mineral trapping is resulted from the chemical reactions between CO_2 and metal ions such as Ca^{2+}, Fe^{2+} and Mg^{2+}, which are rich in the surrounding rocks. As the carbonates produced precipitate in the rock pores the reactions slow down over time. As a result, it takes 10–10,000 years to saturate the pores. Meanwhile, the chemical reactions take place in the pores of the cap rock, improving the integrity of cap rock over time.

12.8.4 Deep Ocean Storage

The ocean naturally traps over 143,000 Gt of CO_2, which is 50 times more than that in the atmosphere. The uptake of CO_2 in ocean has been increasing over the past centuries as a result of the increasing atmospheric CO_2 concentration. However, it still can hold much more! A comprehensive documentation of deep ocean CO_2 storage is given by Caldeira et al. in the form of a SRCSS Special Report to IPCC [32] (Chap. 6, Ocean Storage).

Similar to what was introduced in mass transfer in absorption (Sect. 2.3.4), natural CO_2 storage in ocean is a chemical absorption process. CO_2 enters ocean through the surface water, where the equilibrium concentration is governed by Henry's law, $x_{CO_2} = P_{CO_2}/H$, where H is the Henry's law constant for CO_2 -sea water system.

The solubility of CO_2 in seawater is not constant everywhere; it depends on the pressure, salinity, pH, and temperature of seawater.

Deeper into the ocean, CO_2 goes through chemical reactions. Several simplified chemical equilibriums define the process as

$$CO_2 + H_2O \leftrightarrow H_2CO_3 \leftrightarrow HCO_3^- + H^+ \leftrightarrow CO_3^{2-} + 2H^+ \tag{12.65}$$

All these species are grouped as dissolved inorganic carbon (DIC). H_2CO_3 also reacts with dissolved limestone in the ocean, which is a result of weathering the earth surface, and the simplified chemical reaction is

$$CaCO_3 + H_2CO_3 \rightarrow Ca^{2+} + 2HCO_3^- \tag{12.66}$$

Besides the inorganic carbons, CO_2 is also converted into particulate organic carbon (POC) and fertilizer by photosynthesis. For example,

$$6CO_2 + 6H_2O + uv \rightarrow C_6H_{12}O_6 + 6O_2 \tag{12.67}$$

$$\begin{aligned} &106CO_2 + 122H_2O + 16HCO_3^- + H_3PO_4 \\ &\rightarrow C_{106}H_{263}O_{110}N_{16}P + 138O_2 \end{aligned} \tag{12.68}$$

One way or another, CO_2 is not saturated in deep ocean as a result of these chemical reactions and biological conversions of CO_2 into organic and inorganic carbon compounds. Therefore, more CO_2 can be stored in deep ocean. However, natural absorption of CO_2 into ocean is a slow process and it cannot catch up with the increase rate of anthropogenic CO_2 in the atmosphere.

This process can be expedited by injecting CO_2 into deep ocean. Based on the properties of liquid CO_2 in ocean, engineered deep ocean CO_2 storage can be achieved by

- direct CO_2 dissolution, and
- liquid CO_2 isolation

The fate of the injected CO_2 depends on the injection depth in the ocean. At a depth of about 500 m (corresponding to the pressure of 4–5 MPa and temperatures in the range of 0–10 °C), CO_2 starts to liquefy and the liquid CO_2 has a density of 860–920 kg/m^3. The deeper into the ocean, the greater liquid CO_2 density as a result of the greater pressure.

According to the liquid CO_2 density relative to the seawater density, the depths can be divided into three zones where CO_2, floats, suspends, and sinks, respectively. As shown in Fig. 12.11, liquid CO_2 density is close to that of seawater at depths of about 2,500–3,000 m. This is also referred to as thermocline. Above this zone, CO_2 density is less than the surrounding seawater. Therefore, the liquid CO_2 droplets rise as a plume. In the transition zone, CO_2 density is nearly close to that of seawater and it suspends in seawater due to neutral buoyance. Further deep into the ocean, CO_2 density surpasses that of the seawater resulting in sinking CO_2 plume. For any one of the three scenarios, most of the released CO_2 will eventually dissolve in the ocean.

The sinking CO_2 droplets eventually reach the bottom of the ocean enabling liquid CO_2 isolation as a liquid lake in an ocean floor depression. Such a CO_2 lake can be made by injecting liquid CO_2 to the ocean floor depression or releasing CO_2 at a depth that is close to the target lake. Settled liquid CO_2 in the lake mixes with

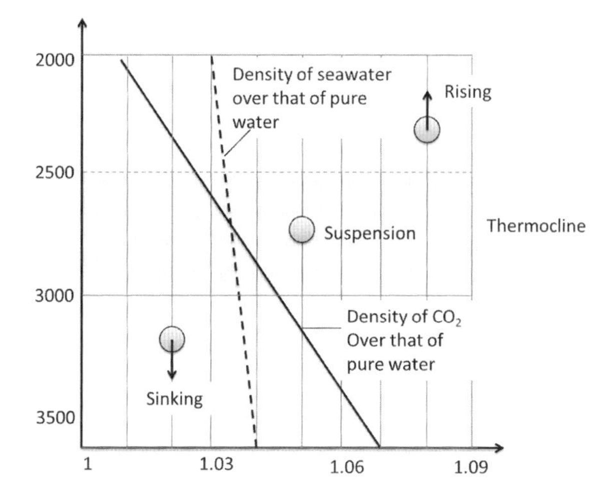

Fig. 12.11 CO_2 density versus ocean depth

the capping sea water due to ocean turbulence. This vertical mixing results in certain amount of CO_2 dissolution. Undisturbed CO_2 lake is expected to have a lifetime of about 10,000 years.

12.8.5 Ecosystem Storage

Like extraterrestrial oceans, terrestrial ecosystems have been an important and long-lasting natural CO_2 sink by photosynthetic reactions in biomass and complex interactions with soils. Unfortunately, recent human activities have reduced its CO_2 storage capacity resulting in increased atmospheric CO_2 concentration. Before the availability of other cost-effective options, ecosystem CO_2 storage is deemed to have a potential to buy human more time and slow down the negative impact of increasing anthropogenic CO_2 emissions.

While the CO_2 storage capacity of terrestrial ecosystems is much less than that of the ocean, it is easy to operate and cost-effective. Ecosystem carbon storage can be achieved by

- Agricultural carbon storage
- Change in land use
- Energy crops production.

12.8.5.1 Agricultural Carbon Storage

Agricultural carbon storage is an approach to additional CO_2 storage in biomass by improved management of vegetation and soil in the ecosystem. Specifically, it can be achieved by conservation tillage, crop selection, rotation and intensified cropping, managing soil biogeochemistry, and manipulation of microbial communities.

Tilling disturbs the soil resulting in reduced CO_2 storage. Therefore, reduced tillage intensity contributes to increased CO_2 storage. It can be achieved by no till, ridge till, minimum till, and mulch till. Conservation tillage aims at the reduction of soil erosion and the maintenance of water-holding capacity of soils. It is expected to restore the loss of CO_2 that was released by historical intensive tillage practice over a short period of time.

Crop selection, rotation, and intensified cropping allow more above-ground biomass, and consequently enable more CO_2 storage. Their net carbon storage effects will be maximized when applied in conjunction with conservation tillage.

Soil organic carbon may depend on the soil biological and chemical properties, for example, pH. Addition of metal oxides such as CaO and MgO is expected to enhance the chemical sorption of CO_2 into the soils. However, this change in pH may also change the behaviors of microorganisms in the soils.

Fungal and bacterial species in the soils play an important role in the release and storage of soil organic carbon as well as the emissions of other GHGs such as N_2O and CH_4. The complexity of microbial community opens another opportunity for potential CO_2 storage and/or reduced GHG emissions by multiplication of the microbial species. Different ongoing research projects are being conducted to evaluate the feasibility of this approach.

12.8.5.2 Change in Land Use

Land on the earth can be grouped into, in the increasing order of carbon storage share, degraded land $<$ croplands $<$ pasture $<$ grass lands $<$ wetlands $<$ forest. Land upgrading in terms of carbon storage capacity is expected to contribute to additional CO_2 storage. IPCC (2000) report on land use, land-use change and forestry estimated that about 1 Gt-C per year could be stored in the short term as a result of the regrowth of perennial vegetation and improvements of land management practices in croplands, grasslands, and forests. For example, change in land use can be achieved by wetland management and restoration and forestry management, afforestation, and reforestation.

Protection and restoration of wetlands presents an opportunity for increased underground organic carbon storage. The residence time of the CO_2 stored in wetland depends on the plant type and the degree of inundation. Since wetland is the second best land in terms of CO_2 storage, only after forest, more CO_2 is released into the atmosphere when wetland is drained for uses as agricultural or urban and industrial development. Unfortunately, recent research shows that it is challenging to restore the lost CO_2 simply by recreating wetlands. More research and development is needed in this area.

The loss of CO_2 storage in forest can be reduced by reducing logging and deforestation. Protection of old trees and regeneration of secondary and degraded forests stand biomass with improved CO_2 storage. Afforestation of croplands is another option at a reasonable cost. The effect can be further extended by replacing soft bushes with standing wood forest.

12.8.5.3 Energy Crop Production

Energy crop production is the option that regenerates biomass from CO_2 in the atmosphere or concentrated CO_2 from carbon capture CO_2 is converted into biomass by photosynthesis and these crops are burned or converted into fuels for combustion, where CO_2 is released again. Energy crops are plants grown specifically for energy production. They are planted and harvested periodically. These energy crops contain oils (soybeans, nuts and grains), sugar (sugar beets), starches (corns, cereal), and lignocellulose in residues or woody biomass.

Technical description of energy production and biofuels from energy crops have been introduced in Sect. 8.4. It is not repeated here.

Energy crops can be classified into herbaceous and short-rotation woody crops [10]. The former is characterized with low lignin content enabling easy delignification and improved accessibility to carbohydrate in lignocellulose for biofuel production. Examples of herbaceous energy crops are

- Perennials like sugarcane, napier grass
- Annuals such as corn, forage sorghum
- Thin-stemmed, warm season perennial, e.g., switch grass

Woody energy crops are fast growing and suitable for use in dedicated supply systems. Desirable candidates are characterized with rapid juvenile growth, wide site adaptability, and pest and disease resistance. They are grown on a sustainable basis and harvested on a rotation of 3–10 years. For example,

- Flowering plants like willow, oak, poplar
- Evergreens like pine, spruce, cedar

Algal biomass is a special energy crop that can be grown in an aqueous environment. It can be cultivated in ponds built beside stationary CO_2 capture sources such as power plants, cement plants, and municipal waste plants. This practice enables high productivity per area of land with low foot prints. In addition, it is expected to produce and harvest biomass continuously with engineered nutrient control. The process allows simple operation with low quality land and water that are unsuitable for other crop growth. Processed algae can be used as a solid fuel or a feedstock for biodiesel. Readers are referred to the literature for advances in algae and energy crop production.

12.9 Environmental Assessment

Although many technologies have been claimed to be effective in reducing carbon emissions, a comprehensive life cycle analysis (LCA) must be conducted before an affirmative conclusion can be made. Since the 1970s, environmental assessment has been developed as a systematic process to identify, analyze, and evaluate the environmental effects of products or activities to ensure that the environmental implications of decisions are taken into account *before* the decisions are made. Environmental assessment allows effective integration of environmental considerations and public concerns into decision-making.

In principle, environmental assessment can be undertaken for individual projects such as a dam, motorway, factory, or a bioenergy plantation; it can also be done at a large scale for plans, programs, and policies. These approaches aim at providing a systematic procedure for identifying potential risks to human health and the environment, and a comparison of the respective risks to alternative options for different environmental compartments (air, soil, water).

LCA is a specific method developed in the 1980s for determining and comparing the potential environmental impacts of product systems or services at all stages in their life cycle—from extraction of resources, through the reduction and use of the product to reuse, recycling or final disposal. It can be applied in strategy formulation, product development, and marketing. The LCA methodology has been developed extensively during the last decade. Moreover, a number of LCA-related standards (ISO 14040-14043) and technical reports have been published by the International Organization for Standardization (ISO) to streamline the methodology.

The LCA approach is quite data-intensive. Not only are direct impacts included, but also those stemming from "upstream" activities such as mining, processing, and transport, as well as the materials (and energy) needed to manufacture all processes. With LCA being developed as a specific assessment methodology to compare products, the formal assessment requirements from the ISO standards for LCA are demanding with regard to time and resources.

References and Further Readings

1. Aboudheir A, Tontiwachwuthikula P, Chakmab A, Idema R (2003) Kinetics of the reactive absorption of carbon dioxide in high CO_2-loaded, concentrated aqueous monoethanolamine solutions. Chem Eng Sci 58(2003):5195–5210
2. Adanez J, Abad A, Garcia-Labiano F, Gayan P, de Diego LF (2012) Progress in chemical-looping combustion and reforming technologies. Prog Energy Combust Sci 38:215–282
3. Almantariotis D, Gefflaut T, Padua AAH, Coxam JY, Costa Gomes MF (2010) Effect of fluorination and size of the alkyl sidechain on the solubility of carbon dioxide in 1-alkyl-3-methylimidazolium bis(trifluoromethylsulfonyl)amide ionic liquids. J Phys Chem B 114:3608–3617

4. Anderson JL, Dixon JK, Brennecke JF (2007) Solubility of CO_2, CH_2, C_2H_6, C_2H_4, O_2 and N_2 in 1-hexyl-3-methylpyridinium bis (trifluoromethylsulfonyl)imide: comparison to other ionic liquids. Acc Chem Res 40:1208–1216
5. Anderson DM, Mauk EM, Wahl ER, Morrill1 C, Wagner AJ, Easterling D, Rutishauser T (2013) Global warming in an independent record of the past 130 years. Geophys Res Lett 40 (1):189–193
6. Anthony JL, Anderson JL, Maginn EJ, Brennecke JF (2005) Anion effects on gas solubility in ionic liquids. J Phys Chem B 109:6366–6374
7. Aspelund A, Mølnvik MJ, de Koeijer G (2006) Ship transport of CO_2 technical solutions and analysis of costs, energy utilization, exergy efficiency and CO_2 emissions. Chem Eng Res Des 84(A9):847–855
8. Bidwe AR, Mayer F, Hawthorne C, Charitos A, Schuster A, Schenecht G (2011) Use of ilmenite as an oxygen carrier in chemical looping combustion-batch and continuous dual fluidized bed investigation. Energy Procedia 4:433–440
9. Boot-Handford ME et al (2014) Carbon capture and storage update. Energy Environ Sci 7:130–189
10. Brown RC (2003) Biorenewable resources, engineering new products from agriculture. Blackwell Publishing, Iowa State Press, Boston
11. Cadena C, Anthony JL, Shah JK, Morrow TI, Brennecke JF, Maginn EJ (2004) Why is CO_2 so soluble in imidazolium-based ionic liquids? J Am Chem Soc 126:5300–5308
12. Carvalho PJ, Coutinho JAP (2010) On the nonideality of CO_2 solutions in ionic liquids and other low volatile solvents. J Phys Chem Lett 1:774–780
13. Chadwick RA, Zweigel P, Gregersen U, Kirby GA, Holloway S, Johannessen PN (2004) Geological reservoir characterization of a CO_2 storage site: the Utsira Sand, Sleipner, northern North Sea. Energy 29:1371–1381
14. Colebrook CF (1939) Turbulent flow in pipes, with particular reference to the transition region between smooth and rough pipe laws. J Inst Civil Eng 11(4):133–156
15. Cuadrat A, Abad A, Garcıa-Labiano F, Gayan P, de Diego LF, Adanez J (2012) Effect of operating conditions in chemical-looping combustion of coal in a 500 Wth unit. Int J Greenhouse Gas Control 6:153–163
16. Danckwerts PV (1979) The reaction of CO_2 with ethanolamines. Chem Eng Sci 34:443–446
17. D'Alessandro DM, Smit B, Long JR (2010) Carbon dioxide capture: prospects for new materials. Angew Chem Int Ed 49(35):6058–6082. doi:10.1002/anie.201000431
18. DOE-NETL (2011) Research and development goals for CO_2 capture technology; Pittsburgh, PA, DOE/NETL-2009/1366
19. DOE-NETL (2012) Techno-economic analysis of CO_2 capture-ready coal-fired power plants, DOE/NETL-2012/1581
20. de Nevers N (2000) Air pollution control engineering. The McGraw-Hill Companies, Inc, New York
21. EIA (Energy Information Administration) (2007) International Energy Outlook (IEO), May 2007
22. EIA (2014) EIA.doe.gov. http://www.eia.gov/oiaf/1605/ggccebro/chapter1.html. Accessed 17 June 2014
23. Freeman S, Davis J, Rochelle GT (2010) Degradation of aqueous piperazine in carbon dioxide capture. Int J Greenhouse Gas Control 4:756–761
24. Grande CA (2012) Advances in pressure swing adsorption for gas separation. ISRN Chem Eng 2012, Article ID 982934. doi:10.5402/2012/982934
25. Glilgen R, Klenrahm R, Wagner W (1992) Supplementary measurements of the (pressure, density, temperature) relation of carbon dioxide in the homogeneous region at temperatures from 220 K to 360 K and pressures up to 13 MPa. J Chem Thermodyn 24:1243–1250
26. Houghton JT et al (eds) (1996) Climate change 1995: the science of climate change, IPCC. Cambridge University Press, Cambridge
27. Houghton J (1997) Global warming: the complete briefing, 2nd edn. Cambridge University Press, Cambridge

28. Houghton JT, Jenkins GJ, Ephramus JJ (eds) (1990) Climate change: the IPCC science assessment. Cambridge University Press, Cambridge
29. Houghton RA, Woodwell GM (1989) Global climate change. Sci Am 260(4):36–44
30. IPCC (The Intergovernmental Panel on Climate Change) (2005) Carbon capture and sequestration report
31. IPCC (2007) Changes in atmospheric constituents and in radiative forcing of the 2007. IPCC Fourth Assessment Report (AR4) by Working Group 1 (WG1). http://www.ipcc.ch/pdf/assessment-report/ar4/wg1/ar4-wg1-chapter2.pdf
32. IPCC (2013) Climate change 2013. In: Solomon S, Qin D, Manning M, Marquis M, Averyt K, Tignor MMB, Miller HL Jr, Chen Z (eds) The physical science basis. Cambridge University Press, New York
33. Ishida M, Zheng D, Akehat T (1987) Evaluation of a chemical-looping-combustion power-generation system by graphic exergy analysis. Energy 12(2):147–154
34. Jerndal E, Mattisson T, Lyngfelt A (2006) Thermal analysis of chemical-looping combustion. Chem Eng Res Des 84:795–806
35. Kim YE, Lim JA, Jeong SK, Yoon YI, Bae ST, Nam SC (2013) Comparison of carbon dioxide absorption in aqueous MEA, DEA, TEA, and AMP solutions. Bull Korean Chem Soc 34 (3):783–787
36. Ko J, Li M (2000) Kinetics of absorption of carbon dioxide into solutions of N-methyldiethanolamine + water. Chem Eng Sci 55:4139–4147
37. Krooss BM, van Bergen F, Gensterblum Y, Siemons N, Pagnier HJM, David P (2002) High-pressure methane and carbon dioxide adsorption on dry and moisture-equilibrated Pennsylvanian coals. Int J Coal Geol 51:69–92
38. Lewis WK, Gilliland ER (1954) US Pat., No. 2,665,972
39. Li K, Yu H, Tade M, Feron P, Yu J, Wang S (2014) Process modeling of an advanced NH_3 abatement and recycling technology in the ammonia-based CO_2 capture process. Environ Sci Technol 48(12):7179–7186
40. Li J, Henni A, Tontiwachwuthikul P (2007) Reaction kinetics of CO_2 in aqueous ethylenediamine, ethyl ethanolamine, and diethyl monoethanolamine solutions in the temperature range of 298–313 K, using the stopped-flow technique. Ind Eng Chem Res 46:4426–4434
41. Li L, Li H, Namjoshi O, Du Y, Rochelle GT (2013) Absorption rates and CO_2 solubility in new piperazine blends. Energy Procedia 37:370–385
42. Liu J, Thallapally PK, McGrail BP, Brown DR, Liu J (2012) Progress in adsorption-based CO_2 capture by metal–organic frameworks. Chem Soc Rev 41:2308–2322
43. Liu Y, Wang Z, Zhou H (2012) Recent advances in carbon dioxide capture with metal-organic frameworks. Greenhouse Gas Sci Technol 2:239–259
44. Ma'mun S, Dindore VY, Svendsen HF (2007) Kinetics of the reaction of carbon dioxide with aqueous solutions of 2-((2-aminoethyl)amino)ethanol. Ind Eng Chem Res 46:385–394
45. Mattisson T, Lyngfelt A, Leion H (2009) Chemical-looping oxygen uncoupling for combustion of solid fuels. Int J Greenhouse Gas Control 3:11–19
46. Nakayama S, Noguchi Y, Kiga T, Miyamae S, Maeda U, Kawai M, Tanaka T, Koyata K, Makino H (1992) Pulverized coal combustion in O_2/CO_2 mixtures on a power plant for CO_2 recovery. Energy Convers Manage 33(5–8):379–386
47. Oke TR (1987) Boundary layer climates, 2nd edn. Methuen, London, p 14
48. Pirngruber GD, Leinekugel-le-Cocq D (2013) Design of a pressure swing adsorption process for postcombustion CO_2 capture. Ind Eng Chem Res 52:5985–5996
49. Ramdin M, de Loos TW, Vlugt TJH (2012) State-of-the-art of CO_2 capture with ionic liquids. Ind Eng Chem Res 51:8149–8177
50. Ryden M, Lyngfelt A, Mattisson T (2011) $CaMn_{0.875}Ti_{0.125}O_3$ as oxygen carrier for chemical-looping combustion with oxygen uncoupling (CLOU)—experiments in a continuously operating fluidized-bed reactor system. Int J Greenhouse Gas Control 5:356–366

51. Seevam P, Race JM, Downie MJ, Hopkins P (2008) Transporting the next generation of CO_2 for carbon capture and storage: the impact of impurities on supercritical CO_2 pipelines. In: 2008 7th international pipeline conference, Calgary, Alberta, Canada, vol 1, pp 39–51. Paper No. IPC2008-64063, 29 Sept–3 Oct 2008
52. Smith JB, Tirpak DA (1990) The potential effects of global climate change on the United States. Hemisphere, New York
53. Torp TA, Gale J (2004) Demonstrating storage of CO_2 in geological reservoirs: the Sleipner and SACS projects. Energy 29:1361–1369
54. Vaidya PD, Kenig EY (2007) CO_2-alkanolamine reaction kinetics: a review of recent studies. Chem Eng Tech 30(11):1467–1474
55. Vega LF, Vilaseca O, Llovell F, Andreu JS (2010) Modeling ionic liquids and the solubility of gases in them: recent advances and perspectives. Fluid Phase Equilib 294:15–30
56. Xu S, Wang Y, Otto FD, Mather AE (1996) Kinetics of the reaction of carbon dioxide with 2-amino-2-methyl-1-propanol solutions. Chem Eng Sci 51(6):841–850
57. Yoon SJ, Lee H (2002) Kinetics of absorption of carbon dioxide into aqueous 2-amino-2-ethyl-1,3-propanediol solutions. Ind Eng Chem Res 41:3651–3656

Chapter 13
Nanoaerosol

Nanomaterials are now widely used in many industries, for example, for improving combustion efficiency, environmental protection, health, and renewable energy production. Once these nanoparticles enter the air, they may result in nano air pollution and have to be monitored and filtered for the protection of the environment and health. Engineering approaches to nano air pollution control is the core part of this chapter. Specially, it will cover the properties of nano air pollution and its implications on air monitoring and air filtration technologies.

13.1 Sources of Nanoaerosol

Nanoaerosols are nanoparticles suspended in a gas. These nanoparticles could be liquid droplets but more often solid particles with at least one dimension being less than 100 nm. Most researchers consider nanoaerosol as another name of ultrafine aerosol or ultrafine particulate matter. There is actually a slight difference between ultrafine aerosol or particulate matter and nanoaerosol. The former is commonly used to describe airborne nanoparticles that are produced incidentally without intension and are suspended in the atmosphere. The latter has a broader coverage including both environmental and engineered nanoparticles in any carrier gas.

Nanoaerosols are produced from various sources intentionally or as a byproduct. Environmental nanoaerosols are produced in the atmosphere by natural nucleation and condensation or incomplete combustion of hydrocarbons. The latter are mostly soot particles; a soot particle is a cluster of nanoparticles between 10 and 100 nm. Engineered nanoaerosols are a result of recent rapid advances in nanotechnology, produced when manufactured nanomaterials become suspended in the air or other carrier gases. These particles usually have complex shapes including sphere, cube, cylinder, flake, crystal, and so on. As introduced in Chap. 4, these different shapes affect their aerodynamics.

Thousands of years ago, the Chinese started collecting soot particles from burnt pine to make high quality ink for fine painting. This may be the earliest engineered nanoaerosol. Nowadays, engineered nanomaterials find more applications to the

Z. Tan, *Air Pollution and Greenhouse Gases*, Green Energy and Technology,
DOI 10.1007/978-981-287-212-8_13

improvement of our quality of life. Many different types of nanosized drugs have been developed and some of them can be aerosolized and delivered by inhalation. Respiratory nanomedicine delivery benefits a patient due to high drug deposition in central and peripheral regions of the lungs [7]. However, human respiratory systems did not seem to be effective in capturing nanoaerosol particles smaller than 40 nm [23]. Drug delivery systems for nanoaerosol are believed to have potential to significantly reduce extrathoracic depositional drug loss [7]. Dosing effect can be improved by charging the particles. Manufacturing cylindrical or tubular shaped nanoaerosol drugs may also help since their actual aerodynamic diameters are increased to micron range and tend to travel only one way into the human respiratory system. On the other hand, manufacturing sub-10 nanoaerosol-based drugs and associated delivery devices are challenging because they agglomerate easily at high concentrations.

Another example is the wide use of nano-silver spray as a disinfectant. During the short life of nanoaerosol after spraying, high dose exposure could result in acute or chronic health effects. The mechanisms of toxicity of silver ions (Ag^+) are well known, but little is known about toxicity nano-silver induced to living organisms. In addition, limited data available has shown that other potentially hazardous and toxic materials are involved in the product and the disinfectant spray could also have negative impacts on the environment and human health.

Nanotechnology market keeps increasing in the past decade to a level of the order of hundreds billion US dollars, according to a report of European Agency for Safety and Health at Work [5]. New market opportunities lead to growing concerns regarding potential risks to the environment and health of human beings. In addition to the safety risk of explosion, some experts suspect that nanomaterials affect health more than microscale materials. At the workplace, workers can be exposed during the production process, use of nanomaterials, transport, storage or waste treatment.

Nanomaterials might also have negative impact on the environment by water and solid contamination, causing indirect impact on people's health. As illustrated in Fig. 13.1, nanoaerosol emitted and formed in the atmosphere eventually falls down to the ground with precipitation. Many experts in nanotechnology believe that it is the next industrial revolution. On the other hand, new technologies come with new risks, which are discovered often later than their benefits.

13.2 Exposure to Nanoaerosol

While the society is enjoying the benefits and excitement brought by nanotechnology, some experts are concerned about its negative impact on human health and the environment. In addition to the nanoaerosol produced in nature and by combustion, recent rapid advances in nanotechnology have outpaced the risk assessment and government regulations in this industry. Nanoaerosol can be more toxic than larger ones of the same material because of their small size, large surface area and great diffusivity. Long-term exposure to nanoaerosol may cause ischemic heart disease, cardiovascular diseases, stroke, chronic bronchitis, asthma, and respiratory tract infections [14].

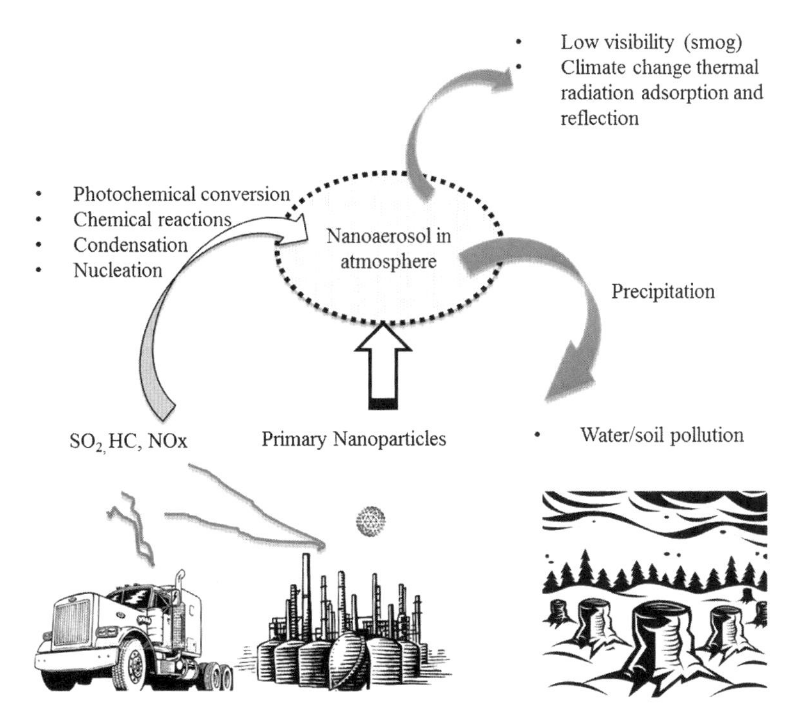

Fig. 13.1 Environmental impact of nanoaerosol

Nanoparticles in certain size range can transport toxic chemicals into the human body through the respiratory system. These chemicals may cause more damage than their counterparts in micron size range. Comparison between the rat's exposure to 250 and 20 nm titanium oxide (TiO_2) aerosol particles of the same weight showed that more 20 nm particles were trapped in the lung, resulting in inflammation [12]. Similar studies showed that exposure to 30-nm Teflon particles, which are considered as inert at micron level, led to acute pulmonary toxicity in rats [27].

Nanoparticles can cause adverse health effects due to the direct action of the particles or to their rate as carriers of toxic elements [61]. Because nanoparticles are not removed from the upper respiratory tract, they are inhaled into the deeper areas. Their rather high deposition (more than 90 %) in the alveolar region or other respiratory tract regions leads to their subsequent entry into the blood stream [7, 58]. The small size and large surface area of nanoparticles enable significant interaction with biological systems. Nanoaerosols may also cause toxic effects on other organs. Nanoaerosols deposited in alveolar region or other respiratory tract region can lead to diffusion into the circulatory system [58], and they can reach the brain and the heart. A direct interaction of particles (or compounds generated by particles) with DNA is considered possible. Therefore, there may be a great risk of carcinogenicity.

The most important toxic effects are induced by nanoaerosol inhalation, and thereby chronic toxicity and carcinogenicity in the lungs [39]. Since the size of nanoaerosol particles is smaller than cells, they can penetrate through the respiratory or integumentary system and leach into the bloodstream. Contrary to neutral nanoaerosols of the same size; charged nanoaerosol particles have a much greater probability of depositing into the lungs [9]. Similar to the effects of fine aerosol, nanoaerosol can cause inflammation, fibrosis, and lung tumors if the exposure at workplaces passes certain threshold and long-lasting. Increased surface area and decreased particle diameter are believed to be the cause of the increased toxicity of granular nanomaterials in the lungs. Lung toxicity, evidenced by inflammation and tissue damage, was also proven to be induced by fibrous nanomaterials such as carbon nanotubes (CNTs). Despite of the evidences provided by tests using rate, mice, and hamsters, not everybody agrees with the mechanisms of tumor development.

Effects on the skin or a relevant skin penetration were not well quantified. However, the barrier function of the skin could be breached by the presence of skin lesions, strong mechanical strain or small-sized nanoparticles (<5–10 nm). For example, a human case of allergic response to nanomaterials was described by a person exposed to dendrimers in Japan [57]. A 22-year old student involved in synthesis of dendrimers developed erythema multiforme-like contact dermatitis on his hands. Conventional treatments and anti-histamines could not stop the disease from progressing to other areas of the body. After 3 weeks of hospitalization, he recovered, but it appeared again when he reentered the same laboratory.

At this moment, there is no legal standard that sets the occupational exposure threshold of nanoaerosol. The development of risk assessment of exposure to nanoaerosol has been limited by the lack of standard methods and compact instrumentation for long term monitoring. Accurate risk assessment requires advanced nanoaerosol sampling and characterization techniques for the analysis of both physical and chemical properties of nanoaerosol. Nonetheless, occupational exposure limits for nanomaterials are set by different organizations. Proposed occupational exposure limits for engineered nanoaerosol are summarized in Table 13.1.

Table 13.1 Proposed occupational exposure limits of nanoaerosols

Nanoaerosol	Occupational exposure limit	Parameters
Titanium dioxide	0.1 mg/m^3	0.1 risk level particles\100 nm
General dust	3 mg/m^3	
Photocopier toner	0.6 mg/m^3	Tolerable risk
	0.06 mg/m^3	2009 acceptable risk
	0.006 mg/m^3	2018 acceptable risk
Biopersistent granular materials (e.g., metal oxides)	20,000 particles/cm^3	Density > 6,000 kg/m^3
	40,000 particles/cm^3	Density < 6,000 kg/m^3
Carbon Nanotubes (CNTs)	0.01 f/cm^3	Exposure risk ratio for asbestos
Fibrous	0.01 f/cm^3	3:1; length 75,000 nm
Multi-walled CNTs	0.0025 mg/m^3	Nanocyl product only [40]

Source Schulte et al. [47]

13.3 Properties of Nanoaerosol

13.3.1 Number and Size of Nanoaerosol Particles

As shown Fig. 13.2, by number, nanoaerosol particles constitute 90 % or more of ambient aerosols, although they only account for a very small fraction of the total mass [48].

The small size of nanoparticles leads to a great surface area to mass ratio and consequently other unique properties of nanoparticles. The high surface area leads to a great surface reactivity. On a nano scale, both classic physics and quantum physics play roles in the interfacial behavior of nanoaerosols.

Some researchers believe that high concentration nanoparticles in the air agglomerate rapidly to form larger particles by chemical bonding and physical reactions [51]. As a result, the lifespan of individual nanoaerosol particles is usually short. Nanoparticles with the sizes of 1–10 nm have the lifespan of a few minutes to hours [2]. Meanwhile larger nanoparticles are formed by agglomeration. As a result, the size distribution of the nanoaerosol in a certain environment may change over time. The agglomeration mechanisms of nanoaerosols are not yet well understood.

On the other hand, well-dispersed nanoaerosols at low concentration may remain airborne for a long period of time; their setting velocity is extremely low because of their small aerodynamic sizes. At low concentration, the chance for nanoaerosol particles to agglomerate is low because of their great impact velocities.

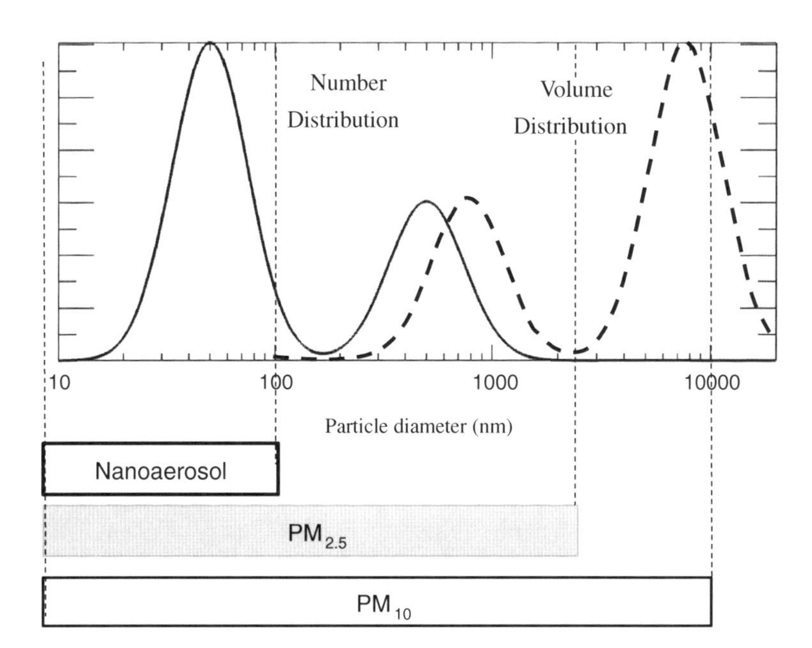

Fig. 13.2 Aerosol number and mass distribution versus size

To take advantage of the unique properties of nanoaerosol, one of the future trends in nanomedicine is targeted drug delivery to the respiratory system by nanoaerosol; it is important to improvement of drug therapies, lung imaging, gene delivery and therapy, tuberculosis diagnosis and treatment. It is important to clearly understand the toxicological effect of inhaled nanoaerosol to address the increasing concerns over potentially harmful public and occupational exposure. And this type of research should be conducted systematically on a global scale.

13.3.2 Noncontinuum Behavior

Nanoaerosol particles are small enough to approach the mean free path of air, which is about 67 nm under standard conditions. For nanoaerosol the continuum assumption is no longer valid and can attain free molecular flow; there is a non-continuum interaction between the particles and the carrier gas. The dimensionless parameter that defines the nature of the aerosol is the Knudsen number, which is the ratio of gas mean free path to particle radius.

$$Kn = 2\lambda/d_p \tag{13.1}$$

where d_p is the particle diameter and λ is the gas mean free path that was introduced in Sect. 2.1.7 above. Under normal conditions the mean free path of the air molecules is 66 nm. Thereby Kn is in the range of 1.32–132 when the diameter of nanoparticle drops from 100 to 1 nm. Now the air-nanoparticle is in noncontinuum regime, Cunningham correction factor, C_c, becomes much more important than micro sized particles.

$$C_c = 1 + Kn\left[1.142 + 0.558\exp\left(-\frac{0.999}{Kn}\right)\right] \tag{13.2}$$

Both Kn and C_c are dimensionless parameters. Since the theoretical value of C_c is always greater than 1, the drag force experienced with slipping effect considered is always smaller than the value calculated with non-slipping assumption. The drag force exerted by the air on the nanoaerosol is calculated by

$$F_D = \frac{3\pi\mu(V_p - V_g)d_p}{C_c} \tag{13.3}$$

Under normal conditions, air flow immediately surrounding a nanoaerosol particle is laminar or in the Stokes regime, although the bulk air flow may be turbulent. In addition, a nanoaerosol particle tends to follow the moving carrier gas and it is very difficult to separate them simply by inertia.

13.3.3 Diffusion of Neutral Nanoaerosol

The diffusion coefficient of uncharged nanoaerosol particles in the air can be determined by the Stokes-Einstein equation described in Eq. (13.4):

$$D_p = \frac{kTC_c}{3\pi \mu d_p} \tag{13.4}$$

where k is the Boltzmann's constant, T the temperature, μ the kinetic viscosity of the carrier gas. For smaller nanoparticles in the size range of 0.5–2 nm, the diffusivity can be calculated using the equation given by Ichitsubo et al. [21] as

$$D_p = \frac{0.815 c_{\mathrm{rms}}}{12\pi N \left(d_g + d_p\right)^2} \sqrt{1 + \frac{M}{M_n}} \tag{13.5}$$

where d_g is the gas molecule diameter (0.37 nm for air), N is the number concentration of gas molecules ($2.45 \times 10^{25}/\mathrm{m}^3$ for air at 293 K and 1 atm), M is the molar weight of the carrier gas (28.82 for air), M_n is the molar weight of nanoaerosol particles, c_{rms} is the root mean square velocity of the carrier gas molecules, which can be determined using Eq. (13.6).

$$c_{\mathrm{rms}} = \left(\frac{3RT}{M}\right)^{1/2} \tag{13.6}$$

13.3.4 Electrical Properties of Nanoaerosol

Nanoaerosol particles are primarily charged by diffusive charging. The number of ions charged to a nanoaerosol particle is calculated using Eq. (13.7) [63]

$$n(t) = \frac{d_p kT}{2e^2 K_E} \ln\left(1 + \frac{d_p K_E \overline{c}_i \pi e^2 N_{i0}}{2kT} t\right) \tag{13.7}$$

where $\overline{c}_i$ is the mean thermal speed of ions (239 m/s at standard conditions $T = 293$ K, $P = 1$ atm), k is Boltzmann constant (1.38×10^{-23} J/K), K_E is a constant of proportionality ($1/4\pi\varepsilon_0 = 9 \times 10^9$ Nm^2/C^2), N_{i0} is ion concentration.

While micron particles may be charged with hundreds of ions, a nanoaerosol particle smaller than 20 nm will probably acquire only a couple of ions; in some cases it will not acquire any. If polydisperse nanoaerosol particles pass through a bipolar charger, two nanoparticles of the same size may obtain different charges [32]. Experimental data show that generally, after charging, sub-20 nm particles carry a negative charge while larger particles carry a positive charge [1].

High concentration of ions and sufficient charging time allow particles to reach maximum charging. The maximum charging by unipolar charging enables the particles to carry ions with same polarity, negative or positive. Charged particles are subjected to electrical forces in an electrical field. The bipolar charging process eventually leads to Boltzmann charge equilibrium if the particles experience sufficient charging time. In such cases, they are considered neutralized. Highly charged particles may be discharged by colliding with ions with different polarity [43].

There are many types of nanoaerosol chargers. Unipolar ions can be produced by unipolar corona discharge, UV charging, carbon fiber ionizer, and separation of ions produced by bipolar ions. Bipolar ions are usually produced by radioactive sources such as Kr^{85} or Po^{210}, soft X-ray, AC corona discharge or dual electrode corona discharge. According to the mechanisms of ion generation, the chargers can be classified into:

- Corona discharge chargers,
- Radioactive chargers, and
- Photoelectric chargers.

Corona discharge is the most commonly used for high ion concentrations [22]. Bipolar charging method has a lower charging efficiency due to particle loss and ions recombination. In the unipolar method, produced ions in the corona charger are moved using the filtered air passing opposite the aerosol flow. Filtered air causes aerosols to be diluted and decreases charging efficiency.

However, nanoparticles may be generated in a corona. Various studies have investigated nanoparticle generation associated with bipolar and unipolar corona chargers [26, 35, 45, 52] and have reported methods to reduce nanoparticle generation in corona charging. One reason for generating nanoparticles by the corona charger is that a corona charger has enough energy to start gas-phase chemical reactions in the charger region, such as forming ozone from oxygen, which may lead to particle generation [45]. Moreover, sputtering of metal from the surface or erosion of the electrodes is another reason which may cause particles to be generated [30].

13.4 Separation of Nanoaerosol from the Air

Separation of general aerosol particles and particulate matter has been introduced in Chap. 6. Among the technologies introduced therein, separation based on inertia in cyclone and gravity settling chamber have almost no effect on nanosized size range. Under normal conditions, nanoaerosol particles follow the air due to their non-continuum behavior.

It is technically challenging to remove nanoparticles from the air by electrostatic precipitators (ESPs) only because nanoaerosol particle charging efficiency is low. Furthermore, extra nanoaerosol particles are likely to be produced in corona chargers, which is a critical component of an ESP. Passing nanoaerosol through a liquid (e.g., water) column can effectively remove the unwanted particles from the air. It may work effective for small quantity of air, however, it becomes costly for large air flow rates.

Filtration of nanoaerosol particles from its carrier gas is important to nanoaerosol sampling and characterization as well as air cleaning. The filtration efficiency for nanoaerosol is also described using Eq. (13.8)

$$\eta_{\text{nano}} = \eta_{\text{ts}}\eta_{\text{ad}} \tag{13.8}$$

13.4.1 Nanoparticle Transport Efficiency

Same as microparticles, nanoparticle transport efficiency (η_{ts}) is calculated based on the single fiber efficiency (η_{sf}) as described in Sect. 6.5:

$$\eta_{\text{ts}} = 1 - \exp\left[\frac{-4\alpha\eta_{\text{sf}}L}{(1-\alpha)\pi d_f}\right] \tag{13.9}$$

where d_f is the diameter of the fiber, L is the thickness of the bulk filter, and α is the solidity of the filter. The single fiber filtration efficiency can still be determined using Eq. (6.8)

$$\eta_{\text{sf}} = 1 - (1-\eta_{\text{it}})(1-\eta_{\text{ip}})(1-\eta_D)(1-\eta_E) \tag{13.10}$$

However, the dominating mechanisms for removing neutral airborne nanoparticles using fibrous filters are Brownian diffusion and interception.

$$\eta_{\text{sf}} = 1 - (1-\eta_{\text{it}})(1-\eta_D) \tag{13.11}$$

There have been several models developed for these mechanisms, and the state of the art is summarized in the paper by Givehchi and Tan [16]. We choose the latest one for each of these two mechanisms. The single fiber efficiency for neutral particles in a slip flow for a Brownian diffusion (η_D) considering slipping effect is described using the following equation [62].

$$\eta_D = 0.84\text{Pe}^{-0.43} \tag{13.12}$$

where Pe is Peclet number is defined using Eq. (13.13)

$$\text{Pe} = \frac{U_0 d_f}{D_p} \tag{13.13}$$

The single fiber filtration efficiency for interception mechanism η_{it} is [41]

$$\eta_{\mathrm{it}} = 0.6\left(\frac{1-\alpha}{Y}\right)\frac{R^2}{1+R}\left(1+\frac{1.996\mathrm{Kn}_f}{R}\right) \tag{13.14}$$

Y is the Kuwabara hydrodynamic factor, defined using Eq. (13.15). It is a function of filter solidity α

$$Y = -\frac{\ln\alpha}{2} - \frac{3}{4} + \alpha - \frac{\alpha^2}{4} \quad \text{for } \mathrm{Kn}_f \ll 1 \tag{13.15}$$

Kn_f is the fiber Knudesn number described in Eq. (13.16)

$$\mathrm{Kn}_f = 2\lambda/d_f \tag{13.16}$$

where d_f is the fiber diameter and R is the interception parameter defined in Eq. (13.17).

$$R = \frac{d_p}{d_f} \tag{13.17}$$

Example 13.1: Nanoaerosol filtration efficiency
A filter is made of fiberglass with a solidity of 5 %, and it is 5-cm thick. The average diameter of the fibers is 5 μm. When the face speed is 0.15 m/s, calculate and plot the fractional transport efficiency as a function of particle aerodynamic diameter in the range of 1–100 nm under standard conditions, by interception and diffusion, respectively.

Solution
In this problem, the following parameters are considered as constant

$$d_f = 5\,\mu\mathrm{m}, \alpha = 0.05, \mu = 1.81 \times 10^{-5}\,\mathrm{Pa.s}$$

$$\mathrm{Kn}_f = \frac{2\lambda}{d_f} = 0.0264$$

$$Y = -\frac{\ln\alpha}{2} - \frac{3}{4} + \alpha - \frac{\alpha^2}{4} = 0.80$$

The following variables can be calculated in an Excel sheet for different particle diameters

$$R = \frac{d_p}{d_f}$$

$$C_c = 1 + \mathrm{Kn}_p \left[1.142 + 0.558 \exp\left(-\frac{0.999}{\mathrm{Kn}_p} \right) \right]$$

$$D_p = \frac{kTC_c}{3\pi\mu d_p}$$

$$\mathrm{Pe} = \frac{U_0 d_f}{D_p}$$

The single fiber filtration efficiency by inertial interception and diffusion, per unit length of fiber is calculated using

$$\eta_{\mathrm{it}} = 0.6\left(\frac{1-\alpha}{Y}\right)\frac{R^2}{1+R}\left(1 + \frac{1.996\mathrm{Kn}_f}{R}\right)$$

$$\eta_D = 0.84\mathrm{Pe}^{-0.43}$$

With the face speed of 0.15 m/s, we can get the single mechanism filtration efficiency for interception, and diffusion. The result is shown in Fig. 13.3.

As seen in Fig. 13.3, diffusion dominates the nanoaerosol particle transport. The transport efficiency by diffusion increases from nearly 0 to 100 % as the particle size decreases from 100 to 1 nm.

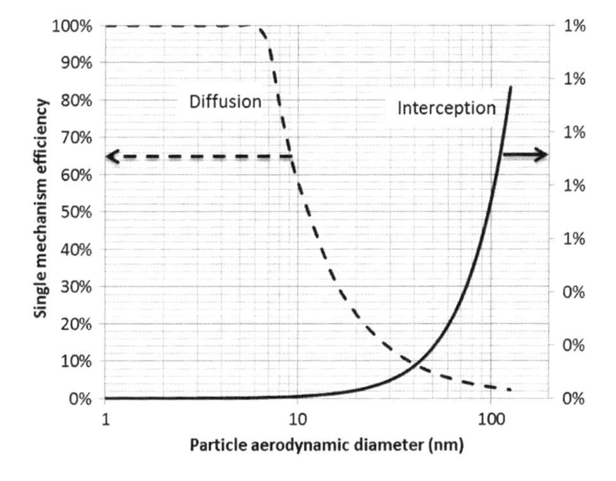

Fig. 13.3 Calculated nanoaerosol filtration efficiency

13.4.2 Adhesion Efficiency and Nanoaerosol Thermal Rebound

In conventional filtration theory, the adhesion efficiency is assumed unity ($\eta_{ad} \equiv 1$). However, it is not certain for nanoaerosol particles. Conventional filtration theory (in Sect. 6.5) indicates that nanoaerosol particle filtration efficiency increases inversely with particle size. Base on this hypothesis, filtration efficiency of nanoaerosol particles can reach 100 % for a properly designed filter. In reality, however, there should be a critical size from which filtration efficiency drops with the decrease of particle diameter (see Fig. 13.4). Otherwise, gas molecules, which are indeed extremely small particles, should be captured by filters resulting in no separation of aerosol particles and the carrier gas. Knowledge of this critical size is important to the design of effective nanoaerosol filters.

When an aerosol particle impacts on a filtration surface, there is an interfacial adhesion force attempting to hold them together. When the adhesion force is strong enough to offset the outgoing momentum at the end of impact, the particle is captured by the filtration surface. It has been well accepted that aerosol particles always stick on the surface in contact. However, this may not be true for nanoaerosol particles because the impact between a solid nanoparticle and a solid surface is most likely elastic because of the small contact area, high speed, and unique properties of nanoaerosol [10]. As a result, nanoparticles may rebound from the filtration surface.

Most researchers (e.g., [60]) assume that the thermal speed of nanoaerosol particles follow the Maxwell–Boltzmann distribution, which is described in Eq. (13.18).

$$f(v_{\mathrm{im}}) = 4\pi v_{\mathrm{im}}^2 \left(\frac{m}{2\pi kT}\right)^{3/2} \exp\left(-\frac{m v_{\mathrm{im}}^2}{2kT}\right) \tag{13.18}$$

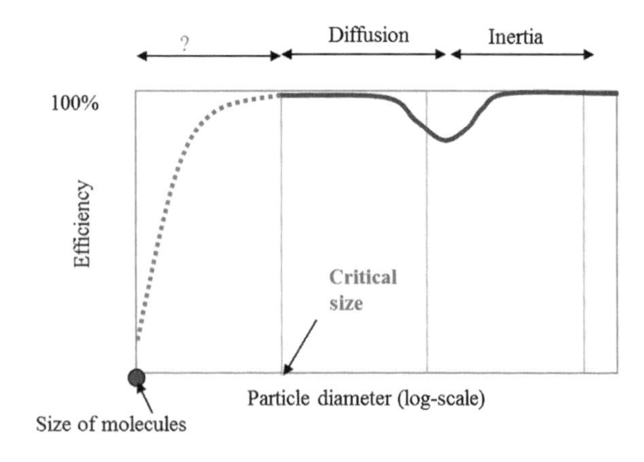

Fig. 13.4 Filtration efficiency vs aerosol particle diameter (not in scale)

where m, k and T denote particle mass, the Boltzmann constant and temperature, respectively. Following the analysis that is similar to the molecular dynamics introduced in Sect. 2.1, the mean impact speed of the nanoaerosol particles is described by Eq. (13.19)

$$\bar{v}_{\text{im}} = \sqrt{\frac{8kT}{\pi m}} \tag{13.19}$$

where the mass (m) in the denominator is now the mass of a single nanoaerosol particle instead of that of a gas molecule. With the mass of a particle, $m = \rho_p \pi d_p^3/6$, Eq. (13.9) becomes

$$\bar{v}_{\text{im}} = \sqrt{\frac{48kT}{\pi^2 \rho_p d_p^3}} \tag{13.20}$$

Equation (13.4) shows that $\bar{v}_{\text{im}} \propto d_p^{-1.5}$. Therefore, the mean thermal impact speed increase dramatically as particle size drops.

We can use Eq. (13.20) to estimate the mean thermal impact speeds of nanoaerosol particles with aerodynamic diameters in the range of 1–100 nm in standard air. For standard air, $T = 293$; particle density $\rho_p = 1{,}000\,\text{Kg}/\text{m}^3$. We can plot the thermal impact speed vs particle diameter in a logarithm scale as in Fig. 13.5.

This assumption may be valid only for dispersed nanoaerosol particles. It has been widely accepted and validated that Maxwell–Boltzmann distribution governs the speed of gas molecules by which the nanoaerosol particles are surrounded, then the motion of the particles are resulted from the impact between the gas molecules and the aerosol particles. The nanoaerosol particles may not move as freely and randomly as the gas molecules due to their inertia. Unlike gas molecules, which do not coagulate to each other upon collision, nanoaerosol particles could agglomerate

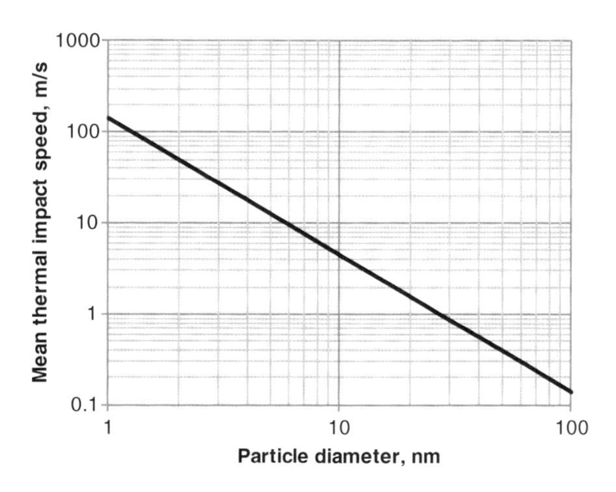

Fig. 13.5 Thermal impact speed of nanoaerosol particle in standard air

and change in size and number concentration. Thus, the assumption of Maxwell–Boltzmann distribution of nanoparticle thermal speed might be valid for diluted cases only. Nonetheless, we have to carry on with the analysis before a better hypothesis is established.

13.4.3 Critical Thermal Speed

The critical particle speed that enables thermal rebound of aerosol particles is a function of adhesion energy (E_{ad}), the coefficient of restitution (e) and particle mass (m).

$$v_{\mathrm{cr}} = \sqrt{\frac{2E_{\mathrm{ad}}}{me^2}} \tag{13.21}$$

The particle critical velocity, above which particle rebounds from the surface, is calculated using Eq. (13.21) and, with $m = \rho_p \pi d_p^3/6$, it becomes

$$v_{\mathrm{cr}} = \sqrt{\frac{12E_{\mathrm{ad}}}{\left(\pi\rho_p d_p^3\right)e^2}} \tag{13.22}$$

where E_{ad} is the adhesion energy, which will be elaborated shortly, e is the coefficient of restitution defined as the particle velocity at rebound over the normal particle velocity at the instant of contact. While intuitively one may assume that $e \approx 1$ for nanoaerosol because of the great rigidity, it is not true. The absolute value is unknown [16]. The coefficient of restitution is dependent on the material of the nanoparticles and the filter surface and the impact velocity of the nanoparticles [3]. For the impact velocities close to the critical velocity, the coefficient of restitution is small, and it leads to small rebound velocities. Molecular dynamics simulation by Ayesh et al. [3] showed that, for solid nanoparticles,

$$e \leq 0.6$$

Unfortunately, the database for coefficient of restitution for nanoparticles is still not well developed yet.

13.4.4 Adhesion Efficiency

Since the particles are considered as being collected when their thermal velocities are below v_{cr}, the fractional adhension efficiency can be mathematically described by

$$\eta_{\mathrm{ad}} = \frac{\int_0^{v_{\mathrm{cr}}} f(v_{\mathrm{im}})\mathrm{d}v_{\mathrm{im}}}{\int_0^{\infty} f(v_{\mathrm{im}})\mathrm{d}v_{\mathrm{im}}} \tag{13.23}$$

With the assumption of Maxwell–Boltzmann distribution for the particle thermal/impact velocity, Eq. (13.23) can be written as, by cancelling the common constants

$$\eta_{\mathrm{ad}} = \frac{\int_0^{v_{\mathrm{cr}}} v_{\mathrm{im}}^2 \exp\left(-\frac{mv_{\mathrm{im}}^2}{2KT}\right)\mathrm{d}v_{\mathrm{im}}}{\int_0^{\infty} v_{\mathrm{im}}^2 \exp\left(-\frac{mv_{\mathrm{im}}^2}{2KT}\right)\mathrm{d}v_{\mathrm{im}}} \tag{13.24}$$

Note that the general integration of

$$\int \left[x^2 \exp(-ax^2)\right]\mathrm{d}x = \frac{\sqrt{\pi}\mathrm{erf}(\sqrt{a}x)}{4a^{1.5}} - \frac{x \times \exp(-ax^2)}{2a} \tag{13.25}$$

where erf is the error function, and erf(0) = 0; erf(∞) = 1. And $x\Delta \exp(-ax^2) \to 0$ when $x \to \infty$, the numerator and the denominator in Eq. (13.24) are, respectively,

$$\int_0^{v_{\mathrm{cr}}} v_{\mathrm{im}}^2 \exp\left(-\frac{mv_{\mathrm{im}}^2}{2KT}\right)\mathrm{d}v_{\mathrm{im}} = \frac{\sqrt{\pi}\mathrm{erf}\left(\sqrt{\frac{m}{2KT}}x\right)}{4\left(\frac{m}{2KT}\right)^{1.5}} - \frac{x\exp\left(-\frac{m}{2KT}x^2\right)}{\frac{m}{KT}} \tag{13.26}$$

$$\int_0^{\infty} v_{\mathrm{im}}^2 \exp\left(-\frac{mv_{\mathrm{im}}^2}{2KT}\right)\mathrm{d}v_{\mathrm{im}} = \frac{\sqrt{\pi}}{4\left(\frac{m}{2KT}\right)^{1.5}} \tag{13.27}$$

Then the adhesion efficiency can be described as

$$\eta_{\mathrm{ad}} = \mathrm{erf}\left(\sqrt{\frac{m}{2KT}}v_{\mathrm{cr}}\right) - \sqrt{\frac{2m}{\pi KT}}v_{\mathrm{cr}} \cdot \exp\left(-\frac{m}{2KT}v_{\mathrm{cr}}^2\right) \tag{13.28}$$

Recall that

$$\bar{v}_{\mathrm{im}} = \sqrt{\frac{8kT}{\pi m}} \tag{13.29}$$

Equation (13.28) becomes

$$\eta_{ad} = \mathrm{erf}\left(\frac{2}{\sqrt{\pi}}\frac{v_{cr}}{\overline{v}_{im}}\right) - \frac{4}{\pi}\frac{v_{cr}}{\overline{v}_{im}}\exp\left[-\frac{4}{\pi}\left(\frac{v_{cr}}{\overline{v}_{im}}\right)^2\right] \tag{13.30}$$

For the ease of presentation, we define an interim term

$$z = \frac{2}{\sqrt{\pi}}\frac{v_{cr}}{\overline{v}_{im}} \tag{13.31}$$

The equation for adhesion energy can be simply presented as

$$\eta_{ad} = \mathrm{erf}(z) - \frac{2z}{\sqrt{\pi}}\exp\left[-z^2\right] \tag{13.32}$$

For the ease of calculation without software, the error function can be approximated with

$$\mathrm{erf}(z) \approx 1 - \frac{1}{\left(1 + a_1 z + a_2 z^2 + a_3 z^3 + a_4 z^4\right)^4} \tag{13.33}$$

where $a_1 = 0.278393$, $a_2 = 0.230389$, $a_3 = 0.000972$, and $a_4 = 0.078108$. The maximum error is 5×10^{-4} (Fortran 77 manual).

13.4.5 Adhesion Energy

Several models of adhesion energy (E_{ad}) were developed before and they were summarized by Givehchi and Tan [16]. As guidance, we will introduce only two of them, the JKR model [24] and the DMT model [11]. These two models complement each other because they represent two extremes in the Tabor parameter spectrum. JKR model is applicable to soft material, large radius, compliant spheres, and large adhesion energy and DMT model is for hard material, small radius with low adhesion energy [33]. The effectiveness of the DMT model has been proven for smaller and stiffer contact solids [44]. However, the main defect in this theory is that it neglects deformations outside the contact area [33].

Consider a nanoparticle deformed on the solid surface in Fig. 13.6. The adhesion energy between the particles and the surface of the filter material can be mathematically calculated based on consideration of elastic or plastic impaction.

$$E_{ad} = \Delta\gamma\pi a^2 \tag{13.34}$$

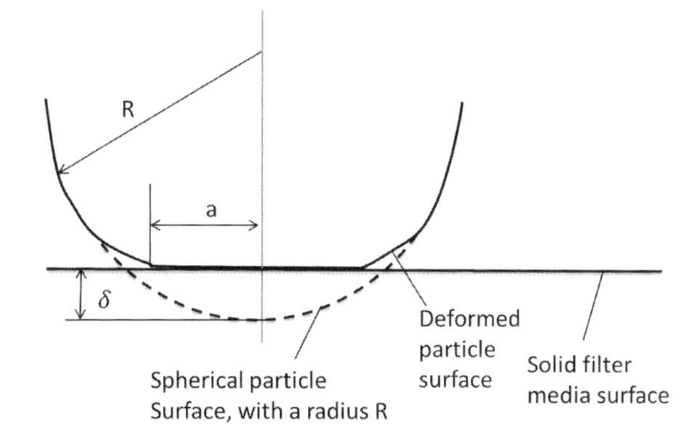

Fig. 13.6 Schematic diagram of adhension energy analysis

where a = the contact radius between particle and filter fiber and πa^2 is the contact area, E_{ad} = the adhesion energy (J), $\Delta\gamma$ = the specific adhesion energy (J/m^2), and $\Delta\gamma$ is a function of the Hamaker constant [11].

$$\Delta\gamma = \frac{H}{12\pi Z_e^2} \tag{13.35}$$

where Z_e = 0.4 nm is the equilibrium distance between the bodies. H = the Hamaker constant between the particle and the filter surface. A great deal of uncertainty is thus related to the determination of the specific adhesion energy especially in the case of nanomaterials.

The Hamaker constant between the particle and the filter surface can be calculated using

$$H = \left(H_p H_f\right)^{1/2} \tag{13.36}$$

Table 13.2 Material properties for thermal rebound calculation

Material	Hamaker constant $H_i \times 10^{19}$ J	Density (kg/m^3)	Mechanical constant $\left(K_i \times 10^{11}\ \text{m}^2/\text{N}\right)$	
			Calculated by Givehchi [16]	Given by Wang and Kasper [60]
Polystyrene	0.79	1,005	10.130	8.86
Glass (Dry)	0.85	2,180	0.443	–
NaCl	0.7	2,165	0.746	2.35
WOx (Tungsten)	1.216	19,250	0.071	–
Steel	2.12	7,840	0.137	0.139
Nickle			–	0.137
Copper	3.3	8,890	0.218	0.216
Fused quartz	0.65		–	0.432

where the subscript p and f stand for particle and filter surface, respectively. A typical Hamaker constant is in the order of $(10^{-19}–10^{-21})$ J. The values for some materials are listed in Table 13.2.

Example 13.2: Specific adhesion energy
Estimate the specific adhesion energy between a NaCl particle and a glass fiber

Solution
For salt particles and a glass fiber filter, $H_p = 7 \times 10^{-20}$ J, $H_f = 8.5 \times 10^{-20}$ J; then the Hamaker constant is

$$H = \left(H_p H_f\right)^{1/2} = \sqrt{7 \times 8.5} \times 10^{-20}\,\mathrm{J} = 7.71 \times 10^{-20}\,\mathrm{J}$$

The specific adhesion energy is

$$\Delta\gamma = \frac{H}{12\pi Z_e^2} = \frac{7.71 \times 10^{-20}\,\mathrm{J}}{12\pi(0.4 \times 10^{-9})^2\,\mathrm{m}^2} = 0.0128\,\mathrm{J/m^2}$$

where $Z_e = 0.4\,\mathrm{nm}$.

By considering the effect of surface adhesion energy and contact pressure inside the contact area, the contact radius between bodies and the adhesion energy are respectively given by the following two equations, ignoring the external force:

$$\begin{cases} a = \left[\dfrac{R^*}{Y^*}(6\Delta\gamma\pi R^*)\right]^{1/3} & \text{(JKR Model)} \\ a = \left[\dfrac{R^*}{Y^*}(2\Delta\gamma\pi R^*)\right]^{1/3} & \text{(DMT Model)} \end{cases} \tag{13.37}$$

where R^* = the characteristic radius of two bodies. In this case, they are considered as the nanoaerosol particle and the filter fiber. And it is defined as

$$\frac{1}{2R^*} = \frac{1}{d_p} + \frac{1}{d_f} \tag{13.38}$$

Y^* is the composite Young's modulus of bodies with the mechanical constant of K_P and K_f

$$Y^* = \frac{4}{3\pi}\left(\frac{1}{K_p + K_f}\right) \tag{13.39}$$

Mechanical constants are also listed in Table 13.2.

Example 13.3: Nanoaerosol adhesion efficiency
Calculate the adhesion efficiency using the DMT model with the following

properties of a glass fiber filter for NaCl nanoparticles 1–100 nm when filter fiber diameter $d_f = 5\,\mu\text{m}$.

Solution
For salt particles and glass fiber filter

$$H_p = 7 \times 10^{-20}\,\text{J}$$

$$H_f = 8.5 \times 10^{-20}\,\text{J}$$

$$K_p = 0.75 \times 10^{-11}\,\frac{\text{m}^2}{\text{N}}$$

$$K_f = 0.443 \times 10^{-11}\,\frac{\text{m}^2}{\text{N}}$$

Then the Hamaker constant

$$H = \left(H_p H_f\right)^{1/2} = \sqrt{7 \times 8.5} \times 10^{-20} J = 7.71 \times 10^{-20}\,\text{J}$$

And the specific adhesion energy is

$$\Delta\gamma = \frac{A_H}{12\pi Z_e^2} = \frac{7.71 \times 10^{-20}\,\text{J}}{12\pi(0.4 \times 10^{-9})^2\,\text{m}^2} = 0.0128\,\text{J}/\text{m}^2$$

The composite Young's modulus of bodies with the mechanical constants of K_p and K_f

$$Y^* = \frac{4}{3\pi}\left(\frac{1}{K_p + K_f}\right) = \frac{4}{3\pi}\left(\frac{1}{0.75 \times 10^{-11} + 0.443 \times 10^{-11}}\right)$$
$$= 3.51 \times 10^{11}(Pa)$$

The characteristic radius of two bodies is

$$R^* = \frac{1}{2}\left(\frac{1}{d_p} + \frac{1}{d_f}\right)^{-1}$$

The impact contact area is determined by

$$a = \left[\frac{R^*}{Y^*}(2\Delta\gamma\pi R^*)\right]^{1/3}$$

Then the adhesion energy is calculated using

$$E_{ad} = \Delta\gamma\pi a^2$$

Then the critical speed and impact speed are calculated using

$$v_{cr} = \sqrt{\frac{12E_{ad}}{\pi\rho_p d_p^3 e^2}}; \quad \bar{v}_{im} = \sqrt{\frac{48kT}{\pi^2\rho_p d_p^3}}$$

Then the adhesion efficiency is determined by

$$z = \frac{2}{\sqrt{\pi}}\frac{v_{cr}}{\bar{v}_{im}}; \quad \eta_{ad} = \text{erf}(z) - \frac{2z}{\sqrt{\pi}}\exp\left[-z^2\right]$$

In calculation using spread sheet, the error function is approximated with

$$\text{erf}(z) \approx 1 - \frac{1}{\left(1 + a_1 z + a_2 z^2 + a_3 z^3 + a_4 z^4\right)^4}$$

where a_1 = 0.278393, a_2 = 0.230389, a_3 = 0.000972, and a_4 = 0.078108.

The results are shown in Fig. 13.7 in terms of penetration efficiency.

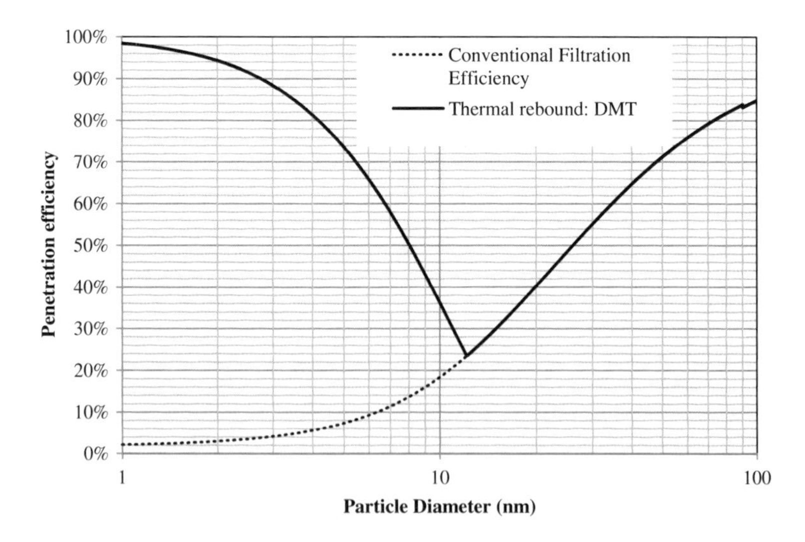

Fig. 13.7 Calculated penetration efficiency using DMT model

13.5 Nanoaerosol Characterization

Nanoaerosol can be sampled on a filter or grid for offline analyses of the morphology and composition of individual particles. The most common offline method is transmission electron microscope (TEM) and energy-dispersive X-ray spectroscopy (EDX). However, the physical and chemical properties may change due to agglomeration and/or chemical reactions during the sampling, transport and offline characterization processes. An online measurement is preferred when it is available.

Online monitoring of nanoaerosol is applied mainly to measure the size distribution. A number of technologies have been developed to measure airborne nanoparticle size distribution for use in a laboratory setting. And they are briefly summarized for guidance as follows.

13.5.1 Scanning Mobility Particle Sizer

Scanning mobility particle sizer (SMPS) employs a differential mobility analyzer (DMA) to classify nanoaerosols based on their electrical mobility after passing through a bipolar charger; the performance of a DMA is mostly limited by the low charging efficiency of sub-20 nm nanoaerosol particles. Classified nanoparticles are sent into a condensation particle counter (CPC), where they grow to 25 μm by condensation in butanol or water vapor. Then these large particles are counted by light scattering technique.

The principles of SMPS were introduced by Wang and Flagan [59]. Particles of different sizes are separated in DMA based on their electrical mobilities that depend on particle size. The electrical mobility Z_p, a measure of the particle's ability to move in an electric field, is defined as

$$Z_p = \frac{neC_c}{3\pi\mu d_p}. \tag{13.40}$$

where n is number of elementary charges on the particle, e is the charge of an ion, C_c is the Cunningham slip correction factor, μ is gas viscosity and d_p is particle diameter.

The polydisperse aerosol enters a bipolar neutralizer in the electrostatic classifier where aerosol particles reach a state of charge equilibrium due to collisions with bipolar ions. Then the charged aerosol enters DMA. The DMA consists of two concentric metal cylinders. The inner cylinder (r_1) is maintained at a controlled negative DC voltage and outer one (r_2) is electrically grounded. Thus, an electric field between the two cylinders is created. The polydisperse aerosol (flow rate Q_p) and sheath air (flow rate Q_{sh}) from the top of the classifier flow down the annular space between the cylinders. Due to the action of the electric field, positively charged particles are attached to the inner electrode while negatively charged particles are attached to the outward one. Uncharged particles are removed with the excess flow. Only particles with the optimum electrical mobility Z_p^* exit the DMA, and

$$Z_p^* = \frac{Q_{\text{sh}}}{2\pi \overline{V} L} \ln\left(\frac{r_2}{r_1}\right) \tag{13.41}$$

where Q_{sh} is sheath air flow rate (equal to excess air flow rate), r_2 is outer radius of annular space, r_1 is inner radius of annular space, $\overline{V}$ is average voltage on the inner collector rod, L is length between the exit slit and polydisperse aerosol inlet.

Combining Eqs. (13.40) and (13.41) leads to the particle diameter measured as a function of the collector rod voltage, number of charges on the particle, classifier flow rate, and geometry of the DMA:

$$d_p = \frac{2(ne)\overline{V} L C_c}{3\mu Q_{\text{sh}} \ln\left(\frac{r_2}{r_1}\right)} \tag{13.42}$$

By changing the voltage supplied to the inner cylinder of the DMA, scanning over the whole particle size interval is possible.

Once the particles are classified according to electrical mobility; their number concentration is measured by CPC. The CPC counts particles with a diameter from a few nanometers to one micrometer. It is very difficult to optically detect submicron particles because they have a diameter that is comparable with or even less than the wavelengths of most lights. To address this challenge, CPC works by passing the aerosol samples though a supersaturated vapor stream; the vapor condenses quickly on the nanosized particles. These large droplets with nanoparticles as seeds inside can be easily detected or counted by optical methods. The pulses of scattered light are collected by a photo detector and converted into electrical pulses. The concentration of particles is obtained from calibration of DC voltage against known concentrations.

The combination of size and number gives us the particle size distribution. To obtain this particle size distribution SMPS requires about 2 min.

13.5.2 Particle Classification by Aerodynamic Particle Focusing

Alternatively, particle can be classified by aerodynamic particle focusing. Under normal condition, nanoaerosol particle follow the air due to its noncontinuum behavior. As introduced in Sect. 4.2.4 above, the inertia of a particle in curvilinear motion is characterized by the Stokes number (Stk), which is defined in Eq. (13.43).

$$\text{Stk} = \frac{\tau U_0}{d_c} = \frac{\rho_p d_p^2 C_c U_0}{18\mu d_c} \tag{13.43}$$

where the characteristic dimension d_c in the above equation depends on the specific application. U_0 is the undisturbed air velocity. In standard air, a particle with Stk $\gg$ 1.0 will continue in a straight line as the fluid turns around the obstacle. But for Stk $\ll$ 1, particles will follow the fluid streamlines closely.

Example 13.4: Stokes numbers of nanoaerosol particles
Estimate the Stokes numbers of nanoaerosol particles with an aerodynamic particle diameter of 100 nm in standard air flowing at 1 m/s perpendicular to a cylinder of diameter 1 cm.

Solution
Given U_0 = 1 m/s, d_c = 1 cm, $\rho_p = 1{,}000$ kg/m^3, and $d_p = 100$ nm

Under standard conditions

$$Kn = 2\lambda/d_p = 2 \times 66/100 = 1.32$$

$$C_c = 1 + Kn\left[1.142 + 0.558\ \exp\left(-\frac{0.999}{Kn}\right)\right]$$
$$= 1 + 1.32\left[1.142 + 0.558\ \exp\left(-\frac{0.999}{1.32}\right)\right] = 4.36$$

$$\text{Stk} = \frac{\rho_p d_p^2 C_c U_0}{18\mu d_c} = \frac{1000 \times (100 \times 10^{-9})^2 \times 4.36 \times 1}{18 \times 1.81 \times 10^{-5} \times 0.01} = 1.34 \times 10^{-5}$$

The extremely small Stokes number indicates that nanoparticles follow air under STD conditions and it is difficult to separate them from the air simply by inertia. However, it is doable under other conditions when the particle viscosity is reduced and the particle Cunningham correction factor is enhanced. Under a carefully engineered condition with low pressure, however, nanoaerosol particles can be separated from the gas phase because of the relatively small amount of molecules surrounding the nanoaerosol particles.

By this approach, nanoaerosol particles can then be focused to a beam after expansion through an orifice, referred to as aerodynamic particle focusing. Aerodynamic focusing of particles has been employed for the accurate particle sizing [13]. As depicted in Fig. 13.8, the trajectories of the aerosol particles depend on their corresponding *Stokes* number. A well designed focusing orifice could isolate particles down to a few nanometers [42]. Large particles cross the center line due to

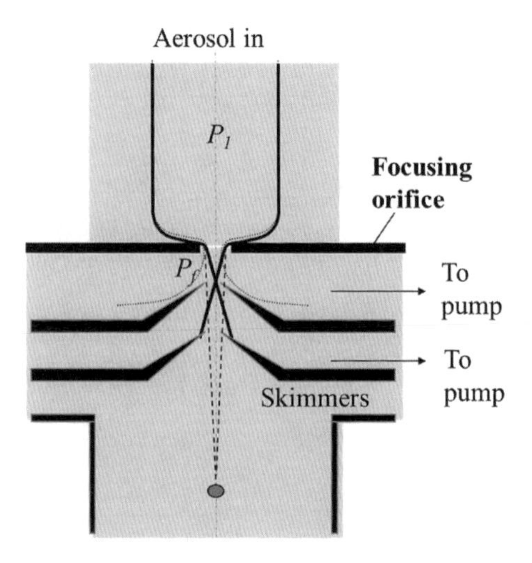

Fig. 13.8 Schematic diagram of aerodynamic partile focusing

great inertia and small ones with low inertia do not cross the center line. The optimally focused particle size by a focusing orifice can be described as [31, 36]

$$d_p^* = \sqrt{\left(d_p^m\right)^2 + (1.657\lambda)^2 - 1.657\lambda} \quad (13.44)$$

where λ is the mean free path of the carrier gas, the maximum size of the focused particles (d_p^m) is a function of critical Stokes number Stk*,

$$d_p^m = \frac{18\mu d_f \mathrm{Stk}^*}{\rho_p v_f} \quad (13.45)$$

where ρ_p is the density of particle, μ is the viscosity of gas, d_f is the focusing orifice diameter and v_f is the average speed in the focusing orifice exit plane. Gas speed in the focusing orifice should reach sonic speed to enable aerodynamic focusing of a certain-size particle. The critical Stokes number Stk* is based on the gas properties at the orifice throat. The value of Stk* has been numerically determined to be between 1 and 2 [31] and experimentally determined to be around 2 [42]. It does not depend on the pressure of the carrier gas.

Substituting Eq. (13.45) into (13.44) leads to

$$d_p^* = \sqrt{\left(\frac{18\mu d_f \mathrm{Stk}^*}{\rho_p v_f}\right)^2 + (1.657\lambda)^2 - 1.657\lambda} \quad (13.46)$$

This equation shows that the optimally focused particle size d_p^* depends on the properties of the gas such as the mean free path (λ) and viscosity (μ) of the carrier gas and the focusing orifice diameter (d_f) the gas viscosity and the gas velocity through the orifice. From an engineering practice point of view, it is challenging to alter the diameter of the focusing orifice or the gas velocity through the orifice. However it is doable to control the size of the optimally focused particles by changing the gas mean free path, which can be done by adjusting the upstream gas pressure.

The optimally focused size also depends on the location of the measurement spot downstream the focusing orifice [13]. For each pressure setting, the focusing is defined by the mean particle size (d_p^*) and the range of particle sizes focused (Δd_p^*).

13.5.3 *Particle Counting by Current Measurement Electrospray Technique*

Aerosol particles can be counted by relating their diameters to the maximum charges of ions. According to what we learned in Sect. 6.3.2, the maximum amount of ions a particle can carry is a function of particle diameter. For nanoaerosol particles, diffusion charging is the dominating mechanism, and the number of ions can be estimated by

$$n(t) = \frac{d_p kT}{2e^2 K_E} \ln\left(1 + \frac{d_p K_E \bar{c}_i \pi e^2 N_{i0}}{2kT} t\right) \tag{13.47}$$

The current by these moving particles with charges is

$$I = N_p Q n e \tag{13.48}$$

This current can be detected by a Faraday cup connected to an electrometer. By combination of aerodynamic particle focusing and current measurement, we can measure the particle size distribution of nanoaerosol particles [54, 56].

The electrometers for the detection of nanosized aerosol particles must be extremely sensitive. The GRIMM model 5.705 electrometer can measure the charge on aerosol particles of the size 0.8–700 nm. Another option is the TSI 3068B electrometer, which measures total net charge on aerosol particles from 2 nm to 5 μm.

Due to the unique properties of nanoaerosol, too much is unknown in this emerging area of research. Unfortunately, there is still a great need to develop stationary or portable instrument to measure particle size distribution that are practical for broad industrial applications.

Online chemical characterization of nanoaerosol is another challenge with significant potential for future research. Compared to size distribution instruments, much less is developed for online analysis of nanoaerosol chemical composition. No instrumentation has been reported aiming at online single particle chemical analysis for sub-10 nm nanoaerosols, which is crucially needed in order to fully understand the mechanisms of secondary aerosol nanoaerosol formation [6]. Currently, it has to be conducted by a combination of offline and online approaches in a statistically significant manner.

13.6 Nanoaerosol Generation

Nanoaerosol particle instruments are calibrated by suppliers or specialized laboratories before they are delivered to end users. In addition to the primary and secondary calibration methods, calibration shall also include a check on the particle number counters's (PNC's) detection efficiency with particles of a known size, often being 23 nm (electrical mobility diameter). Proper selection of the test aerosol particles is essential to the PNC calibration [15]. However, PNCs from different manufacturers are calibrated using different aerosol materials. For example, emery oil from TSI and NaCl from GRIMM. When they are used for industrial nanoaerosol, such as diesel soot, a material with similar behavior with diesel soot should be used too for the calibration.

An ideal aerosol generator is expected to be able to produce a constant and reproducible output of stable aerosol particles with adjustable size and concentration distributions. There are many nanoaerosol generators developed and commercially available based on different mechanisms. Typical ones are summarized as follows.

13.6.1 Evaporation–Condensation Technique

This method can be used for the generation of solid nanoaerosol particles like NaCl, C40 (tetracontane), silver or Tungsten. Nanoparticles are produced based on the principle of evaporation and condensation on nuclei. As shown in Fig. 13.9, the bulk nanoaerosol material is placed in a ceramic crucible (shown) or a ceramic heater container (not shown), where the bulk material is heated to its boiling point. A small flow is introduced into the heater, above the bulk material, to displace the concentrated vapor. The hot vapor is delivered to an area where it is mixed with the cool carrier air to enable condensation.

The output is polydisperse nanoaerosol where particles can be as small as 2 nm [46], which is the starting point of nucleation [29]. The nanoaerosol particle sizes can be varied by controlling one or both of the following factors.

- the crucible air flow rate (to control the vapor feeding rate)
- the carrier air flow (to control the subsequent cooling rate).

13.6.2 Electrospray Technique

This method is employed mainly for the generation of nanosized liquid droplets. As shown in Fig. 13.10, the bulk liquid material (e.g., Emery 3004 or PAO 4 cSt) is feed through a capillary into a container or suspended through a capillary tube from the container. By applying an electrical field to liquid at the capillary tip, the liquid

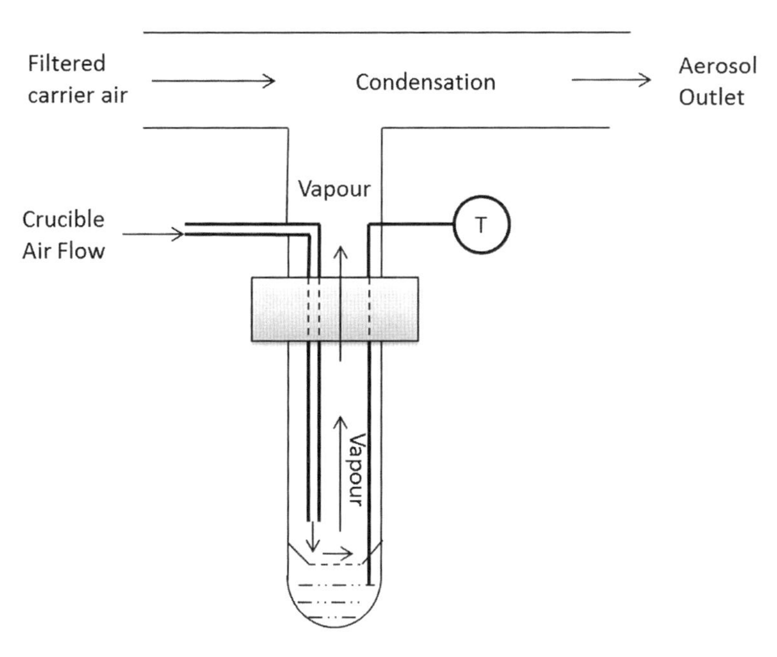

Fig. 13.9 Evaporation–condensation technique for nanoaerosol generation [15]

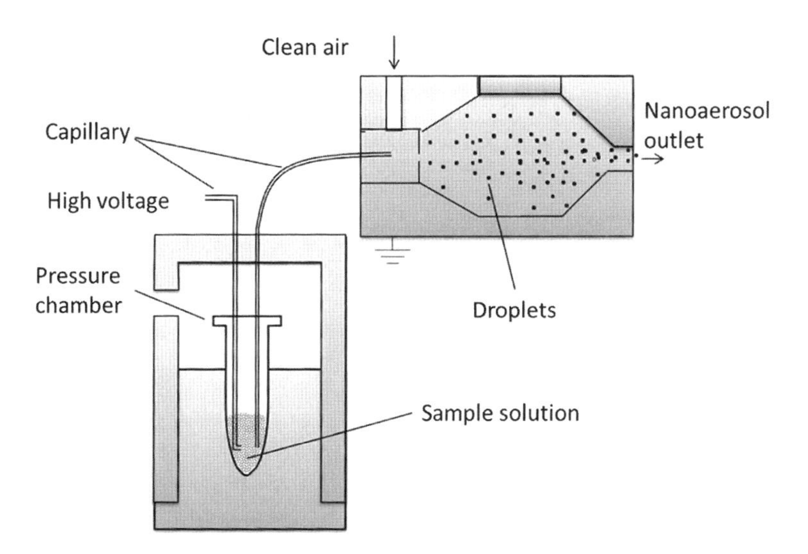

Fig. 13.10 Electrospray technique for nanoaerosol generation [15]

is drawn from the tip into a conical jet to mix with the clean air to produce nanoaerosol. These nanoparticles are then neutralized by an ionizer. Meanwhile, Air and CO_2 are merged with the droplets, and the liquid evaporates [15].

The output is neutralized monodisperse aerosol particle that is free of solvent residue. The particles are supposed to be spherical in shape. The sizes can be downs to a few nanometers. The factors that affect the concentration and the size distribution of nanoaerosols include solution concentration, humidity, and pressure. More about this technique can be found in literature (e.g., [37]).

13.6.3 Soot Nanoaerosol Particles

A properly designed combustion process can be used to generate soot particles, and it can also be called a soot generator. A soot generator uses a diffusion flame to form soot particles during pyrolysis (Fig. 13.11). A gaseous fuel is preferred for easy operation. Within the soot generating burner the flame is mixed with quenching gas at a definite flame height, resulting in a soot particle flow. Extra air is introduced then to dilute the soot particle stream.

The output aerosol particle sizes are controlled by means of varying fuel and its flow rate (in the order of ml per minute), air flow rate (e.g., 200 ml/min) and the inert quenching gas (say, Nitrogen) flow rate (e.g., 1 l/min) and further dilution air stream flow rate (1 l/min) [15]. Sufficient quenching ensures that stable soot particle output in terms of size and concentration. The generated aerosol particle diameter can be down to a few nanometers and within a range of 10^6–10^7 particle/cm^3 are diluted by quench gas and as an option, subsequently by adding dilution air.

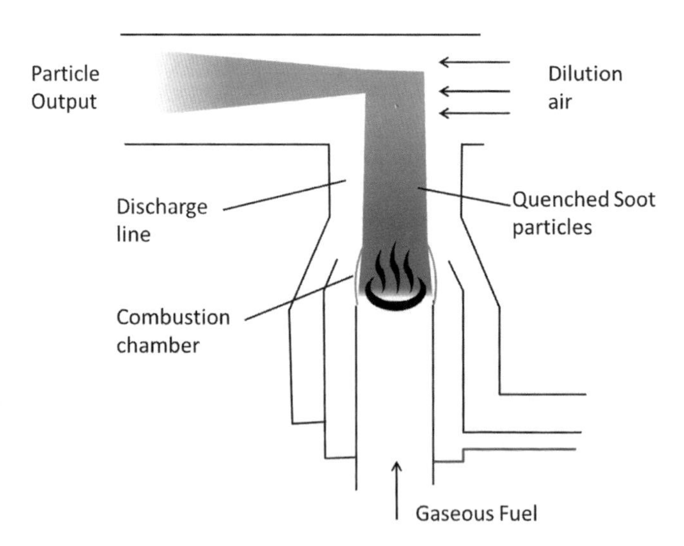

Fig. 13.11 Combustion-based soot particle generation [15]

References and Further Readings

1. Alonso M, Hernandez-Sierra A, Alguacil F (2002) Diffusion charging of aerosol nanoparticles with an excess of bipoloar ions. J Phys A Math Gen 35:6271–6280
2. Anastasio C, Martin S (2001) Atmospheric nanoparticles. In: Banfield JF, Navrotsky A (eds) Nanoparticles and the environment. Reviews in mineralogy & geochemistry 44. Mineralogical Society of America, Chantilly, VA, pp 293–349
3. Ayesh AI, Brown SA, Awasthi A, Hendy SC, Convers PY, Nichol K (2010) Coefficient of restitution for bouncing nanoparticles. Phys Rev B: Condens Matter 81:195422–195426
4. Biris AS, De S, Mazumder MK, Sims RA, Buzatu DA, Mehta R (2004) Corona generation and deposition of metal nanoparticles on conductive surfaces and their effects on the substrate surface texture and chemistry. Part Sci Technol 22:405–416
5. Brun E (ed) (2012) Risk perception and risk communication with regard to nanomaterials in the workplace: European risk observatory literature review, European agency for safety and health at work (EU-OSHA). Publications Office of the European Union, Luxembourg
6. Bzdek B, Pennington M, Johnston M (2012) Single particle chemical analysis of ambient ultrafine aerosol: a review. J Aerosol Sci 52:109–120
7. Castellano P, Ferrante1 R, Curini1 R, Canepari S (2009) An overview of the characterization of occupational exposure to nanoaerosols in workplaces. J Phys Conf Ser 170:012009
8. Chang J, Lawless PA, Yamamoto T (1991) Corona discharge processes. IEEE Trans Plasma Sci 19:1152–1166
9. Cohen B, Xiong J, Fang C, Li W (1998) Deposition of charged particles on lung airways. Health Phys 74:554–560
10. Dahneke B (1971) The capture of aerosol particles by surfaces. J Colloid Interface Sci 37:342–353
11. Derjaguin BV, Muller VM, Toporov YP (1975) Effect of contact deformations on the adhesion of particles. J Colloid Interface Sci 53:314–326
12. Ferin J, Oberdoerster G, Penney DP, Soderholm SC, Gelein R, Piper HC (1990) Increased pulmonary toxicity of ultrafine particles? 1. Particle clearance, translocation, morphology. J Aerosol Sci 21:381–384
13. Fernández de la Mora J, Riesco-Chueca P (1988) Aerodynamic focusing of particles in a carrier gas. J Fluid Mech 195:1–21
14. Ferreira AJ, Cemlyn-Jones J, Cordeiro C (2013) Nanoparticles, nanotechnology and pulmonary nanotoxicology. Pneumologia 19:28–37
15. Giechaskiel B, Alessandrini S, Forni F, Carriero M, Krasenbrink A, Spielvogel J, Gerhart C, Wang X, Horn H, Southgate J, Jörgl H, Winkler G, Jing L, Kasper M (2008) Calibration of PMP condensation particle number counters: effect of material on linearity and counting efficiency, EUR 23495 EN—2008. European Commission Joint Research Centre, Institute for Environment and Sustainability. doi:10.2788/95549, http://ies.jrc.ec.europa.eu/
16. Givehchi R, Tan Z (2014) An overview of airborne nanoparticle filtration and thermal rebound theory. Aerosol Air Qual Res 14:45–63
17. Hakim LF, Portman JL, Casper MD, Weimer AW (2005) Aggregation behavior of nanoparticles in fluidized beds. Powder Technol 160:149–160
18. Hautanen J, Kilpelainen M, Kauppinen EI, Jokiniemi J, Lehtinen K (1995) Electrical agglomeration of aerosol particles in an alternating electric field. Aerosol Sci Technol 22:181–189
19. Hernandez-Sierra A, Alguacil FJ, Alonso M (2003) Unipolar charging of nanometer aerosol particles in a corona ionizer. J Aerosol Sci 34:733–745
20. Hinds W (1999) Aerosol technology: properties, behavior, and measurement of airborne particles, 2nd edn. Wiley-Interscience, NJ
21. Ichitsubo H, Hashimoto T, Alonso M, Kousaka Y (1996) Penetration of ultrafine particles and ion clusters through wire screens. Aerosol Sci Technol 24:119–127

22. Intra P, Tippayawong N (2011) An overview of unipolar charger developments for nanoparticle charging. Aerosol Air Qual Res 11:187–209
23. Jaques P, Kim C (2000) Measurement of total lung deposition of inhaled ultrafine particles in healthy men and women. Inhal Toxicol 12:715–731
24. Johnson L, Kendall K, Roberts A (1971) Surface energy and the contact of elastic solids. Proc R Soc Lond A 324:301–313
25. Jouve G, Goldman A, Goldman M, Haut C (2001) Surface chemistry induced by air corona discharges in a negative glow regime. J Phys D Appl Phys 34:218
26. Kimoto S, Mizota K, Kanamaru M, Okuda H, Okuda D, Adachi M (2009) Aerosol charge neutralization by a mixing-type bipolar charger using corona discharge at high pressure. Aerosol Sci Technol 43:872–880
27. Kittelson DB (1998) Engines and nanoparticles: a review. J Aerosol Sci 29:575–588
28. Kulmala M, Mordas G, Petaja T, Gronholm T, Aalto PP, Vehkamaki H, Hienola AI, Herrmann E, Sipila M, Riipinen I, Manninen HE, Hameri K, Stratmann F, Bilde M, Winkler PM, Birmili W, Wagner PE (2007) The condensation particle counter battery (CPCB): a new tool to investigate the activation properties of nanoparticles. J Aerosol Sci 38:289–304
29. Kulmala M, Riipinen I, Sipilä M, Manninen HE, Petäjä T, Junninen H, Dal Maso M, Mordas G, Mirme A, Vana M, Hirsikko A, Laakso L, Harrison RM, Hanson I, Leung C, Lehtinen KEJ, Kerminen V-M (2007) Toward direct measurement of atmospheric nucleation. Science 318:89–92
30. Liu BYH, Pui DYH, Kinstley WO, Fisher WG (1987) Aerosol charging and neutralization and electrostatic discharge in clean rooms. J Environ Sci 30:42–46
31. Mallina R, Wexler A, Johnston M (1999) High speed particle beam generation: simple focusing mechanisms. J Aerosol Sci 30:719–738
32. Marlow W, Brock J (1975) Calculations of bipolar charging of aerosols. J Colloid Interface Sci 51:23–31
33. Maugis D (2000) Contact, adhesion and rupture of elastic solids. Springer, Berlin
34. McPhee J (2011) The curve of binding energy: a journey into the awesome and alarming world of Theodore B. Taylor. Macmillan, London
35. Medved A, Dorman F, Kaufman SL, Pöcher A (2000) A new corona-based charger for aerosol particles. J Aerosol Sci 31:S616–S617
36. Middha P, Wexler A (2003) Particle focusing characteristics of sonic jets. Aerosol Sci Technol 37:907–915
37. Morozov VN (2011) Generation of biologically active nano-aerosol by an electrospray-neutralization method. J Aerosol Sci 42(5):341–354
38. Mouret G, Chazelet S, Thomas D, Bemer D (2011) Discussion about the thermal rebound of nanoparticles. Sep Purif Technol 78:125–131
39. NIOSH (2005) National Institute for Occupational Safety and Health, NIOSH current intelligence bulletin: evaluation of health hazard and recommendations for occupational exposure to titanium dioxide (2005)
40. Nanocyl (2009) Responsible care and nanomaterials case study nanocyl. Presentation at European responsible care conference, Prague, 21–23 Oct 2009. http://www.cefic.be/Files/Downloads/04_Nanocyl.pdf
41. Payet S, Boulaud D, Madelaine G, Renoux A (1992) Penetration and pressure drop of a HEPA filter during loading with submicron liquid particles. J Aerosol Sci 23:723–735
42. Phares D, Rhoads K, Wexler A (2002) Performance of a single-ultrafine-particle mass spectrometer. Aerosol Sci Technol 36:583–592
43. Pui D, Fruin S, McMurry P (1988) Unipolar diffusion charging of ultrafine aerosols. Aerosol Sci Technol 8:173–187
44. Rahmat M, Ghiasi H, Hubert P (2012) An interaction stress analysis of nanoscale elastic asperity contacts. Nanoscale 4:157–166
45. Romay FJ, Liu BYH, Pui DYH (1994) A sonic jet corona ionizer for electrostatic discharge and aerosol neutralization. Aerosol Sci Technol 20:31–41

46. Scheibel HG, Porstendorfer J (1983) Generation of monodisperse Ag and NaCl aerosols with particle diameters between 2 and 200 nm. J Aerosol Sci 14:113–126
47. Schulte PA, Murashov V, Zumwalde R, Kuempel ED, Geraci CL (2010) Occupational exposure limits for nanomaterials: state of the art. J Nanopart Res 12:1971–1987
48. Seinfeld J, Pandis S (2006) Atmospheric chemistry and physics, Chapter 8. Properties of the atmospheric aerosol, 2nd edn. Wiley-Interscience, Hoboken, pp 350–395
49. Severac C, Jouve G, Goldman A, Goldman M (2004) Chemistry on the low field anodes and cathodes of air corona gaps. J Appl Phys 95:3297–3303
50. Singh Y, Javier JRN, Ehrman SH, Magnusson MH, Deppert K (2002) Approaches to increasing yield in evaporation/condensation nanoparticle generation. J Aerosol Sci 33:1309–1325
51. Stahlmecke B, Wagener S, Asbach C, Kaminski H, Fissan H, Kuhlbusch T (2009) Investigation of airborne nanopowder agglomerate stability in an orifice under various differential pressure conditions. J Nanopart Res 11:1625–1635
52. Stommel YG, Riebel U (2005) A corona-discharge-based aerosol neutralizer designed for use with the SMPS-system. J Electrostat 63:917–921
53. Tabor D (1977) Surface forces and surface interactions. J Colloid Interface Sci 58:2–13
54. Tan Z, Wexler A (2007) Fine particle counting with aerodynamic particle focusing and corona charging. Atmos Environ 41:5271–5279
55. Tan Z (2013) Nanoaerosol. In: Li D (ed) Encyclopedia of microfluidics and nanofluidics, 2nd edn. Springer, New York
56. Tan Z, Givehchi R, Saprykina A (2014) Submicron particle sizing by aerodynamic dynamic focusing and electrical charge measurement, Particuology. doi:10.1016/j.partic.2014.01.002
57. Toyama T, Matsuda H, Ishida I, Tani M, Kitaba S, Sano S, Katayama I (2008) A case of toxic epidermal necrolysis-like dermatitis evolving from contact dermatitis of the hands associated with exposure to dendrimers. Contact Dermat 59(2):122–123
58. Tsai C, White D, Rodriguez H, Munoz C, Huang C, Tsai C, Barry C, Ellenbecker M (2012) Exposure assessment and engineering control strategies for airborne nanoparticles: an application to emissions from nanocomposite compounding processes. J Nanopart Res 14:989
59. Wang SC, Flagan RC (1990) Scanning electrical mobility spectrometer. Aerosol Sci Technol 13:230–240
60. Wang H, Kasper G (1991) Filtration efficiency of nanometer-size aerosol particles. J Aerosol Sci 22:31–41
61. Wang J, Pui DYH (2011) Characterization, exposure measurement and control for nanoscale particles in workplaces and on the road. J Phys Conf Ser 304. doi:10.1088/1742-6596/304/1/012008
62. Wang J, Chen DR, Pui DYH (2007) Modeling of filtration efficiency of nanoparticles in standard filter media. J Nanopart Res 9:109–115
63. White H (1951) Particle charging in electrostatic precipitation. Trans Am Inst Electr Eng 70:1186–1191
64. Zhang L, Ranade MB, Gentry JW (2002) Synthesis of nanophase silver particles using an aerosol reactor. J Aerosol Sci 33:1559–1575

Chapter 14
Indoor Air Quality

The greatest impact of air pollution on our daily life is not outdoor, but rather indoor because we spend most of our time in indoor environments or built environments. Air pollutants at the ground level atmosphere enter indoor environments through windows, doors, cracks, and mechanical systems.

Indoor air quality (IAQ) engineering becomes more and more important to modern society. And it is another relatively new topic related to air emissions and it cannot be missed in a book like this for completeness. This chapter starts with a brief introduction of indoor air quality (IAQ) and the sources and effects of indoor air pollution leading to the engineering approaches to the reduction of indoor air pollution. The approaches include source control, exhaust by hood, dilution by fresh air (ventilation and/or in-duct air cleaning), and indoor air cleaning by air recirculation.

14.1 Introduction

Indoor air quality (IAQ) is the quality of air in an indoor environment. The IAQ indicators include air temperature, air relative humidity, and concentrations of air pollutants. Temperature and relative humidity aim at primarily thermal comfort and the others mainly health.

Indoor air quality is important for various reasons. The first and most obvious one is that the majority of the people in a modern society spend most of their time in indoor environments. On average, people in the industrialized countries like USA spend 90 % or more of their lifetime indoors. Certain group of people, such as young children and seniors, may spend more time indoors than the regular working force. While the number may be lower in less developed nations, it could also be even higher in cold areas like Canada and Russia.

A few decades ago, however, few people realized that there are air pollutants in indoor environment, and that they have great negative impact on the health and performance of the occupants. Since the early 1970s, there has been a constant

Z. Tan, *Air Pollution and Greenhouse Gases*, Green Energy and Technology,
DOI 10.1007/978-981-287-212-8_14

increase in the awareness of indoor air pollution with the advances of research and education in related field.

Increasing attention to indoor air quality was largely attributed to the awareness of poor health associated with the poor indoor environment. According to the studies sponsored by the US EPA, concentrations of indoor air pollutants may be 2–5 times, and sometimes, over 100 times higher than their outdoor counterparts, and indoor air pollution is one of the top five environmental risks to public health, comfort, and performance [1]. The situation is believed to be worsening with reduced ventilation rate for energy conservation and increased use of chemically formulated household products, and emissions from electronic office products.

Indoor air pollution has a profound impact on the quality of life and the economy. The National Institute for Occupational Safety and Health (NIOSH) ranked occupational lung dysfunctions (including lung cancer, pneumonoconioses, and occupational asthma) as the top occupational diseases and injuries. Lung dysfunction is undoubtedly related to the indoor air quality that people are exposed to. In some extreme case, indoor air pollutant could cause death.

Illnesses related to indoor air pollution can be classified into two categories:

- Sick building syndrome (SBS)
- Building related illness (BRI)

SBS is defined as the discomfort or sickness associated with poor indoor air quality without clear identification of the source substances. These symptoms could be, but are not limited to, irritation to eyes, noses, or throat, fatigue, and nausea. BRI is defined as a recognized disease caused by known agents that can be clinically identified. Examples of SRI symptoms include asthma, legionella, hypersensitivity, and humidifier fever. Obviously, the difference between the SBS and BRI is whether the causes of the sickness can be diagnosed clinically. Most of the time, the general public do not differentiate them, and they both are often referred to as sick building syndrome. Approximately one million buildings in the United States are sick buildings with 70 million occupants [2].

It is challenging to list all the sources of indoor air pollutants. Nonetheless, they can be grouped into outdoor and indoor sources. Outdoor air contaminants can enter indoor environments through HVAC systems, building envelopes, or even windows and doors. Indoor sources include combustion sources, such as smoking and cooking, operation of equipment, such as printers and computers, and biological sources, such as plants, animals, and human beings.

An incomplete list of sources of indoor air pollutants (excluding those with outdoor origins) is shown in Table 14.1. The list cannot be complete because of the nearly infinite contaminants in the air and multiple sources of each pollutants. Therefore, this table shall be used for guidance only.

Asbestos is a set of six naturally occurring silicate minerals used commercially for their desirable physical properties. They all have in common their eponymous asbestiform habit: long (roughly 1:20 aspect ratio), thin fibrous crystals.

The prolonged inhalation of asbestos fibers can cause serious illnesses including malignant lung cancer, mesothelioma, and asbestosis. Asbestosis is most closely

Table 14.1 Incomplete list of sources of indoor air pollutants

Contamination (Examples)	Typical sources
Asbestos	• Insulation materials • Ceiling and floor tiles
Combustion related contaminants	• Open fire cooking • Heating • Tobacco smoking • Incent • Candle
Formaldehyde	• Engineered board: Dry wall, interior-grade plywood, cabinetry and furniture, foam insulation, fabrics • Building materials: Adhesives, glues, furniture finishing, sealants, paints, stains, varnishes, wood preservatives, new carpet dyes and fibers, plastics
Biological contaminants (Allergen, mold, dust mite)	• Indoor plants • Animals • Human beings • Bedding for animals • Wet or damp materials
Radon	• Soil • Rock • Basement • Some building materials
Nanoaerosol	• Cooking • Printer • Photocopy machine • Nanospray
None-combustion particulates	• Dust • Hair • Skin
Volatile organic compounds (VOCs)	• Daily consumable products • Building materials

associated with the surface area rather than the diameters of inhaled asbestos; Mesothelioma is most closely associated with numbers of asbestos that are longer than 5 μm and thinner than 0.1 μm or so; lung cancer is most closely associated with those longer than about 10 μm and thicker than about 0.15 μm [3]. This does not mean other asbestos fibers can be considered nonhazardous because all asbestos can induce pathological responses and may contribute to the development of asbestos-related diseases [4].

The trade and use of asbestos have been restricted or banned in many jurisdictions. However, asbestos is not a health concern until it becomes airborne and enters the respiratory system. Care must be executed when renovating asbestos based old buildings.

Formaldehyde has been produced by catalytic oxidation of methanol for over two centuries [5]. It is widely used for the production of resins. Formaldehyde-based resins are used as adhesives in the manufacturing of construction materials, such as plywood, particle-board, and moulding materials. They are also used for furniture and other wood products. It is also a raw material for surface coatings, leather, rubber, and cement industries. Other indoor sources include stonewool and glasswool mats in insulating materials [6].

It is not clear how many people are occupationally exposed to formaldehyde worldwide. However, there are three types of occupations that are of high risk. The first is those working in the production of aqueous solutions of formaldehyde and their downstream chemical industries such as the synthesis of resins. The second group is related to its release from formaldehyde-based resins during the manufacture of wood products, textiles, synthetic vitreous insulation products, and plastics. Last but not the least, people are exposed to the pyrolysis or combustion of organic matter, e.g., in engine exhaust gases.

Indoor volatile organic compounds (VOCs) are emitted by many indoor materials. They can be generally grouped into two, being building materials and daily consumable products. Building materials include adhesives, glues, furniture finishing, sealants, paints, stains, varnishes, wood preservatives, new carpet dyes and fibers, and plastics. The daily consumable products include air fresheners, perfumes, hairsprays, hair gel, cleaning solvents, shoes and fabrics, automotive products, and contaminated water in sewage and sink.

Indoor air quality is a comprehensive topic that requires multiple disciplines, including, almost all branches in modern science and engineering. In this chapter, we focus on the engineering basics and technologies for indoor air quality control.

14.2 Threshold Limit Values

Threshold limit values (TLVs) refer to the upper limit of the concentrations of indoor air pollutants under which it is believed to be safe for all working occupants without impacting their health. The TLVs are established for different substances mainly for protecting the occupants' health with some tolerance of the discomfort such as irritation, narcosis, nuisance, or some stress. Examples of the health impairments that the TLVs are set against include compromise physiological function, adversely affect reproduce developmental processes, and shorten life expectancy.

There are two categories of TLVs. One is issued by governmental regulatory agencies and enforced by law; the other is published as recommended guidelines by scientific communities. For example, in the US, the former is defined by the Environmental Protection Agency (EPA) and the Occupational Safety and Health Agency (OSHA) of the United States; while the later can be set by America Conference of Governmental and Industrial Hygienists (ACGIH), American

Society of Heating, Ventilating, and Air-conditioning Engineering (ASHRAE) or even individual researchers.

The recommended TLVs are not the fine lines between safe and dangerous concentrations, nor are they a relative index of toxicity. Serious adverse health effects do not necessarily take place as a result of exposure to the indoor air pollutants above their TLVs. It is the best practice to maintain concentrations of all indoor air pollutants as low as practical.

According to the exposure period, TLVs can also be classified into the following three categories,

(1) Time-Weighted Average Threshold Limit Value (TLV-TWA),
(2) Short-Term Exposure Limit Threshold Limit Value (TLV-STEL), and
(3) Threshold Limit Value—Ceiling (TLV-C).

TLV–TWA is the time-weighted average concentration based on 8 h workday and a 40 h workweek, when all workers may be repeatedly exposed, day after day. The second one, STEL is defined as a 15 min TWA exposure that should not be exceeded at any time during a workday even if the 8 h TWA is within the TLV-TWA. TLV-STEL is the concentration to which it is believed that workers can be exposed continuously for a short period of time without suffering from (1) irritation, (2) chronic or irreversible tissue damage, or (3) narcosis of sufficient degree to increase the likelihood of accidental injury, impair self-rescue, or materially reduce work efficiency. STELs are recommended only where toxic effects have been reported from high short-term exposures in either humans or animals. There should be at least 60 min between successive exposures in this range. An averaging period other than 15 min may be recommended when this is warranted by observed biological effects.

Example TLVs of typical indoor air pollutants are listed in Table 14.2. More data can be found in literature such as [7] guidelines for indoor air quality. American Conference of Governmental Industrial Hygienists (ACGIH) also publishes annual threshold limit values for chemical substances and physical agents and biological exposure indices. Note that many air pollutants do not have all the three TLVs. It is important to observe whether any one of these types of TLVs is exceeded, a potential hazard from that contaminant is presumed to exist.

14.2.1 Normalized Air Contaminant Concentration

In most indoor environments, more than one air pollutants exist simultaneously. For example, carbon dioxide and airborne particles are always present in a typical working environment. When multiple pollutants exist, the effects of some on human health may be independent of others. In this case, TLVs for individual contaminants can be used to determine the indoor air quality. In general, however, the combined effects of multiple air pollutants on the health and comfort are considered greater than the summation of all individual effects. In order to quantify the combined effect, normalized concentration, C_N, should be used [2]. Mathematically, it is,

$$C_N = \sum \frac{C_i}{TLV_i} \tag{14.1}$$

where C_i and TLV_i are the concentration and threshold limit value of the ith air pollutant of concern, respectively. The equation shows that the normalized contaminant concentration is a dimensionless parameter as long as the units in the nominator and denominator match for each air pollutant counted.

When C_N is less than unity, the air quality is considered acceptable. Otherwise, the air quality needs improvement by reducing the concentrations of one or multiple air pollutants. In reality, however, it is very challenging to enforce the normalized concentration, because there is always a chance that it is greater than unity, provided the list of the pollutants is long enough.

Example 14.1: Threshold limit value
In a welding shop, the measured concentrations of CO, CO_2 and welding fumes are 10, 1,500 ppmv and 3.5 mg/m^3, respectively, all below the recommended TLV-TWA. Is this working environment safe to the workers daily based on normalized concentration?

Solution
TLV-TWA can be found in Table 14.2. And they are listed as follows

Table 14.2 Threshold values of typical indoor air pollutants in work places

Particulate pollutant		TWA (mg/m^3, except for asbestos)	
Asbestos		0.1 fiber/ml	
Coal dust, anthracite		0.4	
Coal dust, bituminous		0.9	
Grain dust (Oat, wheat, barley)		4.0	
Graphite (non fiber)		2.0	
Iron oxide particles and fume, inhalable		5.0	
Lead		0.05	
Welding fumes		5.0	
Gases	Formula	TWA (ppmv)	STEL (ppmv)
Ammonia	NH_3	25	35
Carbon dioxide	CO_2	5,000	30,000
Carbon monoxide	CO	25	
Formaldehyde	HCHO	–	0.3
Hydrogen sulfide	H_2S	10	15
Methanol	CH_3OH	200	250
Ozone	O_3	0.05–0.2	–

Sources ACGIH [11], WHO [7]

Pollutant	TLV-TWA	Measured concentration	C_i/TLV_i
CO	25	10	0.40
CO_2	5,000	1500	0.30
Welding fumes	5	3.5	0.70

Then the normalized concentration can be calculated as follows

$$C_N = \sum \frac{C_i}{TLV_i} = 0.40 + 0.30 + 0.70 = 1.4 > 1.0$$

Since the $C_N > 1$, it is not safe for the workers to be there daily (8 h).

14.2.2 Clean Room

Clean room is a special indoor environment where the number of airborne particles is controlled to avoid contamination of the products. The cleanness of the clean room is specified by the International Standard Organization (ISO) Standard ISO 146441-1. The upper limit of particle number concentration (number/m^3) is defined as,

$$c_p^* = 10^N \left(\frac{0.1}{d_p^*} \right)^{2.08} \tag{14.2}$$

where c_p^* = upper limit of particle number concentration, number/m^3
N = the clean room class number (1, 2,…, 9)
d_p^* = threshold particle diameter in μm, and they are 0.1, 0.2, 0.3, 0.5, 1 and 5 μm.

According to ISO 146441-1, in a class N cleanroom, the number concentration of particles greater than d_p^* cannot exceed c_p^*. To help appreciate the cleanness of a cleanroom, one can compare it with a typical room in an office building, which is close to a Class 9 clean room [2].

Example 14.2: Cleanroom class calculation
An ISO 146441-1 Class 2 cleanroom is 3 m high with a total floor area of 100 m^2, what is the maximum amount of particles that are larger than 100 nm in diameter?

Solution Substitute

$$d_p^* = 100 \text{ nm} = 0.1 \text{ μm}; \quad N = 2$$

into Eq. (14.2), we get

$$c_p^* = 10^N \left(\frac{0.1}{d_p^*}\right)^{2.08} = 10^2 \left(\frac{0.1}{0.1}\right)^{2.08} = 100/\mathrm{m}^3$$

Since the volume of the room is 300 m^3, the total number of particles larger than 100 nm in diameter cannot exceed 30,000 in total. That is an extremely clean room!

For the ease of practice, the cleanroom classes and their corresponding maximum particle number concentrations are computed using Eq. (14.2) and they are summarized in Table 14.3. In this table, the values are the maximum concentrations for the particles size or greater. The greater the class number, the higher allowable particle concentration can be.

For a given class, all limits for all particles sizes must be satisfied. For a Class 1 clean room, the concentration of particles greater than 0.2 μm should be 2 or less particle/m^3. Meanwhile, the particles ≥0.1 μm should not exceed 10 #/m^3.

Cleanroom is important to the quality of many industries, such as medical, manufacturing, and packaging. Cleanrooms were first developed for aerospace applications to manufacture and assemble satellites, missiles, and aerospace electronics. Most applications involve clean airspaces of large volumes with cleanliness levels of ISO Class 9 or cleaner. Recent advances in electronics industries continue to drive the design of cleanrooms. Most recently designed semiconductor clean rooms are ISO Class 5 or cleaner. In addition, preparation of pharmaceutical, biological, and medical products all require clean airspaces to control airborne bacteria and viruses to prevent contamination. Furthermore, some operating rooms in hospital may be classified as cleanrooms, but their primary function is to limit particular types of contamination rather than the quantity of particles.

Table 14.3 Cleanroom classes

ISO class number (N)	Maximum particle number concentration (#/m^3) for particle size $c_p^* = 10^N \left(\frac{0.1}{d_p^*}\right)^{2.08}$					
	0.1 μm	0.2 μm	0.3 μm	0.5 μm	1 μm	5 μm
ISO Class 1	10	2				
ISO Class 2	100	24	10	4		
ISO Class 3	1,000	237	102	35	8	
ISO Class 4	10,000	2,370	1,020	352	83	
ISO Class 5	100,000	23,700	10,200	3,520	832	29
ISO Class 6	1,000,000	237,000	102,000	35,200	8,320	293
ISO Class 7				352,000	83,200	2,930
ISO Class 8				3,520,000	832,000	29,300
ISO Class 9				35,200,000	8,320,000	293,000

14.3 IAQ Control by Ventilation/Dilution

Indoor air quality (IAQ) can be effectively controlled by HVAC systems. The acronym HVAC stands for heating, ventilation and air conditioning. A HVAC system provides treated outdoor air to the indoor environment and deliver it to the point of interest through ducts. As shown in Fig. 14.1, outdoor air enters the system through the air intake followed immediately the air cleaning devices, most likely a filter. As needed, air can be heated or cooled depending on the weather condition. Moisture can be added or removed as well. The air after filtration and conditioning is then distributed to multiple points in the building through supply air duct. After mixing within the rooms, air pollutants will be taken out of the room and become return air. Meanwhile, the temperature and humidity of the indoor air are adjusted to a comfort level. A portion of the return air, after treatment, is merged with the intake air in order to save energy. The rest will be discharged to the atmosphere through the exhaust.

14.3.1 Minimum Ventilation Rate

Ventilation controls the indoor air quality by bringing fresh air into an indoor environment to supply or reduce the heat and moisture and to dilute gaseous and

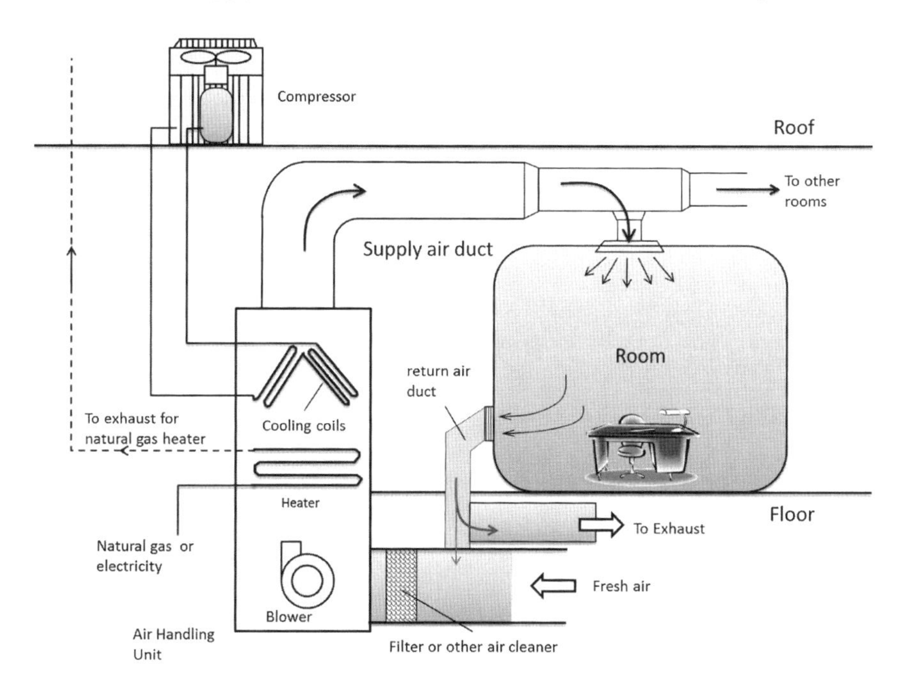

Fig. 14.1 HVAC system with internal air circulation and cleaning

particulate pollutants. In general, three types of variables are of concern, and can be controlled by ventilation system for an indoor environment:

- Temperature,
- Relative humidity, and
- Air pollutants.

Ideally a minimum ventilation rate must be maintained in order to control all these parameters at desired levels for an indoor environment with minimum energy consumption. However, practically the minimum ventilation rates of many buildings are based on the temperature control to reduce energy consumption. In some special buildings such as an animal holding room in a cold climate, ventilation may be based on relative humidity control. Air pollution levels are controlled by ventilation in very few indoor environments such as a welding workshop, which may be near the working station.

As a starting point of the analysis, consider a control volume, where temperature, relative humidity, and air pollutant levels are uniform. This assumption is acceptable if the control volume is small enough or the error is acceptable. This control volume could be an entire room or a zone within. When the control volume is a room, the air within is assumed completely mixed. Admittedly, this is a bold assumption, but it has been widely used in guiding the HVAC industry.

As shown in Fig. 14.2, intake supply air enters the room at a volumetric flow rate of air Q_s. The volume of the pace is V. The total enthalpy, moisture content, and pollutant concentration are denoted as h_s, w_s, and c_s, respectively. In a real HVAC system there is also recirculating air that contains the air within the space of concern. It is herein considered part of the supply air, the property of which can be calculated. Air exits the room at a volumetric flow rate of air, Q_e, and the total enthalpy, moisture content, and pollutant concentration are h_e, w_e, and c_e, respectively. Within the room, the total sensible heat transfer rate is $\dot{q}$, the moisture

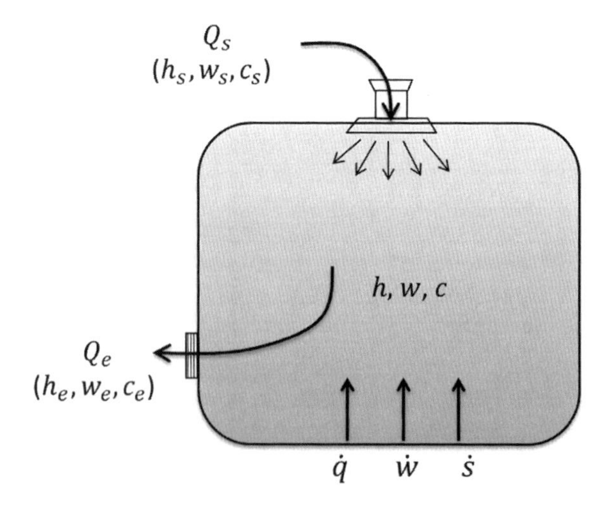

Fig. 14.2 Simplified heat and mass transfer model for a ventilated space

production rate is $\dot{w}$, and the pollutant production rate is $\dot{s}$. The densities of the air at the intake, indoor space, and the exit are denoted as ρ_s, ρ, and ρ_e, respectively. Since it is assumed that indoor air is completely mixed, the properties of the exhaust air are the same as those of the room air. For example,

$$\begin{aligned} v &= v_e \\ h &= h_e \\ c &= c_e \end{aligned} \tag{14.3}$$

Then, mass balance for dry air leads to,

$$\frac{d}{dt}\left(\frac{V}{v}\right) = \frac{Q_s}{v_s} - \frac{Q_e}{v_e} \tag{14.4}$$

where v is specific volume of moist air, which is defined herein as the volume of the moist air (dry air plus the water vapor) containing one unit of mass of "dry air" (m^3 air/kg dry air). As such the unit of both sides of the equation is kg dry air/s and that of Q's is m^3 air/s.

Equation (14.4) shows that the difference between the mass flow rates of the supply dry air and the mass flow rate of exhaust dry air equals the change of mass of dry air within the airspace. In this equation, *we use specific volume* of the air instead of the density because specific volume the of the air defines the volume of the mixture (dry air plus the water vapor) containing one unit of mass of "dry air". The SI unit of specific volume of air is (m^3 air per kilogram of dry air). Note that it is different from the inverse of air density.

$$\frac{1}{v} \neq \rho \tag{14.5}$$

The unit of air density is kg of air/m^3 air. Similarly, we can describe the energy balance based on the sensible heat as follows

$$\frac{d}{dt}\left(\frac{Vh}{v}\right) = \dot{q} + \frac{Q_s}{v_s}h_s - \frac{Q_e}{v_e}h_e \tag{14.6}$$

where h = sensible heat for supply air and exhaust air (kJ/kg of *dry* air). It implies that the difference between the sensible heat of the supply air and that of the exhaust air equals is the total sensible heat production plus the change of sensible heat in the room air.

Similarly, the moisture mass balance and a specific pollutant mass balance can be described using the following two equations, respectively,

$$\frac{\mathrm{d}}{\mathrm{d}t}\left(\frac{Vw}{v}\right) = \dot{w} + \frac{Q_s}{v_s}w_s - \frac{Q_e}{v_e}w_e. \tag{14.7}$$

$$\frac{\mathrm{d}}{\mathrm{d}t}(Vc) = \dot{s} + Q_s c_s - Q_e c_e. \tag{14.8}$$

where w is humidity ratio (kg of water vapor per kg of *dry* air), $\dot{w}$ is the water vapor production rate in kg/s, c is the pollutant concentration in air (kg of pollutant per m^3 of air), $\dot{s}$ is the mass production rate of the particulate pollutant in *kg/s*, Subscripts '*s*' and '*e*' stand for supply and exhaust air, respectively.

If the unit of air pollutant concentration c is in ppmv, the unit conversion is necessary for accurate calculation of ventilation rate with a unit of $(\mathrm{m^3 air/s})$.

The total sensible heat transfer rate, $\dot{q}$ in Eq. (14.6), of the ventilated airspace is the sum of all heat loss or gain within the airspace, including the sensible heat production rate by occupants and indoor equipment (e.g. stove, lights), and the heat transfer through building envelope. $\dot{q}$ can be positive or negative, which indicates heating load or cooling load. Moisture production rate $\dot{w}$ in Eq. (14.7) depends on the status of the indoor sources, for example, exhale of human beings or the capacity of a working humidifier. For the occupants (human or animals) related data can be found in handbooks and/or standards such as those published by ASHRAE [8].

It is relatively challenging to determine the pollutant production rate, $\dot{s}$ in Eq. (14.8), which varies dramatically for different pollutants and sources. Among all the typical air pollutants in indoor environments, carbon dioxide can often be determined indirectly from the load of the occupants because the increase of CO_2 level in an indoor environment is mainly a metabolic product. From large scale statistical analyses, it has been determined that the carbon dioxide production rate of average human and animals is *1* m^3 *of* CO_2 *per 24,600 kJ* of total heat production (THP) under normal indoor condition [8]. Converting to the mass production rate of carbon dioxide in kg/s leads to

$$\dot{s}_{\mathrm{CO_2}} = \rho_{\mathrm{CO_2}} \frac{THP}{24{,}600}. \tag{14.9}$$

where $\dot{s}_{\mathrm{CO_2}}$ = the mass production rate of CO_2 (kg/s), $\rho_{\mathrm{CO_2}}$ = the density of carbon dioxide (1.83 kg/m^3 at 20 °C and 1 atm), *THP* = the total heat production of occupants (kJ/s). Note that the total heat includes the sensible heat and latent heat, whereas the heat production $\dot{q}$ in Eq. (14.6) only includes the sensible heat.

For steady state operation, the rate of mass of dry air entering (Q_s/v_s) equals to that exiting (Q_e/v_e) the room. Energywise, the total sensible heat contained in the supply air plus the net sensible heat transfer rate, must be equal to the total sensible heat removed by the exhaust air. With this minimum requirement, the $\mathrm{d}/\mathrm{d}t$ terms are zeros and aforementioned equations become, respectively,

$$\frac{\mathrm{d}}{\mathrm{d}t} = 0 \tag{14.10}$$

$$\frac{Q_s}{v_s} = \frac{Q_e}{v_e}. \tag{14.11}$$

$$\dot{q} + \frac{Q_s}{v_s} h_s = \frac{Q_e}{v_e} h_e. \tag{14.12}$$

$$\dot{w} + \frac{Q_s}{v_s} w_s = \frac{Q_e}{v_e} w_e. \tag{14.13}$$

$$\dot{s} + Q_s c_s = Q_e c_e. \tag{14.14}$$

Solving these equations leads to the relationship between volumetric ventilation rates and sensible heat balance (or temperature balance) control at steady state

$$\frac{\dot{q}}{h_e - h_s} = \frac{Q_s}{v_s} = \frac{Q_e}{v_e}. \tag{14.15}$$

By similar approaches, we can get the moisture balance ventilation requirement at steady state. Combination of the conservation of mass equation and the conservation of moisture equation gives volumetric ventilation rates for moisture (or relative humidity) balance control at steady state

$$\frac{\dot{w}}{w_e - w_s} = \frac{Q_s}{v_s} = \frac{Q_e}{v_e}. \tag{14.16}$$

Applying the mass balance to pollutant balance one can get the pollutant balance ventilation requirement. At steady state,

$$\dot{s} = Q_e c_e \left(1 - \frac{v_s c_s}{v_e c_e}\right) = Q_s c_s \left(\frac{v_e c_e}{v_s c_s} - 1\right). \tag{14.17}$$

The analysis above defines different ventilation rates for the control of temperature, moisture, and a specific air pollutant.

There is a minimum ventilation rate that is required to maintain an acceptable indoor environmental condition (temperature, humidity, and a specific pollutant concentration). This minimum value must be the *greatest value* of the *six* ventilation rates for balances of sensible heat, moisture, and the given pollutant defined by Eqs. (14.15) to (14.17). Otherwise, at least one of the requirements cannot be met. In reality, the HVAC equipment of a building has an upper limit in capacity and there is a maximum ventilation rate. Therefore, we have to be careful in selecting the right HVAC equipment.

14.3.2 Psychrometric Chart

In order to calculate the ventilation rates above, we can no longer assume that air is always dry because moisture is a major concern in ventilation. The physical and thermodynamic properties of the moist air at sea level can be found in psychrometric chart. The ASHRAE-style psychrometric chart, shown Figure A.2, was pioneered by Willis Carrier in 1904 [9]. It depicts the parameters that are needed for our calculation above such as dry bulb temperature (T), humidity ratio (w), relative humidity (ϕ), specific sensible enthalpy (h), specific volume(v).

Example 14.3: Ventilation rate calculation
There are 20 people in a dinning room, each produces 200 W total heat. The carbon dioxide concentration in the supply air is 500 ppmv. Assume the supply air temperature is 15 °C and relative humidity 50 %. The room air is to be maintained at 22 °C and 60 % relative humidity. If the required maximum CO_2 concentration in the room is 1000 ppmv, estimate the minimum ventilation rate based on the CO_2 concentration.

Solution
Based on the supply air temperature of 15 °C and 50 % relative humidity we can get the specific volume of supply air using a psychrometric chart,

$$v_s = 0.822\,\text{m}^3/\text{kg dry air}$$

The specific volume of exhaust air at 22 °C and 60 % relative humidity is

$$v_e = 0.855\,\text{m}^3/\text{kg dry air}$$

The total heat production rate by 20 people in the dining room is,

$$THP = 20 \times 200\,\text{W} = 4000\,\text{W}\,or\,4\,\text{kJ/s}.$$

Then the mass production rate of CO_2 is estimated using Eq. (14.9)

$$\dot{s}_{CO_2} = \rho_{CO_2}\frac{THP}{24{,}600} = 1.83 \times \frac{4}{24{,}600} = 2.98 \times 10^{-4}(\text{kg/s})$$

Then the minimum ventilation rate fan is determined using Eq. (14.17)

$$\dot{s}_{CO_2} = Q_e c_e\left(1 - \frac{v_s c_s}{v_e c_e}\right) = Q_s c_s\left(\frac{v_e c_e}{v_s c_s} - 1\right)$$

Since $v_s < v_e$, the minimum ventilation rate should be calculated based on the exhaust air. However, we need to convert the unit of concentration c from ppmv to kg CO_2/m^3 air.

$$c_e = \frac{1{,}000\,\text{m}^3\,\text{CO}_2 \times 1.83\,\text{kg CO}_2/\text{m}^3\,\text{CO}_2}{10^6\text{m}^3\,\text{air}} = 1.83 \times 10^{-3}\text{kgCO}_2/\text{m}^3\text{air}$$

The exhaust ventilation rate is

$$Q_e = \frac{\dot{s}_{\text{CO}_2}}{c_e\left(1 - \frac{v_s c_s}{v_e c_e}\right)}$$

$$= \frac{2.98 \times 10^{-4}\,\text{kg CO}_2/\text{s}}{\left(\frac{1000\,\text{m}^3\,\text{CO}_2 \times 1.83\,\text{kg CO}_2/\text{m}^3\text{CO}_2}{10^6\text{m}^3\,\text{air}}\right)\left(1 - \frac{0.822}{0.855} \times \frac{500}{1000}\right)}$$

$$= 0.31\ \text{m}^3\ \text{air/s}$$

14.4 Indoor Air Cleaning Model

Similar to the air dispersion models for ambient air pollutants, we can develop models for indoor air dispersion. The simplest is box model as introduced in Sect. 11.1. We consider an indoor space of concern as a box, which seems to be more reasonable than being treated for an ambient environment. Most models assume the indoor air is uniform for the ease of analysis [10]. This is called compete mixing model. Zonal models are available too, but we will not discuss them in this book. Instead, we will introduce an indoor air cleaning model as follows [10].

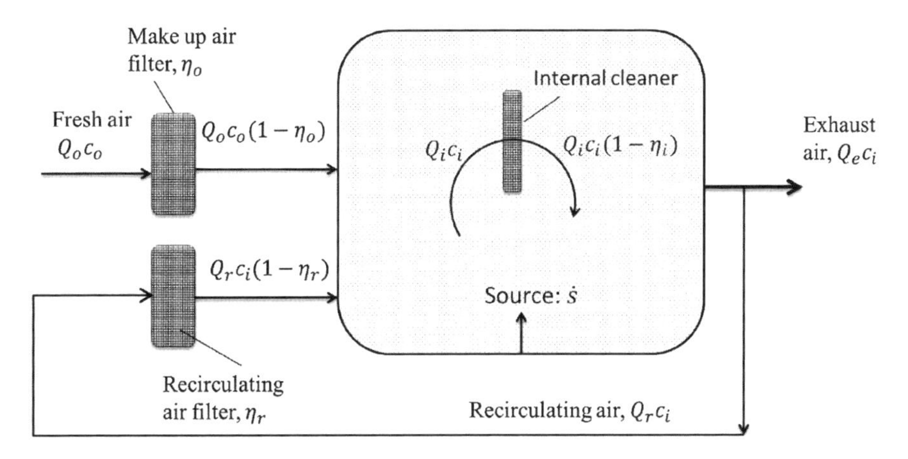

Fig. 14.3 Schematic diagram for the ventilation model with indoor air cleaning

Consider the schematic diagram in Fig. 14.3, where air flow through the building envelop is ignored. There is also an internal air cleaner to help clean the air. The mass balance of an air pollutant leads to

$$V\frac{dc_i}{dt} = [Q_o c_o(1-\eta_o) + Q_r c_i(1-\eta_r) + \dot{s}] - (Q_e c_i + Q_i c_i \eta_i) \tag{14.18}$$

where Q_o, Q_r, Q_i, and Q_e are the volumetric flow rates of fresh air, recirculating air, internal air cleaning, and exhaust air, respectively. V is the volume of the space of concern, and the total area within the space is A. The internal source rate with respect to the air pollutant is $\dot{s}$. At any moment the indoor air concentration is c_i and that in the fresh air is c_o. The last term $(Q_e c_i + Q_i c_i \eta_i)$ stands for the removal rate of the pollutant from the room air.

Since air pollutant is the main concern in this analysis, the change of moisture level in the air can be ignored; then the mass balance of air leads to

$$Q_o = Q_e \tag{14.19}$$

Then Eq. (14.18) becomes,

$$-V\frac{\mathrm{d}c_i}{\mathrm{d}t} = [Q_o - Q_r(1-\eta_r) + Q_i\eta_i]c_i - [Q_o c_o(1-\eta_o) + \dot{s}]. \tag{14.20}$$

If we assume only c_i and t are variables, then integration leads to

$$c_i = \left[c_{i0} - \frac{Q_o c_o(1-\eta_o) + \dot{s}}{Q_o - Q_r(1-\eta_r) + Q_i\eta_i}\right]\exp\left[-\frac{Q_o - Q_r(1-\eta_r) + Q_i\eta_i}{V}t\right] + \frac{Q_o c_o(1-\eta_o) + \dot{s}}{Q_o - Q_r(1-\eta_r) + Q_i\eta_i}. \tag{14.21}$$

It can be presented in a simple form as

$$c_i(t) = (c_{i0} - c_{i\infty})\exp\left(-\frac{t}{\tau}\right) + c_{i\infty} \tag{14.22}$$

where $C_i = C_{i0}$ at $t = 0$

$$c_{i\infty} = \frac{Q_o c_o(1-\eta_o) + \dot{s}}{Q_o - Q_r(1-\eta_r) + Q_i\eta_i} \tag{14.23}$$

$$\tau = \frac{V}{Q_o - Q_r(1-\eta_r) + Q_i\eta_i} \tag{14.24}$$

In the above equation, $c_{i\infty}$ is the steady state concentration in the indoor space when $t \to \infty$, and τ can be considered as a time constant that is used to characterize the transient concentration in the space.

Example 14.4: Transient indoor air pollutant concentration
Consider a commercial kitchen with a volume of 80 m^3 where natural gas is used as a cooking fuel. Assume that after cooking the indoor fume concentration is 1,000 $\mu g/m^3$. The HVAC system works with a fresh air flow rate of 10 m^3/s and an internal recirculating air flow rate of 1/3 of the fresh air flow rate. Ignore the outdoor and indoor source other than cooking. The intake and recirculating filter efficiencies are both 90 % and the internal air cleaner has a flow rate of 1 m^3/s and an efficiency of 95 %. Plot the indoor fume concentration over time.

Solution
The problem is simplified by ignoring the outdoor and indoor sources other than cooking. After cooking, $\dot{s} = 0$, and $c_0 = 0$ in Eq. (14.23). This simplification leads to

$$c_{i\infty} = \frac{Q_o c_o (1 - \eta_o) + \dot{s}}{Q_o - Q_r(1 - \eta_r) + Q_i \eta_i} = 0$$

Then, Eq. (14.22) becomes

$$c_i(t) = c_{i0} \exp\left(-\frac{t}{\tau}\right)$$

where the time constant is calculated using Eq. (14.24):

$$\tau = \frac{V}{Q_o - Q_r(1 - \eta_r) + Q_i \eta_i} = \frac{800}{10 - \frac{10}{3}(1 - 0.9) + 1 \times 0.95} = 75.35(s)$$

Then we have

$$c_i(t) = 1000 \exp\left(-\frac{t}{75.35}\right) (\mu m/m^3)$$

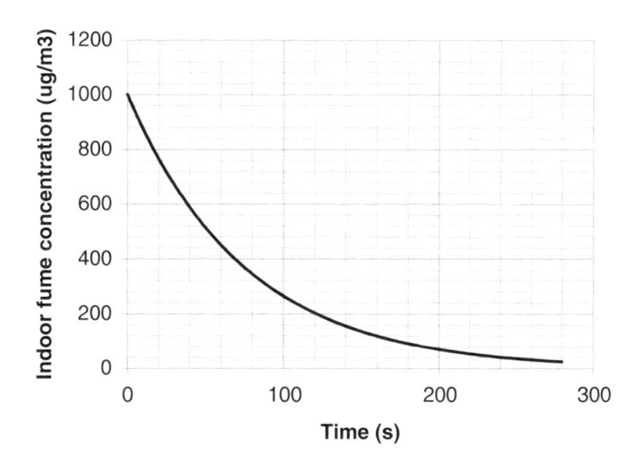

Fig. 14.4 Calculated indoor air pollutant concentration over time

The plot over time is shown in Fig. 14.4.

In addition to the analytical models above, there are a variety of computation fluid dynamics (CFD) models that can be employed for the predication of indoor air quality and indoor air pollutant concentrations. Readers are recommended to learn these specialized models from other publications including those listed in the Reference section [10–20].

14.5 Practice Problems

1. Minimum ventilation requirement is the ________ value of the ventilation requirements based on sensible heat balance, moisture balance, and pollutant balance.

   a. maximum
   b. minimum
   c. average
   d. arbitrary

2. While you were cooking using gas burners in your kitchen, the measured concentrations of CO and CO_2 were 10 and 3500 ppmv, respectively. Is the air quality in your cooking environment acceptable based on these two pollutants only? (Given: the TLV's for CO and CO_2 are 25 and 5000 ppmv, respectively).

   a. Yes, because the CO concentration is below the corresponding TLV.
   b. No, because the CO_2 concentration is below the corresponding TLV.
   c. Yes, because the normalized TLV is below unit.
   d. No, because the normalized TLV is above unit.

3. In a Class 1 clean room, the measured total particle concentration is 9 particles/m^3 for particles larger than 0.1 μm and 3 particle/m^3 for particles larger than 0.2 μm. Does this cleanroom meet the Class 1 criteria?
4. In a large auditorium with a ventilation rate of 6 m^3/s, the carbon dioxide concentrations in the exhaust air and supply air are 800 and 400 ppmv, respectively. The specific volume of exhaust and supply air are 0.82 and 0.78 m^3/kg dry air, respectively. The building has an air volume of 8,000 m^3, and the air inside is assumed to be completely mixed. In a power failure, all exhaust fans are stopped, and there are 1000 people inside, each generates a heat of 200 W. Assume there is no air exchange between the inside and outside of the auditorium. How long it will take the CO_2 to reach the ceiling threshold value of 5,000 ppmv after the power failure.
5. There are 30 students in a classroom, each student produce 120 W of sensible heat and 100 g of water vapour per hour (or 70 W latent heat). The room air temperature is 20 °C and the specific volume and a relative humidity of 50 %.

The carbon dioxide concentration of the supply air is 400 ppmv. Supply air has a temperature of 10 °C a relative humidity of 70 %. Heat loss through the building shelter is 2 kW and heat gain from equipment and lights are 1.5 kW. It is desired to maintain an indoor carbon dioxide concentration of less than 800 ppmv, relative humidity of less than 50, and room temperature of 20 °C, what should the minimum ventilation be for the air?

6. Consider a tire center with a volume of 1,000 m^3 where tire continuously emit VOC at a rate of 100 μg/s. Assume at time zero, indoor VOC concentration is 1,000 μg/m^3. The HVAC system works at a fresh air flow rate of 100 m^3/min and an internal recirculating flow rate of 1/3 of that of the intake fresh air. An internal air cleaner has a flow rate of 0.1 m^3/s; ignore the outdoor source of VOC. The fresh air is free of VOC and recirculating activated carbon filter efficiency is the same as that of the internal air cleaner, which is 90 %. Plot the indoor VOC concentration over time.

References and Further Readings

1. US EPA (2005) IAQ tools for schools: actions to improve indoor air quality, EPA 402-F-05-016. http://www.epa.gov/iaq/schools/actions_to_improve_iaq.html
2. Zhang Y (2004) Indoor air quality engineering. CRC Press, FL, USA
3. Lippmann M (1988) Asbestos exposure indices. Environ Res 46:86–106
4. Dodson RF, Atkinson MAL, Levin JL (2003) Asbestos fiber length as related to potential pathogenicity: a critical review. Am J Ind Med 44:291–297
5. Estévez Sánchez AM, Fernández Tena A, Márquez Moreno MC (1989) Oxidation of methanol to formaldehyde on iron-molybdenum oxide catalysts, with and without chromium as a promoter. React Kinet Catal Lett 38(1):193–198
6. Gerberich HR, Seaman GC (2004) Formaldehyde. In: Kroschwitz JI, Howe-Grant M (eds) Kirk-Othmer encyclopedia of chemical technology, vol 11, 5th edn. Wiley, New York, pp 929–951
7. WHO (2010) WHO guidelines for indoor air quality: selected pollutants
8. ASHRAE (2013) ASHRAE handbook: fundamentals. ASHRAE (American Society of Heating, Refrigerating and Air-Conditioning Engineers), Atlanta
9. Gatley DP (2004) Psychrometric chart celebrates 100th anniversary. ASHRAE J 46(11):16–20
10. Shair FH, Heitner KL (1974) Theoretical model for relating indoor pollutant concentrations to those outside. Environ Sci Technol 8(5):444–451
11. ACGIH (2013) TLVs and BEIs, publication #0113, American conference of governmental industrial hygienists, Cincinnati
12. Feng Z, Long Z, Chen Q (2014) Assessment of various CFD models for predicting airflow and pressure drop through pleated filter system. Build Environ 75:132–141
13. Chen C, Lin C-H, Long Z, Chen Q (2014) Predicting transient particle transport in enclosed environments with the combined CFD and Markov chain method. Indoor Air 24:81–92
14. Liu W, Wen J, Lin C-H, Liu J, Long Z, Chen Q (2013) Evaluation of various categories of turbulence models for predicting air distribution in an airliner cabin. Build Environ 65:118–131

15. Xue Y, Zhai Z, Chen Q (2013) Inverse prediction and optimization of flow control conditions for confined spaces using a CFD-based genetic algorithm. Build Environ 64:77–84
16. Jin M, Zuo W, Chen Q (2013) Simulating natural ventilation in and around buildings by fast fluid dynamics. Numer Heat Transf Part A Appl 64(4):273–289
17. Chen C, Liu W, Li F, Lin C-H, Liu J, Pei J, Chen Q (2013) A hybrid model for investigating transient particle transport in enclosed environments. Build Environ 62:45–54
18. Wang H, Chen Q (2012) A new empirical model for predicting single-sided, wind-driven natural ventilation in buildings. Energy Build 54:386–394
19. Zhu Y, Zhao B, Zhou B, Tan Z (2012) A particle resuspension model in ventilation ducts. Aerosol Sci Technol 46:222–235
20. Zhao B, Chen C, Tan Z (2009) Modeling of ultrafine particle dispersion in indoor environments with an improved drift flux model. J Aerosol Sci 40(1):29–43

Chapter 15
Air Monitoring

As mentioned in Chap. 1, there are two approaches to implement the air emission standards. One is best technology approach and another is air emission monitoring. Air emission monitoring from a source is also referred to as source test, which provides technical information for judging the relative importance of a given source contribution of pollutants. It is also a measure to determine whether a pollution control device installed is working effectively as expected. Sometimes, the analysis of sources of air emissions, such as boilers, incinerators, and diesel engines, also provides complementary information to their energy efficiencies. After all, energy production and the environmental pollution are closely related. The source test results are used in the design and tuning of control methods as well as a legal tool in the case of a violation of an air pollution control regulation.

It is unusual to take only one sample and make a claim because it is solely by luck one can achieve this goal. Actual engineering applications are much more complicated. In order to obtain this accurate information, multiple samples and measurements must be taken followed by engineering statistical analysis to quantify both emission rates and their uncertainties.

Reference Methods have been developed by government agencies or professional societies to guide these practices. In the United States, for example, all reference methods for stationary source tests are available to the public for free on the web site of Emission Measurement Center (http://www.epa.gov/ttn/emc/). These Reference Methods specify the procedures and certified equipment for measuring the constituents of emissions. They also describe the principles, applicability, and presentation of the test results.

15.1 Flow Rate and Velocity Measurement

There are a number of flow rate measurement technologies and related instruments; one of them is Pitot tube, named after Henri Pitot in 1732. It is widely used in engineering practices because of its simplicity, accuracy, reliability, and cost-effectiveness.

Z. Tan, *Air Pollution and Greenhouse Gases*, Green Energy and Technology,
DOI 10.1007/978-981-287-212-8_15

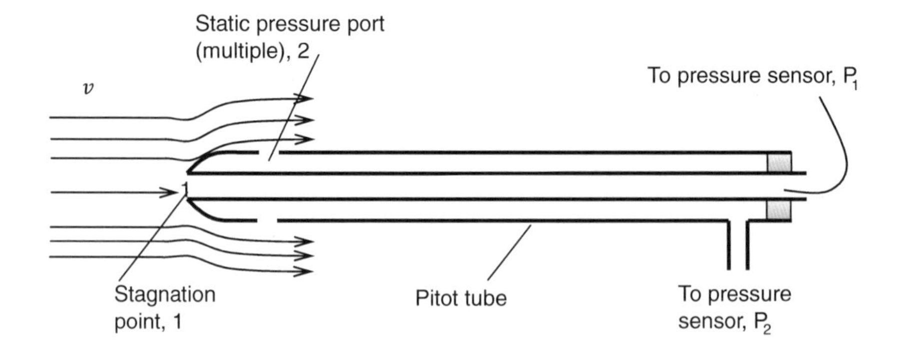

Fig. 15.1 *L-shaped* Pitot tube

A standard Pitot tube is an L-shaped tube depicted in Fig. 15.1. When it is located in a pipe, streamlines connect the stagnation point and the multiple static taps, which are small holes evenly distributed along the circumference of the outer tube.

Applying Bernoulli's Equation (Eq. 2.74) to the stagnation point, where $u_1 = 0$, and the static taps, we can get

$$\Delta P = P_1 - P_2 = \frac{1}{2}\rho u_2^2 \tag{15.1}$$

and the fluid speed near the static taps is

$$u = u_2 = \left(\frac{2\Delta P}{\rho}\right)^{1/2} \tag{15.2}$$

where ΔP is the reading from the differential manometer at the other end of the Pitot tube. The corresponding air volume flow rate is calculated as

$$Q = Au. \tag{15.3}$$

15.2 Source Sampling

As introduced above, a representative sample is the key to the accurate measurement of air emissions from a source. Different source test methods have been developed for a variety of realistic environments. In the USA, for example, most states rely upon US EPA Reference Methods, Methods 1–8, for characterization of the gas flow and specific air pollutants from a stationary source. They are listed as follows:

Method 1: Sample and velocity transverse for stationary sources
Method 2: Determination of stack gas velocity and volumetric flow rates using Type S Pitot tube,
Method 3: Gas analysis for the determination of dry molecular weight
Method 4: Characterizing moisture content in stack gases
Method 5: Determination of particulate matter emission from stationary sources
Method 6: Determination of sulfur dioxide emissions from stationary sources
Method 7: Determination of nitrogen oxide emissions from stationary sources
Method 8: Determination of sulfuric acids and sulfur dioxide emissions from stationary sources

These methods set the foundation for other subsequent methods. The first four methods are supportive but necessary in that they provide information for other methods or the calculation of the results. The other four are for the determination of different air emissions as explicitly indicated by their titles. Method 1 is the first step for any source testing; it is concerned with assessing the suitability of the sampling site and determining the sampling points. The principles behind these Methods will be introduced shortly.

These sampling methods are similar to each other with exchangeable equipment; this similarity allows a single sampling train to be used by multiple methods by adding or removing impingers, filters, or other appropriate devices. For example, Method 5 is typically combined with Methods 2, 3, and 4 in order to determine gas velocity, molecular weight and moisture content, which are required to calculate a particulate emission rate. Method 5 and 8 can also be combined by adding additional impingers and an in-stack filter to the probe on the Method 5 sampling train.

Figure 15.2 shows a schematic diagram of a typical sampling train for Method 5, which is available online (e.g., http://www.epa.gov for free use). A typical sampling train starts with a heated probe, or a hollow glass tube that is inserted into the stack or duct and the last component is a pump drawing gases through the system. Other pieces of equipment include filters and impingers to capture different air pollutants of concern. A gas meter measures the flow rate of sampled gas and the stack gas velocity at the sampling point is measured using a Pitot tube with a manometer (Sect. 15.1). Finally, the sample exhaust is discharged into the atmosphere through the orifice, which is used for sample train flow adjustments at the by-pass valve.

15.2.1 Isokinetic Sampling

Isokinetic sampling has to be executed for particulate matter related source tests. Under isokinetic condition, the gas velocity entering the probe is the same as the bulk air velocity.

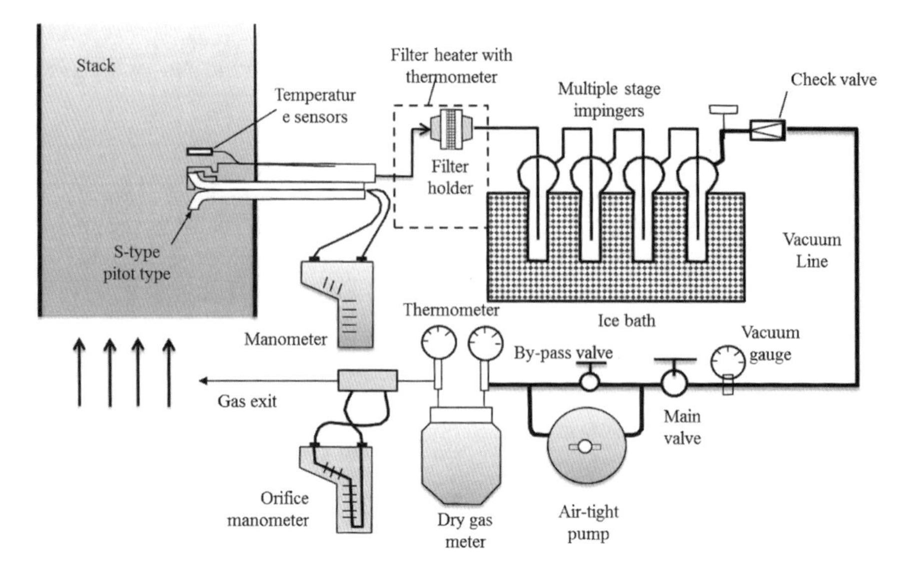

Fig. 15.2 Sampling train for particulate matter (US EPA Method 5)

$$U_s = U_0 \tag{15.4}$$

Under isokinetic condition, as illustrated in Fig. 15.3, particles of different sizes follow the air stream and enter the probe as if there were no probe in the area.

Sampling with velocity U_s lower or higher than the gas stream velocity U_0 is called anisokinetic sampling. They are depicted in Fig. 15.4. Anisokinetic sampling results in errors in the particulate concentration measurements. When the sampling velocity is greater than the gas stream velocity ($U_s > U_0$), it is called over isokinetic; gas from regions not directly in front of it is drawn into the probe and the converging streamlines results in loss by separation of large particles. These particles are

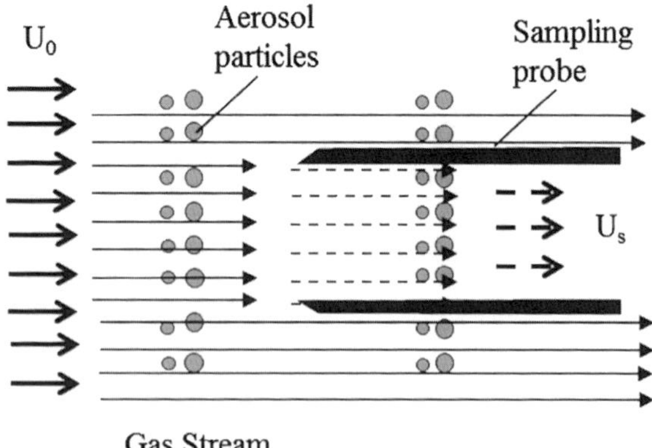

Fig. 15.3 Isokinetic sampling

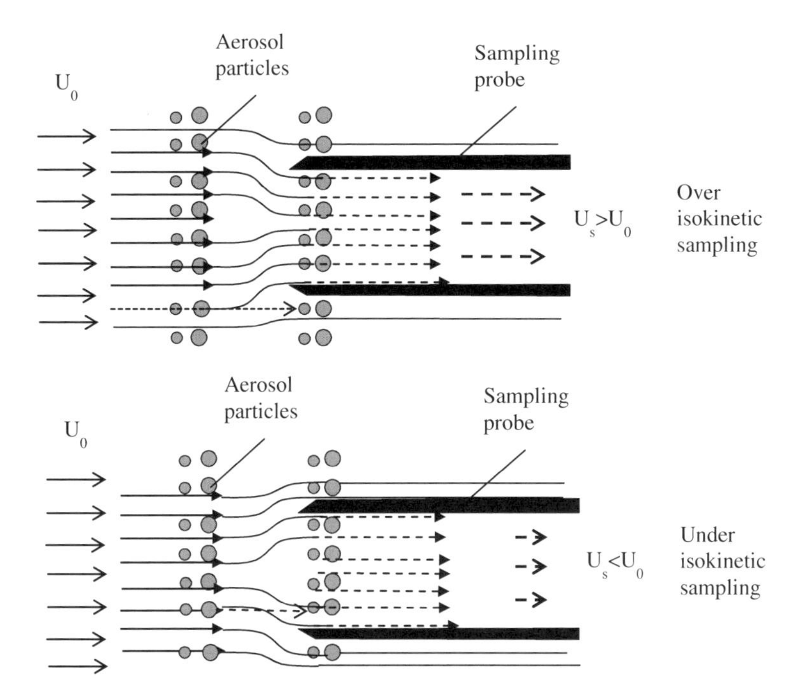

Fig. 15.4 Anisokinetic sampling by velocity mismatch

excluded from the sample, and consequently this practice underestimates the concentrations of large particles in the gas stream. On the other hand, small particles tend to follow the gas streamlines, and their concentrations are over estimated. Overall, over isokinetic sampling results in the underestimation of the particulate mass concentration because large particles contribute much more to the mass than smaller ones. An alternative way to understand this process is that the particulate matter is entering the sampling probe diluted by over sampling of the gas into the probe.

When the sampling velocity is less than that of the gas stream, it results in under isokinetic sampling. Because of the lower sampling velocity, some gas bypasses the probe as if it were an obstruction. As a result, a smaller volume of gas is taken into the probe. However, the large particles in the gas stream still enter the probe due to their great inertia and the diverging gas streamline. The sample is also biased with overestimated large particle concentrations.

The sampling efficiency is a ratio of the sampled concentration to that in the bulk air stream. That is,

$$\eta_s = C_s/C_0 \tag{15.5}$$

In reality, it is very challenging to achieve 100 % isokinetic sampling and most reference methods consider an isokinetic sampling efficiency in the range of ±10 %

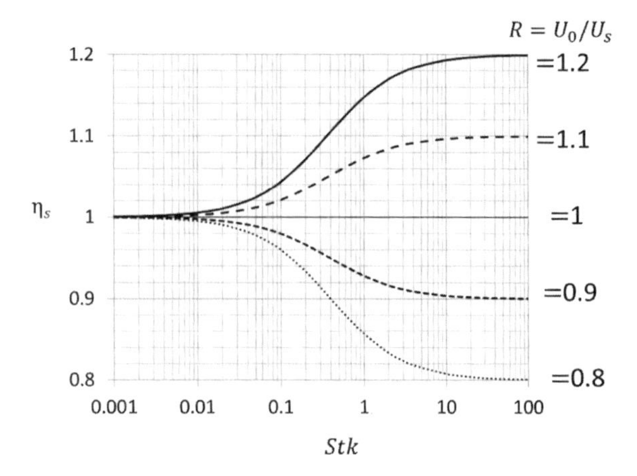

Fig. 15.5 Sampling efficiency η_s versus R and Stk (velocity mismatch only)

acceptable. Note that isokinetic sampling is not necessary if only gaseous pollutants are of concern.

The sampling efficiency can be calculated in terms of particle *Stokes* number, the sampling velocity and the bulk air velocity.

$$\eta_s = 1 + (R - 1)\beta(\text{Stk}, R) \tag{15.6}$$

$$\text{where} \quad \text{R} = \text{U}_0/\text{U}_\text{s} \tag{15.7}$$

and $\beta(\text{Stk}, R)$ is a function of particle Stokes number and the velocity ratio. Belyaev and Levin [1] gave a practical formula that matched the experimental data corresponding to $0.18 < \text{Stk} < 6.0$ and $0.16 < \text{R} < 5.5$.

$$\beta(\text{Stk}, R) = 1 - \frac{1}{1 + (2 + 0.617/R)\text{Stk}} \tag{15.8}$$

Figure 15.5 is produced using Eq. (15.8), the calculation results show that $\eta_s \rightarrow 1$ when $\text{Stk} \rightarrow 0.01$ and $\eta_s \rightarrow R$ when $\text{Stk} \rightarrow \infty$. $\eta_s \approx 1 \pm 0.05$ for $\text{Stk} < 0.1$ and η_s becomes little dependent on Stk. When $\text{Stk} > 10$ the equation can be practically simplified as $\eta_s = R$

$$\begin{aligned} \eta_s &\approx 1 \quad \text{for Stk} < 0.1 \\ \eta_s &\approx R \quad \text{for Stk} > 10 \end{aligned} \tag{15.9}$$

15.2.2 Effect of Misalignment

The preceding analysis is based on the assumption that the sampling velocity and the bulk air velocity have the same direction. In practice, it implies the perfect

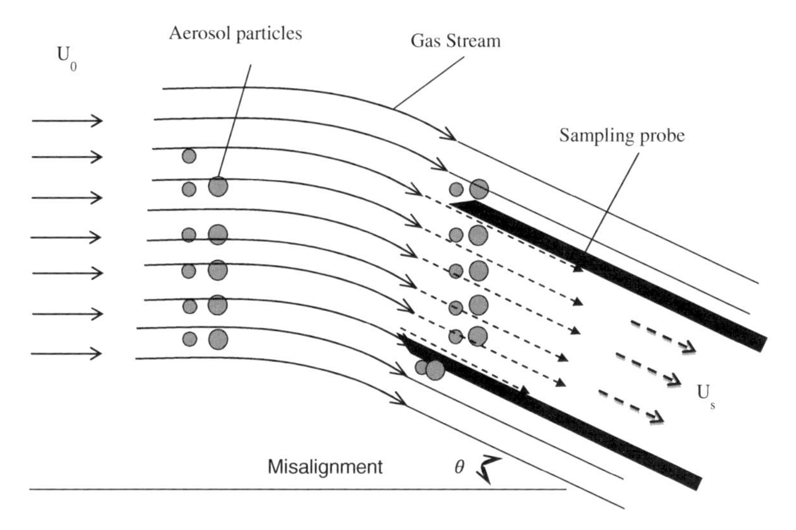

Fig. 15.6 Misalignment in air sampling

alignment between the sampling probe and the bulk air flow direction. As seen in Fig. 15.6, misalignment between the probe and the bulk air velocity may also lead to sampling error.

Empirical equations were given by Durham and Lundgren [5] to quantify the misalignment effect for anisokinetic sampling as follows:

$$\eta_s = 1 + (\cos\theta - 1)\beta'(\mathrm{Stk}', \theta) \quad \text{for R} = 1 \tag{15.10}$$

where $\beta'(\mathrm{Stk}', \theta)$ is a function of Stokes number and the angle of misalignment,

$$\beta'(\mathrm{Stk}', \theta) = 1 - \frac{1}{1 + 0.55\ \mathrm{Stk}'\exp(0.25\ \mathrm{Stk}')} \tag{15.11}$$

and

$$\mathrm{Stk}' = \mathrm{Stk} \times \exp(0.022\theta) \quad \text{for}\, 0 \leq \theta < \pi/2 \tag{15.12}$$

Since $\cos\theta < 1$, this equation indicates that misalignment of the probe always results in the underestimation of the particle concentration.

Figure 15.7 is produced using Eq. (15.10) above for $R = 1$. It shows that the sampling efficiency is always less than unity when $\theta \neq 0$; again, the most dramatic change in sampling efficiency corresponds to the range of $0.1 < \mathrm{Stk} < 10$.

When there are both misalignment and mismatch in air velocities, Eq. (15.13) can be used for estimating the sampling efficiency.

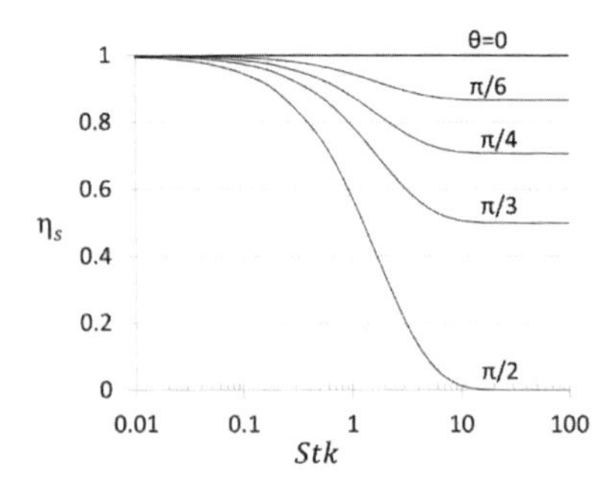

Fig. 15.7 Misalignment sampling efficiency when $R = 1$

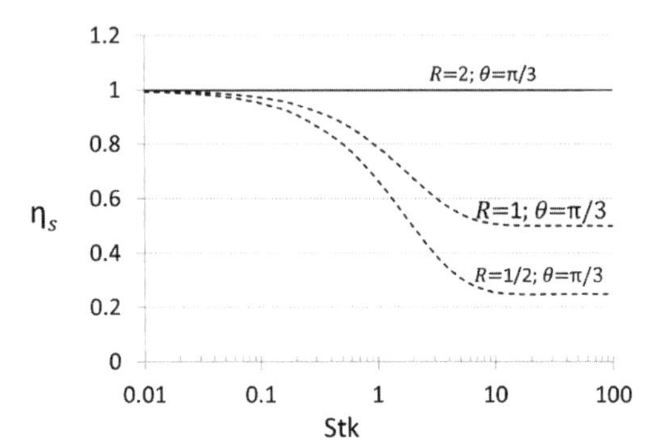

Fig. 15.8 Misalignment sampling efficiency when $R \neq 1$ ($\theta = \frac{\pi}{3}$)

$$\eta_s = 1 + (R\cos\theta - 1)\beta'(\text{Stk}', \theta)\frac{\beta(\text{Stk}', R)}{\beta(\text{Stk}', R = 1)} \tag{15.13}$$

However, it is useful to realize that the sampling efficiency is always 1 when $R\cos\theta = 1$, regardless of the value of $\beta'(\text{Stk}', \theta)\beta(\text{Stk}', R)/\beta(\text{Stk}', R = 1)$. This simple relationship is of practical use; the error introduced by misalignment can be compensated by increasing R value to such a level as $R = 1/\cos\theta$. This concept is also illustrated by a sample calculation of the sampling efficiency shown in Fig. 15.8, which shows that $\eta_s = 1$ when $R = 2$ and $\cos\theta = 1/2$.

15.2.3 Multiple Sampling Locations

Air in the stack is not as uniform as described in the theoretical analysis above; it is turbulent and is not characterized as parabolic velocity profile. Therefore, it is very difficult for a practitioner to determine one sampling point that can represent the average velocity of the entire stack. As a rule of thumb, air sampling in duct should be conducted at multiple points at the transverse plan perpendicular to the duct axis. A common practice in determining the sampling points is the *equal area method*.

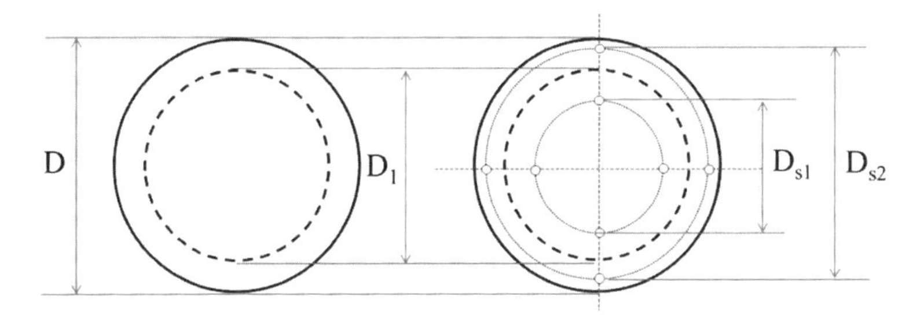

Fig. 15.9 Equal area method for a circular duct (*Left* equal area determination. *Right* sampling point determination)

Sampling points are distributed over equal areas of the cross section of the stack or duct.

A typical stack is circular. Consider a circular stack with an inner diameter of D, its cross section area can be easily determined as $A = \pi D^2/4$. If samples are to be taken, for whatever reasons, from two equal areas as shown in Fig. 15.9, the area of the inner circle (dashed line) is $A/2$. In this case, $\pi D_1^2/4 = A/2 = \pi D^2/8$ and $D_1 = D/\sqrt{2}$. However, the sampling points are NOT to be located on the thick dashed circle defined by D_1. Rather there will be eight sampling points evenly distributed on two other circles indicated by the dotted lines defined by D_{s1} and D_{s2} in Fig. 15.9. These dotted lines divide the inner circle and the *O*-ring into equal areas again. Then four evenly distributed sampling points on each dotted line are determined finally.

A reader may have noticed that the total stack area was actually divided into four equal ones by the dashed circle and two dotted circles. The ultimate goal of this practice is to determine the sampling points, which is defined by the imaginary sampling circles (dotted ones). A fast approach to this can bypass the step for determining the diameters of the circles (dashed lines). For the above practice, the stack is first divided into four instead of two equal areas, then the sampling points are located every other circles starting from the innermost.

Despite the tedious work, one can determine the corresponding sampling points for any stack sampling following the same principles introduce above. Consider a general case where the stack with an inner diameter of D and an area of A is to be divided into n equal annular areas (with diameters of D_i) and the corresponding $4n$ sampling points are determined as follows. In order to determine the sampling points without calculating the n equal areas, treat it as if the total area were to be divided into $2 \times n$ equal annular areas, then the diameter ith circle (D_{si}) counting from the inner-most outward is defined by

$$D_{si}^2 = \frac{i}{2n} D^2 \qquad (i = 1, 3, \ldots 2n - 1) \tag{15.14}$$

On each sampling circle, there are four sampling points evenly distributed.

Similarly, the circles that divide the total area into n equal areas can be determined with i being even numbers, although they are of no practical use.

$$D_i = D\sqrt{i/2n} \quad \text{where } i = 2, 4 \ldots 2n \tag{15.15}$$

Example 15.1: Sampling location in stack
A stack with an inner diameter of 1.2 m is subject to source emission tests. Determine the sampling points by dividing the stack area into three equal sub-areas.

Solution
In this problem, $n = 3$ and $D = 1.2$ m are given. The sampling points are on the circles with diameters of $D_{si} = D\sqrt{i/2n}$ where $i = 1, 3$ and 5

$$D_{s1} = 1.2 \times \sqrt{1/6} = 0.4899(m)$$

$$D_{s2} = 1.2 \times \sqrt{3/6} = 0.8485(m)$$

$$D_{s3} = 1.2 \times \sqrt{5/6} = 1.0954(m)$$

These sampling points are depicted in Fig. 15.10, where readers can determine the equal area circle D_i by using Eq. (15.15).

In addition, samples should be taken at each of the 4n sampling points for equal duration, typically being an hour or so. For statistical analysis of the data, at least three replicates are recommended for each sample. This means at least $3 \times 4 \times n$ samples are taken for one complete test. Usually averaged results are reported with standard deviation to quantify the variations at the corresponding sampling points. As a result, source tests are time-consuming and a complete stack test may last for many hours to days.

In a real source test, the stack cross section area is often divided into more than two equal areas. The number of points is determined following local standards. In general, more sampling points are needed if particulate matter is of interest. Less sampling points are required for gaseous pollutants because the gases in typical stacks are well mixed by turbulence.

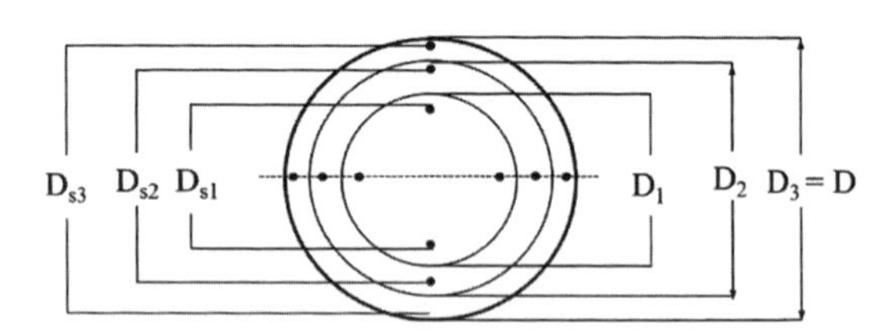

Fig. 15.10 Sampling points determined by equal area method for a circular duct

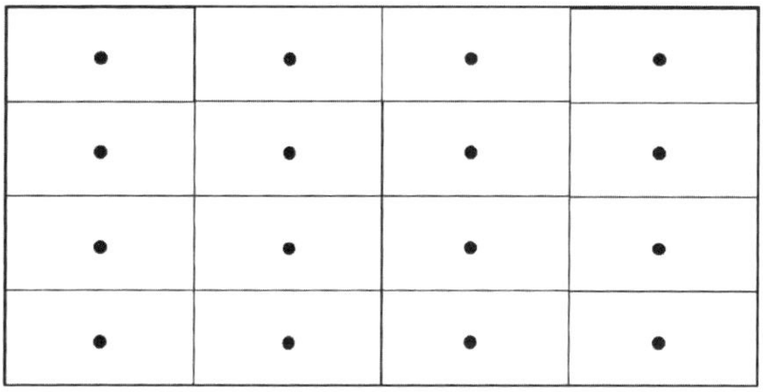

Fig. 15.11 Equal area method for a rectangular duct

Equal area method is also applicable to rectangular duct, where the determination of the sampling points is much simple. The duct perimeter is first divided into a number of girds and the sampling points are chosen at the center of each grid (Fig. 15.11). The more sampling points the more representative samples can be obtained. However, one has to compromise for time and costs.

15.3 Collection of Air Pollutant Samples

After sampled stack gas is extracted from the bulk air stream into the probe nozzle, this air sample is further drawn through a sampling train or a series of leak-proof equipment components configured to capture different air pollutants of concern. As seen in Fig. 15.2 (Method 5), water vapor and condensable are best captured by condensation or bubbling the sample gas through chilled impingers, which are sealed glass vessels. Liquid reagents may be added into the impingers to absorb some gases. Solids are likely captured on heated filters, which are connected to the probe with interconnecting glassware. Solids can also be captured by liquids in impingers.

The mass of the target air pollutant separated from the sample gas can be determined by the weights before and after sampling. Filters are weighed and reagent volumes in the impingers are measured prior to use. After each run, the content of each sampling train component is carefully recovered to a sealed vessel followed by weighing the solid loaded filters or liquids from the impingers. For some tests, the samples collected may be evaluated in a laboratory using scientific equipment.

15.4 Data Analysis and Reporting

Regardless of the end use of the data, the immediate objective of source tests is to obtain reliable and representative information of the air emissions and the rates of emissions into the atmosphere.

In general, air emission monitoring is completed by gas sampling followed by gas characterization. In addition, the properties of total gas flow from the source must also be determined accurately before calculating the corresponding air emission rates. If a gas sample with a sampling flow rate of Q_s is taken from a continuous emission source with a total flow rate of Q, typical value of Q_s is much less than Q, $Q_s \ll Q$. Characterization of the sample leads to the concentration of certain air pollutant in the sample as C_s. If this sample can be used to represent the total gas, one can claim that the concentration of the corresponding air pollutant in the main stream C_0 is the same as c_s.

Now consider a simplified case as follows. The total gas flow rate discharged from a stack to the atmosphere is Q (m^3/s), and a sample is taken with a volume flow rate of Q_s (m^3/s) for a period of time, t(s). Analysis of this sample, either in situ or in a laboratory setting, shows that the mass of air pollutant of interest is m (kg). With this information, one can calculate the mass concentration of the air pollutant in the sample $c = m/Q_s t$ (kg/m^3). If the sample is representative, it can be used to represent the concentration of the air pollutant in the stack: $C = c$. Then the mass flow rate, $\dot{m}$, of this air pollutant in the stack is determined as

$$\dot{m} = \frac{Qm}{Q_s t}(\mathrm{kg/s}) \tag{15.16}$$

The emission rate of the air pollutants can be calculated with the data obtained from the samples. However, the samples taken from different stacks are different in temperature, pressure, and moisture contents. The calculated results shall be corrected to a standard temperature and pressure and expressed on a dry basis. This standardized measure allows the source test results under different conditions to be comparable to regulatory standards or the results of other similar tests.

A common standard for emission test report is to correct the concentrations against 50 % excess air, 7 or 12 % of CO_2, or 7 % of O_2. Take 12 % of CO_2 as an example, the calculated emission rate is converted using the equation that follows. Actual procedure is much more complex, and readers are encouraged to seek local guidelines for more information.

$$C_{\mathrm{std}(12\,\%\,\mathrm{CO_2})} = C_s \times \frac{12\,\%}{100\,\% \times y_{\mathrm{CO_2}}} \tag{15.17}$$

where y_{CO_2} is the mole fraction of CO_2 in the stack gas measured (e.g., following US EPA Method 3).

15.5 Continuous Emission Monitoring and Opacity Measurement

Manual source test is a time-consuming and labor-intensive process, and there has been a growing interest in a more efficient, automatic approach. Advances in instrumental analyzers have enabled continuous emission monitoring (CEM) with modern instruments that can quickly identify many gas compounds at very low concentrations. CEM also allows monitoring of peak emissions.

There are many types of CEM instruments and systems for continuous monitoring of air emissions from stationary sources. They can be classified into extractive and In situ systems. Extractive systems can be further categorized as either dilution or direct. In situ analyzers directly measure source-level concentrations by changing the original gas properties by removal of particulate or interfering vapors or gases.

There are two types of in situ analyzers: path and point measurements. Either single or double path CEM instruments perform measurement across the entire duct or stack at certain elevation, whereas point measurement is completed by a sensor at a specific point or over a path of only a few centimeters distance. Typical in situ CEM systems can tolerate harsh environment at most locations and require fewer accessories such as probe heaters, gas conditioning systems, tubing and pumps than their extractive predecessors [7].

The 1970s saw advancements in German optical systems, which have shown great promise for measuring gas opacity, an indicator of the air emission rate [7]. The opacity of the stack gas is a function of light transmission through the plume and is determined by

$$OP = (1 - I/I_0) \times 100 \tag{15.18}$$

where OP = percent opacity, I = light flux leaving the plume, and I_0 = incident light flux.

Opacity is more lenient than the corresponding mass emission standards. The USEPA developed Method 9 to quantify the opacities of the plumes emitted from stationary point sources (http://www.epa.gov/ttn/emc/promgate/m-09.pdf). Recently, Du et al. [3] reported a method to quantify the plume opacity by digital photography and Du et al. [4] published a paper for fugitive fume detection by opacity.

Most air pollutants can now be continuously monitored by different instruments. The selection of the instruments is important to the design of the monitoring system. The choice of instrumentation for an air monitoring depends on type of pollutants and expected concentrations, accuracy required by local air quality standards, budget and personnel, etc. In the USA, CEM is required by the Clean Air Act Amendments to monitor SO_2, NO_x, CO, CO_2, opacity, total hydrocarbons, and total reduced sulfur (TRS). All the instruments are required for calibration before being used on site.

15.6 Ambient Air Quality Monitoring

Ambient air quality monitoring data is used to determine local air quality index and the levels of air pollution in the region, to establish and evaluate air pollution control measures, and to evaluate air dispersion models or some of their parameters. Sometimes it is also used to warn the citizens when the air pollution is hazardous.

Ambient air quality monitoring network in most regions employs continuous monitoring of outdoor air quality. These networks measure the amount of gaseous air pollutants and particulate matter at multiple locations to produce statistically correct data.

Like air emission monitoring systems, there are two types of ambient air quality monitoring systems: integrated and continuous monitoring systems. In integrated monitoring, samples are collected using standard devices over a certain time interval. The samples collected are analyzed in a laboratory.

A continuous air quality monitoring system constitutes both air sampling and pollutant characterization devices in one system. Samples are taken and analyzed continuously. Air pollutant concentrations are nearly instantaneously displayed or continuously recorded by a data logging mechanism. The system can also be programmed to take samples and display results over certain interval, say every 10 min or 2 h. The intervals set by the operator should follow the guideline set in the corresponding air quality standards. The data can also be sent to a central location for further analysis via wired or wireless network. On the other hand, this type of system requires sophisticated instruments and well-trained professionals to operate and to maintain the system.

An air quality monitoring network can also be mobile. Mobile air quality monitoring networks are set in trailers, automobiles, or other mobile units. They take samples and display data on schedule among selected locations. Consequently, data are collected in time sequence. The mobile air quality monitoring and sampling features its flexibility in air quality measurements and it provides a much higher resolution in the geographical variations if the air monitoring practice lasts long enough to generate meaningful data.

References and Further Readings

1. Belyaev S, Levin L (1974) Techniques for collection of representative aerosol samples. Aerosol Sci 5:325–338
2. Bowers K (1998) Continuous emission monitors compliance assistance program. Air Resources Board, California Environmental Protection Agency, p 95
3. Du K, Rood MJ, Kim B, Kemme M, Franek B, Mattison K (2007) Quantification of plume opacity by digital photography. Environ Sci Technol 41:928–935
4. Du K, Shi P, Rood M, Wang K, Wang Y, Varma R (2013) Digital optical method to quantify the visual opacity of fugitive plumes. Atmos Environ 77:983–989

5. Durham MD, Lundgren DA (1980) Evaluation of aerosol aspiration efficiency as a function of Stokes' number, velocity ratio and nozzle angle. J Aerosol Sci 11:179–188
6. Glasastone S, Laidler KJ, Eyring H (1941) The theory of rate processes. McGraw-Hill Book Co., New York
7. Jahnke J (2000) Continuous emission monitoring, 2nd edn. Wiley, New York
8. Lodge JP Jr (ed) (1989) Methods of air sampling and analysis, Interscience committee of A&WMA, ACS, AIChE, APWA, ASME, AOAC, HPS, and ISA. Lewis Publishers Inc., Chelsea, MI
9. Siegmund C, Goldstein HL (1976) Influence of heavy fuel oil composition and boiler combustion conditions on particulate emissions. Environ Sci Technol 10(12), 1110–1116

Appendix

Table A.1 Molar weights of typical substances

Substance	Chemical formulae	Molar weight (kg/kmole)
Hydrogen	H_2	2.01
Water	H_2O	18.02
Hydrogen sulfide	H_2S	34.08
Hydrogen chloride	HCl	36.46
Carbon	C	12.011
Methane	CH_4	16.04
Ethane	C_2H_6	30.07
Propane	C_3H_8	44.1
Ethylene	C_2H_4	28.05
Acetylene	C_2H_2	26.04
Carbon monoxide	CO	28.01
Carbon dioxide	CO_2	44.01
Oxygen	O_2	32
Nitrogen	N_2	28.01
Nitric oxide	NO	30.01
Nitrogen dioxide	NO_2	46.01
Ammonia	NH_3	17.03
Ammonium sulfate	$(NH_4)_2SO_4$	132.14
Nitrous oxide	N_2O	44.01
Nitric acid	HNO_3	63.01
Sulfur	S	32.07
Sulfur dioxide	SO_2	64.06
Sulfur trioxide	SO_3	80.06
Sulfuric acid	H_2SO_4	98.08
Calcium	Ca	40.08
Calcium oxide (lime)	CaO	56.08

(continued)

Z. Tan, *Air Pollution and Greenhouse Gases*, Green Energy and Technology,
DOI 10.1007/978-981-287-212-8

Table A.1 (continued)

Substance	Chemical formulae	Molar weight (kg/kmole)
Calcium carbonate (lime stone)	$CaCO_3$	100.09
Calcium sulfite	$CaSO_3$	120.14
Calcium sulfate	$CaSO_4$	136.14
Calcium hydroxide	$Ca(OH)_2$	74.09
Sodium carbonate (soda)	Na_2CO_3	105.99
Sodium bicarbonate	$NaHCO_3$	84.01
Sodium sulfite	Na_2SO_3	126.04
Sodium sulfate	Na_2SO_4	142.04
Sodium hydroxide	NaOH	40.00

Table A.2 Quantities and constants

Properties of Air at 20 °C and 1 atm	
Density	1.205 kg/m^3
Viscosity	1.81×10^{-5} Pa s
Mean free path	0.066 μm
Molecular weight	28.86 g/mole
Specific heat ratio, C_p/C_v	1.40
Diffusion coefficient	0.19 cm^2/s
Gas constant for dry air when pressure is in Pa and mass in kg, R	287.055 J/ kg.K
Constants	
Avogadro's number, N_a	6.022×10^{23} /mole
Boltzmann's constant, $k = R/N_a$	1.38×10^{-23} m^2 kg/s.K
Gravitational acceleration at sea level, g	9.81 m/s
Universal gas constant	
When pressure in Pascal, R	8.314 J/mole.K
When pressure in atm and volume in cm^3, R	82.1 $atm.cm^3/mole.K$

Table A.3 Conversion factors

Metric	Imperial to metric	Metric to imperial
Length		
1 km = 1,000 m 1 hm = 100 m 1 m = 100 cm 1 m = 1,000 mm 1 m = 1,000,000 μm 1 μm = 1,000 nm 1 nm = 10 Å	1 mile = 1.609 km 1 foot = 0.3048 m 1 in = 2.54 cm	1 km = 0.54 miles 1 m = 3.2808 feet 1 cm = 0.394 in
Area		
1 m^2 = 100 dm^2 1 dm^2 = 100 cm^2 1 cm^2 = 100 mm^2	1 ft^2 = 0.0929 m^2 1 in^2 = 645.16 mm^2	1 km^2 = 0.3861 $mile^2$ 1 m^2 = 10.7639 ft^2 1 mm^2 = 0.0016 in^2
Volume		
1 m^3 = 1,000 l 1 l = 1 dm^3 1 l = 1,000 cc 1 l = 1,000 ml	1 gallon = 4.5461 l 1 US gallon = 3.785 l 1 pint = 0.5683 l 1 cu. in = 16.3871 cm^3	1 l = 0.22 gal. 1 l = 0.2642 US gal. 1 m^3 = 219.969 gal. 1 m^3 = 35.3147 $feet^3$
Mass/Weight		
1 ton = 1,000 kg 1 kg = 1,000 g 1 g = 1,000 mg	1 ton = 1.016 tons 1 lb. = 0.4536 kg 1 oz. = 28.3495 g *1 US ton = 0.9072* tonnes	1 ton = 0.9842 ton 1 ton = 1.1023 US ton 1 kg = 2.2046 lb. 1 kg = 35.274 oz.
Flow rate		
1 m^3/s = 1,000 l/s 1 m^3/s = 3,600 m^3/h	1 ft^3/min= 4.72 × 10^{-4} m^3/s	1 m^3/s = 2118.88 ft^3/min
Pressure		
1 kPa = 1,000 Pa 1 Pa = 1 N/m^2	1 atm = 760 mm Hg 1 atm = 101.325 kPa 1 atm = 1.01 × 10^6 dyn/cm^2 1 atm = 1.01325 bar	1 atm = 14.7 psia
Concentration		
1 mg/m^3 = 1 μg/l 1g/cm^3 = 1,000 kg/m^3 1g/cm^3 = 1,000 g/l x% = 10,000x ppmv	1 grain/ft^3 = 2.29 g/m^3	1 g/m^3= 0.44 grain/ft^3
Viscosity		
1 Pa.s = 1 N.s/m^2 1 g/cm.s = 0.1 Pa.s 1 g/cm.s = 0.1 N.s/m^2	1 poise = 1 g/cm.s 1 dyn.s/cm^2 = 0.1 Pa.s 1 dyn.s/cm^2 = 0.1 N s/m^2	1 g/cm.s = 1 dyn.s/cm^2
Energy		
1 cal = 4.19 J 1 kW.h = 3,600 kJ	1 Btu = 0.000293 kW.h	1 kW.h = 3412.14 Btu

(continued)

Table A.3 (continued)

Metric	Imperial to metric	Metric to imperial
Power		
1 W = 1 J.s 1 W = 3.6 kJ/h 1 kW = 3600 kJ/h	1 hp = 746 W 1 Btu/h = 0.2987 W	1 W = 3.412 Btu/h 1 kW = 3412.14 Btu/h 1 kW = 1.3405 hp
Temperature		
K = °C + 273	°F = 1.8 °C + 32	1 °C = 5/9 × (°F – 32)

Table A.4 Approximate thermodynamic data for species

Species	Name	Δh_f^0 (298 K) (J/mol)	$C_p = a + bT$ (J/mol K)	
			a	b
C	Carbon, monatomic	716,033	20.5994	0.00026
C(s)	Graphite (ref.)	0	14.926	0.00437
CH_3	Methyl	145,896	42.8955	0.01388
CH_4	Methane	–74,980	44.2539	0.02273
CO	Carbon monoxide	–110,700	29.6127	0.00301
CO_2	Carbon dioxide	–394,088	44.3191	0.00730
C_2H_4O	Ethylene oxide	–52,710	70.1093	0.03319
H_2	Hydrogen (ref.)	0	27.3198	0.00335
H_2O	Water vapor	–242,174	32.4766	0.00862
H_2O_2	Hydrogen peroxide	–136,301	41.6720	0.01952
H_2S	Hydrogen sulfide	–20,447	35.5142	0.00883
H_2SO_4	Sulfuric acid vapor	–741,633	101.7400	0.02143
H_2SO_4	Sulfuric acid liquid	–815,160	144.0230	0.02749
NH_3	Ammonia	–45,965	38.0331	0.01593
NO	Nitric oxide	90,421	30.5843	0.00278
NO_2	Nitrogen dioxide	33,143	43.7014	0.00575
NO_3	Nitrogen trioxide	71,230	61.1847	0.00932
N_2	Nitrogen (ref.)	0	29.2313	0.00307
N_2O	Nitrous oxide	82,166	44.9249	0.00693
O_2	Oxygen (ref.)	0	30.5041	0.00349
O_3	Ozone	142,880	46.3802	0.00553
S (*g*)	Sulfur, gas	279,391	22.4619	–0.0004
S (*l*)	Sulfur, liquid	1,425	29.5005	0.00976
S (*s*)	Sulfur, solid (ref.)	0	13.9890	0.02191
SO_2	Sulfur dioxide	–297,269	45.8869	0.00574
SO_3	Sulfur trioxide	–396,333	62.1135	0.00877

Table A.5 Higher and lower heating values of typical fuels

Fuels	LHV (MJ/kg)	HHV (MJ/kg)
Gaseous Fuels at 0 °C and 1 atm		
Natural gas	47.141	52.225
Hydrogen	120.21	142.18
Liquid Fuels		
Crude oil	42.686	45.543
Conventional gasoline	43.448	46.536
Reformulated or low-sulfur gasoline	42.358	45.433
U.S. conventional diesel	42.791	45.766
Low-sulfur diesel	42.612	45.575
Petroleum naphtha	44.938	48.075
NG-based FT naphtha	44.383	47.654
Methanol	20.094	22.884
Ethanol	26.952	29.847
Butanol	34.366	37.334
Acetone	29.589	31.862
Liquefied petroleum gas (LPG)	46.607	50.152
Liquefied natural gas (LNG)	48.632	55.206
Dimethyl ether (DME)	28.882	31.681
Dimethoxy methane (DMM)	23.402	25.670
Methyl ester (biodiesel, BD)	37.528	40.168
Fischer-Tropsch diesel (FTD)	43.247	45.471
Renewable Diesel I (SuperCetane)	43.563	46.628
Renewable Diesel II (UOP-HDO)	43.979	46.817
Renewable Gasoline	43.239	46.314
Liquid Hydrogen	120.07	141.80
Methyl tertiary butyl ether (MTBE)	35.108	37.957
Ethyl tertiary butyl ether (ETBE)	36.315	39.247
Tertiary amyl methyl ether (TAME)	36.392	39.322
Butane	45.277	49.210
Isobutane	44.862	49.096
Isobutylene	44.824	48.238
Propane	46.296	50.235
Solid Fuels		
Coal (wet basis)	22.732	23.968
Bituminous coal (wet basis)	26.122	27.267
Coking coal (wet basis)	28.610	29.865
Farmed trees (dry basis)	19.551	20.589

(continued)

Table A.5 (continued)

Fuels	LHV (MJ/kg)	HHV (MJ/kg)
Herbaceous biomass (dry basis)	17.209	18.123
Corn stover (dry basis)	16.370	17.415
Forest residue (dry basis)	15.402	16.473
Sugar cane bagasse	15.058	16.355
Petroleum coke	29.505	31.308

Source GREET, The Greenhouse Gases, Regulated Emissions, and Energy Use In Transportation Model, GREET 1.8d.1, developed by Argonne National Laboratory, Argonne, IL, released August 26, 2010. Used with permission

Table A.6 VOC-AC adsorption coefficients

Formula	Name	a	b	d
$CBrCl_3$	Bromotrichloromethane	1.39842	0.23228	−0.02184
$CBrF_3$	Bromotrifluoromethane	−1.46247	0.58361	−0.01044
CBr_2F_2	Dibromodifluoromethane	0.82076	0.30701	−0.01384
CBr_3F	Tribromofluoromethane	−1.43748	0.55503	−0.00450
CCl_2F_2	Dichlorodifluoromethane	−0.07350	0.40145	−0.01404
CCl_2O	Phosgene	−0.64469	0.60428	−0.02986
CCl_3F	Trichlorofluoromethane	0.17307	0.40715	−0.01915
CCl_3NO_2	Chloropicrin	1.26745	0.20841	−0.01288
CCl_4	Carbon Tetrachloride	1.07481	0.28186	−0.02273
$CHBr_3$	Tribromomethane	1.73184	0.19948	−0.02246
$CHCl_3$	Chloroform	0.67102	0.36148	−0.02288
CHN	Hydrogen Cyanide	−4.39245	1.08948	−0.00740
CH_2BrCl	Bromochloromethane	0.61399	0.41353	−0.02531
CH_2BrF	Bromofluoromethane	0.45483	0.36332	−0.01606
CH_2Br_2	Dibromomethane	1.08376	0.37211	−0.03238
CH_2Cl_2	Dichloromethane	−0.07043	0.49210	−0.02276
CH_2I_2	Diiodomethane	1.94756	0.14984	−0.01947
CH_2O	Formaldehyde	−2.48524	0.69123	−0.00375
CH_2O_2	Formic Acid	−1.77731	1.09503	−0.06354
CH_3Br	Methyl Bromide	−1.23835	0.78564	−0.05521
CH_3Cl	Methyl Chloride	−1.91871	0.62053	−0.00549
CH_3Cl_3Si	Methyl Trichlorosilane	1.07198	0.24275	−0.01911
CH_3I	Methyl Iodide	0.73997	0.32985	−0.01330
CH_3NO	Formamide	1.30981	0.25274	-
CH_3NO_2	Nitromethane	−0.32847	0.70602	−0.05111
CH_4	Methane	−4.31008	0.77883	−0.00628
CH_4Cl_2Si	Methyl Dichlorosilane	0.73271	0.29305	−0.01822
CH_4O	Methanol	−1.96739	0.82107	−0.01393

(continued)

Table A.6 (continued)

Formula	Name	a	b	d
CH_4S	Methyl Mercaptan	−1.12288	0.60573	−0.02094
CH_5N	Methylamine	−1.93548	0.64710	−0.01057
CN_4O_8	Tetranitromethane	1.49047	0.18181	−0.01894
CO	Carbon Monoxide	−5.18782	0.90121	−0.01358
COS	Carbonyl Sulfide	−1.42882	0.51061	0.00028
CO_2	Carbon Dioxide	−3.65224	0.80180	−0.00328
CS_2	Carbon Disulfide	−0.18899	0.47093	−0.01481
$C_2Br_2F_4$	1,2-Dibromotetrafluoroethane	0.90388	0.25693	−0.00974
C_2ClF_5	Chloropentafluoroethane	0.08264	0.34756	−0.01343
$C_2Cl_3F_3$	1,1,2-Trichlorotrifluoroethane	1.27368	0.18656	−0.01231
C_2Cl_4	Tetrachloroethylene	1.40596	0.20802	−0.02097
$C_2Cl_4F_2$	1,1,2,2-Tetrachlorodifluoroethane	1.37307	0.17625	−0.01465
$C_2HBrClF_3$	Halothane	0.92405	0.31204	−0.02004
C_2HCl_3	Trichloroethylene	1.02411	0.29929	−0.02539
C_2HCl_3O	Dichloroacetyl Chloride	1.23647	0.26219	−0.02596
C_2HCl_3O	Trichloroacetaldehyde	1.17362	0.26971	−0.02513
C_2HCl_5	Pentachloroethane	1.64566	0.13515	−0.01572
$C_2HF_3O_2$	Trifluoroacetic Acid	−0.12577	0.59373	−0.03445
C_2H_2	Acetylene	−2.24177	0.82454	−0.03390
$C_2H_2Br_4$	1,1,2,2-Tetrabromoethane	-	-	-
$C_2H_2Cl_2$	1,1-Dichloroethylene	0.48740	0.33282	−0.01622
$C_2H_2Cl_2$	cis-1,2-Dichloroethylene	0.47567	0.39061	−0.02554
$C_2H_2Cl_2$	trans-1,2-Dichloroethylene	0.47567	0.39061	−0.02554
$C_2H_2Cl_2O_2$	Dichloroacetic Acid	1.69237	0.09630	-
$C_2H_2Cl_4$	1,1,1,2-Tetrachloroethane	1.44097	0.19166	−0.01995
$C_2H_2Cl_4$	1,1,2,2-Tetrachloroethane	1.52322	0.17848	−0.02019
C_2H_3Cl	Vinyl Chloride	−0.98889	0.66564	−0.04320
C_2H_3ClO	Acetyl Chloride	0.03627	0.45526	−0.02093
$C_2H_3ClO_2$	Methyl Chloroformate	0.41186	0.42776	−0.02776
$C_2H_3Cl_3$	1,1,1-Trichloroethane	0.97331	0.28737	−0.02277
$C_2H_3Cl_3$	1,1,2-Trichloroethane	1.17163	0.27791	−0.02746
C_2H_3N	Acetonitrile	−0.79666	0.63512	−0.02598
C_2H_3NO	Methyl Isocyanate	−1.07579	0.85881	−0.06876
C_2H_4	Ethylene	−2.27102	0.61731	−0.01467
$C_2H_4Br_2$	1,1-Dibromoethane	1.37260	0.25671	−0.02516
$C_2H_4Br_2$	1,2-Dibromoethane	1.44231	0.25500	−0.02666
$C_2H_4Cl_2$	1,1-Dichloroethane	0.54485	0.36091	−0.02192
$C_2H_4Cl_2$	1,2-Dichloroethane	0.55343	0.37072	−0.02161
$C_2H_4Cl_2O$	Bis(chloromethyl)ether	0.95599	0.33784	−0.03200
$C_2H_4F_2$	1,2-Difluoroethane	−3.97902	2.51862	−0.31617

(continued)

Table A.6 (continued)

Formula	Name	a	b	d
C_2H_4O	Acetaldehyde	−1.17047	0.62766	−0.02475
C_2H_4O	Ethylene Oxide	−2.42379	0.94878	−0.04062
$C_2H_4O_2$	Acetic Acid	−0.05553	0.68410	−0.06071
$C_2H_4O_2$	Methyl Formate	−0.99586	0.61693	−0.01847
C_2H_4S	Thiacyclopropane	0.02258	0.45520	−0.02154
C_2H_5Br	Bromoethane	0.31783	0.43549	−0.03072
C_2H_5Cl	Ethyl Chloride	−0.50828	0.50364	−0.02179
C_2H_5ClO	2-Chloroethanol	0.74164	0.46933	−0.05158
C_2H_5I	Ethyl Iodide	1.00356	0.32123	−0.02405
C_2H_5N	Ethyleneimine	−1.16912	0.91238	−0.07400
C_2H_5NO	N-Methylformamide	1.23333	0.21723	-
$C_2H_5NO_2$	Nitroethane	0.44968	0.49708	−0.04612
C_2H_6	Ethane	−2.40393	0.68107	−0.01925
C_2H_6O	Ethanol	−0.51153	0.67525	−0.04473
C_2H_6OS	Dimethyl Sulfoxide	1.24042	0.31302	−0.04768
$C_2H_6O_2$	Ethylene Glycol	1.40474	0.18738	−0.02663
$C_2H_6O_4S$	Dimethyl Sulfate	1.34617	0.21539	−0.02336
C_2H_6S	Dimethyl Sulfide	0.48472	0.37358	−0.02770
C_2H_6S	Ethyl Mercaptan	0.00552	0.40506	−0.01802
$C_2H_6S_2$	Dimethyl Disulfide	0.75878	0.35928	−0.02953
C_2H_7N	Dimethylamine	−1.22492	0.63962	−0.03266
C_2H_7NO	Monoethanolamine	1.21569	0.21994	-
$C_2H_8N_2$	Ethylenediamine	0.56504	0.46307	−0.04789
C_3H_3Cl	Propargyl Chloride	0.27135	0.40480	−0.02135
C_3H_3N	Acrylonitrile	0.07669	0.49986	−0.03500
C_3H_3NO	Oxazole	0.63350	0.30620	−0.02350
C_3H_4	Methylacetylene	−2.52865	1.74715	−0.21635
$C_3H_4Cl_2$	2,3-Dichloropropene	0.95417	0.30034	−0.02614
C_3H_4O	Acrolein	−0.29632	0.49437	−0.02471
C_3H_4O	Propargyl Alcohol	0.22971	0.57711	−0.05441
$C_3H_4O_2$	Acrylic Acid	0.75549	0.47108	−0.05615
$C_3H_4O_3$	Pyuvic Acid	1.07410	0.41414	−0.05768
C_3H_5Br	3-Bromo-1-Propene	0.84815	0.32392	−0.02398
C_3H_5Cl	3-Chloropropene	0.32792	0.36553	−0.01853
C_3H_5ClO	alpha-Epichlorohydrin	0.83203	0.38983	−0.03932
$C_3H_5ClO_2$	Methyl Chloroacetate	1.07657	0.32514	−0.03617
$C_3H_5ClO_2$	Ethyl Chloroformate	0.94901	0.32529	−0.03201
$C_3H_5ClO_3$	1,2,3-Trichloropropane	1.47241	0.18136	−0.02165
C_3H_5I	3-lodo-1-Propene	1.33634	0.24222	−0.02271
C_3H_5N	Propionitrile	0.05925	0.51747	−0.03781

(continued)

Table A.6 (continued)

Formula	Name	a	b	d
C_3H_5NO	Hydracrylonitrile	1.50994	0.11037	-
C_3H_5NO	Lactonitrile	1.44156	0.12689	-
C_3H_6	Propylene	−0.93674	0.57775	−0.03853
$C_3H_6Cl_2$	1,1-Dichloropropane	0.95379	0.28791	−0.02487
$C_3H_6Cl_2$	1,2-Dichloropropane	0.98872	0.28700	−0.02571
$C_3H_6Cl_2$	1,3-Dichloropropane	1.10340	0.27837	−0.02824
$C_3H_6Cl_2$	2,2-Dichloropropane	0.85314	0.29432	−0.02255
C_3H_6O	Acetone	−0.14546	0.47497	−0.02286
C_3H_6O	Allyl Alochol	0.32390	0.49368	−0.04370
C_3H_6O	n-Propionaldehyde	0.05519	0.49738	−0.04331
C_3H_6O	1,2-Propylene Oxide	−0.42829	0.53858	−0.02757
C_3H_6O	1,3-Propylene Oxide	−0.50421	0.51872	−0.02296
$C_3H_6O_2$	Ethyl Formate	0.12618	0.42260	−0.02090
$C_3H_6O_2$	Methyl Acetate	0.13314	0.42849	−0.02188
$C_3H_6O_2$	Propionic Acid	0.77846	0.44570	−0.05209
$C_3H_6O_2S$	3-Mercaptopropionic Acid	1.68823	0.05916	-
$C_3H_6O_3$	Lactic Acid	1.60722	0.09225	-
$C_3H_6O_3$	Methoxyacetic Acid	1.61885	0.08873	-
C_3H_6S	Thiacyclobutane	0.67420	0.37225	−0.03151
C_3H_7Br	1-Bromopropane	0.83601	0.32406	−0.02407
C_3H_7Br	2-Bromopropane	0.81137	0.31043	−0.02155
C_3H_7Cl	Isopropyl Chloride	0.31428	0.34779	−0.01661
C_3H_7Cl	n-Propyl Chloride	0.40133	0.34678	−0.01931
C_3H_7I	Isopropyl Iodide	1.26456	0.24157	−0.02122
C_3H_7I	n-Propyl Iodide	1.30623	0.24227	−0.02250
C_3H_7N	Allylamine	0.16250	0.39815	−0.02105
C_3H_7N	Propyleneimine	0.06919	0.43529	−0.02293
C_3H_7NO	N,N-Dimethylformamide	0.90253	0.37875	−0.04523
$C_3H_7NO_2$	1-Nitropropane	0.91328	0.34648	−0.03730
$C_3H_7NO_2$	2-Nitropropane	0.83248	0.35732	−0.03608
C_3H_8	Propane	−0.79460	0.49029	−0.02398
C_3H_8O	Isopropanol	0.27183	0.46419	−0.03682
C_3H_8O	n-Propanol	0.38644	0.48033	−0.04505
C_3H_8O2	2-Methoxyethanol	0.74339	0.41792	−0.04536
$C_3H_8O_2$	Methylal	0.19079	0.38167	−0.01775
$C_3H_8O_2$	1,2-Propylene Glycol	1.48275	0.11594	-
$C_3H_8O_2$	1,3-Propylene Glycol	1.58563	0.08395	-
C_3H_8S	n-Propylmercaptan	0.59031	0.31407	−0.02190
C_3H_8S	Isopropyl Mercaptan	0.55779	0.31539	−0.02051
C_3H_8S	Ethyl-Methyl-Sulfide	0.62830	0.31889	−0.02320

(continued)

Table A.6 (continued)

Formula	Name	a	b	d
C_3H_9N	n-Propylamine	0.05768	0.34918	−0.01241
C_3H_9N	Isopropylamine	0.07464	0.37106	−0.01568
C_3H_9N	Trimethylamine	−0.09422	0.32583	−0.00337
C_3H_9NO	1-Amino-2-Propanol	1.25496	0.27456	−0.04254
C_3H_9NO	3-Amino-1-Propanol	1.53733	0.08156	-
C_3H_9NO	Methylethanolamine	1.14745	0.30208	−0.04243
$C_3H_9O_3P$	Trimethyl-Phosphite	1.00568	0.21001	−0.01402
$C_3H_9O_4P$	Trimethyl Phosphate	1.48463	0.16933	−0.02290
$C_3H_{10}N_2$	1,2-Propanediamine	0.90237	0.31904	−0.03400
C_4H_4O	Furan	0.04084	0.40613	−0.01620
$C_4H_4O_2$	Diketene	0.87430	0.37094	−0.03962
C_4H_4S	Thiophene	0.80753	0.32166	−0.02654
C_4H_5Cl	Chloroprene	0.72957	0.29786	−0.02111
C_4H_5N	trans-Crotonitrile	0.70791	0.37284	−0.03756
C_4H_5N	cis-Crotonitrile	0.58131	0.39427	−0.03590
C_4H_5N	Methacrylonitrile	0.46655	0.38890	−0.03042
C_4H_5N	Pyrrole	0.83128	0.38413	−0.04217
C_4H_5N	Vinylacetonitrile	0.61844	0.40844	−0.04032
$C_4H_5NO_2$	Methyl Cyanoacetate	1.56587	0.09143	-
C_4H_6	1,3-Butadiene	−0.03359	0.34764	−0.01297
C_4H_6	Dimethylacetylene	−0.06673	0.39387	−0.01524
C_4H_6	Ethylacetylene	−0.02918	0.33636	−0.01056
$C_4H_6Cl_2$	1,3-Dichloro-trans-2-Butene	1.30208	0.19939	−0.02091
$C_4H_6Cl_2$	1,4-Dichloro-cis-2-Butene	1.40119	0.17876	−0.02041
$C_4H_6Cl_2$	1,4-Dichloro-trans-2-Butene	1.40904	0.18120	−0.02179
$C_4H_6Cl_2$	3,4-Dichloro-1-Butene	1.23394	0.21476	−0.02135
C_4H_6O	trans-Crotonaldehyde	0.68353	0.36560	−0.03400
C_4H_6O	2,5-Dihydrofuran	0.33990	0.40041	−0.02451
C_4H_6O	Methacrolein	0.43461	0.37019	−0.02474
$C_4H_6O_2$	gamma-Butyrolactone	1.29434	0.29719	−0.04658
$C_4H_6O_2$	cis-Crotonic Acid	1.30871	0.25008	−0.03752
$C_4H_6O_2$	Methacrylic Acid	1.23099	0.27648	−0.03903
$C_4H_6O_2$	Methyl Acrylate	0.45869	0.32104	−0.02001
$C_4H_6O_2$	Vinyl Acetate	0.61067	0.34797	−0.02595
$C_4H_6O_3$	Acetic Anhydride	1.07388	0.31083	−0.03575
C_4H_7N	n-Butyronitrile	0.64311	0.38787	−0.03822
C_4H_7N	Isobutyronitrile	0.56697	0.38807	−0.03531
C_4H_7NO	3-Methoxypropionitrile	1.13283	0.28534	−0.03732
C_4H_8	1-Butene	0.07313	0.32701	−0.01452
C_4H_8Br	1,2-Dibromobutane	1.69234	0.12766	−0.01497

(continued)

Table A.6 (continued)

Formula	Name	a	b	d
C_4H_8Br	2,3-Dibromobutane	1.68176	0.12916	−0.01492
$C_4H_8Cl_2$	1,4-Dichlorobutane	1.38278	0.17796	−0.02030
C_4H_8O	n-Butyraldehyde	0.45056	0.37372	−0.02689
C_4H_8O	Isobutyraldehyde	0.39315	0.36715	−0.02379
C_4H_8O	1,2-Epoxybutane	0.36719	0.37654	−0.02360
C_4H_8O	Methyl Ethyl Ketone	0.46525	0.37688	−0.02801
C_4H_8O	Ethyl Vinyl Ether	0.33311	0.33471	−0.01711
C_4H_8O	Tetrahydrofuran	0.29856	0.35648	−0.01550
$C_4H_8O_2$	Isobutyric Acid	1.14021	0.29004	−0.03833
$C_4H_8O_2$	n-Butyric Acid	1.22589	0.26481	−0.03737
$C_4H_8O_2$	1,4-Dioxane	0.66781	0.36208	−0.03034
$C_4H_8O_2$	Ethyl Acetate	0.63612	0.34441	−0.02691
$C_4H_8O_2$	Methyl Propionate	0.64273	0.34862	−0.02767
$C_4H_8O_2$	n-Propyl Formate	0.65855	0.34340	−0.02750
$C_4H_8O_2S$	Sulfolane	1.77762	0.00118	0.00005
C_4H_8S	Tetrahydrothiophene	0.93777	0.33197	−0.03877
C_4H_9Br	1-Bromobutane	1.16698	0.24380	−0.02270
C_4H_9Br	2-Bromobutane	1.13872	0.24481	−0.02203
C_4H_9C	n-Butyl Chloride	0.80024	0.29114	−0.02370
C_4H_9Cl	sec-Butyl Chloride	0.75046	0.29132	−0.02212
C_4H_9Cl	tert-Butyl Chloride	0.68673	0.28529	−0.01944
C_4H_9N	Pyrrolidine	0.60693	0.36363	−0.03004
C_4H_9NO	N,N-Dimethylacetamide	1.20026	0.25124	−0.03263
C_4H_9NO	Morpholine	1.00673	0.30572	−0.03294
C_4H_{10}	n-Butane	0.03071	0.34304	−0.01596
C_4H_{10}	Isobutane	−0.01676	0.33495	−0.01274
$C_4H_{10}O$	n-Butanol	0.89881	0.32534	−0.03648
$C_4H_{10}O$	sec-Butanol	0.76814	0.34611	−0.03478
$C_4H_{10}O$	Diethyl Ether	0.23477	0.36044	−0.02236
$C_4H_{10}O$	Methyl-Propyl-Ether	0.36764	0.32893	−0.01787
$C_4H_{10}O$	Methyl Isopropyl Ether	0.36373	0.31940	−0.01647
$C_4H_{10}O$	Isobutanol	0.84818	0.33155	−0.03559
$C_4H_{10}O_2$	1,3-Butanediol	-	-	-
$C_4H_{10}O_2$	1,4-Butanediol	-	-	-
$C_4H_{10}O_2$	2,3-Butanediol	1.50642	0.09239	-
$C_4H_{10}O_2$	t-Butyl Hydroperoxide	1.08563	0.26496	−0.03035
$C_4H_{10}O_2$	1,2-Dimethoxyethane	0.74981	0.31330	−0.02616
$C_4H_{10}O_2$	2-Ethoxyethanol	1.07911	0.27792	−0.03199
$C_4H_{10}O_4S$	Diethyl Sulfate	1.64797	0.05805	-
$C_4H_{10}S$	n-Butyl Mercaptan	0.98086	0.24388	−0.02251

(continued)

Table A.6 (continued)

Formula	Name	a	b	d
$C_4H_{10}S$	Isobutyl Mercaptan	0.93709	0.24802	−0.02179
$C_4H_{10}S$	sec-Butyl Mercaptan	0.92287	0.24856	−0.02146
$C_4H_{10}S$	tert-Butyl Mercaptan	0.84380	0.24937	−0.01939
$C_4H_{10}S$	Diethyl Sulfide	0.95993	0.24465	−0.02195
$C_4H_{10}S$	Isopropyl-Methyl-Sulfide	0.92769	0.24689	−0.02132
$C_4H_{10}S$	Methyl-Propyl-Sulfide	0.97217	0.24453	−0.02229
$C_4H_{10}S_2$	Diethyl Disulfide	1.42594	0.12564	−0.01373
$C_4H_{11}N$	n-Butylamine	0.64570	0.31857	−0.02570
$C_4H_{11}N$	Isobutylamine	0.60137	0.31538	0.02369
$C_4H_{11}N$	sec-Butylamine	0.57706	0.31254	−0.02250
$C_4H_{11}N$	tert-Butylamine	0.50036	0.30334	−0.01897
$C_4H_{11}N$	Diethylamine	0.54770	0.30799	−0.02107
$C_4H_{11}NO$	Dimethylethanolamine	1.18381	0.23249	−0.02857
$C_4H_{12}Si$	Tetramethylsilane	0.70867	0.23089	−0.01505
$C_4H_{13}N_3$	Diethylene Triamine	1.54270	0.06479	-

Source Yaws CL, Bu L, Nijhawan S (1995) Determining VOC adsorption capacity. Pollut Eng 27(2):34–37. http://www.pollutionengineering.com/

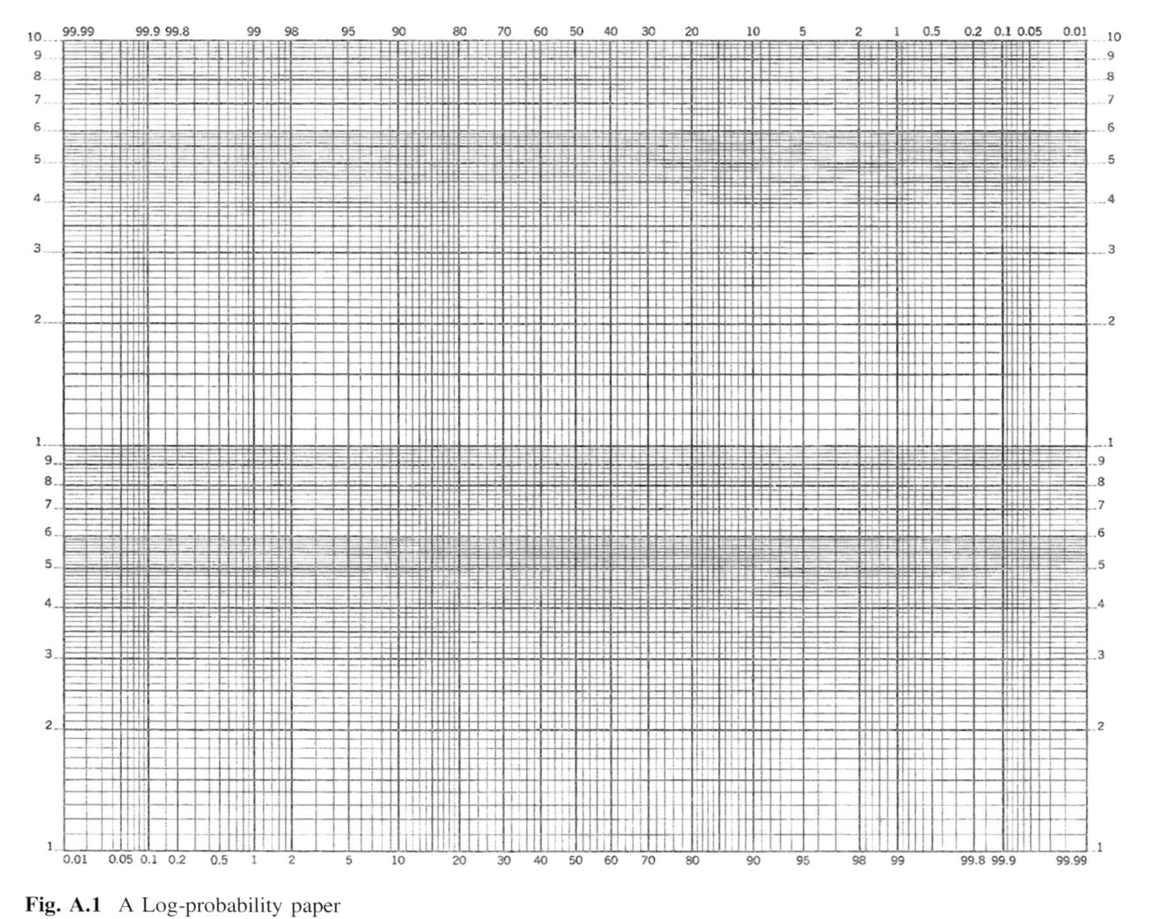

Fig. A.1 A Log-probability paper

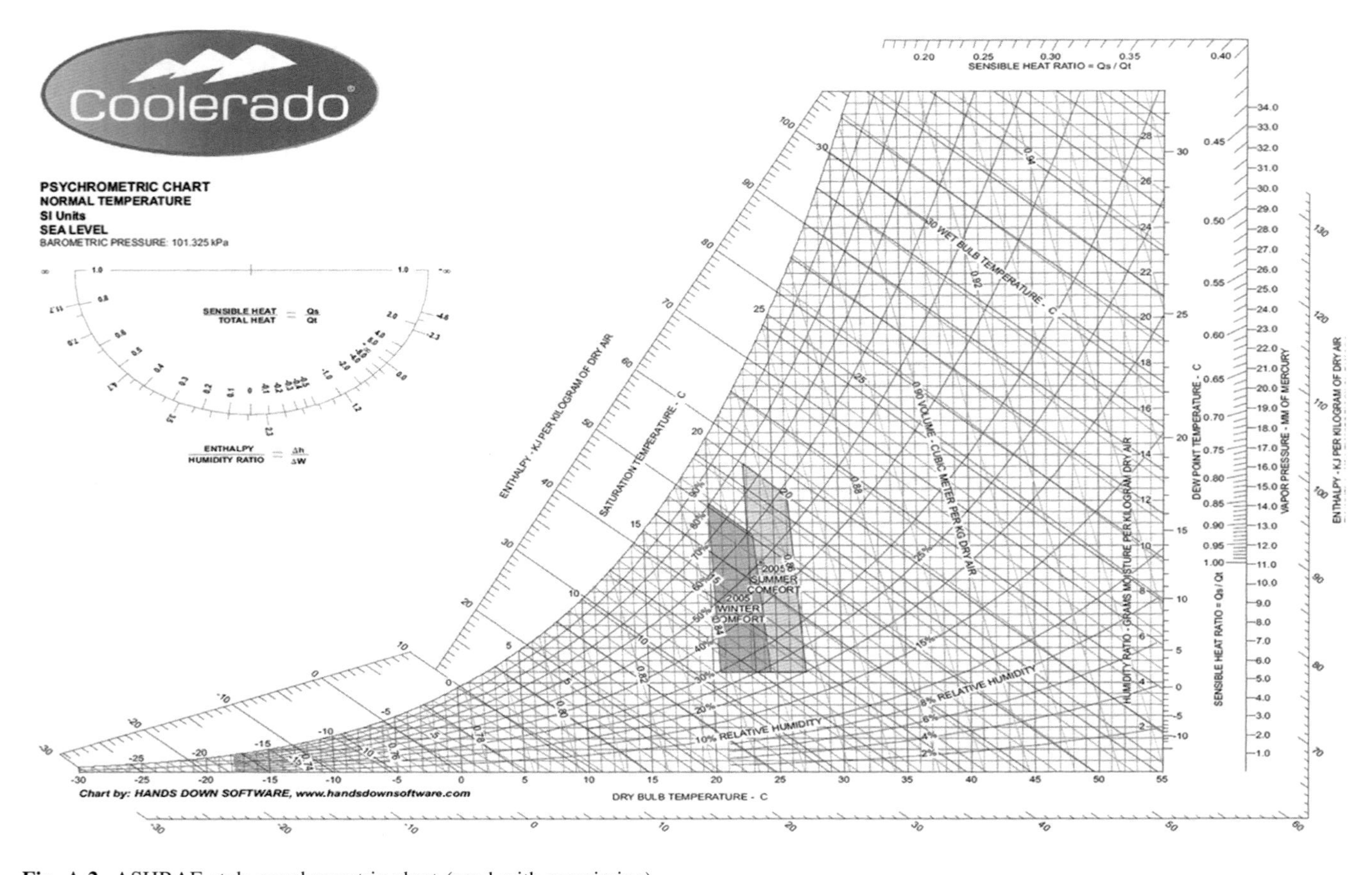

Fig. A.2 ASHRAE-style psychrometric chart (used with permission)

Index

Z. Tan, *Air Pollution and Greenhouse Gases*, Green Energy and Technology,
DOI 10.1007/978-981-287-212-8

Printed by Printforce, the Netherlands